The Road from Nanomedicine to Precision Medicine

Part B

Endorsed by
American Society for Nanomedicine
European Foundation for Clinical Nanomedicine
European Society for Nanomedicine

Jenny Stanford Series on Nanomedicine

Series Editor
Raj Bawa

Titles in the Series

Vol. 1
Handbook of Clinical Nanomedicine: Nanoparticles, Imaging, Therapy, and Clinical Applications
Raj Bawa, Gerald F. Audette, and Israel Rubinstein, eds.
2016
978-981-4669-20-7 (Hardcover)
978-981-4669-21-4 (eBook)

Vol. 2
Handbook of Clinical Nanomedicine: Law, Business, Regulation, Safety, and Risk
Raj Bawa, ed., Gerald F. Audette and Brian E. Reese, asst. eds.
2016
978-981-4669-22-1 (Hardcover)
978-981-4669-23-8 (eBook)

Vol. 3
Immune Aspects of Biopharmaceuticals and Nanomedicines
Raj Bawa, János Szebeni, Thomas J. Webster, and Gerald F. Audette, eds.
2018
978-981-4774-52-9 (Hardcover)
978-0-203-73153-6 (eBook)

Vol. 4
The Road from Nanomedicine to Precision Medicine
Shaker A. Mousa, Raj Bawa, and Gerald F. Audette
2020
978-981-4800-59-4 (Hardcover)
978-0-429-29501-0 (eBook)

Vol. 5
Current Issues in Medicine, Pharma, and Biotech
Raj Bawa and Gerald F. Audette, eds.
2021

Vol. 6
Drug Delivery at the Nanoscale: A Guide for Scientists, Physicians, and Lawyers
Raj Bawa, ed.
2022

Jenny Stanford Series on Nanomedicine Vol. 4

The Road from Nanomedicine to Precision Medicine

Part B

edited by

Shaker A. Mousa, PhD, MBA
Chairman & Executive Vice President, The Pharmaceutical Research Institute,
Rensselaer, New York, USA
Professor of Pharmacology, Albany College of Pharmacy and Health Sciences,
Albany, New York, USA

Raj Bawa, MS, PhD
Patent Agent, Bawa Biotech LLC, Ashburn, Virginia, USA
Vice President and Chief IP Officer, Guanine Inc., Rensselaer, New York, USA
Scientific Advisor, Teva Pharmaceutical Industries, Ltd., Israel

Gerald F. Audette, PhD
Associate Dean of Faculty, Faculty of Science,
Associate Professor, Department of Chemistry,
York University, Toronto, Canada

JENNY STANFORD
PUBLISHING

Published by

Jenny Stanford Publishing Pte. Ltd.
Level 34, Centennial Tower
3 Temasek Avenue
Singapore 039190
Email: editorial@jennystanford.com
Web: www.jennystanford.com

Note from the Series Editor and the Publisher

Extensive efforts have been made to make the information provided herein as accurate and as up-to-date as possible. However, the accuracy and completeness of this information cannot be guaranteed. It is important to note that knowledge and best practices in the various fields represented in this book (medicine, drug delivery, nanomedicine, precision medicine, genomics, bioinformatics, biologics, tissue engineering, patent law, FDA law, regulatory science, toxicology, pharmaceutical sciences, etc.) are constantly evolving. As new research and experience broaden our knowledge base, changes in research methods, legal and business practices, diagnostics, assays, tools and techniques, medical formulations, and/or treatments may become necessary. Therefore, it is imperative that the reader does not solely rely on the information presented herein and always carefully reviews product labels, warnings, data, and directions before using or consuming any formulation, device or drug product. For additional information about a product, please contact the manufacturer, FDA, inventor, physician, pharmacist, or other licensed health-care professional, as appropriate. Similarly, careful evaluation of any procedures, manufacturing steps, legal ideas, medical protocols, regulatory guidances, or assays described herein is warranted. To the fullest extent of the law, neither the publisher nor the editors or authors make any representations or warranties, express or implied, with respect to the information presented in this book, for its use or misuse, or interpretation thereof. In this regard, they assume no liability for any injury and/or damage to persons or property as a matter of product liability, negligence, or otherwise.

A catalogue record for this book is available from the Library of Congress and the British Library.

ISBN: 978-981-4800-59-4 (Set)
ISBN: 978-0-429-29501-0 (Set) (eBook)
ISBN: 978-981-4800-99-0 (Part B) (Hardcover)

Printed in Canada

This book is dedicated to my wife and children for their love, support, and encouragement.

—Shaker A. Mousa

Dedicated with love to the three most amazing women in my life:
my mother, my wife, and my mother-in-law.

—Raj Bawa

This book is dedicated to my wife and son, for their love and support.

—Gerald F. Audette

Cover Image

A depiction of nano-robots in the human cardiovascular system. Such nano-combination products (NCPs), currently at the R&D phase, could be commercialized in the coming decades. They could serve multiple functions of diagnostic evaluation, drug delivery, and assessment of therapy—all within the same nanoengineered combination product. As technological advances continue to merge product types (drug, biologic, or device), I expect that more products will fall into the category of combination products. The FDA and the EMA are certain to struggle with appropriate regulatory pathways for such highly integrated combination products at the interface of the three product domains.

—Dr. Raj Bawa, Series Editor

Also Read

Handbook of Clinical Nanomedicine. Vol. 1. Nanoparticles, Imaging, Therapy, and Clinical Applications

Raj Bawa, PhD, Gerald F. Audette, PhD, and Israel Rubinstein, MD (Editors)

978-981-4669-20-7 (Hardcover), 978-981-4669-21-4 (eBook)
1662 pages

This handbook (55 chapters) provides a comprehensive roadmap of basic research in nanomedicine as well as clinical applications. However, unlike other texts in nanomedicine, it not only highlights current advances in diagnostics and therapeutics but also explores related issues like nomenclature, historical developments, regulatory aspects, nanosimilars and 3D nanofabrication. While bridging the gap between basic biomedical research, engineering, medicine and law, the handbook provides a thorough understanding of nano's potential to address (i) medical problems from both the patient and health provider's perspective, and (ii) current applications and their potential in a healthcare setting.

"Dr. Bawa and his team have meticulously gathered the distilled experience of world-class researchers, clinicians and business leaders addressing the most salient issues confronted in product concept development and translation."

Gregory Lanza, MD, PhD
Professor of Medicine and Oliver M. Langenberg Distinguished Professor
Washington University Medical School, USA

"This is an outstanding, comprehensive volume that crosscuts disciplines and topics fitting individuals from a variety of fields looking to become knowledgeable in medical nanotech research and its translation from the bench to the bedside."

Shaker A. Mousa, PhD, MBA
Vice Provost and Professor of Pharmacology
Albany College of Pharmacy and Health Sciences, USA

"Masterful! This handbook will have a welcome place in the hands of students, educators, clinicians and experienced scientists alike. In a rapidly evolving arena, the authors have harnessed the field and its future by highlighting both current and future needs in diagnosis and therapies. Bravo!"

Howard E. Gendelman, MD
Margaret R. Larson Professor and Chair
University of Nebraska Medical Center, USA

"It is refreshing to see a handbook that does not merely focus on preclinical aspects or exaggerated projections of nanomedicine. Unlike other books, this handbook not only highlights current advances in diagnostics and therapies but also addresses critical issues like terminology, regulatory aspects and personalized medicine."

Gert Storm, PhD
Professor of Pharmaceutics
Utrecht University, The Netherlands

Handbook of Clinical Nanomedicine. Vol. 2. Law, Business, Regulation, Safety, and Risk

Raj Bawa, PhD (Editor), Gerald F. Audette, PhD, and Brian E. Reese, PhD, MBA, JD (Assistant Editors)

978-981-4669-22-1 (Hardcover), 978-981-4669-23-8 (eBook)
1448 pages

This unique handbook (60 chapters) examines the entire "product life cycle," from the creation of nanomedical products to their final market introduction. While focusing on critical issues relevant to nanoproduct development and translational activities, it tackles topics such as regulatory science, patent law, FDA law, ethics, personalized medicine, risk analysis, toxicology, nano-characterization and commercialization activities. A separate section provides fascinating perspectives and editorials from leading experts in this complex interdisciplinary field.

"The distinguished editors have secured contributions from the leading experts in nanomedicine law, business, regulation and policy. This handbook represents possibly the most comprehensive and advanced collections of materials on these critical topics. An invaluable standard resource."

Gregory N. Mandel, JD
Peter J. Liacouras Professor of Law and Associate Dean
Temple University Beasley School of Law, USA

"This is an outstanding volume for those looking to become familiar with nanotechnology research and its translation from the bench to market. Way ahead of the competition, a standard reference on any shelf."

Shaker A. Mousa, PhD, MBA
Vice Provost and Professor of Pharmacology
Albany College of Pharmacy, USA

"The editors have gathered the distilled experience of leaders addressing the most salient issues confronted in R&D and translation. Knowledge is power, particularly in nanotechnology translation, and this handbook is an essential guide that illustrates and clarifies our way to commercial success."

Gregory Lanza, MD, PhD
Professor of Medicine and Oliver M. Langenberg Distinguished Professor
Washington University Medical School, USA

"The title of the handbook reflects its broad-ranging contents. The intellectual property chapters alone are worthy of their own handbook. Dr. Bawa and his coeditors should be congratulated for gathering the important writings on nanotech law, business and commercialization."

Richard J. Apley, JD
Chief Patent Officer
Litman Law Offices/Becker & Poliakoff, USA

"It is clear that this handbook will serve the interdisciplinary community involved in nanomedicine, pharma and biotech in a highly comprehensive way. It not only covers basic and clinical aspects but the often missing, yet critically important, topics of safety, risk, regulation, IP and licensing. The section titled 'Perspectives and Editorials' is superb."

Yechezkel (Chezy) Barenholz, PhD
Professor Emeritus of Biochemistry and Daniel Miller Professor of Cancer Research
Hebrew University-Hadassah Medical School, Israel

Immune Aspects of Biopharmaceuticals and Nanomedicines

Raj Bawa, János Szebeni, Thomas J. Webster and Gerald F. Audette (Editors)
978-981-4774-52-9 (Hardback), 978-0-203-73153-6 (eBook)
1038 pages

The enormous advances in the immunologic aspects of biotherapeutics and nanomedicines in the past two decades has necessitated an authoritative and comprehensive reference source that can be relied upon by immunologists, biomedical researchers, clinicians, pharmaceutical companies, regulators, venture capitalists, and policy makers alike. This text provides a thorough understanding of immunology, therapeutic potential, clinical applications, adverse reactions, and approaches to overcoming immunotoxicity of biotherapeutics and nanomedicines. It also tackles critical, yet often overlooked topics such as immune aspects of nano-bio interactions, current FDA regulatory guidances, complement activation-related pseudoallergy (CARPA), advances in nanovaccines, and immunogenicity testing of protein therapeutics.

"This outstanding volume represents a review of the various effects of biopharmaceuticals and nanomedicines on the immune system: immunotherapy, vaccines, and drug delivery; challenges and overcoming translational barriers stemming from immunotoxicity; strategies to designing more immunologically friendly formulations."

África González-Fernández, PhD, MD
Professor of Immunology and President of the Spanish Society of Immunology,
University of Vigo, Spain

"For those who are specialists, and for those interested in a broader understanding of biologics and nanomedicines, this is a superb book, with internationally accomplished contributors. It serves both as a reference and as a practical guide to the newest advances in these important fields. Highly recommended!"

Carl R. Alving, MD
Emeritus Senior Scientist, Walter Reed Army Institute of Research, Silver Spring, Maryland, USA

"A skillfully produced book that addresses an often-missed topic: immune aspects of biologicals and nanoscale therapeutics, with an emphasis on clinical relevance and applications."

Rajiv R. Mohan, PhD
Professor and Ruth M. Kraeuchi Missouri Endowed Chair Professor,
University of Missouri, Columbia, USA

"An indispensable masterpiece! It represents a rich source of information on interactions of biologics and nanodrugs with the immune system—all critical for medical applications. Volume 3, once again, achieves the series' high standards."

László Rosivall, MD, PhD, DSc Med, Med habil.
Széchenyi Prize Laureate and Professor, Faculty of Medicine, Semmelweis University,
Budapest, Hungary

American Society For Nanomedicine

The American Society for Nanomedicine (ASNM) (https://www.nanomedus.org) is a nonprofit, professional medical organization based in Ashburn, Virginia, USA. It was founded in 2008 by Dr. Raj Bawa of Bawa Biotech LLC and Dr. Esther Chang of Georgetown Medical Center. The ASNM comprises members drawn from diverse fields, including medicine, law, nanotechnology, pharma, biotech, engineering, and biomedical sciences with the common goal of advancing nanomedicine research to benefit global health. These goals are achieved through an open forum of ideas and collaborative efforts as well as close cooperation with our partner organizations. Since its inception, the ASNM has organized major international conferences.

Specifically, the vision of the ASNM includes

- promoting research related to all aspects of nanomedicine and providing a forum through scientific meetings for the presentation of basic, clinical, and population-based research;
- promoting and facilitating the formal training of physicians, basic medical scientists, engineers, molecular biologists, statisticians, and allied healthcare providers in nano-related medical research and education;
- encouraging primary and secondary preventive measures and nano-based technologies to reduce the incidences of various diseases;
- facilitating the establishment of programs and policies that can better serve early diagnosis.

CLINAM

The European Foundation for Clinical Nanomedicine (https://www.clinam.org), founded in 2007 by Beat Löffler and Patrick Hunziker, is an organization based in Basel, Switzerland. Its primary mission is to advance medicine to the benefit of individuals and society through the application of nanoscience and targeted medicine. Aiming at prevention, diagnosis, and therapy, it supports clinically focused research and the interaction and information flow between clinicians, researchers, and the public. The major goal is to support the development and application of nanomedicine and targeted medicine and having in scope all nanomedicine-related fields. The foundation runs a lab, creates an annual summit for clinical nanomedicine, and established the *European Journal of Nanomedicine* (now PRNANO online). The 12th CLINAM Summit will be held May 17–20, 2020, and is titled "Clinical Nanomedicine and the Impact of Digitalization and Artificial Intelligence for Precision Medicine—The Technologies for Diagnosis and & Therapy in Personalized Medicine." The Summit traditionally brings together over 500 participants from more than 40 countries.

CLINAM founded the European Society for Nanomedicine (ESNAM), which has more than 1,000 members today. ESNAM was the driving force for the formation of the International Society for Nanomedicine, (ISNM) which brings together members from Japan, Korea, USA, Canada, Europe, South America, Australia, Africa, and India. CLINAM organizes worldwide summer schools. The next one is planned in Asia for the autumn of 2020. During the summit, CLINAM also hosts satellite meetings. For the past five years, the International Pharmaceutical Regulators Programme (IPRP), a meeting for global regulatory authorities, is also held. This group uses the CLINAM platform to provide statements on the global cooperation to come to an optimal framework for regulatory matters pertaining to nanomedicine and precision medicine.

About the Editors

 **Shaker A. Mousa, PhD, MBA, FACC, FACB, FAHA, FNAI,** is an endowed chair, tenured professor of pharmacology at Albany College of Pharmacy in Albany, New York, USA. He also serves as the executive vice president and chair of the Pharmaceutical Research Institute in Rensselaer, New York, USA. Dr. Mousa is the founder of several spin-off pharmaceutical and biotechnology companies. Previously, for two decades, he was a senior principal research scientist and a research fellow at DuPont Pharmaceuticals and Imaging Co., DuPont Merck, and DuPont Pharmaceuticals Company. He is a recipient of several national and international awards, including the 2017 Kuwait Foundation for Advancement in Sciences Laureate for Applied Sciences in Medicine.

Among Dr. Mousa's professional accomplishments are his contributions to several patents, specifically to the discovery and development of novel anti-platelet, anti-thrombotic therapies, noninvasive myocardial perfusion, and thrombus imaging agents. His work is reported in over 1,000 peer-reviewed publications and he holds over 400 US and international patents. He contributed to the discovery and development of the following products/clinical candidates: Cardiolite® (Tc-99m sestamibi, non-invasive myocardial perfusion imaging agent), Marluma (for breast cancer detection), DMP444 (Tc-99m platelet GPIIb/IIIa antagonist for non-invasive thrombus imaging agent for venous and arterial thromboembolic disorders), Roxifiban (DMP754, oral anti-platelet/anti-thrombotic agent for the prevention and treatment of coronary, carotid and peripheral artery thromboembolic disorders). Dr. Mousa is involved in the discovery of novel site directed anti-α_v/β_3 for tumor radiotherapy, imaging, and novel pharmacological aspects of heparins and non-anticoagulant heparins. His main research interests are in drug discovery and drug development arena as well as bringing novel concepts from the bench-to-the-bedside and *vice versa* via key enabling technologies/strategies, including nanotechnology, biotechnology, stem cell, and pharmacotherapy.

Dr. Mousa received his PhD in pharmacology from Ohio State University (USA), completed postdoctoral research in cardiovascular pharmacology at the University of Kentucky (USA), and earned his MBA from Widener University (USA). He is an elected Fellow of the American College of Cardiology (FACC), a Fellow of the National Academy of Clinical Biochemistry (FACB), and fellow of the American Heart Association (FAHA). In 2018, he was elected as a fellow of the National Academy of Inventors (FNAI). He is a member of several national and international societies and serves on several NIH study sections, US Department of Defense, and other funding agencies. In addition, he is the editor-in-chief of *Biomedicines* and serves on the editorial boards of several high impact scientific and medical journals.

 **Raj Bawa, MS, PhD,** is president of Bawa Biotech LLC, a biotech/pharma consultancy and patent law firm based in Ashburn, Virginia, USA that he founded in 2002. Trained as a biochemist and microbiologist, he is an inventor, entrepreneur, professor, and registered patent agent licensed to practice before the US Patent & Trademark Office. He is currently a scientific advisor to Teva Pharmaceutical Industries, Ltd. (Israel), a visiting research scholar at the Pharmaceutical Research Institute of Albany College of Pharmacy (Albany, New York), and vice president and chief intellectual property officer at Guanine, Inc. (Rensselaer, New York). He has previously served as a principal investigator of SBIRs as well as a reviewer for both the NIH and NSF. Currently, he is principal investigator of a CDC grant to develop an assay for Carbapenem-resistant *Enterobacteriaceae*. In the 1990s, Dr. Bawa held various positions at the US Patent & Trademark Office, including primary examiner from 1996–2002. Previously, he was an adjunct professor at Rensselaer Polytechnic Institute in Troy, New York from 1998–2018, where he received his doctoral degree in three years (biophysics/biochemistry). He is a life member of Sigma Xi, co-chair of the nanotech and precision medicine committees of the American Bar Association (2015–20) and founding director of the American Society for Nanomedicine (founded in 2008). He has authored over 100

scientific and legal publications, co-edited four texts, and serves on the editorial boards of a dozen peer-reviewed journals, including serving as an associate editor of *Nanomedicine* (Elsevier). Some of Dr. Bawa's awards include the Innovations Prize from the Institution of Mechanical Engineers, UK (2008); Appreciation Award from the Undersecretary of Commerce, Washington, DC (2001); the Key Award from Rensselaer's Office of Alumni Relations (2005); and Lifetime Achievement Award from the American Society for Nanomedicine (2014).

 **Gerald F. Audette, PhD,** is associate dean of faculty in the faculty of science, an associate professor of chemistry, and member of the Centre for Research on Biomolecular Interactions at York University in Toronto, Canada. His research focuses on the correlation between protein structure and biological activity of proteins involved in bacterial conjugation, in particular the type 4 secretion system from the conjugative F-plasmid of *Escherichia coli*. In addition, his research targets the type IV pilins and associated assembly systems from several bacterial pathogens and is exploring the adaptation of these protein systems for applications in bionanotechnology and nanomedicine. Dr. Audette is the co-editor of volumes 1–4 of the *Jenny Stanford Series on Nanomedicine* and is a subject editor of structural chemistry and crystallography for the journal *FACETS*.

Contents

List of Corresponding Authors

Angelina Angelova Institut Galien Paris-Sud, CNRS UMR 8612, Univ Paris-Sud, Univ Paris-Saclay, 5 rue Jean-Baptiste Clément, F-92296 Châtenay-Malabry, France, Email: angelina.angelova@u-psud.fr

Alejandro Baeza Dpto. Materiales y Produccion Aeroespacial, ETSI Aeronautica y del Espacio, Universidad Politecnica de Madrid, 28040 Madrid, Spain, Email: alejandro.baeza@upm.es

Raj Bawa Bawa Biotech LLC, 21005 Starflower Way, Ashburn, VA 20147, USA, Email: bawa@bawabiotech.com

Neil A. Belson Law Office of Neil A. Belson, LLC, 6225 Hampstead Ct., Port Tobacco, MD 20677, USA, Email: nabelsonlaw@hotmail.com

David M. Berube PCOST, 1070 Partners Way, Campus Box 7565, Suite 5100, Hunt Library, North Carolina State University, Raleigh, North Carolina 27606-7565, USA, Email: dmberube@ncsu.edu

Rakesh Biswas Department of Internal Medicine, Kamineni Institute of Medical Sciences, Narketpally 508254, Telangana, India, Email: rakesh7biswas@gmail.com

Albert Boretti Department of Mechanical and Aerospace Engineering, Benjamin M. Statler College of Engineering and Mineral Resources, West Virginia University, Morgantown, WV 26506, USA, Email: a.a.boretti@gmail.com

James A. L. Brown Department of Biological Sciences, University of Limerick, Limerick, V94 T9PX, Ireland, Email: James.brown@ul.ie

Moses Sing Sum Chow College of Pharmacy, Western University of Health Sciences, Pomona, CA 91766, USA, Email: mchow@westernu.edu

Paul J. Davis The Pharmaceutical Research Institute, Albany College of Pharmacy and Health Sciences, One Discovery Drive, Rensselaer, NY 12144, USA, Email: pdavis.ordwayst@gmail.com

Adam K. Ekenseair Department of Chemical Engineering, 313 Snell Engineering Center, Northeastern University, 360 Huntington Avenue, Boston, MA 02115-5000, USA, Email: a.ekenseair@neu.edu

Kadija Ferryman Data & Society Research Institute, 36 West 20th St., 11th Floor, New York, NY 10011, USA, Email: ferryk18@newschool.edu

Mario Ganau Suite 2204 – 70 Temperance St., Toronto, Ontario M5H0B1, Canada, Email: mario.ganau@alumni.harvard.edu

David T. Harris AHSC Biorepository, 1501 N Campbell Avenue, AHSC 6122, University of Arizona, Tucson, AZ 85724, USA, Email: davidh@email.arizona.edu

Xiao Hu Department of Physics and Astronomy, Rowan University, 201 Mullica Hill Road, Glassboro, NJ 08028, USA, Email: hu@rowan.edu

Satoshi Inoue Department of Systems Aging Science and Medicine, Tokyo Metropolitan Institute of Gerontology, 35-2 Sakae-cho, Itabashi-ku, Tokyo 173-0015, Japan, Email: sinoue@tmig.or.jp

Richard S. Isaacson Department of Neurology, Weill Cornell Medicine, 525 East 68th Street, Room F-616, New York, NY 10065, USA, Email: rii9004@med.cornell.edu

Yann Joly Centre of Genomics and Policy, Faculty of Medicine, McGill University, 740 Dr. Penfield Avenue, #5101, Montréal H3A0G1, Quebec, Canada, Email: yann.joly@mcgill.ca

Alexandre R. Loukanov Division of Strategic Research and Development, Graduate School of Science and Engineering, Saitama University, Shimo-Ohkubo 255, Sakura-ku, Saitama 338-8570, Japan, Email: loukanov@mail.saitama-u.ac.jp

Jonathan Lovell Department of Biomedical Engineering, University at Buffalo, State University of New York, 201 Bonner Hall, Buffalo, NY 14260, USA, Email: jflovell@buffalo.edu

Maria Teresa Di Martino Department of Experimental and Clinical Medicine, Magna Graecia University, Salvatore Venuta University Campus, Viale Europa, 88100 Catanzaro, Italy, Email: teresadm@unicz.it

Shaker A. Mousa The Pharmaceutical Research Institute, Albany College of Pharmacy and Health Sciences, 1 Discovery Drive, Suite 238, Rensselaer, NY 12144, USA, Email: shaker.mousa@acphs.edu

Ajit Narang Small Molecule Pharmaceutics Department, Genentech, Inc., One DNA Way, South San Francisco, CA 94080, USA, Email: narang.ajit@gene.com

Filomena Rossi Department of Pharmacy and CIRPeB, Università Federico II, 80134 Naples, Italy, Email: filomena.rossi@unina.it

Philip Serwer Department of Biochemistry and Structural Biology, The University of Texas Health Science Center, 7703 Floyd Curl Drive, San Antonio, Texas 78229-3900, USA, Email: serwer@uthscsa.edu

Joachim Storsberg Fraunhofer Institute for Applied Polymer Research, Geiselbergstraße 9, Potsdam D-14476, Germany, Email: joachim.storsberg@iap.fraunhofer.de

Vladimir P. Torchilin Center for Pharmaceutical Biotechnology & Nanomedicine, Bouvé College of Health Sciences, 206TF, Northeastern University, 360 Huntington Avenue, Boston, MA 02115, USA, Email: v.torchilin@neu.edu

Thomas J. Webster Department of Chemical Engineering, 313 Snell Engineering Center, Northeastern University, 360 Huntington Ave, Boston, MA 02115, USA, Email: th.webster@neu.edu

Marc S. Williams Genomic Medicine Institute, Geisinger, 100 Academy Drive, Danville, PA 17822, USA, Email: mswilliams1@geisinger.edu

Lei Zheng Departments of Oncology and Surgery, Sidney Kimmel Comprehensive Cancer Center, Johns Hopkins University School of Medicine, 1650 Orleans Street, CRB1 3rd Floor, Baltimore, MD 21287, USA, Email: lzheng6@jhmi.edu

Roy C. Ziegelstein Johns Hopkins University School of Medicine, Baltimore, MD 21205, USA, Email: rziegel2@jhmi.edu

Preface

I spent about two decades in drug discovery and drug development (DDD) at DuPont and DuPont Merck Pharmaceuticals, where we utilized various technology platforms to accelerate drug discovery and improve the probability of success. Although we made significant advances in achieving FDA-approved novel diagnostics and therapeutics that fulfilled unmet medical needs in various diseases, overall it remains a high-risk business that takes a long time to move a concept from the lab to the clinic. In 2002, I decided to transition from pharma to academia where I currently reside in my present position. Throughout my career, I have searched for enabling technologies that would shorten the time to DDD, mitigate the risk, and extend product life cycle. I found that the integration of nanotechnology and biotechnology ("Nanobiotechnology" or "Nanomedicine") fulfilled that promise and even facilitated the road towards precision medicine, the subject matter of this timely volume.

The overarching theme for this volume (36 chapters), accurately labeled *The Road from Nanomedicine to Precision Medicine* relates to the impact of nanomedicine along the continuum toward precision medicine. This specific volume of the outstanding book series was inspired by the first international nanotechnology conference held at my institution in Albany, New York, in August 2015, and numerous speakers at that conference have graciously contributed. This volume aims to define the terminology and nomenclature most appropriately employed in nanomedicine and explores the impact of nanotechnology on drug delivery utilizing passive and/or active targeting facilitating the road toward precision medicine. While not explicitly organized into discrete sections, this volume can be separated into several themes, each addressing a different aspect.

The first theme of this volume sets the stage, giving some general introductory aspects of drug delivery and nanodrugs, and providing a spectacular overview of precision medicine. The chapters in this theme, Chapters 1, 2, 19, 25, 32, and 36, highlight the origins of what is now known as precision medicine and how it arose from nanomedicine to much more personalized medicine through big data, artificial intelligence, data analytics, and the role of genomics, proteomics, and the human genome project. Also included in this theme are discussions on the ethical, legal, and social issues associated with both nanomedicine and personalized medicine. In addition, these chapters provide an overview of the major ethical issues raised by pharmacogenomics and personalized medicine. In this context, some issues will likely be easily resolved and disappear over time. Others may persist, and new ones will probably emerge within the next few years. Consequently, it is critical to continuously revisit the practices and impacts of personalized approaches to medicine.

Following this setting of the stage, the second theme, covered in Chapters 10, 11 18, 20, 21, 26, 27, 29, 30, and 31, explores the definition of the person within the personalized/precision medicine context. Within this theme are chapters pertaining to key aspects of personalized approaches to Alzheimer's disease and cancer as well as the use of genomic and proteomic profiling in approaches to oncology. For instance, Chapter 21 discusses a case-based blended learning ecosystem (CBBLE) employing omics-driven diagnostics to optimize precision medicine and reduce over diagnosis and overtreatment. And Chapter 30 discusses the concept of "Personomics" where the person is central to their care, and knowing your patient (now at a genomic level as well as through personal interactions) is something that physicians are coming back to in their approach to medicine. Clearly, a deeper understanding of the patient is a central and critical component to any precision/personalized treatment regime.

The third theme covers what might be most often though of when considering precision medicine, that is the therapeutics themselves as well as their delivery. Chapters 4 and 5 provide a discussion of multifunctional gold nanoparticles and the impact of nanoparticle size and structure on their functionality, respectively, and Chapter 13 covers the use of cerium oxide nanoparticles as a reactive oxygen-modulating drug. More recent developments in nanomedicine are the use of proteins, peptides, viruses, hydrogels, and other water-soluble polymers as drug delivery vehicles for precision medicine—all these key areas within nanomedicine are extensively covered in this volume. Many of these systems are self-assembling and biocompatible, making them exciting novel approaches to challenging disorders.

Our fourth theme covers topics generally considered as those involved in the detection/diagnosis of disease and monitoring of personalized therapeutic approaches, as well as novel tools and nanomachines. Starting this theme off is Chapter 8, where Drs. Christian Schmidt and Joachim Storsberg discuss the roles of nanomaterials in therapy, diagnostics, theranostics, lab-on-a-chip and other biomedical approaches. In Chapter 9, Rosa and colleagues explore the application of nano-magnetic resonant imaging (Nano-MRI), a novel tool for personalized diagnoses and detection. This novel approach is followed on by the discussion of using porphyrins as novel contrast agents and photosensitizers by Huang and Lovell. This theme is rounded out with three chapters on artificial nanomachines/nanorobotics for personalized healthcare (Chapter 14), nanomotors for the detection of nucleic acids, proteins, and pollutants (Chapter 15), and magneto-responsive biomaterials for tissue engineering (Chapter 16); there is clearly a large array of approaches and potential for precision medicine.

The fifth and final theme provides a view of how the American regulatory framework is approaching personalized medicine. Two chapters (Chapters 3 and 22) from the US Food and Drug Administration explore the agency's views on personalized and precision medicine, genetic variant databases (a key to personalized medicine), and companion diagnostics. The success of many personalized medicines depends on a person's individual genetic makeup, and bringing new medical approaches and products to treat a disease often requires synergistic approaches within an existing regulatory framework to ensure patient safety, the hallmark of any drug regulatory system.

I express my sincere gratitude to the authors, coeditors, and reviewers for their excellent effort in undertaking this project with great enthusiasm. I especially thank the other two editors, Dr. Raj Bawa and Dr. Gerald Audette, for meticulously reviewing various chapters of this book. I also thank Mr. Stanford Chong and Ms. Jenny Rompas of Jenny Stanford Publishing for commissioning us to edit this volume. Mr. Arvind Kanswal of Jenny Stanford Publishing and the staff at Bawa Biotech LLC are acknowledged for their valuable assistance with publication coordination.

Shaker A. Mousa, MBA, PhD
October 5, 2019
Albany, New York

Chapter 18

Proteomic Profiling and Predictive Biomarkers in Neuro-Traumatology and Neuro-Oncology

Mario Ganau, MD, PhD,[a] Nikolaos Syrmos, MD, MSc,[b] Marco Paris, MD,[c] and Lara Prisco, MD, MSc[d]

[a]*Toronto Western Hospital, University of Toronto, Canada*
[b]*Faculty of Health Sciences, Aristotle University of Thessaloniki, Greece*
[c]*National Hospital for Neurology and Neurosurgery, University College London, UK*
[d]*John Radcliffe Hospital, Oxford University, UK*

Keywords: nanomedicine, biomarkers, biosignature, proteomics, genomics, metabolomics, lab-on-a-chip, biosensors, disease profile, quantitative neuroscience, preventive medicine, prognosis, central nervous system, neuro-traumatology, traumatic brain injury, spinal cord injury, cerebrospinal fluid, neurocognitive disorders, degenerative neuropathologies, neuro-oncology, brain tumors, gliomas, microRNA, systems biology, nanoneurosurgery, point-of-care diagnostic testing

18.1 Biomarkers and Quantitative Neuroscience

Biomarkers are quantitative biological signatures of several physiological states or pathological conditions. As such, they are

The Road from Nanomedicine to Precision Medicine
Edited by Shaker A. Mousa, Raj Bawa, and Gerald F. Audette
Copyright © 2020 Jenny Stanford Publishing Pte. Ltd.
ISBN 978-981-4800-59-4 (Hardcover), 978-0-429-29501-0 (eBook)
www.jennystanford.com

commonly used to provide information about the risk of developing specific diseases, the likelihood and rapidity of progression, and the prediction of their outcome [1]. In clinical practice, a biomarker of any type may be used individually or in combination with another biomarker: The two or more (i.e., a profile of data gathered from imaging, genomics, and proteomics testing) are sometimes referred to as a biosignature. As a general rule, a composite measure such as a biosignature can significantly enhance the sensitivity and specificity of diagnostic protocols compared to that of each measure alone.

In quantitative neuroscience, identifying suitable biomarkers is pivotal to streamlining the clinical screening for early and ultra-early diagnosis of degenerative diseases, including cancers. Noteworthy, attempts to identify biosignatures to monitor also the evolution of traumatic injuries, specifically those to the brain and spine, currently represent the latest frontier in neuro-traumatology. Similarly, neuro-oncology is focusing on the markers of spreading of neoplastic cells and recurrence of disease following initial treatment to optimize the management of many primary and secondary brain malignancies.

Furthermore, in the era of preventive medicine what has garnered considerable interest is the possible role of biomarkers not only during the phases of early and ultra-early diagnosing and monitoring of diseases, but more importantly, in research settings. Biomarkers can help scientists with measurable parameters to test the efficacy and safety of experimental protocols. In this regard, the pharmaceutical industry is now looking at a peculiar category of quantitative indicators known as surrogate biomarkers: a specific group of intrinsic laboratory markers that can be objectively measured and evaluated as an indication of normal or pathologic biologic processes or used as pharmacologic responses to a therapeutic intervention [2]. Surrogate biomarkers can, in fact, be used as endpoints when the pathophysiology of the disease and the mechanisms of action of the intervention are thoroughly understood. In those instances, they can actually serve as potential substitutes for clinically meaningful endpoints in clinical trials [3].

In this chapter, we offer an overview of how improvements in the state of the art of many methodologies for genomics, proteomics, and metabolomics testing are rapidly finding a place in modern medicine. In particular, we review the latest discoveries in quantitative neuroscience—specifically those that, through nanotechnology, hold the promise to foster the field of preventive and personal medicine in neuro-traumatology and neuro-oncology—and finally forecast how system biology will further enhance neurosurgery in the realm of diagnostics at the nanoscale.

18.2 Nanomedicine and Its Contribution to Genomics, Proteomics, and Metabolomics

Diagnostics, one of the building blocks of nanomedicine, relies on methodologies that are empowering clinicians with information coming from accurate biosensors: devices that combine a biological component with a physicochemical detector used for the recognition of an analyte [4]. The improved applicability and sensitivity of modern biosensors allowed molecular diagnostics to rapidly move beyond genomics to proteomics and to identify a disease based on their typical post-translational modifications [5]. The proteome and secretome, by definition, are dynamic, changing both in physiological and in pathological conditions; the ultimate goal of determining them is therefore to characterize the flow of information within the cells, through the intercellular protein circuitry that regulates the extracellular microenvironment. Indeed, the study of proteomics and molecular biomarkers already allowed us to identify direct or indirect predictive factors and to determine which affected pathway has more chances to be a selective therapeutic target. Eventually the goal of nanodiagnostics became the study of single cells or the identification of single proteins in a variety of biologic samples, such as serum, plasma, cerebrospinal fluid (CSF), urine, and cell extracts.

Whereas traditional readout systems (i.e., ELISA) require large volumes that ultimately dilute the specimens to generate detectable signals at a cost of reduced sensitivity to the picomolar range (pM, 10^{-12} M/L), with single-protein or single-molecule essays, the focus has shifted on the presence or absence of the protein or molecule of interest and the related signal (see Fig. 18.1), at very low, femtomolar concentrations (fM, 10^{-15} M/L) [6–9]. To support the development of those analytical methods for the detection of relevant biomarker at a micro- or nanometer scale, scientists pursued innovative techniques able to reduce the quantities of biological specimens required for diagnostic assays, while scaling down the minimum amount of DNA or proteins that can be directly detected, and acquired a more comprehensive knowledge regarding the over-expression or under-expression of certain proteins and their physiological or pathological correlations.

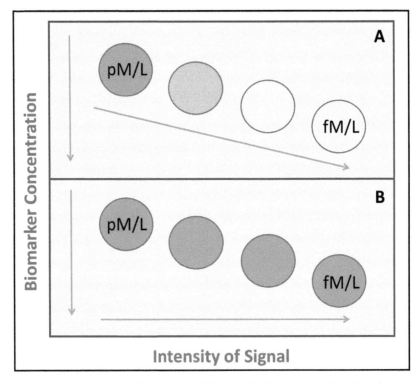

Figure 18.1 Accuracy of biosensors. The sensitivity of conventional readout (A) is proportional to biomarker concentration; as such, its detection fades beyond pM/L. Single-molecule imaging (B) instead is not concentration dependent, and therefore provides positive detection even at much smaller volumes in the range of fM/L.

On one hand, the miniaturization supported by nanotechnology-based methodologies allowed for the manufacture of portable, hand-held, implantable, or even injectable devices; as a result of their minute size, these devices need less sample or reagent for analysis or operation, resulting in enhanced cost- and time-effectiveness [4]. For instance, by harnessing the ability to precisely and reproducibly actuate fluids and manipulate bioparticles, such as DNA, cells, and molecules, micro- and nano-fluidics revolutionized the chemical and biological analysis of those biological substrates by replicating laboratory bench-top technology on a miniature chip-scale device.

A wide range of laboratory and consumer biotechnological applications, from genetic and proteomic analysis kits to cell culture and manipulation platforms allowing *in vitro* analyses of established inflammatory, viral, or oncological biomarkers, enable scientists to predict the behavior of cells under various exogenous stimuli [10]. Furthermore, point-of-care diagnostic testing, which enables testing directly at the patient's bedside, permits physicians to diagnose pathological conditions more rapidly than conventional lab-based testing. By using these devices to reduce the time to diagnosis, the physician is able to make better patient management decisions, leading to improved patient outcomes and reduce the overall cost of care.

On the other hand, advances in microelectronics and biosensor tools have been instrumental in facilitating the development of these diagnostic devices. Various platforms were developed to allow for the simultaneous real-time evaluation of a broad range of disease markers by non-invasive techniques. Among them two classes of microtechnological devices developed since the early 1990s, such as microarray DNA chip and microfluidics systems for lab-on-a-chip diagnostics, found their full application following further miniaturization at the nanoscale [4]. Several techniques from the field of nanotechnology are nowadays available for the miniaturization and biofunctionalization of diagnostic surfaces which result promising in the screening armamentarium for molecular analyses. Indeed, many of them appear particularly suitable for a high-sensitivity determination of panels of biomarkers [11].

18.3 Highlights on Predictive Biomarkers for Neuro-Traumatology

Traumatic brain injury (TBI) and spinal cord injury (SCI) represent a critical worldwide health problem; despite remarkable advances in medical and surgical management, their diagnosis and prognosis remain a major challenge of modern healthcare systems [12,13].

With an estimated 10 million people affected annually by TBI worldwide, it is predicted by the World Health Organization (WHO) that by the year 2020, TBI will surpass many diseases to become the third leading cause of global mortality and disability [14]. In recent years neuro-traumatologists attempted to elucidate the complex mechanisms of TBI progression and to seek acute and chronic biomarkers helpful in optimizing TBI prognosis and management [15]. Given the commonalities between TBI and degenerative diseases such as Alzheimer and other forms of dementia, one fruitful research approach focused on progressive white matter degeneration [16]. As such, the mechanisms involved in myelin loss, delayed microvascular damage, and appearance of focal microbleeds, that are temporally and regionally associated with punctuate blood–brain barrier breakdown, were better understood [17]. Furthermore, four relatively early stages of TBI resulted particularly productive in terms of biomarker discovery: oxidative stress, neuroinflammation, apoptosis, and autophagy [18]. Overall, the above approach allowed to theorize and, later on, experimentally confirm several putative glial and inflammatory biomarkers, which appear to be early (within 24 h from injury) or chronically (over the first 3 months from injury) upregulated. Of note, most of them can be derived from brain tissue, blood, and CSF at time of surgical intervention, or during invasive intracranial monitoring through external ventricular drains or parenchymal microdialysis [19].

For instance, S-100β, a calcium-binding protein retrievable in both serum and CSF, is now considered an established marker for severe TBI [20–22]. Similarly, tau proteins seem to correlate with elevated intracranial pressure, an acute symptom reflective of medically refractory TBI, whereas phosphorylated tau can be found in serum of patients who suffered severe TBI up to several months after injury [23, 24]. However, due to the complexity and heterogeneity of TBI, multiple proteins identified as good candidates to monitor its evolution actually present as perturbed networks; as such, investigations on animal models of TBI are now trying to attempt a more comprehensive system biology approach as the next step in studying the many pathological processes involved [25]. This will likely serve as a new ground to better predict the short- and long-term outcome of patients with very different conditions such as moderate concussion versus devastating brain injuries.

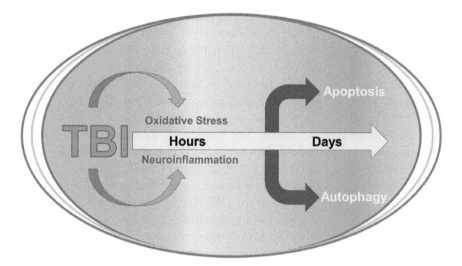

Figure 18.2 Pivotal temporal events following TBI. The upregulation and downregulation of several proteins involved in the processes of oxidative stress, neuro-inflammation, apoptosis, and autophagy are those providing the most putative prognostic biomarkers in neuro-traumatology.

Recently, the area of neuroproteomics was used also as a potential tool in biomarker discovery for SCI, which is considered among the most frequent causes of mortality and morbidity in every medical care system around the world [26]. The incidence of SCI in the United States alone is estimated to be around 11,000 new cases per year, with a prevalence of more than 200,000 individuals in the general population [27]. Hence, there is growing interest in using some of the new diagnostic tools previously mentioned to seek a broader view of the dynamics of genes and proteins involved in the various stages of acute and chronic SCI. Interestingly, in the field of SCI, the goal was never limited to the diagnostic and prognostic aspects pertaining to this pathology; instead several groups attempted to understand the mechanism behind cell regeneration after injury, as well as therapeutic targets for chronic pain and spasticity [28–32].

So far, the most common models of SCI used in genomic and proteomic investigations involved, obviously, contusive injuries, rather than those transecting the spinal cord. In SCI, similarly to TBI, the genes involved in inflammation and immune cell recruitment resulted upregulated in the acute phase post-injury. Specifically, transcriptional changes in inflammation related molecules (i.e., cyclooxygenase 2, interleukins, and tumor necrosis factor α) are known to increase at early time points (between 30 min and 6 h) post-injury; this is followed by immediate activation of cell cycle genes involved in cell damage and death [26]. Also, another common finding seems to be the downregulation of ion channels involved in cell excitability (i.e., N-methyl-D-aspartate, glutamate receptors, and potassium ion channels) which prelude to cytoskeletal destruction (i.e., catabolism of neurofilaments and microtubules associated proteins that reflect the early tissue loss at the site of spinal contusion) [33]. Finally, during the subacute period following SCI, the overall pattern of gene expression changes dramatically from a damage-dominant transcriptional profile to one of active repair involving cellular proliferation and migration, whereas in the chronic phase, markers of matrix and blood vessel remodeling, antioxidant action, and blood–spinal cord barrier re-establishment (i.e., angiopoietin, glial fibrillary acidic protein, glutathione S-transferase) markedly increased [26].

Ultimately, rather than targeting the SCI transcription levels, microRNA array technology was used as a new, powerful tool to reveal the consequences of injury at the gene modulation level [34, 35]. Briefly, microRNAs are a class of RNA molecules implicated in a wide range of gene regulation mechanisms. Their role in SCI can be classified into three categories: (1) upregulation, (2) downregulation, and (3) an early upregulation at 4 h followed by downregulation at 1 and 7 days post-SCI. An initial bioinformatics analysis indicated that the potential targets for miRNAs altered after SCI include genes involved in the inflammation, oxidation, and apoptosis [34]. Soon, this approach could represent a useful way to determine pathologic biosignatures (i.e., abnormal expression of miRNAs coupled with microsegmental neuroradiological imaging of the spinal cord) and contribute not only to the monitoring of SCI but also to the development of therapeutic interventions, through the analysis of the clinical response to innovative regenerative drugs currently tested in randomized controlled trials, such as the Rho inhibitor known as Cethrin [37].

18.4　Why Progress in Neuro-Oncology Relies on Biomarkers?

Whereas the diagnostics in neuro-traumas investigate a biological system that implies a physiological state at time of injury, the tumor microenvironment represents a much more complex and heterogeneous system, consisting of intricate interactions between the tumor cells and its neighboring non-cancerous stromal cells [38]. On one hand, the principal stromal cells in the tumor niche consist of endothelial cells, macrophages, immune cells, and fibroblasts; on the other hand, the tumor cells and their progenitor stem cells show unique behaviors due to variation in genetic and environmental factors, with dramatic implications in the pathogenic conditions of brain cancers. For these reasons, efforts in neuro-oncology, and in particular those revolving around gliomas, the most common and aggressive primary brain tumors, continue to focus on elucidating the complex molecular mechanisms underlying the pathophysiology of brain cancers, with the aim to define sensitive and specific biomarkers exploitable in clinical settings [39, 40].

Highlighting the issue of inter- and intra-tumor heterogeneity, many single-cell analysis techniques, such as cell-based, nucleic acid-based, protein-based, metabolite-based, and lipid-based, emerged over the past decade as an important approach to detecting variations in morphology and genetic or proteomic expression within the tumor niche [41].

The demand for parallel, multiplex analysis of protein biomarkers from very small biospecimens obtained at the time of surgery, or through blood/CSF samples during follow-up, represented for years an increasing trend. To date, one example of this approach to multiplex protein assays relies on the DNA-directed-immobilization technique, where spatially encoded antibody microarrays (capitalizing on the chemical robustness of DNA oligomer strands and on the reliable assembly of DNA-labeled antibodies via complementary hybridization) allow the capturing of multiple proteins of interest in a cell lysate all at once [10]. Furthermore, in the past five years, the miniaturization of conventional techniques led to the development of DNA barcode-type arrays at 10 times higher density than standard spotted microarrays, potentially enabling high-throughput and low-cost measurements. The versatility of barcode immunoassay is demonstrated by the ability to stratify glioma patients via multiple measurements of a dozen blood protein biomarkers for each patient [42]. In a research setting, this technique was further implemented by coupling it with the immobilization of living cells, with an outlook for multiplex assay of cytoplasmic proteins. By this method, it was possible to detect simultaneously not only enzymes, such as phospho-extracellular signal regulated kinase (ERK), but also receptors, such as EGFR, both key nodes of the PI3K signaling pathway, remarkably upregulated in high-grade gliomas (such as aggressive glioblastomas), at concentrations of 10 to 1 ng/ml [43, 44].

Of note, another recent achievement was the interrogation of cross talk between signaling pathways within a cellular population as a paradigm to understand the overall tumor architecture, so that now we have a broader view of how cell–cell contacts, and soluble factors signaling, influence the interaction between glioma cells and the surrounding environment. In particular, experiments conducted on PI3K signaling in a model of glioma cancer (U87 cells line), as a function of cell–cell separation, demonstrated that the expression of EGFR in a subpopulation of cells represents a trigger for parenchymal invasion, whereas its expression in the majority or the entirety of the tumor would not enhance tumorigenicity but would instead result in a self-inhibiting state [45].

With specific regard to the forthcoming steps of multiplexing analysis of gliomas, the concomitant detection of key biomarkers such as p53, nestin, EGF-R, OLIG2, and PTEN and the increase in the statistical value of assays obtained from homogeneous cell populations (perhaps sorted according to their positivity for CD 133, a marker of glioma stem cells) will result in an enhanced recognition of recurrent tumors, or the newly discovered glioma circulating tumor cells [40–46].

18.5 System Biology and the Future of Neurosurgery in the Realm of Nanometers

Nanotechnology-based approaches are being extensively explored to discover, identify, and quantify clinically useful cellular signatures for early detection, diagnosis, and prognosis of any pathology. The study of the brain as the main focus of neuroscience poses significant challenges for protein analysis. Apart from the large number of proteins spanning a wide dynamic range in abundance, brain function is governed by a multidimensional interplay between gene–protein, protein–protein, and protein–environment interactions, that are tightly regulated both spatially and temporally [47]. As previously described, specimens important for neuroscience research include brain tissue, blood/serum/plasma, and CSF; each of these has specific limitations, which makes their analysis difficult. Namely, the high-molecular complexity of the brain necessitates complex proteomic analysis in order to investigate neuronal processes of clinical interest, and the related proteins governing those processes in physiological and pathological conditions.

As shown in this chapter, the high-throughput nature of genomic, proteomic, and metabolomics studies has generated so far a huge volume of information mostly in a hypothesis-driven manner. This, however, anticipated the next wave of clinical research: In the near future, in fact, those data will probably acquire new meanings as systems biology is paving the way to understand the dynamics that govern protein networks and regulate the progression of the neurologic pathologies of neurosurgical interest discussed here.

Systems biology is one of the newest domains in biological science, and its aim is to better investigate complex biological processes. Briefly, it represents a mathematical model capable of predicting the altered processes or functions of complex biological systems under normal or perturbed conditions [48]. As such, by using tailored algorithms, it is theoretically possible to extract a vast number of statistically significant set of proteins, altered in different functional pathways pertaining to a given disease. The suitability of those sets of proteins as possible biomarkers is then further tested by interpreting their biological role from gene activation to protein expression, and therefore finding meaningful relationships among genes, proteins, cell processes, and diseases. Eventually, system biology allows analyzing changes in gene expression across thousands of transcripts and visualizing the interplay between proteins of interest as pattern profiles for the understanding of CNS neuronal and glial injury. To this extent, the objective is not just to identify single protein candidates but to reveal how those biomarkers interact with each other at subcellular levels, generating a global understanding of neural biological entities, such as receptors, synapses, mitochondria, and organelle structures.

Despite in their infancy, targeted studies coupling high-throughput data from genomics, proteomics, and metabolomics, with the bioinformatics power of a system biology approach might potentially provide insights on many more neurologic diseases [49]. Hence, in the near future, scientists will be hopefully able to identify new key players, propose novel drug targets, and provide clinicians and surgeons with more accurate biosignatures in a non-speculative manner.

In conclusion, because of their clinical importance, the generation at the nanoscale of new materials suitable for diagnostic purposes has always been central in the conversation about nanomedicine. However, given the excellent opportunity provided by diagnostics at the nanoscale, and the unmet dream to predict the occurrence of diseases and somehow prevent them, the mathematical models required to analyze the big data produced by those platforms will acquire even more importance.

Abbreviations

CNS: central nervous system

CSF: cerebrospinal fluid

EGFR: epidermal growth factor receptor

ELISA: enzyme-linked immunosorbent assay

ERK: phospho-extracellular signal regulated kinase
OLIG2: oligodendrocyte lineage transcription factor 2
PTEN: phosphatase and tensin homolog
SCI: spinal cord injury
TBI: traumatic brain injury
WHO: World Health Organization

Disclosures and Conflict of Interest

The opinions and perspectives here reflect the current views of the authors. The authors declare that they have no conflict of interest and have no affiliations or financial involvement with any organization or entity discussed in this chapter. No writing assistance was utilized in the production of this chapter and the authors have received no payment for its preparation.

Corresponding Author

Dr. Mario Ganau
Suite 2204 – 70 Temperance St.
Toronto, Ontario M5H0B1
Canada
Email: mario.ganau@alumni.harvard.edu

References

1. Biomarkers Definitions Working Group (2001). Biomarkers and surrogate endpoints: Preferred definitions and conceptual framework. *Clin. Pharmacol. Ther.*, **69**(3), 89–95.

2. Colburn, W. A. (2000). Optimizing the use of biomarkers, surrogate endpoints, and clinical endpoints for more efficient drug development. *J. Clin. Pharmacol.*, **40**(12 Pt 2), 1419–1427.

3. Prentice, R. L. (2009). Surrogate and mediating endpoints: Current status and future directions. *J. Natl. Cancer Inst.*, **101**(4), 216–217.

4. Ganau, M., et al. (2015). Diagnostic challenges of nanomedicine. In: Escoffier, L., Ganau, M., Wang, J., eds. *Commercializing Nanomedicine: Industrial Applications, Patents and Ethics*, Pan Stanford, Singapore, pp. 21–49.

5. Krutzik, P. O., et al. (2004). Analysis of protein phosphorylation and cellular signaling events by flow cytometry: Techniques and clinical applications. *Clin. Immunol.*, **110**(3), 206–221.

6. Lequin, R. M. (2005). Enzyme immunoassay (EIA)/enzyme-linked immunosorbent assay (ELISA). *Clin. Chem.*, **51**(12), 2415–2418.

7. Engvall, E., Perlmann, P. (1971). Enzyme-linked immunosorbent assay (ELISA). Quantitative assay of immunoglobulin G. *Immunochemistry*, **8**(9), 871–874.

8. Koppelman, S. J., et al. (2004). Detection of soy proteins in processed foods: Literature overview and new experimental work. *J. AOAC Int.*, **87**(6), 1398–1407.

9. Rissin, D. M., et al. (2010). Single-molecule enzyme-linked immunosorbent assay detects serum proteins at subfemtomolar concentrations. *Nat. Biotechnol.*, **28**(6), 595–599.

10. Ganau, M., et al. (2015). A DNA-based nano-immunoassay for the label-free detection of glial fibrillary acidic protein in multicell lysates. *Nanomedicine*, **11**(2), 293–300.

11. Ganau, L., et al. (2018). Understanding the pathological basis of neurological diseases through diagnostic platforms based on innovations in biomedical engineering: New concepts and theranostics perspectives. *Medicines (Basel)*, **5**(1), e22.

12. Prisco, L., et al. (2012). Early predictive factors on mortality in head injured patients: A retrospective analysis of 112 traumatic brain injured patients. *J. Neurosurg. Sci.*, **56**(2), 131–136.

13. Ganau, M., Fehlings, M. G. (2017). Clinical and health policy-related challenges of pediatric spinal cord injuries. *Neurol. India*, **65**(3), 475–476.

14. Mathers, C. D., Loncar, D. (2006). Projections of global mortality and burden of disease from 2002 to 2030. *PLoS Med.*, **3**(11), e442.

15. Ganau, M., Prisco, L. (2013). Comment on "neuromonitoring in traumatic brain injury". *Minerva Anestesiol.*, **79**(3), 310–311.

16. Ganau, M., et al. (2018). Delirium and agitation in traumatic brain injury patients: An update on pathological hypotheses and treatment options. *Minerva Anestesiol.*, **84**(5), 632–640.

17. Blyth, B. J., et al. (2009). Validation of serum markers for blood-brain barrier disruption in traumatic brain injury. *J. Neurotrauma*, **26**(9), 1497–1507.

18. Ganau, M., et al. (2018). Current and future applications of biomedical engineering for proteomic profiling: Predictive biomarkers in neuro-traumatology. *Medicines (Basel)*, **5**(1), 19.

19. Thelin, E. P., et al. (2014). Microdialysis monitoring of CSF parameters in severe traumatic brain injury patients: A novel approach. *Front. Neurol.*, **5**, 159.

20. Korfias, S., et al. (2007). Serum S-100B protein monitoring in patients with severe traumatic brain injury. *Intensive Care Med.*, **33**(2), 255–260.

21. Thelin, E. P., et al. (2014). Secondary peaks of S100B in serum relate to subsequent radiological pathology in traumatic brain injury. *Neurocrit. Care*, **20**(2), 217–229.

22. Goyal, A., et al. (2013). S100b as a prognostic biomarker in outcome prediction for patients with severe traumatic brain injury. *J. Neurotrauma*, **30**(11), 946–957.

23. Zemlan, F. P., et al. (2002). C-tau biomarker of neuronal damage in severe brain injured patients: Association with elevated intracranial pressure and clinical outcome. *Brain Res.*, **947**(1), 131–139.

24. Rubenstein, R., et al. (2015). A novel, ultrasensitive assay for tau: Potential for assessing traumatic brain injury in tissues and biofluids. *J. Neurotrauma*, **32**(5), 342–352.

25. Kobeissy, F. H., et al. (2016). Neuroproteomics and systems biology approach to identify temporal biomarker changes post experimental traumatic brain injury in rats. *Front. Neurol.*, **7**, 198.

26. Guingab-Cagmat, J. D., et al. (2010). Neurogenomics and neuroproteomics approaches to studying neural injury. In: Fehlings, M. G., Vaccaro, A. R., Boakye, M., Rossignol, S, Ditunno, J. F., Burns, A. S., eds. *Essentials of Spinal Cord Injury*, Thieme, New York-Stuttgart, pp. 605–618.

27. Ravenscroft, A., et al. (2000). Chronic pain after SCI. A patient survey. *Spinal Cord*, **38**(10), 611–614.

28. Katano, T., et al. (2006). Proteomic identification of a novel isoform of collapsin response mediator protein-2 in spinal nerves peripheral to dorsal root ganglia. *Proteomics*, **6**(22), 6085–6094.

29. Redell, J. B., et al. (2009). Traumatic brain injury alters expression of hippocampal microRNAs: Potential regulators of multiple pathophysiological processes. *J. Neurosci. Res.*, **87**(6), 1435–1448.

30. Ding, Q., et al. (2006). Proteome analysis of up-regulated proteins in the rat spinal cord induced by transection injury. *Proteomics*, **6**(2), 505–518.

31. Kang, S. K., et al. (2006). Proteomic analysis of injured spinal cord tissue proteins using 2-DE and MALDI-TOF MS. *Proteomics*, **6**(9), 2797–2812.

32. Kunz, S., et al. (2005). Comparative proteomic analysis of the rat spinal cord in inflammatory and neuropathic pain models. *Neurosci. Lett.*, **381**(3), 289–293.

33. Park E., et al. (2004). The role of excitotoxicity in secondary mechanisms of spinal cord injury: A review with an emphasis on the implications for white matter degeneration. *J. Neurotrauma*, **21**(6), 754–774.

34. Nakanishi, K., et al. (2010). Responses of microRNAs 124a and 223 following spinal cord injury in mice. *Spinal Cord*, **48**(3), 192–196.

35. Izumi, B., et al. (2011). MicroRNA-223 expression in neutrophils in the early phase of secondary damage after spinal cord injury. *Neurosci. Lett.*, **492**(2), 114–118.

36. Liu, N. K., et al. (2009). Altered microRNA expression following traumatic spinal cord injury. *Exp. Neurol.*, **219**(2), 424–429.

37. Forgione, N., Fehlings, M. G. (2014). Rho-ROCK inhibition in the treatment of spinal cord injury. *World Neurosurg.*, **82**(3–4), e535–539.

38. Ganau, L. (2018). Predicting complexity of tumor removal and postoperative outcome in patients with high-grade gliomas. *Neurosurg. Rev.*, **41**(1), 371–373.

39. Ganau M. (2014). Tackling gliomas with nanoformulated antineoplastic drugs: Suitability of hyaluronic acid nanoparticles. *Clin. Transl. Oncol.*, **16**(2), 220–223.

40. Ganau, M., et al. (2018). How nanotechnology and biomedical engineering are supporting the identification of predictive biomarkers in neuro-oncology. *Medicines (Basel)*, **5**(1), e23.

41. Ganau, L., et al. (2015). Management of gliomas: Overview of the latest technological advancements and related behavioral drawbacks. *Behav. Neurol.*, **2015**, 862634.

42. Fan, R., et al. (2008). Integrated barcode chips for rapid, multiplexed analysis of proteins in microliter quantities of blood. *Nat. Biotechnol.*, **26**(12), 1373–1378.

43. Shi, Q., et al. (2012). Single-cell proteomic chip for profiling intracellular signaling pathways in single tumor cells. *Proc. Natl. Acad. Sci. U.S.A.*, **109**(2), 419–424.

44. Shin, Y. S., et al. (2010). Chemistries for patterning robust DNA microbarcodes enable multiplex assays of cytoplasm proteins from single cancer cells. *ChemPhysChem*, **11**(14), 3063–3069.

45. Wang, J., et al. (2012). Quantitating cell–cell interaction functions with applications to glioblastoma multiforme cancer cells. *Nano Lett.*, **12**(12), 6101–6106.

46. Ganau, M., et al. (2017). Enhancing contrast agents and radiotracers performance through hyaluronic acid-coating in neuroradiology and nuclear medicine. *Hell. J. Nucl. Med.*, **20**(2), 166–168.

47. Filiou, M. D., et al. (2012). To label or not to label: Applications of quantitative proteomics in neuroscience research. *Proteomics*, **12**(4–5), 736–747.

48. Longtin, R. (2005). An integrated approach: Systems biology seeks order in complexity. *J. Natl. Cancer Inst.*, **97**(7), 476–478.

49. Vizirianakis, I. S., Fatouros, D. G. (2012). Personalized nanomedicine: Paving the way to the practical clinical utility of genomics and nanotechnology advancements. *Adv. Drug Deliv. Rev.*, **64**(13), 1359–1362.

Chapter 19

Nanomedicine and Personalized Care: Facts and Fiction

David M. Berube, PhD[a] **and Emily Winderman, PhD**[b]

[a]*Department of Communication and PCOST, North Carolina State University, Raleigh, North Carolina, USA*
[b]*Department of Communication, University of Minnesota, Twin Cities, Minnesota, USA*

Keywords: especially tasty fruit, nanomedicine, Nanomedicine Initiative, National Nanotechnology Initiative, nanotechnology, personalized medicine, individualized medicine, precision medicine, hyperbole, Nanotechnology Alliance for Cancer, post-truth, team science, President Donald Trump, Senator Hillary Clinton, translational science, bench to bedside, post-truth atmosphere

Nanotechnology-enabled medicine or nanomedicine is one of the most popular applications of nanotechnology. In 2013, Over 240 nanomedicine products were identified either in use or in trials [1]. Public Communication of Science and Technology (PCOST) continues to investigate the linkage between claims about nanomedicine and the broader implications associated with those claims: personalized and individualized medicine. Berube has made three arguments about nanomedicine to date. In this chapter, he and a colleague add a fourth.

The Road from Nanomedicine to Precision Medicine
Edited by Shaker A. Mousa, Raj Bawa, and Gerald F. Audette
Copyright © 2020 Jenny Stanford Publishing Pte. Ltd.
ISBN 978-981-4800-59-4 (Hardcover), 978-0-429-29501-0 (eBook)
www.jennystanford.com

First, nanomedicine like most of nanotechnology has been hyperbolized. To characterize the National Nanotechnology Initiative as something truly groundbreaking, proponents exaggerated the role nanoscience may play in medicine. In turn, media amplify their messages: "Media have risen to help the public along by framing even the dullest findings as monumental with life-saving and life extending qualities." In addition, "[d]iscoveries in the lab are reported through university and industry publicity offices emphasizing the implications of findings. When a researcher has produced findings, department chairpersons and heads contact public relations professionals at a university to release an announcement. University public relations departments are notorious for exaggerating the accomplishments of their researchers as well as hyperbolizing the implications of their discoveries" [2].

Second, nanomedicine combined with health-related phobias transforms nanomedicine related nanotechnology into "especially tasty fruit." Much like the concept of "low hanging fruit," this intrinsic favorability of nanomedicine to various publics suggests nanomedicine—among nearly all other applications of nanotechnology—has the potential to transfer its high positive valence to other applications of nanotechnology becoming a hallmark of the entire industry and a variable to rehabilitate the industry should some crisis with nanotechnology surface [3].

Third, nanomedicine has a very special audience: the patient and their supportive intimates and family. "The 'non-patient' public may want more choice. Given the luxury of critical reflection animated by claims of counter productivity (toxic side effects), evidence of morbidity and mortality, and in some cases, eco-toxicity, non-governmental organizations and civil advocacy groups may challenge claims from researchers and developers." Under high stress circumstances brought on by medical issues, patients react as unique audiences. "When looking into the health care industry, the public is privileged by critical distance. However, once the public becomes 'patients' the dynamics change. As 'patients,' the public finds themselves reoriented from a 'critical' setting into a 'care' setting." Despite the nurturing quality of "care" settings, they can be profoundly alienating [4]. In general, nanotechnology was hailed as the next industrial revolution. More recently, there has been an effort to rebrand it as "green."

Nanomedicine was similarly hyperbolized. Two predictions from 2005 have been recognized as being not only off mark, but hyperbolizing.

The first hyperbolic claim came in 2005, at the 4th annual Nano Business conference at New York Marriott's Financial Center [5]. Jeff Jaffe, then the president of Bell Labs Research, provided a vision of a ubiquitous, virtually invisible nanotech-enabled communication network that would offer global "secure and instantaneous" conversation, as if people were in the same room.

The second claim came from Dr. Anna Baker of the National Cancer Institute, where she served as deputy director for Strategic Scientific Initiatives. For several years, she implemented multi/trans-disciplinary programs including the Nanotechnology Alliance for Cancer. Dr. Barker wowed a gathered audience when she spoke of the grand challenge to "*do away with pain and suffering caused by cancer*" by 2015 [5]. She is the current director of Transformative Healthcare Knowledge Networks and co-director, Complex Adaptive Systems, at the School of Life Sciences at Arizona State University (ASU). Of course, in 2019, we know that Baker's prediction was off the mark.

As social scientists and rhetorical critics, we are interested not only in what has been said, but also why what had been said could have been so "off the mark" by someone whom we assume handled a large sum of grant resources if not the meta-phorical keys to the country's cancer labs. While a determination of her motivations is mostly guesswork, we can name some socio-politically grounded assumptions that we will examine in more detail below.

In the post-truth political world in which we live, the rhetorical function of truth has been tarnished. During our last election, PolitiFact scored US President Donald Trump alone with 174 false and "pants on fire" (zero evidence) statements [6]. Truth doesn't seem to matter much anymore. We have false news events posted on the Internet with sometimes disastrous consequences. For instance, one false news event was associated with a shooter visiting a pizzeria to investigate a child sex ring allegedly run by Senator Hillary Clinton and John Podesta [7]. Conspiracy generated by false statements and false news posts feed into the affirmation bias whereby people tend to believe something as true if it is consistent with what they already feel or believe is true. The public is highly vulnerable to false claims especially when they do not have the capacities to separate fact from fiction and when the false claims are consistent with what they already believe is true. The post-truth world is a world of fiction: claims that seem plausible are not based in facts. The shooter referenced above is a twisted example of the consequences of post-truth fiction.

While in most cases, falsities and hyperbole do not lead to violence and death, they do lead to misunderstanding and wrongful decision making. In some cases, we will argue exaggeration can be dangerous for promoting false promises in regards to the realities of our current technological and medical capabilities. In other words, when researchers promise groundbreaking advances in nano-enabled personalized medicine without considering the economic and communicative constraints preventing that actualization, they risk misleading the public and exacerbating this recently legitimated 'post-truth' cultural climate.

To support our argument, this chapter covers four concepts: a brief review of the prominence of nanomedicine research in the USA among some of the top health and medicine academic research institutions; a critical analysis of personalized and individualized care within the current health care economy; an assessment of how effective nanomedicine can be at implementing individualized care to the public; and finally, an review of some of the dangers associated with hyperbole and exaggeration in promoting public programs involving science and technology in our current climate.

19.1   The Nanomedicine Community

There is the National Network of Nanomedicine Development Centers (https://commonfund.nih.gov/nanomedicine/fundedresearh).   The Nano-medicine Initiative applies an engineering approach to the study of cellular and subcellular systems in an effort not only to understand, but to precisely control molecular complexes that operate at the nanoscale [8].

There are many outstanding nanomedicine centers in the US. For purposes of illustration, the following surface regularly in the literature:

- Johns Hopkins Center (JHC) for Nanomedicine at the Wilmer Eye Institute, which claims six start-ups, including Kala Pharmaceuticals, Graybug, Theraly Pharmaceuticals and its own Theraly Fibrosis, Ashvattha Therapeutics and Orpheris. A highlight involves work on fibrosis. There are currently no effective anti-fibrotic drugs available for patients. The dendrimer-platform technologies targeting neuroinflammation is being commercialized and moved towards clinical trials [9].

- UCSB Center for Nanomedicine (CNM) in Santa Barbara and Goleta is leading research into sepsis, diabetes, cancer and auto-immune diseases. UC San Diego's work in sepsis is very interesting. According to their Web site and CNM publication records, CNM scientists have discovered a receptor system in the body that modulates sepsis. This receptor system controls blood coagulation and thrombosis thereby promoting host survival of pneumococcal sepsis. We have further learned how to manipulate this receptor system to reduce inflammation and increase survival in both Gram-positive and Gram-negative sepsis. There have not been any effective drugs developed to treat sepsis for decades, and the pharmaceutical industry has mostly dropped all sepsis research and development programs. CNM discoveries of a novel receptor system in the body that modulates sepsis by controlling blood protein aging and turnover has provided a new class of molecules for targeted therapies for sepsis [10].

- UC San Diego's Center for Excellence in Nanomedicine and Engineering (CNME). The tissue engineering work with computer-assisted, light-based 3D nanopatterning of hydrogels, which could allow uniform vascularization of cellular constructs. They claim such a development would represent a major step forward, as the non-uniform vascularization resulting from endothelial cell infiltration may not support extended culture or survival of implants [11].

Cancer is the focus of much work in nanomedicine. The National Cancer Institute and its NCI Alliance for Nanotechnology in Cancer have been making promises about nanomedicine for some time [12] and many centers across the country have reported similar promises.

Here are a few illustrations.

- The UCSB teams report their scientists have devised multiple nanoparticle approaches that enable cancer drugs, for example, to home to tumors and subsequently to enter into tumors with increased infiltrating capabilities thereby exposing more of the tumor to drugs that can eliminate cancer cells. The ability to target drugs and imaging agents to diseases such as cancer is one of the next major steps required to ultimately focus new drugs and treatments to diseased tissues while sparing normal healthy tissues and thereby reducing side effects.

- One of the foremost nanomedicine/cancer centers is the Texas Center for Cancer Nanomedicine. It is developing and applying a diverse array of nano-platforms for new therapeutics, methodologies for reliable monitoring of therapeutic efficacy, early detection approaches from biological fluids and advances in imaging, and cancer-prevention protocols for ovarian and pancreatic cancers [13].

Finally, many large hospital and care networks have opened Departments of Nanomedicine as well, far too many to list here.

19.2 Nanomedicine and Personalized Care

The claims associating nanomedicine and personalized care are extensive and what follows hardly exhaust the developments and projections made by researchers in nanomedicine.

Nanomedicine-based tests and assays may add to the list of laboratory tests that can improve care-giving. Potentially, some of these tests could provide highly individualized information. The theory works this way [14]:

> *The promise of personalized nanomedicine was based on the understanding that each individual possesses a unique genetic profile predisposing him or her to respond to therapies differently.*

Now, armed with the predictive power of in silico *models of patient populations, whole-genome testing, clinically qualified biomarkers that can assess individual responses to therapies, and other tools, the biomedical field is poised for significant advances and benefits to individuals rather than to their population mean.*

For example, Glinky wrote that "laboratory tests for measurements of microRNAs and other classes of small noncoding RNAs in archived, formalin-fixed, paraffin-embedded human samples with sufficient specificity and sensitivity has significantly limited the development of clinically relevant noncoding RNA–based diagnostic and therapeutic applications" [15]. His commentary discussed research advances that allow him to conclude: "These exciting studies will facilitate the conclusive, evidence-based interrogation of the molecular and genetic mechanisms of disease states and enable unequivocal validation of diagnostic and therapeutic noncoding RNA targets" [15].

19.3   Challenges Involved in Personalized Medicine

Despite the notable promises professed for nanomedicine's therapeutic implications—Personalized Medicine and Precision Medicine—there are notable constraints that prevent the actualization of widespread implementation. Challenges currently exist in the generation, management, and deployment of relevant data to accommodate healthcare professionals and to reach patients who would benefit from personalized treatment programs.

Personalized medicine has been hailed as a welcome advancement for diseases such as heterogeneous forms of cancer, chronic illnesses, smoking cessation [16], and mental illness [17]. The practice of mapping trends over large genomic data sets is replacing observational data [18]. Yet, it is data (and its utility, applicability, accessibility, and management) that also presents one of the most profound challenges for a future of precision-based medical care. Until researchers can come to terms with data, personalized medicine will remain an implication—not a reality.

A first challenge is generating large sets of data for a comparatively small sample size. Because researchers have embraced the complexity of genomic data to map the structure of heterogeneous diseases [19], an abundance of data emerges from a relatively small sample of patients [20]. As Zakim and Schwab note, high costs of data collection limit the diversity of phenotypes in clinical studies [18]. A large-scale review of personalized treatment in colorectal cancers [21] observed that despite promising to advance the mapping of biomarkers, "further validation is required because of contradictory study reports, small cohort sizes, inconsistent detection of genes and enzyme activity and variations in data analysis." Despite what the authors name as "promising" futures, they still reported a need to "validate more biomarkers through large prospective trials" [21]. This is one way that personalized medicine has hit roadblocks in actualization. The sheer amount of data generated is remarkable when considering that it reflects a small sample of subjects being studied. This "big p, small n" problem potentially undermines the goal of surveying complex disease profiles [22].

We should remain suspicious of hyperbolic claims of broad trans-disease applicability for personalized medicine when the number of subjects studied remains small. For example, Zarrinpar et al.'s [23] pilot prospective randomized clinical study tested tacrolimus drug efficacy in post-liver transplant patients and included eight participants. Four participants received a parabolic personalized dosing based on a second-order algebraic equation. The other four, the control group, were treated by their physician's observational guidance. The authors noted that the equation reduced the need for physician recalibration of dosing and celebrated the possible applications of algebraic drug administration to cancer, infectious disease, and cardiovascular medicine. While the authors tested the efficacy of the equation on animal models, their optimism should raise questions about the equation's translational and predictive power across different classes of disease [23].

Conversely, the abundance of data that can be collected from a single person has also been hailed as an advance in precision medicine, but this too raises concerns about replicability. Schnork argued in *Nature* that one-person trials ("N-of-1") could be advantageous to examine multiple factors that contribute to disease onset. Individual physicians have certainly taken on this task in an ad hoc manner, as when a patient is prescribed a drug to treat a symptom for which it has not yet been approved (typically defined as "compassionate use.") Schnork points to "N-of-1" population aggregation to reach the best of both worlds: limiting negative side effects from large-scale clinical studies and aggregating individuated insights. Unsurprisingly, the cost of care is the major obstacle to implementing "N-of-1" and individual trials would have to be implemented in routine clinical care to be effective [24]. Herein lies the possibility for nanotechnology. "Nanotechnology enabled carrier and reporter systems may assist in the collection of patient-specific data noninvasively by giving access to a previously inaccessible spatiotemporal resolution" [25]. Technologies designed to collect and synthesize vast amounts of data are crucial to implementing laboratory successes in larger populations, particularly as the wealth of data that can be extracted from a single individual grows.

Without improved abilities to generate, analyze, and store data, the enterprise of personalized medicine risks contributing to a multi-tiered global health system with profound access issues. Although a hallmark of personalized medicine lies in the curation of unique treatments for unique individuals facing unique iterations of disease, the generation, storage, and funding for computational resources are lacking. Such financial constraints restrict the reach of personalized medicine to those with resources living in the wealthiest of countries. Alyass et al. conclude: "Major investments need to be made in the fields of bioinformatics, biomathematics, and biostatistics to develop translational analyses of omics data and make the best use of high-thoroughput technologies" [22]. Referencing the "big-p, small-n" issue, visualization technologies generate massive amounts of data applicable to either a single person or a small population, requiring copious economic and technological resources for the benefit of a few.

19.4   Translational Science Problems

Perhaps some of the nanomedicine hyperbole has emerged from grand promises made at the national level. Although former President Obama announced that he would direct his administration to invest significant resources into personalized medicine (what he named "precision medicine") in his 2015 State of the Union Address, there are significant roadblocks to actualizing this investment for improved patient outcome. Scientific translation between domains of knowledge production, or what the 1968 *New England Journal of Medicine* named as the "bench-to-bedside interface" presents personalized medicine with constraints that require interdisciplinary teams to work together effectively to get the insights generated by physician scientists into the hands of clinicians [26]. This requires not only a mechanism for transferring technical insights from research studies, but a plan for clinicians to engage with their patients on timely basis when survival is measured in weeks and months, not years. Unfortunately, much of the work to this end completed between 2009 and today reflects personalized medicine being "lost in translation" [27].

Scientists and clinicians working within personalized medicine research agree that there is a gap between the insights generated in the laboratory and its impact for people living with a given disease. Malentecchi et al. sent a survey to hospital directors in European academic institutions. Eighty-seven percent of surveyed participants agreed. "Cooperation and collaboration between health-care professionals is becoming a pressing need to consolidate a personalized medicine approach" [28, p. 984]. After surveying the field, the authors concluded that a "landslide breakthrough" in translational research was necessary for proper implementation of the research advances into clinical practice. They further explain that this goal is unobtainable without "communication among professionals, doctors, and patients" [28, p. 981]. Communication is pivotal amongst scientists and practitioners even when they share a level of technical expertise. Joly et al. remind us that "the perception of personalized care, in primary care...differs among healthcare professionals and patients" [29, pp. 406–407]. When researchers, hospital clinicians, and primary care professionals are not speaking the same language, the likelihood that life-prolonging insights will make its way to patients in need is low.

While communication plays a pivotal role in translating laboratory insight into clinical practice, advances have been primarily in genomic database and biobanking. In 2009, Madhavan et al. discussed the emergence of Rembrandt, a new database that would integrate, distribute, and analyze complex genomic insights to bridge the gap between the "bench to bedside" [30, p. 157]. While the authors praised the possibility of a genomics database to permit discoveries, they admitted that searching for the right data means that they are effectively looking for a "needle in a haystack" [30, p. 158]. The authors note that the goal of Rembrandt was not necessarily producing numbers, but rather insight into the problem at hand. While this is a worthy goal, it does produce pause towards its ultimate efficacy if it does not include the necessary communicative training in translating the proverbial hay to find the needle.

To pursue the goals of translational science beyond establishing biobanks and genomic databases, personalized and precision medicine advocates need to take seriously the insights that have been generated in *The Science of Team Science* [31]. Team Science seeks to understand how collaboration across multiple areas of expertise can productively occur. Team science training exceeds simply sharing information; it focuses on the communicative practices and policy initiatives that allow insights to translate from "bench to bedside" when experts are siloed in their own daily practices of knowledge production and implementation. Personalized medicine stakeholders need to be integrating the following team-based concerns into their translational considerations: collaboration, including satisfaction, impact and mutual trust, and integration [32]. Each of these frameworks indicates that there must be critical interrogation of the interactions between professionals trained in unique modes—be it laboratory science or clinical practice. Without attending to the human dimensions of translation, we risk missing very real obstacles to integrating personalized medicine—the scientists and clinicians themselves. After all, while there may well be the technologies present to permit cross-pollination of ideas, without the communicative collaboration, we risk burying the needle in the haystack even further.

All told, personalized medicine—while promising—presents notable constraints in translation. This is not a problem that in a post-truth environment we can simply wish away with improved technological capacity. Our technological progress is only as useful to helping patients as the ability for knowledge-generating experts to commune for the benefit of many, not just a few.

19.5 The Economy of Personalized Care

Despite the hype, nanomedicine will not change this commercial form of distribution of medicine and treatments. Worse, the distribution will continue to be concentrated in rich countries and for the upper classes [33, p. 117].

The nursing community has discussed personalized care for some time. The ability to present oneself, to be with a patient in a way that acknowledges shared humanity, is the basis of much of nursing practice [34]. This approach seems fundamental to nursing but we want to use a different definition: *Individualized and personalized care involves providing medical assistance to a patient that has been designed to resolve an ailment or injury maximizing the individualized therapeutic variables each individual patient brings to the case instant.*

Is personalized care possible? The idea that nanomedicine can produce diagnostic and therapeutic technologies that can be effective in treating individuals whose ailments have individual characteristics is commendable. Unsurprisingly, most care regimens we used today do not work most of the time. Indeed, the failure rates are often very high yet still profitable for the health care industries from pharmaceuticals through broad hospital networks. Put simply, selling drugs and procedures that do not work remains profitable.

In an era of debates over hospital care expenses and shortfalls in practicing physicians in some specialty fields, we hear about telemedicine whereby a patient may use a kiosk to begin a dialogue with a caregiver sharing symptoms and concerns. In turn, the caregiver triages the patient to a health care regimen that may involve an actual visit with a nurse practitioner or a doctor, but in some cases, it may involve self-treatment regimens.

Before we begin to examine some of the claims made by researchers in nanomedicine, it is important to examine the economy of individualized and personalized care. For example, it is unsurprising that the pharmaceutical industry profits on the significant failure of drugs designed for a general population. Failure rates can be very high. The public has become quite accustomed to switching from one drug to another one that does the trick. High failure rates are profitable. While in the therapeutic care field that may not be the case, high risks especially economic ones can reduce the incentive to practice in a specific field of practice. High failure rates lead to liability issues and overall competency concerns that can impact a caregiver and associated institutions.

The economy of nanomedicine and its high costs at least in the earliest applications does not fare well for populations on the margins of society.

> *Even in the wealthiest regions of the world, the cost of novel medicines and medical devices are frequently prohibitive, and access to them is limited to the most privileged or denied for all because they are too costly to bring to market. Nanomedicines comprising an active "conventional" pharmaceutical agent, such as doxorubicin and paclitaxel, plus a vectoring nanoparticle, are typically more expensive per dose or per unit mass of the active agent than the naked drug by itself [35].*

The pharmaceutical industry is directed to the rich. Following Forbes, the world's 10 best selling drugs in 2005 were all for rich patients (high cholesterol, heartburn, high blood pressure, schizophrenia, and depression) [35].

The entire concept of personalized treatment may be an obstacle for indigent patients altogether. "Medicines or treatments that cannot be distributed en masse constitute an additional difficulty in getting treatment to those with lower purchasing capacity or who live in locations where there is less medical infrastructure" [36, p. 117]. Indigent populations do not have the capacity to generate the types of information that make personalized therapies effective. It is a fundamental problem of access.

According to the World Health Organization (WHO), by 2002, 80% of the world drug market was concentrated in North America, Europe, and Japan, a geographic area where only 19% of the world population lives. However, 90% of the burden of disease can be found in poor countries, where patients do not have the purchasing capacity to buy medicine [36].

While a 2012 unpublished dissertation [37] suggests a net savings from reduction of side effects and other benefits using quality-of-life years analysis, the costs up front will be higher and could be much higher than the author had previously imagined.

Individuals who may benefit the most from individualized medicine and personalized care will be those who have the means to pay for it.

19.6   Dangers of Exaggeration and Hyperbole

Can exaggeration kill you? Not usually. Most of the rhetoric of exaggeration is treated as falsehood and disposed appropriately into one of the circular files we empty regularly. Whether the file is out short-term memory or the trash bin, exaggeration generally has little serious implications.

While excess is seldom wise, hyperbole is more than excessive exaggeration and it can be unintentional and intentional. As a trope, hyperbole is used effectively when both the speaker/writer and the audience/readership understand they are speaking and hearing exaggerated rhetoric. Indeed, it is intended to work that way, just like the tropes of sarcasm and irony. However, when the audience/reader interprets hyperbole as fact, we get misinformation. When the speaker or writer knows hyperbole will be misinterpreted by the audience or reader, we get mal-information. Misinformation is countered by refutative information. We correct what we understand wrongly with more information.

Mal-information is another problem altogether. Mal-information can be found in conspiracy arguments and conspiracy arguments are the germ of "fake news," a recurring phenomenon in social media. A conspiracy argument involves an arguably false major premise and a truly false minor premise. Returning to our earlier example of fake news:

Major Premise: Senator Hillary Clinton and John Podesta can't be trusted.

Minor Premise: Senator Hillary Clinton and John Podesta run a child porn ring.

Much like rebutting rumors, the problem is the arguer needs to attack and rebut both premises while the defense involves using one premise to bolster the other. This is incredibly frustrating and mostly unproductive.

A second powerful reason mal-information is so powerful is because it generally makes a claim within a set of plausible scenarios that some audiences are predisposed to believe as true. Misinformation is consistent with what already can be found in the wheelhouse of beliefs some publics believe is true. We tend to believe true what we are prepared to believe is true. Affirmation or confirmation bias involves many behaviors, including searching, interpreting, favoring, and recalling information in a way that confirms one's preexisting beliefs or hypotheses, while giving disproportionately less consideration to alternative possibilities [38]. Once you believe something, most anything that aligns with this belief is embraced as believable as well.

We have panoply of beliefs and attitudes that help us organize the complex components of the world in which we live. Susan Fiske called us "cognitive misers" [39]. We spend little time analyzing concepts unless they meet strict requirements: high salience and high exigence [40]. The concepts are important to us and there is some phenomenon or interpretation thereof which is significantly and substantially important. Otherwise, we interpret concepts using what Kahneman calls System 1 reasoning [41] and what Chaiken calls heuristically [42].

Another example: In the early years of this millennium, mal-information generated by anti-genetically modified seeds opponents led to some countries (Malawi, Mozambique, Zambia, and Zimbabwe) rejecting genetically modified related food aid under PL 480, the US Food for Peace program. The result was malnutrition and likely starvation on some scale. According to Vidal [43], anti-GM "groups are putting millions of lives at risk in a despicable way." While some academics have argued the provision of GMO food aid had little to do with starvation [44], most agricultural sciences disagree. While research needs to be done to evidence the direct linkage between anti-GMO information and malnutrition/starvation, the hypothesis is worth contemplating.

Yet another example: In 2012, mal-information was released that indicated comments about the dangers of nanoparticles in sunscreens in Australia resulted in Australians using less sunscreens in general. Suggesting that an anti-nanoparticle public interest group (Friends of the Earth Australia) was misdirecting the public, advocates asserted the claims by the interest group were having a negative effect on health and well-being. First, it is important to note that in 2008, Berube published a piece claiming the Friends of the Earth Australia publication against nanoparticles in sunscreens was suspect [45], a team from PCOST investigated the data that had been generated on sunscreen use and the proximate relationships this may have had to overall sunscreen use by sunbathers and could find no significant effect. Clearly, the hypothesis that mal-information was dangerous in this case instant could not be sustained.

The challenge for the debunked hyperbole is the overwhelming amount of hyperbole and the inability of fact-checkers to have the temporal and financial capacity to investigate each case. Furthermore, we have no protocol in place to help us decide which cases need to be investigated.

19.7 Conclusions

Argument by hyperbole is problematic for many reasons and can be dangerous to public life. The public is unprepared for a truth-free society where anyone can make any claim they want wherein the "fact police" determine what faction deserves to be rebutted and debunked. One of the foundations of debate is that asserting something is true is much easier than proving the assertion is incorrect.

In the current post-truth atmosphere, it might behoove advocates in science and technology, and in this case nano-medicine, to take great precaution against exaggerating their claims. Concerns have been expressed elsewhere in the overclaims of nanomedicine [46]. Overclaims and hyperboles can have real effects.

For example, Baker's claim on cancer may come back to haunt government agencies like the NIH and the NCI when they attempt to increase appropriations for nanomedicine research. Ultimately, we know the public is unprepared to vet claims about complicated science and technology. After the last election season, it is reasonable to suspect this problem may extend to many members of the political elite and the media.

While grants demand broader implications and public relation officers want great claims that capture headlines, learning to temper overclaims can reduce difficult arguments down the line. To date, "the promise and hype of personalized medicine has outpaced its evidentiary support. In order to achieve favorable coverage and reimbursement and to support premium prices for personalized medicine, manufacturers will need to bring better clinical evidence to the marketplace and better establish the value of their products" [47].

Disclosures and Conflict of Interest

This article was supported in part by a grant from the National Science Foundation. ECCS-1542015. NNCI: North Carolina Research Triangle Nanotechnology Network. RTNN. All opinions expressed are the authors and do not necessarily reflect those of the National Science Foundation, North Carolina State University or the University of Minnesota. Both authors report no conflicts of interest.

Corresponding Author

Dr. David M. Berube
PCOST, 1070 Partners Way
Campus Box 7565
Suite 5100, Hunt Library
North Carolina State University
Raleigh, North Carolina 27606-7565, USA
Email: dmberube@ncsu.edu

References

1. Etheridge, M. L., Campbell, S. A., Erdman, A. G., Haynes, C. L., Wolf, S. M., McCullough, J. (2013). The big picture on nanomedicine: The state of investigational and approved nanomedicine products. *Nanomedicine*, **9**, 1–14.

2. Berube, D. M. (2006). *Nanohype: The Truth behind the Nanotechnology Buzz*, Prometheus Books, Amherst, NY.

3. Berube, D. M. (2009). Public acceptance of nanomedicine: A personal perspective. *WIREs Nanomed. Nanobiotechnol.*, **1**, 2–5.

4. Berube, D. M. (2016). The audience is the message: Nanomedicine as apotheosis or *damnatio memoria*. In: Bawa, R., Audette, G. F., and Reese, B. E., eds. *Handbook of Clinical Nanomedicine: Law, Business,*

Regulation, Safety, and Risk, Pan Stanford Publishing, Singapore, pp. 1117–1140.

5. Business Wire (2005). NanoBusiness 2005 concludes fourth successful conference in New York City with record number of sponsors and attendees. Available at: http://www.businesswire.com/news/home/20050607006026/en/NanoBusiness-2005-Concludes-Fourth-Successful-Conference-York (accessed on January 24, 2019).

6. Politifact (2016). Donald Trump's file. Available at: http://www.politifact.com/personalities/donald-trump/ (accessed on January 24, 2019).

7. Kang, C., Goldman, A. (2016). In Washington pizzeria attack, fake news brought real guns. *New York Times*, December 5. Available at: http://www.nytimes.com/2016/12/05/business/media/comet-ping-pong-pizza-shooting-fake-news-consequences.html?_r=0 (accessed on January 24, 2019).

8. NIH Completes Formation of National Network of Nanomedicine Centers. Available at: https://www.nih.gov/news-events/news-releases/nih-completes-formation-national-network-nanomedicine-centers (accessed on January 24, 2019).

9. Commercialization. Available at: http://cnm-hopkins.org/commercialization/ (accessed on January 24, 2019).

10. Sepsis. Available at: https://www.cnm.ucsb.edu/research/sepsis (accessed on January 24, 2019).

11. Center of Excellence for Nano-Medicine and Engineering (CNME). Available at: https://iem.ucsd.edu/centers/cnme-nano-medicine-engineering.html (accessed on January 24, 2019).

12. Main page. Available at: https://www.cancer.gov/ (accessed on January 24, 2019).

13. Center Overview. Available at: https://nano.cancer.gov/action/programs/uthsc/index.asp (accessed on January 24, 2019).

14. Ferrari, M., Philibert, M. A., Sanhai, W. R. (2009). Nanomedicine and society. *Clin. Pharmacol. Ther.*, **85**(5), 466–467.

15. Glinsky, G. V. (2013). RNA-guided diagnostics and therapeutics for next-generation individualized nanomedicine. *J. Clin. Invest.*, **123**(6), 2350–2352.

16. Lerman, C., Schnoll, R. A., Hawk Jr, L. W., Cincirpini, P., George, T. P., Wileyto, E. P., Swan, G. E., Benowitz, N. I., Heitjan, D. F., Tyndale, R. F. (2015). A randomized placebo-controlled trial to test a genetically

informed biomarker for personalizing treatment for tobacco dependence. *Lancet Respir. Med.*, **3**(2), 131–138.

17. Simon, G. E., Perlis, R. H. (2010). Personalized medicine for depression: Can we match patients with treatments?. *Am. J. Psychiatry*, **167**(12), 1445–1455.

18. Zakim, D., Schwab, M. (2015). Erratum: Data collection as a barrier to personalized medicine. *Trends Pharmacol. Sci.*, **36**, 68–71.

19. West, M., Ginsburg, G. S., Huang, A. T., Nevins, J. R. (2006). Embracing the complexity of genomic data for personalized medicine. *Genome Res.*, **16**, 559–566.

20. Beim, P. Y., Elashoff, M., Hu-Seliger, H. H. (2013). Personalized reproductive medicine on the brink: Progress, opportunities, and challenges ahead. *Reprod. Biomed. Online*, **27**, 611–623.

21. Patil, H., Saxena, S. G., Barrow, C. J., Kanwar, J. R., Kepat, A., Kanwar, R. K. (2017). Chasing the personalized medicine dream through biomarker validation in colorectal cancer. *Drug Discovery Today*, **22**(1), 111–119.

22. Alyass, A., Turcotte, M., Meyre, D. (2015). From big data analysis to personalized medicine for all: Challenges and opportunities. *BMC Med. Genomics*, **8**(1), 33.

23. Zarrinpar, A., Lee, D., Silva, A., Datta, N., Kee, T., Eriksen, C., Weigle, K., Agopian, V., Kaldas, F., Farmer, D., Wang, S. E., Busuttil, R., Ho, C., Ho, D. (2016). Individualizing liver transplant immunosuppression using a phenotypic personalized medicine platform. *Sci. Transl. Med.*, **8**(333), 1–13.

24. Schnork, N. J. (2015). Time for one person trials. *Nature*, **520**(7549), 609–611.

25. Herrman, I. K., Rosslein, M. (2016). Personalized medicine: The enabling role of nanotechnology. *Nanomedicine*, **11**(1), 1–3.

26. Lehrer, R. I. (1968). Phagocytes and the 'bench-bedside' interface. *N. Engl. J. Med.*, **278**, 1014–1016.

27. Hulot, J. (2008). Pharmacogenomics and personalized medicine: Lost in translation? *Genome Med.*, **2**(3), 1.

28. Malentacchi, F., Mancini, I., Brandslund, I., Vermeersch, P., Schwab, M., Marc, J., van Schaik, R. H., Siest, G., Theodorsson, E., Pazzagli, M., Di Resta, C. (2015). Is laboratory medicine ready for the era of personalized medicine? A survey addressed to laboratory directors of hospitals/academic schools of medicine in Europe. *Clin. Chem. Lab. Med.*, **53**(7), 981–988.

29. Joly, P. B., Kaufmann, A. (2008). Lost in translation? The need for 'upstream engagement' with nanotechnology on trial. *Sci. Cult.*, **17**(3), 225–247.

30. Madhavan, S., Zenklusen, J. C., Kotliarov, Y., Sahni, H., Fine, H. A., Buetow, K. (2009). Rembrandt: Helping personalized medicine become a reality through integrative translational research. *Mol. Cancer Res.*, **7**(2), 157–167.

31. Falk-Krzesinski, H. J., Contractor, N., Fiore, S. M., Hall, K. L., Kane, C., Keyton, J., Klein, J. T., Spring, B., Stokols, D., Trochim, W. (2011). Mapping a research agenda for the science of team science. *Res. Eval.*, **20**(2), 145–158.

32. Mâsse, L. C., Moser, R. P., Stokols, D., Taylor, B. K., Marcus, S. E., Morgan, G. D., Hall, K. L., Croyle, R. T., Trochim, W. M. (2008). Measuring collaboration and transdisciplinary integration in team science. *Am. J. Preventive Med.*, **35**(2), S151–S160.

33. Hall, R. M., Sun, T., Ferrari, M. (2012). A portrait of nanomedicine and its bioethical implications. *J. Law Med. Ethics*, **40**(4), 763–779.

34. Herper, M., Kang, P. (2006). The world's ten best selling drugs, *Forbes*, March 22. Available at: http://www.forbes.com/2006/03/21/pfizer-merck-amgen-cx_mh_pk_0321topdrugs.html (accessed on January 24, 2019).

35. Invernizzi, N., Foladori, G. (2006). Nanomedicine, poverty and development. *Development*, **49**(4), 114–118.

36. Lichtenburg, F. (2005). Pharmaceutical innovation and the burden of disease in developing and developed countries. *J. Med. Philos.*, **30**, 663–690.

37. Bosetti, R. (2012). Cost effectiveness of cancer nanotechnology. *Dissertation*, D/2012/2451/8, Hasselt University, Belgium.

38. Plous, S. (1993). *The Psychology of Judgment and Decision Making*, McGraw-Hill, Inc., NY.

39. Fiske, S., Taylor, S. (1991). *Social Cognition*, McGraw-Hill, Inc., NY.

40. Bitzer, L. (1968). The rhetorical situation. *Philos. Rhetoric*, **1**, 1–14.

41. Kahneman, D. (2011). *Thinking, Fast and Slow*, Farrar, Straus & Giroux, NY.

42. Chaiken, S. (1980). Heuristic and systematic information processing and the use of source versus message cues in persuasion. *J. Pers. Soc. Psychol.*, **39**, 752–766.

43. Vidal, J. (2002). US dumping unsold GM food on Africa. The Guardian, October 7. Available at: https://www.theguardian.com/science/2002/oct/07/gm.famine (accessed on January 24, 2019).

44. Zerbe, N. (2004). Feeding the famine? American food aid and the GMO debate in Southern Africa, *Food Policy*, **29**, 593–608.

45. Berube, D. M. (2008). Rhetorical gamesmanship in the nano debates over sunscreens and nanoparticles. *J. Nanopart. Res.*, **10**, 23–37.

46. Duncan, R., Gaspar, R. (2011). Nanomedicine(s) under the microscope. *Mol. Pharmacol.*, **8**, 2101–2141.

47. Meckley, L. M., Neumann, P. J. (2010). Personalized medicine: Factors influencing reimbursement. *Health Policy*, **94**(2), 91–100.

Supplemental Readings

Allon, I., et al. (2017). Ethical issues in nanomedicine: Tempest in a teapot? *Med. Health Care Philos.*, **20**(1), 3–11.

Bawa, R., Johnson, S. (2007). The ethical dimensions of nanomedicine. *Med. Clini. N. Am.*, **91**(5), 881–887.

Halappanavar, S., Vogel, U., Wallin, H., Ysuk, C. L. (2017). Promise and peril in nanomedicine: The challenges and needs for integrated systems biology approaches to define health risk. *WIRES Nanomed. Nanobiotechnol.*, **10**(1), e1465.

Hermann, I., Rosslein, M. (2016). Personalized medicine: The enabling role of nanotechnology. *Nanomedicine*, **11**(1), 1–3.

Kaushik, A., Jayan, R. D., Nair, M., eds. (2017). *Advances in Personalized Nanotherapeutics*, Springer.

Mohammed, W., et al. (2018). Potential application of nanotechnology in health care: An insight. *Nanoscale Rep.*, **1**(2), 1–8.

Nouri, M., Lopez, J. (2017). Nanomedicine and personalised medicine: Understanding the personalisation of health care in the molecular era. *Sociol. Health Illn.* **39**(4), 547–565.

Sahakyan, N., et al. (2017). Personalized nanoparticles for cancer therapy: A call for greater precision. *Anti-Canc. Agents Med. Chem.*, **17**(8), 1033–1039.

Shi, J., Kantoff, P. W., Wooster, R., Farokhzad, O. C., et al. (2017). Cancer nanomedicine: Progress, challenges and opportunities. *Nat. Rev.: Cancer.*, **17**, 20–37.

Tinkle, S., et al. (2014). Nanomedicines: Addressing the scientific and regulatory gap. *Ann. N. Y. Acad. Sci.*, **1313**, 35–56.

Ziegelstein, R. C. (2017). Personomics: The missing link in the evolution from precision medicine to personalized medicine. *J. Pers. Med.*, **7B**(4), 11.

Chapter 20

Precision Immuno-Oncology: Prospects of Individualized Immunotherapy for Pancreatic Cancer

Jiajia Zhang, MD, MPH,[a,b,c] Christopher L. Wolfgang, MD, MS, PhD,[a,b,c] and Lei Zheng, MD, PhD[a,b,c]

[a]*Departments of Oncology and Surgery, Sidney Kimmel Comprehensive Cancer Center, Johns Hopkins University School of Medicine, Baltimore, Maryland, USA*
[b]*Bloomberg-Kimmel Institute for Cancer Immunotherapy, Baltimore, Maryland, USA*
[c]*Pancreatic Cancer PMCoE Program, Johns Hopkins University School of Medicine, Baltimore, Maryland, USA*

Keywords: pancreatic cancer, pancreatic ductal adenocarcinoma, precision medicine, immunotherapy, immune checkpoint, tumor microenvironment, biomarker, tumor-infiltrating lymphocytes, mutational burden, mismatch repair deficiency, chromosomal chaos, neoantigens, neoepitopes, stromal matrix, CD40/CD40L, RAS/MAPK activation, focal adhesion kinase, DNA repair, immune tolerance

20.1   Introduction

Pancreatic cancer is the fourth leading cause of cancer death for both men and women, with an annual incidence of approximately

The Road from Nanomedicine to Precision Medicine
Edited by Shaker A. Mousa, Raj Bawa, and Gerald F. Audette
Copyright © 2020 Jenny Stanford Publishing Pte. Ltd.
ISBN 978-981-4800-59-4 (Hardcover), 978-0-429-29501-0 (eBook)
www.jennystanford.com

53,000 new cases in the United States, of whom 43,000 are expected to die [1]. Despite a better understanding of tumor biology and optimization of current treatment modalities, 5-year survival rate of pancreatic cancer is only 5–6% [2]. These sobering results have spawned the efforts spent on developing novel therapies to improve the treatment outcomes. Immunotherapy, which targets cancer cells by augmenting the immune system, has become a game changer in modern cancer cares. Emerging immunotherapeutics including immune checkpoint blockade antibodies and CAR T cell therapies have led to durable response among responsive patients. However, challenges remain as only an objective response rate of 10–30% was observed among those receiving single agent immunotherapy. There is a growing need for individualized medicine solutions to guide patient selection by predicting treatment response, to spare patients from ineffective treatment, and also to avoid toxicity associated with immunotherapy.

With rapid advancement in the technology of next generation sequencing and novel bioinformatics platforms, molecular and genetic profiling of tumors has become an integral venue to guide personalized cancer cares. Opportunities for precision medicine have been expanded to the field of immune-oncology. Integrating immunotherapy with precision medicine by leveraging molecular, genomic, cellular, clinical, behavioral, physiological, and environmental parameters to tailor immunotherapy options has generated enormous interests. PD-L1 status, mutation burden and neoantigen load has been shown in various cancer types to predict positive response to immune checkpoint inhibitors. More challenges lie in validations of the clinical values of these biomarkers in selecting the patients for immunotherapy. Particular opportunities and challenges of personalized immunotherapy exist for pancreatic cancers.

20.2 Overview of Immune-Biology of Pancreatic Cancer

Cancer immunotherapy is based on the exquisite specificity of both antibodies and T cells to differentiate the subtle differences between cancer and normal cells and thus mediate a response against tumor cells. To trigger an effective killing of cancer cells, a series of stepwise events must be initiated and allowed to proceed and be expanded iteratively [3, 4]: (1) release of tumor specific or associated antigens; (2) antigen presentation (dendritic cells/APCs); (3) priming and activation of T cells; (4) trafficking of T cells to tumors (CTLs); (5) infiltration of T cells into tumors; (6) recognition of cancer cells by T cells; (7) killing of cancer cells. Each step of anti-tumor immune response is characterized by the coordination of numerous factors, with stimulatory factors promoting immunity and inhibitory factors reducing immune activity or keeping the process in check. Therefore, cancer immunotherapy has been attempted by targeting each of the rate-limiting steps. Over the last decade, researches have suggested an immunosuppressive TME (Fig. 20.1) as the fundamental basis for most of the rate-limiting steps of an effective anti-tumor immune response in pancreatic cancer [5, 6].

Pancreatic cancer bears unique immunologic hallmarks. With a low-moderate mutational burden, pancreatic cancer cells are less immunogenic and reside within a dense stromal environment. The stromal matrix, which comprised of cellular and acellular components, such as fibroblasts, myofibroblasts, pancreatic stellate cells, immune cells, blood vessels, extracellular matrix and soluble proteins such as cytokines and growth factors, contributes to tumor growth and promotes metastasis [7]. In line with the excessive desmoplasia, another conundrum of pancreatic cancer is a deficiency of vasculature, leading to impaired perfusion and drug delivery. The unique TME of pancreatic cancer gives rise to a series of challenges along multiple steps of anti-tumor immune response: a lack of strong cancer antigens or epitopes recognized by T cells (Step 1), minimal activation of cancer-specific T cells (Step 3), poor infiltration of T cells into tumors (Step 5), downregulation of the major histocompatibility complex on cancer cells (Step 6), and immunosuppressive factors and cells in the tumor microenvironment (Step 2–7). As a consequence, T cells are largely excluded from the immediate TME, which may, at least partially, explain the unresponsiveness to checkpoint blockade therapy (including anti-PD-1/PD-L1 and anti-CTLA-4) [8]. An individual pancreatic cancer patient can have a deficiency in one or multiple step(s) of anti-tumor immune response. Thus, a personalized medicine approach will ultimately be required for effective cancer immunotherapy.

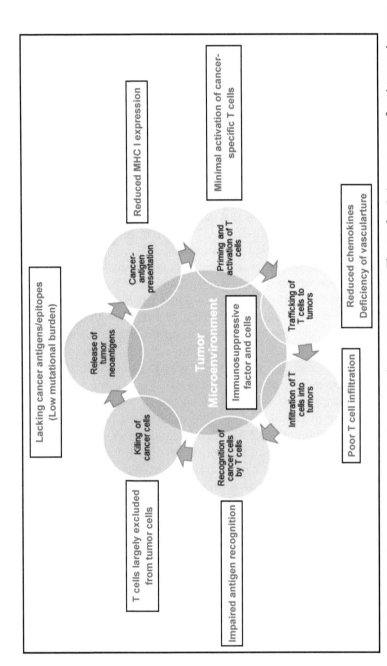

Figure 20.1 Deficiencies in the tumor microenvironment of pancreatic cancer. The mechanistic processes of anti-tumor immune response are delineated in the diagram. The deficiencies in tumor microenvironment that lead to the failure of anti-tumor immune response are highlighted in red.

20.3 Biomarkers Identification

20.3.1 Tumor Microenvironment

20.3.1.1 PD-L1 expression

The PD-1/PD-L1 pathway regulates the balance between the stimulatory and inhibitory signals needed for effective immune responses [9]. In tumors, upregulation of PD-L1 on cancer cells creates an "immune shield" to protect against immune attack from T cells and contributes to the development of T-cell exhaustion [10]. This lays the foundation of checkpoint blockade of the PD-1-PDL1 pathway pathway to unleash the effector functions of T cells and reinvigorate the killing of tumor cells, and also highlights the value of PD-L1 expression as a predictive biomarker for this class of therapy. For tumors reported with clinical response to the anti-PD-1/PD-L1 therapies including melanoma, renal cell carcinoma (RCC), non-small cell lung cancer (NSCLC) and bladder cancer, the range of PD-L1 expression falls from 14% to 100% [11–13]. In the KEYNOTE-010 study, the magnitude of benefit with pembrolizumab was associated with the levels of PD-L1 expression, with increased survival benefits in patients with PD-L1 expression ≥ 50% of tumor cells (regardless of the staining intensity with the 22C3 clone) [14]. A further study restricting patient recruitment among PD-L1 ≥ 50% showed a significantly improved progression free survival and subsequently established pembrolizumab as a first-line treatment for metastatic NSCLC with PD-L1 expression ≥ 50% [14]. By contrast, a clinical trial based on PD-L1 selection of ≥5% showed no survival advantage of nivolumab as the first line treatment compared to the standard-of-care chemotherapy [15]. Of note, differences in the PD-L1 expression threshold may not be the single factor contributing to the divergent results of the above two clinical trials; other factors such as difference in assays of assessing PD-L1 expression and imbalance in baseline patient characteristics may also have contributed to the difference in the clinical trial results.

In PDACs, reports of PD-L1 expression vary from 12–90% [16–19], suggesting that we do not have a consensus on the expression of PD-L1 [20, 21]. Multiple factors may have contributed to the wide range of reported expression rates including the specificity of the staining methodology, the difference in the PD-L1 positivity cutoff, the difference between primary and metastatic lesions, and the inclusion of tumor vs. immune cells in quantifying PD-L1 expression [22]. However, one important factor may be the effect of the prior treatment on the expression of PD-L1 [20, 22]. Thus, extreme cautions should be taken in investigating the predictive role of PD-L1 expression for immunotherapy response in pancreatic cancer. In fact, none of the pancreatic cancer patients without mismatch repair deficiency has responded to the single-agent anti-PD-1 antibody treatment.

20.3.1.2 Tumor-infiltrating lymphocytes

Tumor-infiltrating lymphocytes (TILs) are the immune cell context direct interacting with the TME and have been shown to act either as a predictive or prognostic factor for treatment in various cancers [21–25]. In a study in colon cancer, the densities of $CD3^+$, $CD8^+$, granulysin, and granzyme B $(GZMB)^+$, and $CD45RO^+$ cells in each tumor region (tumor center and invasive margin) were shown to be prognostic [23]. In melanoma, a significant correlation was observed between the presence of both TILs and B7-H1 expression in the tumor microenvironment and the response to checkpoint blockade [26]. Nevertheless, only having positive PD-L1 expression and TILs is not sufficient for pancreatic cancer responding to anti-PD-1 therapies. In our study evaluating 24 pancreatic ductal adenocarcinomas from patients who received neoadjuvant GVAX vaccination, although essentially all the tumors have induction of TILs and PD-L1 expression, the survival of patients is correlated with the infiltration of myeloid cells [27]. Therefore, only a comprehensive characterization of the tumor-infiltrating immune cells would adequately support the precision medicine practice for pancreatic cancer.

20.3.2 Biomarkers in the Genome

20.3.2.1 Mutations and mutation burden

Known as a genetic disease, cancer is initiated by mutations that activate oncogenic drivers and eventually turn normal cells into cancer cells through activation of genes that promote proliferation or suppression of genes that modulate apoptosis. In many cancers, oncogenesis is accompanied by the accumulation of mutations, which lead to generation of neoantigens that capable of eliciting potent T cell responses and drive response to current immunotherapies [4]. A growing body of evidence in the past five years have suggested that the overall mutational burden is a predictive biomarker of checkpoint blockade therapies [28–31]. For example, in tumors with higher spectrum of somatic mutational burdens, such as melanoma and non–small cell lung cancers, treatment with anti–PD-1 blockade has resulted in significantly improved survival outcome [14, 32, 33]. Moreover, recent research identified truncal mutations, which arise early in oncogenesis and are shared by almost all of the cancer cells, are more likely to elicit a potent anti-tumor response compared to those arise later and shared by only a subgroup of cancer cells (branch mutation) [4]. In solid tumors with mismatch repair (MMR) deficiency, higher mutational burdens were seen [34, 35]. In a Phase II clinical trial of progressive metastatic carcinoma with or without MMR deficiency, whole-exome sequencing revealed a significantly increased somatic mutations per tumor (mean, 1782 vs. 73) in MMR–deficient tumors as compared with MMR–proficient tumors. This corresponds to a remarkable increase in immune-related objective response rate (40% vs. 0%) and prolonged immune-related progression-free survival rate in MMR-deficient tumors vs. MMR-proficient tumors (78% vs. 11%) [35]. In the subsequent expanded cohort which included 86 advanced MMR-deficient patients across 12 different tumor types, objective radiographic responses were observed in 53% of patients and complete responses were achieved in 21% of patients [36]. This leads to the recent approval of the microsatellite instability high condition, the phenotype of mismatch repair deficiency, as a pan-cancer biomarker for immune checkpoint blockade therapies and set a stage for the development of precision immuno-oncology. Another potential genetic biomarker is associated with point mutations affecting DNA replication—polymerase epsilon (POLE) or polymerase delta (POLD1), which have been reported to exhibit some of the highest mutational burdens identified to date and render patients with exceptionally mutated (ultramutated) cancers [37–40]. Improved survival has been observed for POLE-mutated tumors in retrospective studies of conventional therapy setting [21]; and a few case reports also reported dramatic responses to immune checkpoint blockade [41, 42].

In pancreatic cancer, the average mutation burden was reported from 26~67 mutations per case [43, 44], representing a relatively low mutational burden compared to those seen in other solid tumors. In general, PDAC with more copy number alterations (indicative of chromosomal instability) exhibited mutations in DNA break repair genes and trended toward poor prognosis [43]. Microsatellite instability (MSI) and POLE/POLD1 mutations are found in 2% and 1–2% of patients with pancreatic cancer, respectively [45]. Favorable outcomes are anecdotally reported for MMR-deficient PDAC [46], suggesting that even lower incidence of MMR deficiency would be found in the PDAC patients who require treatment. A retrospective cohort study of resected PDAC (154 in discovery cohort and 95 in replication cohort) clustered PDAC into five predominant mutational subtypes: age related, double-strand break repair, mismatch repair, and one with unknown etiology. Those with higher frequency of somatic mutations and tumor-specific neoantigens corresponded to double-strand break repair and mismatch repair subtypes, which were found to have higher expression of antitumor immunity, including activation of CD8+ T lymphocytes and overexpression of regulatory molecules (CTLA-4, PD-1, and indolamine 2,3-dioxygenase 1 [IDO1]) [47]. Moreover, a significant number of germline mutations that are associated with genes for DNA repair were found in the double-strand break repair subtype including BRCA1, BRCA2, PALB2, and ATM. These gene mutations render individuals susceptible to pancreatic cancer and are potential biomarkers for response to targeted therapy or immunotherapy. Ongoing clinical trials are testing poly (ADP-ribose) polymerase (PARP) inhibitors in the small fraction of patients (<10%) with mutations in BRCA2 or other DNA repair genes.

Although high mutation burden is a predictive biomarker for favorable response, individual mutations may also act as a predictor of inferior anti-cancer immune response, particularly those cancer-associated driver mutations. Mutations activating the MAP kinase pathway, for example, the KRAS and BRAF mutations, may attenuate T-cell recognition by down-regulation of the major histocompatibility complex (MHC) class I antigen-processing factors [48–51]. Notably, KRAS and BRAF is mutated in 90–95% and 3% of PDAC, respectively [43]. These alterations result in a decrease of T-cell ligands for antigen presentation (Step 2) and thus attenuating anti-tumor inflammatory response. Reduced expression of the MHC class I antigen-processing factors enables cancer cell more likely to escape from immune surveillance and attack, and poses great challenges for effective immunotherapy. Oncogenic RAS has also been shown to upregulate expression of immunomodulatory cytokines, such as IL-8 and GM-CSF [52], which subsequently induces the infiltration of myeloid derived suppressor cells (MDSC) [53]. Other mutations can also attribute to immune resistance. For example, β-2-microglobulin (B2M) and Janus kinases (JAK1 and JAK2) mutations have been reported to attribute to primary and secondary resistance to the PD-1 blockade therapy [54, 55]. Truncating mutations of B2M lead to loss of surface expression of MHC class I molecules. Loss-of-function mutations of AK1/2 result in a lack of response to the interferon gamma signaling. Recognizing that B2M and JAK1/2 mutations would lead to lack of response to PD-1 blockade therapy, it has been suggested that these genes be incorporated in target gene sequencing panels to help select patients for precision cancer treatments. To validate the predictive value of genomic biomarkers with a low incidence, basket trials that incorporate precision medicine into hypothesis-driven clinical trials have emerged as a feasible solution [56]. In these clinical trials, tumors are classified based on genetic alterations instead of tumor histology. These clinical trials scale the number of patients screened for multiple genetic alterations in tumors through a large consortium or multi-institution collaboration and assign patients to a treatment arm corresponding to one particular actionable mutation.

20.3.2.2 Chromosomal Chaos

Aneuploidy, also known as somatic copy number alterations (SCNAs), is characterized by the presence of a chaotic chromosomal environment with abnormal number of chromosomes and chromosomal segments and has been proposed to drive tumorigenesis in various cancer types [57]. In pancreatic cancer,

DNA aneuploidy is an independent factor of poor prognosis [58]. Recently, aneuploidy was found to be associated with decreased expression of cytokines responsible for tumor destruction (IFN-γ, IL-1A, IL-1B, and IL-2) [59]. In this study, compared to the mutation number, the level of SCNAs showed a stronger correlation with the cytotoxic immune signature in most of the tumor types examined, even in those whose mutation numbers positively correlated with the SCNA levels. Clinical validation of the predictive role of aneuploidy in melanoma patients treated with anti–CTLA-4 revealed that high SCNA levels were associated with a poorer response [59].

20.4 Immunotherapeutic Targets for Pancreatic Cancer

20.4.1 Immune Checkpoints

Immune checkpoint inhibitors, including anti-CTLA4, anti-PD-1, and anti-PD-L1 antibodies, are effective as single agents in immune-sensitive cancers like melanoma, renal cell carcinoma and NSCLC, but lack efficacy in immune-quiescent or resistant cancers such as pancreatic cancer [60–62]. In a phase II trial, Ipilimumab was administered to 27 patients with locally advanced or metastatic pancreatic cancer. Unfortunately, no significant improvement in survival was observed [63]. As discussed previously, PD-L1 was rarely expressed on untreated pancreatic cancer. Therefore, combination strategies to leverage the potentials of other therapies to turn tumors from immunologically "cold" to "hot" are a key to achieve the response to immune checkpoint blockade.

20.4.2 Activating or Introducing Cancer Antigen-Specific T Cells

20.4.2.1 Tumor Vaccines

Over the last decade, great efforts have been made to explore the role of vaccines in enhancing the immunogenicity of cancer cells and boosting the T cell response in pancreatic cancer. At the center of vaccine design is the delivery of tumor antigen to antigen presenting cells (APCs), which plays a major role orchestrating the immune response [64]. Early studies with single agent tumor vaccines targeting cancer associated antigens (TSAs) showed improved immune profiles, but have been largely unsuccessful in inducing a positive clinical response [65]. On the one hand, lack of a strong immunogenicity of TSAs may have decreased the efficiency of the vaccine. On the other hand, the excessive immune suppressive TME also contributes significantly to the unresponsiveness. However, our group has raised the "Priming" hypothesis and demonstrated the role of cancer vaccine in priming the TME for immune checkpoint blockade therapies. This evidence supports the combinatorial immunotherapy strategies by providing T cells to immunologic "cold" TME followed by removing the inhibitory signals in the TME [66–69]. GVAX, which composed of two granulocyte-macrophage colony-stimulating factor–secreting allogeneic pancreatic tumor cell lines, induces T-cell immunity to cancer antigens, including mesothelin [70]. In a clinical trial of 59 patients with resectable pancreatic cancer treated preoperatively with GVAX, infiltration of T cells and development of tertiary lymphoid structures in the TME were observed in the tumors resected 2 weeks following the treatment of GVAX, suggesting that the combination therapy converted a "non-immunogenic" neoplasm into an "immunogenic" neoplasm. In addition, the vaccine therapy induces the adaptive immune resistance mechanisms, including the PD-1–PD-L1 pathway, indicates that the vaccine approach could prime pancreatic cancer for immune checkpoint and other immunomodulatory therapies [17]. This was also evidenced by another trial in which favorable objective responses were observed among metastatic PDAC patients treated with the combination of GVAX and Ipilimumab compared to ipilimumab alone [71]. Therefore, combinatorial therapy of vaccines with immunomodulatory therapies (e.g., Checkpoint blockade) or other treatment modifying TME would be the best strategy.

20.4.2.2 Neoantigens and Neoepitopes

The recognition that neoepitope-specific T cells underlies the immune response to checkpoint inhibitors have pushed neoantigens to the wavefront. Indeed, as tumor neoantigens are non-self to the host immune system, they are less likely to induce immune tolerance or trigger autoimmunity [72]. As discussed previously, the fact that mutation/neoantigen burden drives immune response and correlates with clinical outcomes of immune checkpoint blockade strongly supports the use of neoantigens for therapeutic intervention. In a mouse model, therapeutic vaccines of synthetic mutant peptides predicted from genomics and bioinformatics approaches showed comparable outcome as checkpoint blockade in sarcomas [73]. Steven Rosenberg and his group reported that a metastatic cholangiocarcinoma patient treated with adoptive transfer of TIL containing mutation-specific polyfunctional T helper-1 cells achieved notable tumor regression. $CD8^+PD-1^+$ neoantigen-specific lymphocytes detected in the peripheral blood in melanoma patients resembled infiltrating $CD8^+PD-1^+$ cells in their tumors, implying that vaccine-induced $CD8^+PD-1^+$ lymphocytes could also traffic into the tumors [74]. These results provide evidence for the rationale to develop personalized therapies using neoantigen-reactive lymphocytes or T cell receptor (TCR) engineered T cells to treat cancer [75]. Adoptive transfer of T cells targeting the mutant KRAS as a cancer-specific neoantigen is also being tested [76].

In the initial step of developing personalized neoantigen vaccine, somatic mutations can be identified by comparing the whole exome sequences and those of matched normal cells [77, 78]. Tumor RNA expression profiling will help to filter in those expressed candidate neoepitopes [79]. Subsequent prediction of putative peptides involves interrogating the class I and class II HLA binding affinities using neoantigen prediction software. The final step requires well-designed clinical study to validate the vaccine before it applied to the patients.

20.4.3 Targeting the Microenvironment

20.4.3.1 Stromal matrix

Stroma cells contribute to the resistance of pancreatic cancer to chemotherapy by forming an insulating matrix around the tumor, squeezing the blood vessels and preventing chemotherapy drugs and immune cells from reaching it (Fig. 20.2).

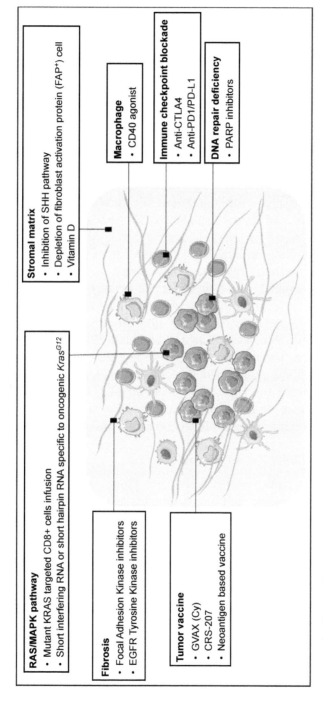

RAS/MAPK pathway
- Mutant KRAS targeted CD8+ cells infusion
- Short interfering RNA or short hairpin RNA specific to oncogenic *Kras*^{G12}

Fibrosis
- Focal Adhesion Kinase inhibitors
- EGFR Tyrosine Kinase inhibitors

Tumor vaccine
- GVAX (Cy)
- CRS-207
- Neoantigen based vaccine

Stromal matrix
- Inhibition of SHH pathway
- Depletion of fibroblast activation protein (FAP$^+$) cell
- Vitamin D

Macrophage
- CD40 agonist

Immune checkpoint blockade
- Anti-CTLA4
- Anti-PD1/PD-L1

DNA repair deficiency
- PARP inhibitors

Figure 20.2 Potential immunotherapeutic targets for pancreatic cancer.

Breaching the significant stroma barriers represents a promising strategy to improve the delivery and efficacy of systemic therapeutics agents. Various attempts have been made to target stromal profibrotic pathways, cytokines and growth factors involved in tumor desmoplasia and angiogenesis. In a mouse PDA model, inhibition of the sonic hedgehog (SHH) signaling pathway could engender a dramatic depletion of stromal components paralleled by an increase in intratumoral vascular density, leading to a significantly enhanced concentration of intracellular metabolite of gemcitabine, transient disease stabilization and a significant prolongation of survival [80]. However, tumors could develop adaptation to chronic SHH inhibition and ultimately resume stromal desmoplasia and hypovascularity [80]. Though disappointing, it provides a proof of principle that disruption of the desmoplastic stroma facilitates the delivery of therapeutic agents into PDAC. Preliminary data in mice have shown that a synthetic form of vitamin D could be used to target the protective matrix, which would increase the delivery of chemotherapy and immune cells into tumor more efficiently [81]. Another study reported depleting carcinoma-associated fibroblasts (CAFs) expressing fibroblast activation protein (FAP) resulted in the immune control of PDAC growth. The depletion of the FAP(+) stromal cell also uncovered the antitumor effects of α-CTLA-4 and α-PD-L1, indicating that the potential therapeutic value of stroma targeting agents in combination with T-cell checkpoint antagonists [8].

20.4.3.2 CD40/CD40L Pathway

The tumor microenvironment in PDAC is predominantly immunosuppressive and exerts potent restrains for antitumor immunity. CD40 agonist antibodies (αCD40) promote APC maturation and enhance macrophage tumoricidal activity [82]. Beatty et al. reported on a small cohort of PDAC patients treated with gemcitabine chemotherapy plus anti-CD40 agonist antibodies and observed tumor regressions in a CD40-dependent mechanism by targeting tumor stroma [83]. More recent studies suggested that CD40 agonists can mediate both T-cell-independent and T-cell-dependent immune mechanisms of tumor regression in pancreatic cancer. The former mechanism involves systemic activation of macrophages that infiltrate the tumor, become tumoricidal, and facilitate the depletion of tumor stroma [84]. The latter finding suggested, when combined with chemotherapy, CD40 blockade can activate T-cell immunity and mediate major tumor regression, but this anti-tumor T-cell response is restrained by suppressive elements in the tumor microenvironment. An early clinical trial testing agonist CD40 monoclonal antibody in combination of gemcitabine was well-tolerated and associated with increased antitumor activity in patients with PDAC [85].

20.4.3.3   RAS/MAPK activation

With the significant role of RAS/MAPK (mitogen-activated protein kinase) pathway in modifying the TME, inhibition of activated RAS downstream pathways may increase anti-tumor immune response and thus therapeutic effect of checkpoint blockade therapy. In melanoma patients, treatment with MAPK (mitogen-activated protein kinase) inhibitors have shown reduced levels of immunosuppressive cytokines and increased TILs [86, 87]. For triple-negative breast cancer (TNBC), the use of MAPK inhibitors has enhanced the responses to immune checkpoint blockade, with an upregulation of MHC I and II on tumor cells, increase in CD8 TILs, and improved prognosis [88]. In pancreatic cancer, most efforts targeting RAS/MAPK activation pathway has been largely unsuccessful. However, simultaneously targeting RAS/MAPK pathway and providing immunotherapy hold a promise to tear down the barriers of immune suppression in the TME and increase the effectiveness of immune-based therapeutic modalities. Nevertheless, additional resistance mechanism may likely need to be identified and overcome.

20.4.3.4   Focal adhesion kinase

Focal Adhesion Kinase (FAK), is a non-receptor tyrosine kinase that plays an critical role in cancer migration, proliferation, and survival [89], and more recently has been found to regulate pro-inflammatory pathway activation and cytokine production [90, 91]. In neoplastic PDAC cells, hyperactivated FAK activity has been shown to orchestrate fibrotic and immunosuppressive TME. In the mouse model of PDAC, elevated FAK activity correlated with high levels of fibrosis and poor CD8+ cytotoxic T cell infiltration in tumors. However, administration of FAK inhibitors was associated with markedly reduced tumor fibrosis and decreased numbers of tumor-infiltrating immunosuppressive cells [92]. Furthermore, FAK inhibitors were shown to render previously unresponsive PDACs sensitive to chemo- and immunotherapy including anti-PD-1 and anti CTLA-4 antibodies in preclinical model of PDACs. These findings have supported the clinical testing of FAK inhibitors in combination with checkpoint inhibitors and chemotherapy for pancreatic cancer treatment.

20.4.4 Targeting DNA Repair Mechanisms

Poly (adenosine diphosphate [ADP]) ribose polymerase (PARP), which represents a powerful machinery for single-strand break repair, orchestrates the DNA damage response (DDR) and the maintenance of genomic stability inhibitors [93, 94]. In murine models, PARP$^{-/-}$ knockout mice demonstrated increased genetic instability, major DNA repair defects as well as hypersensitivity to alkylating agents and radiotherapy [95]. A synthetic lethal therapeutic effect was also proved for patients with compromised ability to repair double-strand DNA breaks by homologous recombination when treated with PARP blockade [96]. Over the past few years, PARP inhibitors (PARPi), which have been tested as monotherapies or in combination with DNA-damaging agents, have shown an efficacy against tumors with defects in DNA repair mechanisms, especially in BRCA1/2 mutation–related cancers [97]. There have been great interests in the combination of immunotherapy with PARP inhibitors based on the rationale that PARPi may increase a synergic therapeutic effect. Treatment using a PARP inhibitor together with immunotherapy of CTLA-4 blockade was shown to induce long-term survival in a BRCA1-deficient ovarian tumor model, with local induction of antitumor immunity and the production of increased levels of interferon-g (IFNγ) in the peritoneal tumor environment. Several clinical trials testing the combination of PARPi and immunotherapy for PDACs are underway.

20.5 Conclusions

The advances in immunogenomics coupled with the recent fundamental advances in the understanding of immune-tumor microenvironment interaction have created opportunities for the development of more effective and personalized immunotherapy approaches for pancreatic cancer patients. Great efforts have focused on identification of biomarkers to facilitate more precise choices of immune modulatory agents. Research-driven drug discovery has led to the emergence of a variety of novel therapeutic targets that are reshaping the pancreatic cancer treatment landscape. Future directions include strategies to pool the right neoantigens/neoepitopes for personalized vaccines, to break host mechanisms of immune tolerance, to enable relevant immune cells to effectively localize to the sites of disease, and to use new drugs such as PARPi to target DNA repair deficiencies. Moreover, studies characterizing PDACs with unique neoantigens and those from long-term survivors might provide new insights into effective treatment strategies for pancreatic cancer [98]. With enormous amount of genomics and immune-biomarker data generated, more efforts should be made to the development of relational databases and bioinformatics platform to empower the practice of precision immuno-oncology. Nevertheless, before unleashing the full power of precision oncology, more studies with robust validation and well-designed clinical trials are warranted to provide evidence to support individualized immunotherapy for pancreatic cancer.

Abbreviations

ADP: adenosine diphosphate
APCs: antigen presenting cells
B2M: β-2-microglobulin
CAFs: carcinoma-associated fibroblasts
CAR: chimeric antigen receptor
CTLA-4: anti-cytotoxic T-lymphocyte antigen 4
DDR: DNA damage response
FAK: focal adhesion kinase

FAP: fibroblast activation protein
IDO1:   indolamine 2,3-dioxygenase 1
IFNγ: interferon gamma
JAK: Janus kinase
MAPK :   mitogen-activated protein kinase
MDSC: myeloid derived suppressor cells
MHC: major histocompatibility complex
MMR: mismatch repair
MSI: microsatellite instability
NSCLC: non-small cell lung cancer
PARP: poly (ADP-ribose) polymerase
PARPi: PARP inhibitors
PD-1:   programmed cell death 1
PDAC: pancreatic ductal adenocarcinoma
POLD: polymerase delta
POLE: polymerase epsilon
RCC: renal cell carcinoma
SCNAs: somatic copy number alterations
SHH: sonic hedgehog
TCR: T cell receptor
TILs: tumor-infiltrating lymphocytes
TME: tumor microenvironment
TNBC: triple-negative breast cancer
TSAs: tumor-specific antigens

Disclosures and Conflict of Interest

This chapter was originally published as: Zhang, J., Wolfgang, C. L., Zheng, L. (2018). Precision immuno-oncology: Prospects of individualized immunotherapy for pancreatic cancer. *Cancers*, **10**, 39., doi:10.3390/cancers10020039, under the Creative Commons Attribution license (http://creativecommons.org/licenses/by/4.0/). It appears here, with edits and updates, by kind permission of the author and the publisher, MDPI (Basel).

This study is partially funded by NIH grants R01 CA169702 (Lei Zheng); R01 CA197296 (Lei Zheng); NIH grant K23 CA148964 (Lei Zheng); the Commonwealth Foundation (Lei Zheng), the Bloomberg-Kimmel Institute for Cancer Immunotherapy (Lei Zheng, Jiajia Zhang), the Viragh Foundation and the Skip Viragh Pancreatic Cancer Center at Johns Hopkins (Lei Zheng); the Sol Goldman Pancreatic Cancer Research Center (Lei Zheng); National Cancer Institute Specialized Programs of Research Excellence in Gastrointestinal Cancers grant P50 CA062924 (Lei Zheng); Sidney Kimmel Comprehensive Cancer Center grant P30CA006973 (Lei Zheng, Christopher L. Wolfgang). We thank Sevier Medical Art for their original design of the cell elements used in Fig. 20.2.

Lei Zheng receives grant supports from Bristol-Meyer Squibb, Merck, iTeos, Amgen, Gradalis, and Halozyme, and receives the royalty for licensing GVAX to Aduro Biotech. Lei Zheng is a paid consultant at Biosynergies and Merrimeck.

Corresponding Author

Dr. Lei Zheng
Departments of Oncology and Surgery
Sidney Kimmel Comprehensive Cancer Center
Johns Hopkins University School of Medicine
1650 Orleans Street, CRB1 3rd Floor
Baltimore, MD 21287, USA
Email: lzheng6@jhmi.edu

References

1. Siegel, R. L., Miller, K. D., Jemal, A. (2017). Cancer statistics, 2017. *CA Cancer J. Clin.*, **67**, 7–30.

2. Seufferlein, T., Mayerle, J. (2016). Pancreatic cancer in 2015: Precision medicine in pancreatic cancer—Fact or fiction? *Nat. Rev. Gastroenterol. Hepatol.*, **13**, 74–75.

3. Chen, D. S., Mellman, I. (2013). Oncology meets immunology: The cancer-immunity cycle. *Immunity*, **39**, 1–10.

4. Chen, D. S., Mellman, I. (2017). Elements of cancer immunity and the cancer—Immune set point. *Nature*, **541**, 321–330.

5. Predina, J., Eruslanov, E., Judy, B., Kapoor, V., Cheng, G., Wang, L.-C., Sun, J., Moon, E. K., Fridlender, Z. G., Albelda, S., et al. (2013). Changes in the local tumor microenvironment in recurrent cancers may explain the failure of vaccines after surgery. *Proc. Natl. Acad. Sci. U.S.A.*, **110**, E415–E424.

6. Gajewski, T. F., Schreiber, H., Fu, Y.-H. (2013). Innate and adaptive immune cells in the tumor microenvironment. *Nat. Immunol.*, **14**, 1014–1022.

7. Feig, C., Gopinathan, A., Neesse, A., Chan, D. S., Cook, N., Tuveson, D. A. (2012). The pancreas cancer microenvironment. *Clin. Cancer Res.*, **18**, 4266–4276.

8. Feig, C., Jonesa, J. O., Kraman, M., Wells, R. J. B., Deonarine, A., Chan, D. S., Connell, C. M., Roberts, E. W., Zhao, Q., Caballero, O. L., et al. (2013). Targeting CXCL12 from FAP-expressing carcinoma-associated fibroblasts synergizes with anti-PD-L1 immunotherapy in pancreatic cancer. *Proc. Natl. Acad. Sci. U.S.A.*, **110**, 20212–20217.

9. Sharpe, A. H., Wherry, E. J., Ahmed, R., Freeman, G. J. (2007). The function of programmed cell death 1 and its ligands in regulating autoimmunity and infection. *Nat. Immunol.*, **8**, 239–245.

10. Patel, S. P., Kurzrock, R. (2015). PD-L1 expression as a predictive biomarker in cancer immunotherapy. *Mol. Cancer Ther.*, **14**, 847–856.

11. Powles, T., Eder, J. P., Fine, G. D., Braiteh, F. S., Loriot, Y., Cruz, C., Bellmunt, J., Burris, H. A., Petrylak, D. P., Teng, S. L., et al. (2014). MPDL3280A (anti-PD-L1) treatment leads to clinical activity in metastatic bladder cancer. *Nature*, **515**, 558–562.

12. Herbst, R. S., Soria, J. C., Kowanetz, M., Fine, G. D., Hamid, O., Gordon, M. S., Sosman, J. A., McDermott, D. F., Powderly, J. D., Gettinger, S. N., et al. (2014). Predictive correlates of response to the anti-PD-L1 antibody MPDL3280A in cancer patients. *Nature*, **515**, 563–567.

13. Dong, H., Strome, S. E., Salomao, D. R., Tamura, H., Hirano, F., Flies, D. B., Roche, P. C., Lu, J., Zhu, G., Tamada, K., et al. (2002). Tumor-associated B7-H1 promotes T-cell apoptosis: A potential mechanism of immune evasion. *Nat. Med.*, **8**, 793–800.

14. Herbst, R. S., Baas, P., Kim, D. W., Felip, E., Pérez-Gracia, J. L., Han, J. Y., Molina, J., Kim, J. H., Arvis, C. D., Ahn, M. J., et al. (2016). Pembrolizumab versus docetaxel for previously treated, PD-L1-positive, advanced non-small-cell lung cancer (KEYNOTE-010): A randomised controlled trial. *Lancet*, **387**, 1540–1550.

15. Carbone, D. P., Reck, M., Paz-Ares, L., Creelan, B., Horn, L., Steins, M., Felip, E., van den Heuvel, M. M., Ciuleanu, T. E., Badin, F., et al. (2017).

First-line nivolumab in stage IV or recurrent non–small-cell lung cancer. *N. Engl. J. Med.*, **376**, 2415–2426.

16. Soares, K. C., Rucki, A. A., Wu, A. A., Olino, K., Xiao, Q., Chai, Y., Wamwea, A., Bigelow, E., Lutz, E., Liu, L., et al. (2015). PD-1/PD-L1 blockade together with vaccine therapy facilitates effector T-cell infiltration into pancreatic tumors. *J. Immunother.*, **38**, 1–11.

17. Lutz, E. R., Wu, A. A., Bigelow, E., Sharma, R., Mo, G., Soares, K., Solt, S., Dorman, A., Wamwea, A., Yager, A., et al. (2014). Immunotherapy converts nonimmunogenic pancreatic tumors into immunogenic foci of immune regulation. *Cancer Immunol. Res.*, **2**, 616–631.

18. Loos, M., Giese, N. A., Kleeff, J., Giese, T., Gaida, M. M., Bergmann, F., Laschinger, M., Büchler, M. W., Friess, H., et al. (2008). Clinical significance and regulation of the costimulatory molecule B7-H1 in pancreatic cancer. *Cancer Lett.*, **268**, 98–109.

19. Lu, C., Paschall, A. V., Shi, H., Savage, N., Waller, J. L., Sabbatini, M. E., Oberlies, N. H., Pearce, C., Liu, K., et al. (2017). The MLL1-H3K4me3 axis-mediated PD-L1 expression and pancreatic cancer immune evasion. *J. Natl. Cancer Inst.*, **109**,djw283.

20. Zheng, L. (2017). PD-L1 expression in pancreatic cancer. *J. Natl. Cancer Inst.*, **109**, djw304.

21. Church, D. N., Stelloo, E., Nout, R. A., Valtcheva, N., Depreeuw, J., ter Haar, N., Noske, A., Amant, F., Tomlinson, I. P., Wild, P. J., et al. (2015). Prognostic significance of POLE proofreading mutations in endometrial cancer. *J. Natl. Cancer Inst.*, **107**(1), 402.

22. Shang, B., Liu, Y., Jiang, S., Liu, Y. (2015). Prognostic value of tumor-infiltrating FoxP3+ regulatory T cells in cancers: A systematic review and meta-analysis. *Sci. Rep.*, **5**, 15179.

23. Galon, J., Costes, A., Sanchez-Cabo, F., Kirilovsky, A., Mlecnik, B., Lagorce-Pagès, C., Tosolini, M., Camus, M., Berger, A., Wind, P., et al. (2006). Type, density, and location of immune cells within human colorectal tumors predict clinical outcome. *Science*, **313**, 1960–1964.

24. Zhang, L., Conejo-Garcia, J. R., Katsaros, D., Gimotty, P. A., Massobrio, M., Regnani, G., Makrigiannakis, A., Gray, H., Schlienger, K., Liebman, M. N., et al. (2003). Intratumoral T cells, recurrence, and survival in epithelial ovarian cancer. *N. Engl. J. Med.*, **348**, 203–213.

25. Fridman, W. H., Pagès, F., Sautès-Fridman, C., Galon, J. (2012). The immune contexture in human tumours: Impact on clinical outcome. *Nat. Rev. Cancer*, **12**, 298–306.

26. Taube, J. M., Anders, R. A., Young, G. D., Xu, H., Sharma, R., McMiller, T. L., Chen, S., Klein, A. P., Pardoll, D. M., Topalian, S. L., et al. (2012). Colocalization of inflammatory response with B7-H1 expression in human melanocytic lesions supports an adaptive resistance mechanism of immune escape. *Sci. Transl. Med.*, **4**, 127ra37.

27. Tsujikawa, T., Kumar, S., Borkar, R. N., Azimi, V., Thibault, G., Chang, Y. H., Balter, A., Kawashima, R., Choe, G., Sauer, D., et al. (2017). Quantitative multiplex immunohistochemistry reveals myeloid-inflamed tumor-immune complexity associated with poor prognosis. *Cell Rep.*, **19**, 203–217.

28. Rizvi, N. A., Hellmann, M. D., Snyder, A., Kvistborg, P., Makarov, V., Havel, J. J., Lee, W., Yuan, J., Wong, P., Ho, T. S., et al. (2015). Mutational landscape determines sensitivity to PD-1 blockade in non–small cell lung cancer. *Science*, **348**, 124–128.

29. McGranahan, N., Furness, A. J., Rosenthal, R., Ramskov, S., Lyngaa, R., Saini, S. K., Jamal-Hanjani, M., Wilson, G. A., Birkbak, N. J., Hiley, C. T., et al. (2016). Clonal neoantigens elicit T cell immunoreactivity and sensitivity to immune checkpoint blockade. *Science*, **351**, 1463–1469.

30. Chen, D. S., Irving, B. A., Hodi, F. S. (2012). Molecular pathways: Next-generation immunotherapy-inhibiting programmed death-ligand 1 and programmed death-1. *Clin. Cancer Res.*, **18**, 6580–6587.

31. Fang, Y., Yao, Q., Chen, Z., Xiang, J., William, F. E., Gibbs, R. A., Chen, C. (2013). Genetic and molecular alterations in pancreatic cancer: Implications for personalized medicine. *Med. Sci. Monit.*, **19**, 916–926.

32. Topalian, S. L., Sznol, M., McDermott, D. F., Kluger, H. M., Carvajal, R. D., Sharfman, W. H., Brahmer, J. R., Lawrence, D. P., Atkins, M. B., Powderly, J. D., et al. (2014). Survival, durable tumor remission, and long-term safety in patients with advanced melanoma receiving nivolumab. *J. Clin. Oncol.*, **32**, 1020–1030.

33. Reck, M., Rodríguez-Abreu, D., Robinson, A. G., Hui, R., Cso" szi, T., Fülöp, A., Gottfried, M., Peled, N., Tafreshi, A., Cuffe, S., et al. (2016). Pembrolizumab versus chemotherapy for PD-L1–positive non-small-cell lung cancer. *N. Engl. J. Med.*, **375**, 1823–1833.

34. Haraldsdottir, S., Hampel, H., Tomsic, J., Frankel, W. L., Pearlman, R., De La Chapelle, A., Pritchard, C. C. (2014). Colon and endometrial cancers with mismatch repair deficiency can arise from somatic, rather than germline, mutations. *Gastroenterology*, **147**, 1308–1316.e1.

35. Le, D. T., Uram, J. N., Wang, H., Bartlett, B. R., Kemberling, H., Eyring, A. D., Skora, A. D., Luber, B. S., Azad, N. S., Laheru, D., et al. (2015). PD-1 blockade in tumors with mismatch-repair deficiency. *N. Engl. J. Med.*, **372**, 2509–2520.

36. Le, D. T., Durham, J. N., Smith, K. N., Wang, H., Bartlett, B. R., Aulakh, L. K., Lu, S., Kemberling, H., Wilt, C., Luber, B. S., et al. (2017). Mismatch repair deficiency predicts response of solid tumors to PD-1 blockade. *Science*, **357**, 409–413.

37. Domingo, E., Freeman-Mills, L., Rayner, E., Glaire, M., Briggs, S., Vermeulen, L., Fessler, E., Medema, J. P., Boot, A., Morreau, H., et al. (2016). Somatic POLE proofreading domain mutation, immune response, and prognosis in colorectal cancer: A retrospective, pooled biomarker study. *Lancet Gastroenterol. Hepatol.*, **1**, 207–216.

38. Palles, C., Cazier, J. B., Howarth, K. M., Domingo, E., Jones, A. M., Broderick, P., Kemp, Z., Spain, S. L., Guarino, E., Salguero, I., et al. (2013). Germline mutations affecting the proofreading domains of POLE and POLD1 predispose to colorectal adenomas and carcinomas. *Nat. Genet.*, **45**, 136–144.

39. Levine, D. A. (2013). The Cancer Genome Atlas Research Network. Integrated genomic characterization of endometrial carcinoma. *Nature*, **497**, 67–73.

40. Rayner, E., van Gool, I. C., Palles, C., Kearsey, S. E., Bosse, T., Tomlinson, I., Church, D. N. (2016). A panoply of errors: Polymerase proofreading domain mutations in cancer. *Nat. Rev. Cancer*, **16**, 71–81.

41. Johanns, T. M., Miller, C. A., Dorward, I. G., Tsien, C., Chang, E., Perry, A., Uppaluri, R., Ferguson, C., Schmidt, R. E., Dahiya, S., et al. (2016). Immunogenomics of hypermutated glioblastoma: A patient with germline POLE deficiency treated with checkpoint blockade immunotherapy. *Cancer Discov.*, **6**, 1230–1236.

42. Bouffet, E., Larouche, V., Campbell, B. B., Merico, D., de Borja, R., Aronson, M., Durno, C., Krueger, J., Cabric, V., Ramaswamy, V., et al. (2016). Immune checkpoint inhibition for hypermutant glioblastoma multiforme resulting from germline biallelic mismatch repair deficiency. *J. Clin. Oncol.*, **34**, 2206–2211.

43. Witkiewicz, A. K., McMillan, E. A., Balaji, U., Baek, G., Lin, W. C., Mansour, J., Mollaee, M., Wagner, K. U., Koduru, P., Yopp, A., et al. (2015). Whole-exome sequencing of pancreatic cancer defines genetic diversity and therapeutic targets. *Nat. Commun.*, **6**, 6744.

44. Biankin, A. V., Waddell, N., Kassahn, K. S., Gingras, M. C., Muthuswamy, L. B., Johns, A. L., Miller, D. K., Wilson, P. J., Patch, A. M., Wu, J., et al.

(2012). Pancreatic cancer genomes reveal aberrations in axon guidance pathway genes. *Nature*, **491**, 399–405.

45. cBioPortal. Available at: http://www.cbioportal.org/ (accessed on January 17, 2019).

46. Banville, N., Geraghty, R., Fox, E., Leahy, D. T., Green, A., Keegan, D., Geoghegan, J., O'Donoghue, D., Hyland, J., Sheahan, K. (2006). Medullary carcinoma of the pancreas in a man with hereditary nonpolyposis colorectal cancer due to a mutation of the MSH2 mismatch repair gene. *Hum. Pathol.*, **37**, 1498–1502.

47. Connor, A. A., Denroche, R. E., Jang, G. H., Timms, L., Kalimuthu, S. N., Selander, I., McPherson, T., Wilson, G. W., Chan-Seng-Yue, M. A., Borozan, I., et al. (2017). Association of distinct mutational signatures with correlates of increased immune activity in pancreatic ductal adenocarcinoma. *JAMA Oncol.*, **3**, 774.

48. Seliger, B., Harders, C., Wollscheid, U., Staege, M. S., Reske-Kunz, A. B., Huber, C. (1996). Suppression of MHC class I antigens in oncogenic transformants: Association with decreased recognition by cytotoxic T lymphocytes. *Exp. Hematol.*, **24**, 1275–1279.

49. Seliger, B., Harders, C., Lohmann, S., Momburg, F., Urlinger, S., Tampé, R., Huber, C. (1998). Down-regulation of the MHC class I antigen-processing machinery after oncogenic transformation of murine fibroblasts. *Eur. J. Immunol.*, **28**, 122–133.

50. Atkins, D., Breuckmann, A., Schmahl, G. E., Binner, P., Ferrone, S., Krummenauer, F., Störkel, S., Seliger, B., et al. (2004). MHC class I antigen processing pathway defects, ras mutations and disease stage in colorectal carcinoma. *Int. J. Cancer*, **109**, 265–273.

51. Bradley, S. D., Chen, Z., Melendez, B., Talukder, A., Khalili, J. S., Rodriguez-Cruz, T., Liu, S., Whittington, M., Deng, W., Li, F., et al. (2015). BRAFV600E co-opts a conserved MHC class I internalization pathway to diminish antigen presentation and CD8+ T-cell recognition of melanoma. *Cancer Immunol. Res.*, **3**, 602–609.

52. Zeitouni, D., Pylayeva-Gupta, Y., Der, C. J., Bryant, K. L. (2016). KRAS mutant pancreatic cancer: No lone path to an effective treatment. *Cancers*, **8**, 45.

53. Erkan, M. (2013). Understanding the stroma of pancreatic cancer: Co-evolution of the microenvironment with epithelial carcinogenesis. *J. Pathol.*, **231**, 4–7.

54. Zaretsky, J. M., Garcia-Diaz, A., Shin, D. S., Escuin-Ordinas, H., Hugo, W., Hu-Lieskovan, S., Torrejon, D. Y., Abril-Rodriguez, G., Sandoval, S.,

Barthly, L., et al. (2016). Mutations associated with acquired resistance to PD-1 blockade in melanoma. *N. Engl. J. Med.*, **375**, 819–829.

55. Shin, D. S., Zaretsky, J. M., Escuin-Ordinas, H., Garcia-Diaz, A., Hu-Lieskovan, S., Kalbasi, A., Grasso, C. S., Hugo, W., Sandoval, S., Torrejon, D. Y., et al. (2017). Primary resistance to PD-1 blockade mediated by JAK1/2 mutations. *Cancer Discov.*, **7**, 188–201.

56. Redig, A. J., Jänne, P. A. (2015). Basket trials and the evolution of clinical trial design in an era of genomic medicine. *J. Clin. Oncol.*, **33**, 975–977.

57. Holland, A. J., Cleveland, D. W. (2009). Boveri revisited: Chromosomal instability, aneuploidy and tumorigenesis. *Nat. Rev. Mol. Cell Biol.*, **10**, 478–487.

58. Sciallero, S., Giaretti, W., Geido, E., Bonelli, L., Zhankui, L., Saccomanno, S., Zeraschi, E., Pugliese, V. (1993). DNA aneuploidy is an independent factor of poor prognosis in pancreatic and peripancreatic cancer. *Int. J. Pancreatol.*, **14**, 21–28.

59. Davoli, T., Uno, H., Wooten, E. C., Elledge, S. J. (2017). Tumor aneuploidy correlates with markers of immune evasion and with reduced response to immunotherapy. *Science*, **355**, eaaf8399.

60. Foley, K., Kim, V., Jaffee, E., Zheng, L. (2016). Current progress in immunotherapy for pancreatic cancer. *Cancer Lett.*, **381**, 244–251.

61. Brahmer, J. R., Tykodi, S. S., Chow, L. Q., Hwu, W. J., Topalian, S. L., Hwu, P., Drake, C. G., Camacho, L. H., Kauh, J., Odunsi, K., et al. (2012). Safety and activity of anti–PD-L1 antibody in patients with advanced cancer. *N. Engl. J. Med.*, **366**, 2455–2465.

62. Brahmer, J. R., Drake, C. G., Wollner, I., Powderly, J. D., Picus, J., Sharfman, W. H., Stankevich, E., Pons, A., Salay, T. M., McMiller, T. L., et al. (2010). Phase I study of single-agent anti-programmed death-1 (MDX-1106) in refractory solid tumors: Safety, clinical activity, pharmacodynamics, and immunologic correlates. *J. Clin. Oncol.*, **28**, 3167–3175.

63. Royal, R. E., Levy, C., Turner, K., Mathur, A., Hughes, M., Kammula, U. S., Sherry, R. M., Topalian, S. L., Yang, J. C., Lowy, I., et al. (2010). Phase 2 trial of single agent ipilimumab (anti-CTLA-4) for locally advanced or metastatic pancreatic adenocarcinoma. *J. Immunother.*, **33**, 828–833.

64. Le, D. T., Jaffee, E. M. (2013). Next-generation cancer vaccine approaches: Integrating lessons learned from current successes with promising biotechnologic advances. *J. Natl. Compr. Cancer Netw.*, **11**, 766–772.

65. Jaffee, E. M., Hruban, R. H., Canto, M., Kern, S. E. (2002). Focus on pancreas cancer. *Cancer Cell*, **2**, 25–28.

66. Keenan, B. P., Saenger, Y., Kafrouni, M. I., Leubner, A., Lauer, P., Maitra, A., Rucki, A. A., Gunderson, A. J., Coussens, L. M., Brockstedt, D. G., et al. (2014). A listeria vaccine and depletion of T-regulatory cells activate immunity against early stage pancreatic intraepithelial neoplasms and prolong survival of mice. *Gastroenterology*, **146**, 1784–1794. E6.

67. Lutz, E. R., Kinkead, H., Jaffee, E. M., Zheng, L. (2014). Priming the pancreatic cancer tumor microenvironment for checkpoint-inhibitor immunotherapy. *Oncoimmunology*, **3**, e962401.

68. Kouo, T., Huang, L., Pucsek, A. B., Cao, M., Solt, S., Armstrong, T., Jaffee, E. (2015). Galectin-3 shapes antitumor immune responses by suppressing CD8+ T cells via LAG-3 and inhibiting expansion of plasmacytoid dendritic cells. *Cancer Immunol. Res.*, **3**, 412–423.

69. Chu, N. J., Armstrong, T. D., Jaffee, E. M. (2015). Nonviral oncogenic antigens and the inflammatory signals driving early cancer development as targets for cancer immunoprevention. *Clin. Cancer Res.*, **21**, 1549–1557.

70. Laheru, D., Lutz, E., Burke, J., Biedrzycki, B., Solt, S., Onners, B., Tartakovsky, I., Nemunaitis, J., Le, D., Sugar, E., et al. (2008). Allogeneic granulocyte macrophage colony-stimulating factor-secreting tumor immunotherapy alone or in sequence with cyclophosphamide for metastatic pancreatic cancer: A pilot study of safety, feasibility, and immune activation. *Clin. Cancer Res.*, **14**, 1455–1463.

71. Le, D. T., Lutz, E., Uram, J. N., Sugar, E. A., Onners, B., Solt, S., Zheng, L., Diaz Jr, L. A., Donehower, R. C., Jaffee, E. M., et al. (2013). Evaluation of ipilimumab in combination with allogeneic pancreatic tumor cells transfected with a GM-CSF gene in previously treated pancreatic cancer. *J. Immunother.*, **36**, 382–389.

72. Yarchoan, M., Johnson, B. A., Lutz, E. R., Laheru, D. A., Jaffee, E. M. (2017). Targeting neoantigens to augment antitumour immunity. *Nat. Rev. Cancer*, **17**, 209–222.

73. Gubin, M. M., Zhang, X., Schuster, H., Caron, E., Ward, J. P., Noguchi, T., Ivanova, Y., Hundal, J., Arthur, C. D., Krebber, W. J., et al. (2014). Checkpoint blockade cancer immunotherapy targets tumour-specific mutant antigens. *Nature*, **515**, 577–581.

74. Gros, A., Parkhurst, M. R., Tran, E., Pasetto, A., Robbins, P. F., Ilyas, S., Prickett, T. D., Gartner, J. J., Crystal, J. S., Roberts, I. M., et al. (2016). Prospective identification of neoantigen-specific lymphocytes in the peripheral blood of melanoma patients. *Nat. Med.*, **22**, 433–438.

75. Tran, E., Turcotte, S., Gros, A., Robbins, P. F., Lu, Y. C., Dudley, M. E., Wunderlich, J. R., Somerville, R. P., Hogan, K., Hinrichs, C. S., et al. (2014). Cancer immunotherapy based on mutation-specific CD4+ T cells in a patient with epithelial cancer. *Science*, **9**, 641–645.

76. Tran, E., Robbins, P. F., Lu, Y. C., Prickett, T. D., Gartner, J. J., Jia, L., Pasetto, A., Zheng, Z., Ray, S., Groh, E. M., et al. (2016). T-cell transfer therapy targeting mutant KRAS in cancer. *N. Engl. J. Med.*, **375**, 2255–2262.

77. Vogelstein, B., Papadopoulos, N., Velculescu, V. E., Zhou, S., Diaz, L. A., Kinzler, K. W. (2013). Cancer genome landscapes. *Science*, **339**, 1546–1558.

78. Tran, E., Ahmadzadeh, M., Lu, Y. C., Gros, A., Turcotte, S., Robbins, P. F., Gartner, J. J., Zheng, Z., Li, Y. F., Ray, S., et al. (2015). Immunogenicity of somatic mutations in human gastrointestinal cancers. *Science*, **350**, 1387–1390.

79. Liu, X. S., Mardis, E. R. (2017). Applications of immunogenomics to cancer. *Cell*, **168**, 600–612.

80. Olive, K. P., Jacobetz, M. A., Davidson, C. J., Gopinathan, A., McIntyre, D., Honess, D., Madhu, B., Goldgraben, M. A., Caldwell, M. E., Allard, D., et al. (2009). Inhibition of hedgehog signaling enhances delivery of chemotherapy in a mouse model of pancreatic cancer. *Science*, **324**, 1457–1461.

81. Sherman, M. H., Ruth, T. Y., Engle, D. D., Ding, N., Atkins, A. R., Tiriac, H., Collisson, E. A., Connor, F., Van Dyke, T., Kozlov, S., et al. (2014). Vitamin D receptor-mediated stromal reprogramming suppresses pancreatitis and enhances pancreatic cancer therapy. *Cell*, **159**, 80–93.

82. Elgueta, R., Benson, M. J., de Vries, V. C., Wasiuk, A., Guo, Y., Noelle, R. J. (2009). Molecular mechanism and function of CD40/CD40L engagement in the immune system. *Immunol. Rev.*, **229**, 152–172.

83. Vonderheide, R. H., Bajor, D. L., Winograd, R., Evans, R. A., Bayne, L. J., Beatty, G. L. (2011). CD40 agonists alter tumor stroma and show efficacy against pancreatic carcinoma in mice and humans. *Science*, **331**, 1612–1616.

84. Vonderheide, R. H., Bajor, D. L., Winograd, R., Evans, R. A., Bayne, L. J., Beatty, G. L. (2013). CD40 immunotherapy for pancreatic cancer. *Cancer Immunol. Immunother.*, **62**, 949–954.

85. Beatty, G. L., Torigian, D. A., Chiorean, E. G., Saboury, B., Brothers, A., Alavi, A., Troxel, A. B., Sun, W., Teitelbaum, U. R., Vonderheide, R. H.,

et al. (2013). A phase I study of an agonist CD40 monoclonal antibody (CP-870,893) in combination with gemcitabine in patients with advanced pancreatic ductal adenocarcinoma. *Clin. Cancer Res.*, **19**, 6286–6295.

86. Wilmott, J. S., Long, G. V., Howle, J. R., Haydu, L. E., Sharma, R. N., Thompson, J. F., Kefford, R. F., Hersey, P., Scolyer, R. A. (2012). Selective BRAF inhibitors induce marked T-cell infiltration into human metastatic melanoma. *Clin. Cancer Res.*, **18**, 1386–1394.

87. Donia, M., Fagone, P., Nicoletti, F., Andersen, R. S., Høgdall, E., Straten, P. T., Andersen, M. H., Svane, I. M. (2012). BRAF inhibition improves tumor recognition by the immune system: Potential implications for combinatorial therapies against melanoma involving adoptive T-cell transfer. *Oncoimmunology*, **1**, 1476–1483.

88. Loi, S., Dushyanthen, S., Beavis, P. A., Salgado, R., Denkert, C., Savas, P., Combs, S., Rimm, D. L., Giltnane, J. M., Estrada, M. V., et al. (2016). RAS/MAPK activation is associated with reduced tumor-infiltrating lymphocytes in triple-negative breast cancer: Therapeutic cooperation between MEK and PD-1/PD-L1 immune checkpoint inhibitors. *Clin. Cancer Res.*, **22**, 1499–1509.

89. Sulzmaier, F. J., Jean, C., Schlaepfer, D. D. (2014). FAK in cancer: Mechanistic findings and clinical applications. *Nat. Rev. Cancer*, **14**, 598–610.

90. Serrels, A., Lund, T., Serrels, B., Byron, A., McPherson, R. C., von Kriegsheim, A., Gómez-Cuadrado, L., Canel, M., Muir, M., Ring, J. E., et al. (2015). Nuclear FAK controls chemokine transcription, tregs, and evasion of anti-tumor immunity. *Cell*, **163**, 160–173.

91. Zhao, X. K., Cheng, Y., Cheng, M. L., Yu, L., Mu, M., Li, H., Liu, Y., Zhang, B., Yao, Y., Guo, H., et al. (2016). Focal adhesion kinase regulates fibroblast migration via integrin beta-1 and plays a central role in fibrosis. *Sci. Rep.*, **6**, 19276.

92. Jiang, H., Hegde, S., Knolhoff, B. L., Zhu, Y., Herndon, J. M., Meyer, M. A., Nywening, T. M., Hawkins, W. G., Shapiro, I. M., Weaver, D. T., et al. (2016). Targeting focal adhesion kinase renders pancreatic cancers responsive to checkpoint immunotherapy. *Nat. Med.*, **22**, 851–860.

93. Durkacz, B. W., Omidiji, O., Gray, D. A., Shall, S. (1980). (ADP-ribose)n participates in DNA excision repair. *Nature*, **283**, 593–596.

94. Dantzer, F., Schreiber, V., Niedergang, C., Trucco, C., Flatter, E., De La Rubia, G., Oliver, J., Rolli, V., Ménissier-de Murcia, J., et al. (1999). Involvement of poly(ADP-ribose) polymerase in base excision repair. *Biochimie*, **81**, 69–75.

95. Shall, S., de Murcia, G. (2000). Poly(ADP-ribose) polymerase-1: What have we learned from the deficient mouse model? *Mutat. Res.-DNA Rep.*, **460**, 1–15.

96. Ashworth, A. (2008). A synthetic lethal therapeutic approach: Poly(ADP) ribose polymerase inhibitors for the treatment of cancers deficient in DNA double-strand break repair. *J. Clin. Oncol.*, **26**, 3785–3790.

97. Michels, J., Vitale, I., Saparbaev, M., Castedo, M., Kroemer, G. (2013). Predictive biomarkers for cancer therapy with PARP inhibitors. *Oncogene*, **33**, 3894–3907.

98. Balachandran, V. P., Łuksza, M., Zhao, J. N., Makarov, V. (2017). Identification of unique neoantigen qualities in long-term survivors of pancreatic cancer. *Nature*, **551**, 512–516.

Chapter 21

Developing a Case-Based Blended Learning Ecosystem to Optimize Precision Medicine: Reducing Overdiagnosis and Overtreatment

Vivek Podder, MBBS,[a] Binod Dhakal, MD,[b]
Gousia Ummae Salma Shaik, MD,[c] Kaushik Sundar, DM,[d]
Madhava Sai Sivapuram, MBBS,[e] Vijay Kumar Chattu, MD,[f]
and Rakesh Biswas, MD[c]

[a]Department of Internal Medicine,
Tairunnessa Memorial Medical College and Hospital, Gazipur, Bangladesh

[b]Division of Hematology/Oncology, Medical College of Wisconsin,
Milwaukee, Wisconsin, USA

[c]Department of Internal Medicine, Kamineni Institute of Medical Sciences,
Narketpally, Telangana, India

[d]Department of Neurology, Rajagiri Hospital, Chunanangamvely, Aluva, Kerala, India

[e]Department of Internal Medicine,
Dr. Pinnamaneni Siddhartha Institute of Medical Sciences and Research Foundation,
Chinaoutapalli, Andhra Pradesh, India

[f]Department of Paraclinical Sciences, Faculty of Medical Sciences,
The University of the West Indies, St. Augustine, Trinidad and Tobago

Keywords: overdiagnosis, overtreatment, case-based blended learning ecosystem, case studies, precision medicine, omics-driven, low resource setting, high resource setting, underdiagnosis, undertreatment, accuracy-driven, case studies, precision lens, critical appraisal, oncogenomics, user-driven health care, optimizing care, online-offline components,

The Road from Nanomedicine to Precision Medicine
Edited by Shaker A. Mousa, Raj Bawa, and Gerald F. Audette
Copyright © 2020 Jenny Stanford Publishing Pte. Ltd.
ISBN 978-981-4800-59-4 (Hardcover), 978-0-429-29501-0 (eBook)
www.jennystanford.com

cartridge-based nucleic acid amplification assay test, tuberculosis, percutaneous coronary intervention, ST-elevation myocardial infarction, technology-driven, health care technology, transparency, patient-centred, evidence-based, self-directed learning

21.1 Introduction

The term "Precision Medicine" was first coined by Clayton Christensen in his book the "Innovator's Prescription", published in 2009 [1]. According to the early definition given by the Institute of Precision Medicine, "Precision medicine is targeted, individualized care that is tailored to each patient based on his or her specific genetic profile and medical history" [2].

While the above definitions allow us to assume that precision medicine is focused on meeting patients' requirements accurately, we need to review the scientific nature of accuracy, precision and their relationship with each other to put things in perspective. This is essential toward optimizing patient requirements and outcomes, minimizing damage to the healthcare ecosystem by reducing under-overdiagnosis and therapy. To quote from Thomas (2014), "The healthcare 'system' is now better understood as an 'ecosystem' of interconnected stakeholders, each one charged with a mission to improve the quality of care while lowering its cost. To ensure patient safety and quality care while realizing savings, these stakeholders are building new relationships—often outside the four walls of the hospital" [3].

We illustrate current concepts borrowed from existing scientific literature around accuracy and precision with Figs. 21.1 and 21.2. We have modified the figure in reference to paper [4] to offer a fresh perspective on accuracy-driven "precision medicine", which is an age-old tool for physicians, currently augmented by technology.

21.1.1 Precision versus Accuracy

We illustrate the above concepts with micro case studies below:

> *"An elderly patient from a country endemic with tuberculosis presented with a chronic cough and weight loss. A lung pathology was detected on imaging that was not amenable to further biopsy*

efforts as a result of unavailable resources. He was started on empirical treatment for tuberculosis after sending a sputum for acid-fast bacillus (AFB) and culture."

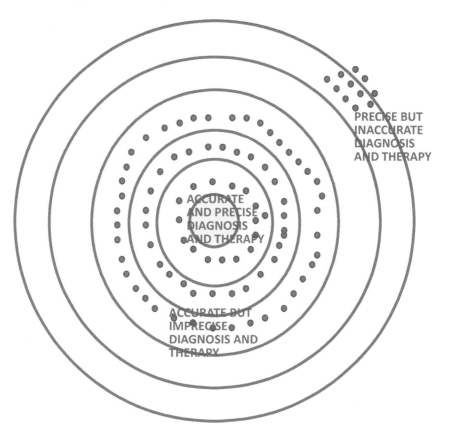

Figure 21.1 Mapping 'Precision' and 'Accuracy' in Medicine (Accuracy—achieving patients' requirement of best outcome, and Precision—the narrowest path to achieve it).

In tuberculosis endemic countries, physicians often treat empirically for tuberculosis in lung pathologies, although lung malignancy is a close differential in such situations. In the above patient's context, physicians were being obviously imprecise in starting treatment for tuberculosis empirically even when the tuberculosis bacilli was undetectable. This is an acceptable standard practice with established protocols for treating sputum-negative tuberculosis utilized globally by many countries that are endemic for tuberculosis.

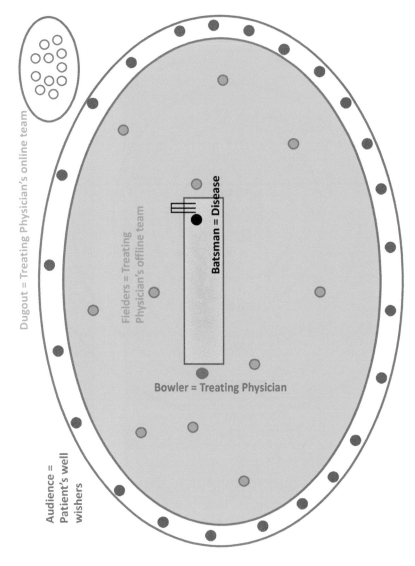

Figure 21.2 Illustrates the analogy of healthcare ecosystem functioning in the form of a cricket match.

Now to illustrate the concepts further, the above case may have the following mentioned outcomes:

(a) The elderly patient's sputum comes out to be positive and once he is begun on antitubercular therapy, he recovers. His cough subsides and weight improves and his sputum culture report that comes after 6 weeks also turns out to be positive for tuberculosis. This is an example of precise and accurate diagnosis and treatment.

(b) The elderly patient's sputum turns out to be negative and yet once he is begun on antitubercular therapy, he recovers. His cough subsides and weight improves and his sputum culture report that comes after 6 weeks turns out to be positive for tuberculosis although his initial sputum smear was negative. This would be an example of initially imprecise, but finally accurate outcomes.

(c) The elderly patient's sputum turns out to be negative and a cartridge-based nucleic acid amplification assay test (CBNAAT) sent at the same time also turns out to be negative for tuberculosis and once he is begun on antitubercular therapy, he does not appear to recover at all. His cough worsens along with his appetite and his weight loss increases. A bronchoscopy with bronchoalveolar lavage is performed and sent again for malignant cytology, AFB, CBNAAT, and culture. His sputum culture report that comes after 6 weeks turns out to be positive for drug-resistant tuberculosis. Although, he receives second-line therapy for his drug-resistant tuberculosis, his condition worsens, and he dies.

The above is an example of a precise approach that still leads to inaccurate patient outcomes. We can be inaccurate in spite of being precise because of the current limitation of information and knowledge that does not always allow us to be accurate. The role of research and learning is to address this limitation and push the boundaries of current knowledge. Precision medicine develops and positively evolves with better research and learning.

(d) The elderly patient's sputum turns out to be negative for tuberculosis and no further tests are done due to lack of resources. He is put on empirical antitubercular therapy, but he does not appear to recover at all. His cough worsens along with his appetite and his weight loss increases. One day he has a sudden shortness of breath and dies. An autopsy reveals bronchogenic carcinoma and pulmonary embolism.

The above is an example of imprecise and inaccurate outcomes. Here again, the obvious precipitants are low resources that prevent further probing toward a precise diagnosis, leading one to resort to imprecise treatment.

Out of the descriptions of a–d above, let's focus on (b), which illustrates imprecise approaches to arrive at accuracy. This is also known as the trial and error method in common parlance. Over the past centuries, medicine has often relied on this imprecise trial and error method, particularly on individual patients before it was replaced by the population-based randomized controlled trial approach to collect generalizable average evidence that is often applied on individual patients. The problem with this current approach is that the requirements of the individual at hand may not always match the average requirements.

Nevertheless, modern-day, evidence-based precision demands that the individual is first viewed through the 'average evidence' lens. Once the individual's response to the 'average' evidence-based available therapy is suboptimal or the therapy itself is unavailable, a therapeutic trial on the individual can be undertaken as a single subject 'n of 1' study design.

Coming back to (b), which is about a situation where empirical therapy is provided to an individual who may have a substantial chance of having an alternative diagnosis, and in (b) that individual turns out to be lucky to have that very diagnosis that was targeted. This imprecise approach may often be overdone in low-resource settings, where one often comes across prescriptions listing out 9–10 medications where the strategy is to target all possible differentials with whatever medicine appears to have the slightest evidence of success. While this form of irrational overuse of medication is a global problem, it appears to be more pronounced in low-resource settings [5]. This problem, a lack of system to make practitioners aware of their follies, is more common in low-resource settings also because of arcane medical education strategies that do not train students in critical appraisal, neither in terms of recognizing nor applying current best evidence. This deficit often stems from the education system's overreliance on rote memorization based on a curriculum that discourages students to ask questions [6]. One solution to the current curricular conundrum is to create more CBBLEs in every medical college, where patient-centered, evidence-based, self-directed learning is the prime emphasis [7].

Behnke L. M. et al. have defined overutilization as *"use of unnecessary care when alternatives may produce similar outcomes, results in a higher cost without increased value"* [8]. Overuse has a huge burden on low- and middle-income countries, where much healthcare is provided by ill-regulated, private providers and with fee-for-service. As a result, healthcare costs significantly increase and potentially harm patients through inappropriate interventions.

While our above description of overuse is true for multiple medication overuse in single patients, some single interventions can be scaled rapidly in larger populations. Current global concerns are mounting over inappropriate interventions such as percutaneous coronary intervention (PCI) for patients who are unlikely to benefit.

Brownlee S. et al. reported that in an Indian second opinion setting, 55% of recommended PCIs were ill-advised [9]. A study in a tertiary care center in India showed that ST-elevation myocardial infarction (STEMI) patients constituted 55% of all inappropriate elective PCIs. These PCI procedures were performed on totally occluded infarct-related vessels after 12 h of symptom onset. The same study also showed that patients with stable angina with single or double vessel disease, low-risk group and sub-optimal medical therapy constituted 45% of all inappropriate elective PCIs [10]. It has been reported that the prevalence of inappropriate PCI procedures is 3.7% in Korea; 12% in the USA; 14% in Germany; 16% in Italy; 22% in Israel and 20% in Spain [9].

Over the last decade, we have adopted an evidence-based precision medicine approach that enables utilizing the best available evidence toward optimizing care for individual patients. Our individual patient requirements have led us to adopt a blended learning platform to enable an informational support for our patients. It also helps medical students to have a platform to help patients locally while learning from global experts in an online ecosystem [11]. Our online learning ecosystem is a community of computer and mobile users comprised of medical students, their physician teachers, other health professionals and patients and their relatives, each one of whom provide inputs (in terms of case-based information) to this ecosystem through devices. They receive learning feedback on the same cases such that their learning outcomes can be potentially translated into patient outcomes. This system, labeled "user-driven healthcare" (UDHC), is not restricted to our CBBLE, but represents an evolving global "phenomenon". Here "improved healthcare is achieved through concerted collaborative learning between multiple users and stakeholders, primarily patients, health professionals and other actors in the caregiving collaborative network across a web interface". This has been described in detail elsewhere [12–18].

21.1.2 Case-Based Blended Learning Ecosystem (CBBLE): Current Case Studies and Implications for Precision Medicine

"The idea of sharing and learning around patients has been alive since the beginning of medicine, when physicians would present their cases to a large audience to primarily learn from the inputs of other physicians." With the invention of the printing press, instead of restricting themselves to verbal face-to-face case presentations, many physicians published their cases in journals and slowly the medical fraternity started naming those published diseases after their first authors. "In this way, case reporting became a gainful activity not only in terms of scientific advancement towards patient benefits, but also as an important instrument of physician fame." We have utilized this case reporting model to help our patients and to train our medical students about disease and patient experience [18, 19].

By reporting cases, this model allows more engagement both from patients and medical students to reach a precise and accurate diagnosis, and also helps as an educational tool.

Our healthcare learning ecosystem is currently based offline in the Kamineni Institute of Medical Sciences, and this offline base keeps shifting with the various university locations in India where our corresponding authors are based for varying numbers of years. The online component of this blended learning ecosystem began on email groups, and then shifted to social media groups such as "Tabula Rasa" [14]. It currently exists in WhatsApp groups with a global audience of medical students and physicians. Gauging by the mentions in our past publications that have flowered through offline inputs processed further online, the students and faculty driving our online discussions range from universities in the US; India; Oxford, UK; Ontario, Canada; Montpellier, France; and Monash, Australia; as well as in the Maldives, Nepal, and Bangladesh.

What follows is a series of case narratives by the physicians in our case-based learning ecosystem and our critical appraisal of them through the "precision lens". We have previously published detailed patient narratives of online patient users of our ecosystem [18], and this article focuses on the physician narratives to provide a qualitative perspective of our workflow.

21.2   Materials and Methods

These case studies were taken from our "case-based blended learning ecosystem" (CBBLE) and were narrated by the physicians, students, and their senior teachers. We illustrated current concepts of scientific accuracy and precision using two analogies. The first was an age-old analogy of "scientific precision" and the second was a creative analogy comparing the medical endeavor of our CBBLE to a game of cricket where health professional bowlers and fielders try to get the batting disease team out in time to win the match. While this is not far from established analogies of "battling" disease, the cricket stadium offers a detailed perspective of our CBBLE framework. A group of our CBBLE online students visited Kamineni Institute of Medical Sciences (KIMS), Narketpally, where its offline component is currently ongoing. They worked with the team of students, physically managing the patients and developing insights into the nature of precision that enabled further analysis of the narratives obtained from our CBBLE students past and present.

This was done to attain our objective of illustrating the role of accuracy-driven "age-old precision medicine" in the framework of our current CBBLE, parallel to current "OMICS-driven" precision medicine narratives, and how we may create a bridge between the two systems.

21.3   Results

21.3.1   Precision Medicine through a Physician Resident's Narrative Lens (in Relatively Low-Resource Settings)

SS is a postgraduate resident physician managing critically ill patients, as well as chronic patients in the outpatient department (OPD) as part of her formal residency program in medicine. She also works part of the time on a thesis that involves clinical decision making around management of thyroid disorders.

What follows is her own narrative of experiences with one recent critically ill patient that she managed onsite, offline along with her other resident colleague as well as other first-year postgraduate colleagues, also known as interns. This was observed in real time by the online members of our CBBLE, who supported with their input once SS shared the patient's deidentified online record on a blog. This was preceded by a snippet of the patient's computed tomography (CT) images, as well as a very brief history of her problem. The online record blog link is accessible in [20].

> *After dinner on my night duty, I rushed to casualty as a patient was brought who needed immediate attention. On going to casualty, I saw that a 60-year-old lady was struggling to breathe and appeared tachypneic. Her % oxygen saturation was only 86% at room air.*

We had to provide her immediate oxygen support which improved her saturation but she was still tachypneic, and opening her mouth to take in the air, basically struggling to take the air in.

After making sure that her saturation was well maintaining I had called the attenders to take a detailed history. Her complaints were not of recent onset. She had complaints since the past 4 months. The lady had been having complaints of burning micturition since 4 months. She had been visiting the hospital often for the above complaints. This time, she had the same complaints of burning micturition, fever, abdominal pain and decreased urine output.

On asking further the attenders related that she hadn't passed urine since 3–4 days. So I had immediately asked for a Foley's catheterization. The moment we had inserted the Foley's the urine was milky white for the initial few minutes. By this time we had got a few of their old prescriptions which showed a urology outpatient card and the urologist made a diagnosis based upon the history of "thin stream of urine" as stricture urethra. Now everything was falling into place. Her persistent complaints from the past few months, her burning micturition, fever.

We got the necessary investigations done, firstly sent a blood and urine culture as I could sense that she may deteriorate. Sent immediately blood for an arterial blood gas (ABG) analysis which showed a metabolic acidosis thus ruling out any lung pathology leading to shortness of breath. She needed dialysis. Blood counts showed elevated leukocytes. Her urine routine microscopy showed plenty of pus cells. Her ultrasonogram (USG) of the abdomen revealed pyonephrosis with dilated ureters.

Had got her dialysis done the next day. In her Foley's interestingly we noted some debris. We had a discussion around it as what it could be. But post dialysis though her acidosis resolved she went into hypotension. We had to start her on inotropic support. Had to monitor her blood pressure (BP) closely to tailor the dose accordingly. We were eagerly waiting for her culture reports as the leucocyte count showed an increasing trend. By this time we had got all her previous outpatient cards which not only revealed her past history and procedures done but were also a representation of the case-based workflow of multiple departments in our rural tertiary medical college hospital.

Her previous outpatient cards revealed that her first visit was in December last year. She had first gone to a gynecologist with complaints of burning micturition, whitish discharge per vaginum and fever. Probably the gynecologist had asked her to void and come as they wanted to examine her per vaginally. But this is where we lost track of the information in her outpatient card progress notes. She then landed up in urology OPD where her outpatient card revealed that she had a thin stream of urine. The diagnosis of stricture urethra was made solely based on this and a dilatation procedure was done, she was catheterized with Foley's and urine culture sensitivity was sent. She was started on antibiotic and asked to turn up after 3 days which she did. She got her Foley's removed and was prescribed the 1st line antibiotic for urinary tract infection (UTI). The urine culture report was, unseen by the care provider, and unasked by the patients. She was prescribed the same 1st line antibiotic as in the first visit. She was then asymptomatic for the next 2 months. When she developed her current symptoms she turned up in the casualty this time.

The organism was the same since the first report, but sadly the culture reports being unseen she was getting the antibiotic for which the organism was never sensitive. The organism had been evolving since then into a more virulent one.

21.3.1.1 Analogy-driven analysis: Retrospective notes from the dugout

We draw below an analogy of our blended ecosystem with a cricket match (Fig. 21.2), where the bowler is the treating physician, along with the rest of the fielders as a part of the treating team. The dugout is the online team sending inputs to the bowler, also known as, the offline treating physician team for taking the wicket of that batsman, also known as, the disease.

The main aim of the entire offline and the online team is to take an accurate wicket using a precise strategy that minimizes overuse and overtreatment and successfully banishes the disease to the pavilion and accurately meets patient requirements. Precision medicine ensures the narrowest path toward accuracy.

The healthcare professional team, as bowlers and fielders, rushed to the casualty (field) to battle the disease of this patient and win the match to meet the patient requirements within a narrowest path possible to attain precision and accuracy without overdiagnosis and overtreatment (Fig. 21.2).

The postmenopausal woman was prescribed first-line antibiotics for UTI without following the urine c/s reports that precisely pointed toward the sensitivity. This permitted the E. coli to strengthen, as the antibiotic in use was a poor match. Moreover, an estrogen ointment was prescribed for local use, although evidence of efficacy for this intervention for UTI prevention or treatment is inconclusive [21]. Lacking adequate diagnosis and follow-up, she returned to casualty with life-threatening symptoms.

The above is an example of an initial accurate managed with an imprecise approach leading to inaccuracy.

The physician resident's narrative, continued:

We got a CT abdomen done for her as the USG abdomen revealed pyonephrosis. We had something very unusual and unexpected in store for us in the CT. She had air pockets in the kidney, in the erector spinae muscle and spinal canal. We decided to change the antibiotic. We were still waiting for the culture report and had started her on a higher antibiotic. But to our dismay, our microbiology department didn't have the sensitivity checking disc for the antibiotic which we had started her on.

21.3.1.2 Online work flow in parallel with the offline component of our CBBLE

At this point, the patient data was shared online in our CBBLE and the online team (analogous to the stadium dugout) swung into action (Figs. 21.3 and 21.4). This enriched the decision-making capacities of the bowling and fielding team by sharing current evidence-based inputs gathered through search engines. This considerably helped the offline treating team (the bowlers and fielders) to intensify their strategy to finally get all the batsmen out. This patient's course was such that one batsman after another tried to score as much as possible, as once she appeared to have recovered from the sepsis, she developed a myocardial infarction. Once that was out, she developed recurrent seizures, which were possibly a result of her uremic encephalopathy. She finally recovered, with all the wickets taken. A graphical representation of the entire hospital course of her illness is shared (Fig. 21.5), here.

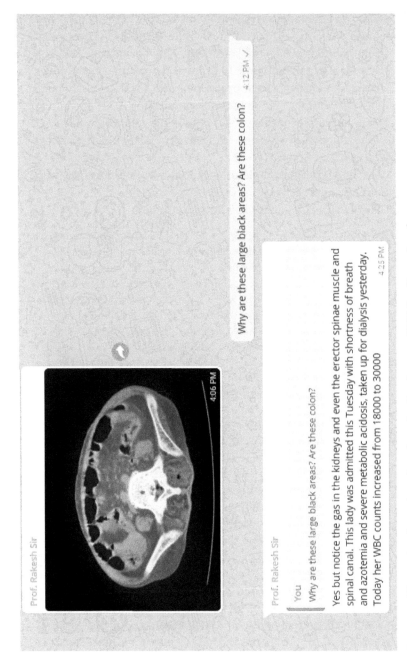

Figure 21.3 Illustrating the patient data shared online into our CBBLE online network.

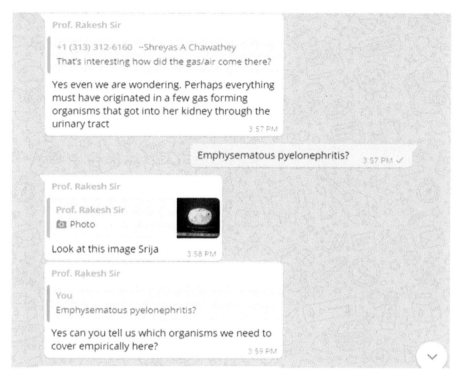

Figure 21.4 Illustrating the patient data shared online into our CBBLE online network.

21.3.2 Precision Medicine through a Neurologist's Narrative Lens (in Relatively High-Resource Settings)

KS has been a member of our CBBLE since 2006, when RB, his offline teacher in his medical college in Bangalore migrated on a teaching assignment to Malaysia and he chose to keep in touch online. Since then, he has completed his postgraduate residency in medicine and fellowship in neurology. He currently practices interventional neurology in a high-resource setting in India, where he often works in the neuro cath lab performing angiography and stenting of cerebral blood vessels.

What follows is a case-based narrative in his voice.

11/4/18	10 IU/L

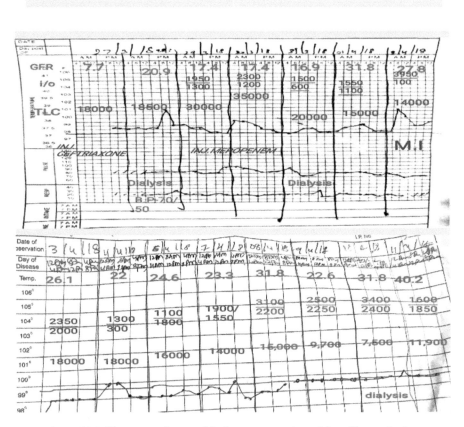

Figure 21.5 Illustrates the graphical representation of her illness during her stay in the hospital.

A 29-year-old male, presented to our Neurology Clinic with his first known episode of generalized tonic-clonic seizures. He denies drug or alcohol usage and his growth and developmental history was normal. Prior to this hospital visit, he had not attended any medical consultation. He was not on any medications for any other ailment. His neurological examination was normal. He was then subjected to an electroencephalography (EEG) and magnetic resonance (MR) imaging of the brain. EEG was normal. However, MR imaging (with contrast) revealed a single ring-enhancing lesion in the left temporal lobe with a visible scolex. A diagnosis of Neurocysticercosis was made. The patient was prescribed Albendazole and steroids for the acute problem and then prescribed T. Levetiracetam 500 mg twice a day for the seizures.

Two weeks into treatment he was brought to the emergency room (ER) in an unresponsive state. His parents found him unresponsive and noticed that he had inadvertently passed urine in his clothes. On examination, his vitals were stable. He had a gaze preference to the left, with paucity of movements on the right side. The possibility of an unwitnessed seizure, Todd's palsy, and postictal confusion were suspected. A repeat MR imaging revealed acute infarct in the left middle cerebral artery (MCA) and left anterior cerebral artery (ACA) territories. MR angiogram revealed a left carotid occlusion. His older MR images were reviewed and showed no evidence of carotid or intracranial vascular disease. He was transferred to the angiography suite and a diagnostic cerebral angiogram was performed. Digital subtraction angiography (DSA) revealed a left carotid T occlusion. A mechanical thrombectomy was successfully performed. However, there was also evidence of left MCA occlusion. Mechanical thrombectomy was repeated, with partial recanalization. Repeat brain imaging revealed significant left hemispheric infarct with evidence of evolving cerebral edema. A decompressive craniectomy was done 3 days after the surgery, repeat imaging revealed persistent cerebral edema and hence he was referred for re-exploration. He underwent left frontal and temporal lobectomy. The patient is slowly recovering with occupational and rehabilitative therapies.

A young stroke workup was considered. Routine blood investigations were normal. Autoimmune markers including antinuclear antibody (ANA) and antineutrophil cytoplasmic antibodies (ANCA) were normal. He had normal homocysteine levels and a normal echocardiogram. Holter monitoring was normal. We screened him for genetic thrombophilic conditions. He was found to have a homozygous MTHFR C677T gene mutation. MTHFR mutation has been previously linked to ischemic strokes.

Two hit hypothesis is well discussed in the field of oncology where a reactive genetic mutation can be triggered by an epigenetic event such as an infection. However, if we are to apply the same principle in this case scenario, our patient had a 'first hit' with a homozygous MTHFR C677T mutation. The 'second hit' may have been the neurocysticercosis infection, followed by the interventions or the resultant systemic distress and functional recovery.

21.3.2.1 Analogy-driven analysis: Retrospective notes from the dugout

In the above game, the bowling and fielding team of KS played with the best possible strategy (precision), but was unable to get the disease out (accuracy), and the game appeared to be drawn, if not lost.

21.3.3 Precision Medicine (in Relatively Low-Resource Settings) through the Lens of a Medical Student's Online Interaction with Our CBBLE

VP is a medical student from Tairunnessa Memorial Medical College and Hospital, Gazipur, Bangladesh, and an active member of our CBBLE who has attended the offline component of our CBBLE, which is also organized in collaboration with BMJ case reports [22]. After finishing his offline elective stint a year back in our host institute, he has subsequently remained active online and regularly collates patient data shared by different members of our CBBLE into deidentified online patient records and ensures that all the signed informed consents from these patients are maintained and preserved in secure locations. The physicians in our CBBLE share their patient data in the hope of obtaining online feedback through conversational engagement with peers along with current best evidence support which VP provides through extensive searches of online and offline electronic resources. The patient narrative below was posted on our WhatsApp discussion forum by a CBBLE member and collated by VP:

> *A 52 year old woman, who was visiting us from North Bengal said she was watching TV in her apartment one day in the summer of 2015 and suddenly noticed that the TV was swinging followed by swinging walls and even the stairs of her apartment when she quickly started running down to safety and amidst all this commotion she noticed a buzz of people shouting earthquake! Earthquake!! She remembers that is the first time she had experienced severe palpitations since then which she complains of repeatedly experiencing over the years. However, she didn't notice the swelling in front of her neck that we noticed. We didn't notice any protruding eyes or tremors etc. USG neck showed nodules in the thyroid that was sent for fine-needle aspiration cytology (FNAC) and her T4 was double the normal value and TSH was very low.*

Toxic adenoma is a toxic thyroid nodule which produces excessive thyroid hormones. They are managed either by prolonged thionamide therapy, surgery or radioiodine ablation. In this non-diabetic patient, a non-velvety, hyperpigmented area involving skin folds was noticed at the back of the neck that was very atypical of acanthosis nigricans. In view of thyroid nodule and hyper-pigmentation in the neck, paraneoplastic acanthosis nigricans was brought into consideration. Later, FNAC report was available which showed features of benign thyroid lesion with foci of mild atypia.

Once the above narrative was posted by one of our CBBLE users feedback started pouring in, in the form of queries and clarifications sought (Fig. 21.6).

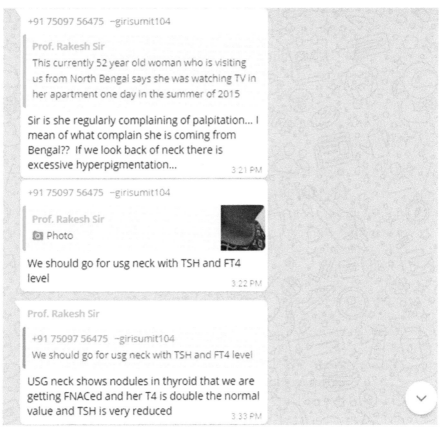

Figure 21.6 Illustrates the feedback and the queries posted from the CBBLE online network.

21.3.3.1 After the discussion, the following questions about the patient were raised

If FNAC is positive, we would know what to do next, but what are the chances of false negativity after FNAC? What is the frequency of having to obtain excision/incision biopsies in such situations? What if we miss an underlying malignancy and go for radioiodine? Based on FNAC findings, how would you decide between surgery and medical management in this patient? These were answered by VP after searching online and offline resources and illustrated by him below along with interactions with the initial CBBLE user who had posted the patient data.

21.3.3.2 If FNAC is positive, we would know what to do next, but what are the chances of false negativity after FNAC?

This made us look up specificity and sensitivity of FNAC in detecting thyroid malignancy. A study showed FNAC is more specific (98%) than sensitive (80%) in detecting malignancy [23]. Here if FNAC is more specific than sensitive so what should we do if the FNAC is negative? There is still 20% chance of its being malignant.

21.3.3.3 What is the frequency of having to obtain excision/incision biopsies in such situations?

The decision to go for excision biopsy primarily for histopathological examination (HPE) to rule out malignancy is often a judgement call. In this context from a precision medicine perspective would a liquid biopsy be feasible or helpful? A feasibility study was conducted by detecting BRAF(V600E) circulating tumour DNA (ctDNA) in the plasma of patients with thyroid nodules to distinguish between benign and malignant nodules. Results showed that BRAF(V600E) ctDNA could distinguish between the two [24].

21.3.3.4 Based on FNAC findings, how would you decide between surgery and medical management in this patient?

The problem with both radioiodine and surgery is that the patient would become hypothyroid eventually and end up in a lifelong dose of maintenance thyroxine. The problem with carbimazole is that we may not be sure of the response.

Eventually, the patient was started on carbimazole after shared decision making with her and her relatives.

21.3.3.5 Analogy-driven analysis: Retrospective notes from the dugout

This game involved a fair amount of precision even in a relatively low-resource setting. It appears to have achieved a fair amount of accuracy, although the game is going to be long, as thyroid-suppressive treatment takes a long time to maintain remission in toxic adenomas. There is always a chance of a malignancy showing up sometime later. This patient was actually referred to our CBBLE from a very distant community, nearly 2000 km away from our tertiary care teaching hospital. VP and RB were already evaluating the patient's inputs posted by one of our CBBLE community health workers, who also resides in the same community 2000 km away, in the patient's village/town. A tentative plan for evaluation was made online, following which the patient was further evaluated in the hospital after she made the long journey. The revised inputs on the patient after evaluation by the hospital residents and faculty were again posted to the online network on WhatsApp, and a final plan of available intervention was drawn. There was a question raised by the team evaluating the patient in the hospital about the uncertainty of future malignancy prediction in this patient when an online CBBLE member suggested that this could become feasible in the near future using BRAF (V600E) ctDNA. At the current point in time, this is out of reach from a rural Indian tertiary teaching hospital perspective, but it is possible that resources and research could make this available even in less than a decade, in the coming years. Until then, we may have to live with the uncertainty posed.

21.3.4 Precision Medicine through a Hematoncologist's Narrative Lens (in Relatively High-Resource Settings)

Author BD is one of the earliest students of our CBBLE from the purely email era, when he published some of his cases even as a graduate student in Manipal College of Medical Sciences, Pokhara, Nepal, way back in 2002. He is currently working at the cutting edge of precision medicine in a high-resource setting at the Medical College of Wisconsin, as a hemato-oncologist focused on multiple myeloma and other plasma cell disorders.

> *In a frigid Wisconsin morning last December, I received a phone call from my friend, whose brother was in the battle for his life against Multiple Myeloma. This patient's disease had progressed after multiple lines of chemotherapy including two stem cell transplants in the span of 4 years. My friend was desperate for a ray of hope to help his brother. He had exhausted all available options, and enrollment in a clinical trial was the only potential solution for his disease. As a myeloma focussed physician, I was aware of a clinical trial that was exploring an experimental agent for the subtype of myeloma he had; myeloma with chromosomal translocation 11 and 14; t(11:14). With ample caution, I routed them towards relevant myeloma clinical trials in their area. Fast forward a few months, my friend called me again, but this time excitement was palpable through the phone. His brother was enrolled in a clinical trial using bcl 2-inhibitor (venetoclax)—a drug active against a particular subtype of multiple myeloma with t(11:14) [25]. This result, a remission, was shared by others and this would open a new era of precision medicine in the therapeutic armamentarium of multiple myeloma.*

21.3.4.1   Analogy-driven analysis: Retrospective notes from the dugout

The patient's disease was difficult to drive out in spite of the best possible efforts by the fielding team over 4 years, and a chemical that targeted a precise molecular pathway in the pathogenesis of the disease offered promise of finally making a dent in the disease.

21.4   Discussion

21.4.1   Current Narratives in Precision Oncogenomics

In 1927, a brilliant physicist from Germany, Werner Heisenberg, introduced a principle, well known as "The Heisenberg Uncertainty

Principle", which would become pivotal for the development of quantum mechanics. The principle asserts the fundamental limit to the precision with which certain pairs of physical properties of a particle can be known. Modern medicine, particularly within the field of oncology, is rapidly moving towards a precision approach, as the molecular underpinnings of disease evolution are made known. As one of the major goals of cancer research is the gaining of understanding of the genetic changes responsible for the establishment of the "cancer clone" and the "key pathways", they could be targeted therapeutically. New insights into human cancers are emerging from basic research, and this has the potential to augment disease diagnostics, therapeutics, and clinical decision-making.

"Precision Medicine"—an abundant term in medical literature—refers to the tailoring of medical treatment guided by genomic or molecular features of the disease and not by the clinicopathological features [26]. Since cancer is a disease of the genome, the field has been the perfect choice to enhance the impact of precision medicine [27]. Because every single cancer patient exhibits a different genetic profile, and that profile can change over time, "tailored" treatment, rather than a "one-size-fits-all" approach, is likely to benefit patients, and hence is an attractive concept. Whether it is a mere concept or realistically assures a better future in oncology continues to remain a debate. The precision medicine approach has transformed the outlook for some deadly cancers. One of the most notable examples is the discovery of bcr-abl gene fusion and the development of imatinib for chronic myelogenous leukemia (CML). CML treated by imatinib resulted in unprecedented results, with a 5-year survival of 90% and some patients even inching towards cure [28]. Other examples include the human epidermal growth factor receptor-2 (HER-2) and development of agents like trastuzumab. Compared to conventional chemotherapy, the addition of this agent has resulted in significant improvement in progression-free survival and reduction of death by 20% [29]. These examples illustrate how the identification of key molecular pathways targeted therapeutically could alter the disease course and result in the desired outcomes. What about other cancers? Has precision medicine delivered its promise in other cancers, as well, and is it a time to celebrate? Or has our approach for more "precision" resulted in more "uncertainty", as described by Heisenberg?

In this context, the design and the results of the SHIVA trial are worth a discussion. It is a phase 2, randomized multicenter trial, which assessed the efficacy of several molecularly targeted therapies based on molecular profiling compared to conventional therapies in patients with advanced cancers [30]. The results showed no improvement in progression-free survival (the primary end-point) with the use of molecularly targeted agents compared to physician's choice of chemotherapy. However, it is important to realize that the majority of the patients in the trial received a hormone modulator or mTOR inhibitor, and thus the justification of the failure of the precision approach based on this limited data is not reasonable. The other was that the study was powered to determine whether the use of an algorithm-based approach to treatment allocation can improve patient outcomes—regardless of the nature of such allocated treatments [30]. In the trial, each patient served as his or her own control in terms of primary end-point assessment. This calls for novel methods of designing the trials, with clinically meaningful endpoints using precision medicine. One of the other concerns has been the possible lack of valid biomarkers. We are experiencing another revolution in personalized treatment, with the introduction of various immune-based approaches. Cancer immunotherapy used to have limited applications, mainly for selected cancers like melanoma and renal cancers and involved the use of interleukin 2 (IL-2). With the dramatic progress in the last few years, this approach has moved into the mainstream in several cancers. Currently, immune-based approaches represent the most exciting area for diseases like melanoma, non-small lung cell cancer, and hematological cancers including multiple myeloma. The immune approach includes both active and passive immunotherapies: monoclonal antibodies, checkpoint inhibitors, and cellular immunotherapy. Recently, Rosenberg et al. demonstrated a new approach to immunotherapy, in which "adoptive transfer of mutant-protein-specific tumour-infiltrating lymphocytes (TILs) in conjunction with interleukin (IL)-2 and checkpoint blockade mediated the complete durable regression of metastatic breast cancer, which is now ongoing for >22 months" [31]. However, not all patients respond to immunotherapy, and there is a variability in response. One of the reasons for the variability in patient response with immunotherapy is the lack of predictive biomarkers. Identifying predictive biomarkers is a challenge in immunotherapy, and the other challenges include cost, toxicity and tumor heterogeneity, which impede the efficacy of immune-based therapy. For checkpoint inhibition, PDL1 has been proposed as a putative biomarker, as pembrolizumab (anti-PD 1) is approved in non-small-cell lung carcinoma (NSCLC) only in patients whose tumor PD-L1 levels are ≥50% [32]. Unfortunately, with the different temporal, spatial and methodological heterogeneity, it remains an unreliable biomarker [27]. The other unreliable biomarker includes the ERCC1 for NSCLC to platinum therapy [33].

Despite the challenges, precision medicine holds a lot of potential in cancer therapy. While we need to conduct well-designed randomized trials to assess broader efficacy of personalized medicine and validate the bio-markers, we can also enjoy the tremendous success of KIT mutations in gastrointestinal stromal tumor (GIST), BRAF (V600E) in melanoma, EGFR, ALK, and ROS1 alterations in NSCLC [34–36]. In a study in which my friend's brother participated, venetoclax monotherapy resulted in unprecedented response rates of 40% in heavily pre-treated patients with multiple myeloma [25]. Biomarker analysis confirmed that response to venetoclax correlated with higher BCL2:BCL2L1 and BCL2:MCL1 mRNA expression ratios, which are predominantly seen in patients with t (11:14). These outcomes show potential in results, for the first time, that pave the way for precision medicine in multiple myeloma with a significant potential to change practice in specific subgroups of patients and the hope is that knowledge in the field will increase to include other subgroups.

Clinical trials are evolving to investigate tumor heterogeneity from patient to patient in their design. The Molecular Analysis for Therapy Choice (NCI-MATCH) is a clinical trial selecting treatments based on genetic features of patients, not traditional tumor histology [32]. Thusfar, 2500 patients in the USA have been enrolled in one of the 24 arms of this trial, representing one half of the recruitment goals [37]. The Molecular Profiling-based Assignment of Cancer Therapy (NCI-MPACT) is another innovative clinical trial to test the hypothesis that targeting an oncogenic driver mutation is more efficacious than not targeting it. NCI-MPACT will recruit advanced cancer patients who have been unresponsive to standard therapeutic options and possess mutations in one of three genetic pathways that include DNA repair, PI3K/mTOR (phosphoinositide-3 kinase/mammalian target of rapamycin), and Ras/Raf/MEK (mitogen-activated protein kinase). The efficacy of diagnosis and therapies using precision medicine could be significantly enhanced, should results deliver the outcomes investigated in these trials.

With the development of novel technologies, it is hoped that understanding of tumor complexity and the immune system will be increased. These will be critical in designing future tailored combination therapies. The recent advances in the development of sequencing technologies have enhanced the ability to sequence cancers at both population and single-cell levels. Diverse mechanisms that lead to disease evolution, disease response, and refractoriness are slowly being understood. These advancements, we hope, will translate to more targeted therapies with better outcomes in patients with cancers in future.

The above discussion by author BD perhaps reflects and echoes the thoughts of many of his peers, who are working to make the dream of precision medicine a reality; while at the same time, this is an evolving area, where there are unseen threats (for patients) and opportunities (for pharma) in terms of aggravating the pandemic of overdiagnosis and overtreatment [38–40].

21.4.2 Age-Old Precision Medicine

The patient of KS presented to a relatively high-resource set-up in India, where the practitioner worked in a highly specialized and narrow domain of precision medicine (interventional neurology) and made the best precise efforts to address the problem in this critically ill patient, but the results were far from accurate. We are not sure of the outcomes that may have been achieved if the same patient had visited a low-resource setting; would the approach have been imprecise but the outcomes serendipitously better? Currently, what are the available strategies for predicting patient outcomes in response to highly specialized precision intervention? Randomized controlled trials have looked at catheter-based interventions in cerebral vessels, and available data supports the use of mechanical thrombectomy from 6 to 24 h for patients with occlusion of the intracranial carotid or proximal middle cerebral artery who present to a stroke center with expertise in both mechanical thrombectomy and automated infarct volume determination using MR imaging or perfusion CT [41]. While the data continues to evolve, there shall remain a grey zone where the push of newer innovations in terms of pharmacological and non-pharmacological devices will encourage overdiagnosis and overtreatment. This appears directly reactionary to combat pre-existing underdiagnosis and undertreatment, particularly in parts of the globe that are plagued by it.

The two patient narratives from relatively low-resource settings naturally reflect the danger of underdiagnosis and undertreatment. This is apparent beginning with the very first narrative, where the urinary tract infection was undertreated due to underdiagnosis. This was again a result of informational discontinuity. This is another malady that our CBBLE is striving to address by tracking each patient record online and encouraging physicians to communicate around their online patient records, which can be accessed through our CBBLE. The solution to this lies in concerted team communication. Ideally, if this patient had first approached our CBBLE, we could have utilized our online network, connected to microbiologists, to collect her results. While this is easier in patients admitted to our hospital, the same may not be true with outpatients. For such situations, we have in the past tried to train community health workers to follow up with our outpatients even after they return to their homes. We have tried to integrate this with rudimentary home health care programs, but we must admit that we have not been able to scale this approach beyond just one or two towns in rural India [42].

Identification, elucidation and correlation of external life event pathways through traditional history taking and internal cellular and molecular event pathways through current precision medicine approaches could serve as a superior predictor of outcomes than the binary preference-bound "patient-related outcome". It can begin with recording and documentation of individual case-based experiences of the external and internal pathways that are otherwise routinely lost to science, and all this data can be harnessed using case-based reasoning (CBR) techniques. Many labels, such as case-based informatics and evidence farming, have been applied to these currently fledgling endeavors [43–46]. Our elective students have begun the process through their case records in their online learning portfolios [47], and in the next few months to years, once the number of cases in our database improves, we shall be able to utilize algorithms for "experiential evidence farming" to demonstrate improved patient outcomes with recording, sharing, reusing and recycling steps in the CBR evidence farming cycle.

21.5 Conclusions

This chapter illustrates how CBBLE can strengthen the bridge between age-old precision approaches with modern omics-driven approaches to deliver precision healthcare and reduce overdiagnosis and overtreatment. The narratives from our CBBLE found that physicians naturally aim for precision and accuracy, although accuracy may remain elusive in spite of our best focused precise approaches. One can considerably increase one's chances of obtaining accuracy if one is regularly supported by a CBBLE that not only offers evidence-based decision tools, but also a regular connection with peers that can go a long way toward improved documentation, transparency and newer learning insights.

We hope to scale the CBBLE approach in the near future to not only reduce overdiagnosis and overtreatment, but also promote transparency, accountability, and innovation toward optimizing solutions for individual patients, integrating age-old as well as technology-driven precision medicine approaches.

Abbreviations

ABG:	arterial blood gas
ACA:	anterior cerebral artery
AFB:	acid-fast bacillus
ANA:	antinuclear antibody
ANCA:	antineutrophil cytoplasmic antibodies

BP: blood pressure
CBBLE: case-based blended learning ecosystem
CBNAAT: cartridge-based nucleic acid amplification assay test
CBR: case-based reasoning
CML: chronic myelogenous leukemia
CT: computed tomography
ctDNA: circulating tumour DNA
DSA: digital subtraction angiography
EEG: electroencephalography
ER: emergency room
FNAC: fine-needle aspiration cytology
GIST: gastrointestinal stromal tumor
HER-2:   human epidermal growth factor receptor-2
HPE: histopathological examination
IL-2:   interleukin 2
MCA: middle cerebral artery
MR: magnetic resonance
NCI-MATCH: National Cancer Institute-Molecular Analysis for Therapy Choice
NCI-MPACT: National Cancer Institute-Molecular Profiling-based Assignment of Cancer Therapy
NSCLC: non-small-cell lung carcinoma
OPD: outpatient department
PCI: percutaneous coronary intervention
STEMI: ST-elevation myocardial infarction
TILs: tumour-infiltrating lymphocytes
TSH: thyroid stimulating hormone
UDHC: user-driven healthcare
USG: ultrasonogram
UTI: urinary tract infection

Disclosures and Conflict of Interest

This chapter was originally published as: Podder, V., Dhakal, B., Shaik, G. U. S., Sundar, K., Sivapuram, M. S., Chattu, V. K., Biswas, R. (2018).

Developing a case-based blended learning ecosystem to optimize precision medicine: Reducing overdiagnosis and overtreatment. *Healthcare*, **6**, 78, doi:10.3390/healthcare6030078, under the Creative Commons Attribution license (http://creativecommons. org/licenses/by/4.0/). It appears here, with edits and updates, by kind permission of the copyright holders.

Conceptualization: R.B., V.P., and M.S.S.; Writing Original Draft Preparation: R.B., V.P., B.D., K.S., G.U.S.S., M.S.S., V.K.C.; Writing-Review and Editing: R.B.; Images: M.S.S. and V.P.

Funding: This research received no external funding.

Acknowledgments: For case 1, we would like to acknowledge faculty and students of Kamineni Institute of Medical Sciences, which was presented by Akhil Sugandhi and Salma to the Department of Microbiology, Gynecology, Hospital Administration, Urology and Emergency Medicine before being finally presented to everyone in the Thursday morning CME for their inputs; Venkat Kishan, Department of Radiodiagnosis for identifying the gas in the kidneys, spinal canal, and erector spinae. We also acknowledge the online members of our CBBLE namely Abhishek Choudary, Monika Pathania, Avinash Kumar Gupta, Srija Katta, Shreyas Chawathey, Ashwini Ronghe, Deepak Badhani, Yogesh Sharma, Kuldeep Gupta, Aadipta Ghosh and all other students in our current offline location for our CBBLE, Kamineni Institute of Medical Sciences (KIMS), Narketpally. We also acknowledge Amy Price, Namrata Dass, Michele Meltzer, Bhavik Shah, Boudhayan Das Munshi, Rajib Sengupta, Amit Taneja, Nidhi Sehgal, Sujoy Dasgupta, Able Lawrence, Rajendra Takhar, Debasish Acharjee, Karthik Balachandran, Binidra Banerjee, Leelavati Thakur, Pradip Kar and other active online members of our CBBLE.

Conflicts of Interest: The authors declare no conflict of interest.

Corresponding Author

Dr. Rakesh Biswas
Department of Internal Medicine
Kamineni Institute of Medical Sciences
Narketpally 508254, India
Email: rakesh7biswas@gmail.com

References

1. Christensen, C. M., Grossman, J. H., Hwang, J. (2009). The innovator's prescription. In: *Soundview Executive Book Summaries*, McGraw-Hill, New York, NY, USA.

2. Institute for Precision Medicine (2015). Available at: http://ipm. Weill. Cornell.Edu/about/definition (accessed on January 2, 2019).

3. Thomas, B., UST Global Health Group. The new healthcare ecosystem: 5 emerging relationships. Available at: https://www. beckershospitalreview.com/hospital-management-administration/ the-new-healthcare-ecosystem-5-emerging-relationships.html (accessed on January 2, 2019).

4. Peter, A. How does accuracy differ from precision in scientific measurements? Socratic questions. Available at: https://socratic.org/ questions/how-does-accuracy-differ-from-precision-in-scientific-measurements (accessed on January 2, 2019).

5. Nagashree, B., Manchukonda, R. S. (2016). Prescription audit for evaluation of present prescribing trends in a rural tertiary care hospital in South India: An observational study. *Int. J. Basic Clin. Pharmacol.*, **5**, 2094–2097.

6. Krishna, P. (1992). Medical education. *Health Millions*, **18**, 42–44.

7. Shankar, P. R., Chandrasekhar, T. S., Mishra, P., Subish, P. (2006). Initiating and strengthening medical student research: Time to take up the gauntlet. *Kathmandu Univ. Med. J.*, **4**, 135–138.

8. Behnke, L. M., Solis, A., Shulman, S. A., Skoufalos, A. (2013). A targeted approach to reducing overutilization: Use of percutaneous coronary intervention in stable coronary artery disease. *Popul. Health Manag.*, **16**, 164–168.

9. Brownlee, S., Chalkidou, K., Doust, J., Elshaug, A. G., Glasziou, P., Heath, I., Nagpal, S., Saini, V., Srivastava, D., Chalmers, K., et al. (2017). Evidence for overuse of medical services around the world. *Lancet*, **390**, 156–168.

10. Patil, D., Lanjewar, C., Vaggar, G., Bhargava, J., Sabnis, G., Pahwa, J., Phatarpekar, A., Shah, H., Kerkar, P. (2017). Appropriateness of elective percutaneous coronary intervention and impact of government health insurance scheme—A tertiary centre experience from Western India. *Indian Heart J.*, **69**, 600–606.

11. Podder, V. A large force of health system—The medical students: Have they been utilized adequately? Available at: https://blogs. Bmj.

Com/case-reports/2017/12/12/a-large-force-of-health-system-the-medical-students-have-they-been-utilized-adequately/ (accessed on January 2, 2019).

12. Price, A., Chandra, S., Bera, K., Biswas, T., Chatterjee, P., Wittenberg, R., Mehta, N., Biswas, R. (2013). Understanding clinical complexity through conversational learning in medical social networks: Implementing user-driven health care. In: *Handbook of Systems and Complexity in Health*, Springer, New York, NY, USA, pp. 767–793.

13. Biswas, R., Sturmberg, J. P., Martin, C. M. (2011). The user driven learning environment. In: *User-Driven Healthcare and Narrative Medicine: Utilizing Collaborative Social Networks and Technologies*, IGI Global, Hershey, PA, USA, pp. 229–241.

14. Purkayastha, S., Price, A., Biswas, R., Ganesh, A. J., Otero, P. (2015). From dyadic ties to information infrastructures: Care-coordination between patients, providers, students and researchers: Contribution of the health informatics education working group. *Yearb. Med. Inform.*, **10**, 68–74.

15. Biswas, R. (2015). Clinical audit and lifelong reflective practice as game changers to integrate medical education and practice. *J. Fam. Med. Prim. Care*, **4**, 476.

16. Price, A. I., Djulbegovic, B., Biswas, R., Chatterjee, P. (2015). Evidence-based medicine meets person-centred care: A collaborative perspective on the relationship. *J. Eval. Clin. Pract.*, **21**, 1047–1051.

17. Price, A., Chatterjee, P., Biswas, R. (2015). Comparative effectiveness research collaboration and precision medicine. *Ann. Neurosci.*, **22**, 127–129.

18. Arora, N., Tamrakar, N., Price, A., Biswas, R. (2014). Medical students meet user driven health care for patient centered learning in clinical medicine. *Int. J. User Driven Healthc. (IJUDH)*, **4**, 7–17.

19. Price, A., Biswas, T., Biswas, R. (2013). Person-centered healthcare in the information age: Experiences from a user driven healthcare network. *Eur. J. Pers. Cent. Healthc.*, **1**, 385–393.

20. Case Based Blended Learning Ecosystem Narratives. Available at: http://cbblenarratives.blogspot.com/2018/05/cbble-narratives.html (accessed on January 2, 2019).

21. Raz, R. (2011). Urinary tract infection in postmenopausal women. *Korean J. Urol.*, **52**, 801–808.

22. BMJ Case Report Student Electives. Available at: https://promotions.bmj.com/jnl/bmj-case-reports-student-electives-2/ (accessed on January 2, 2019).

23. Basharat, R., Bukhari, M. H., Saeed, S., Hamid, T. (2011). Comparison of fine needle aspiration cytology and thyroid scan in solitary thyroid nodule. *Pathol. Res. Int.*, **2011**, 754041.

24. Patel, K. B. (2015). Detection of circulating thyroid tumor DNA in patients with thyroid nodules, Electronic Thesis and Dissertation Repository, London, ON, Canada 3644. https://ir.lib.uwo.ca/etd/3644/ (accessed on January 2, 2019).

25. Kumar, S., Kaufman, J. L., Gasparetto, C., Mikhael, J., Vij, R., Pegourie, B., Benboubker, L., Facon, T., Amiot, M., Moreau, P., et al. (2017). Efficacy of venetoclax as targeted therapy for relapsed/refractory t (11; 14) multiple myeloma. *Blood*, **130**, 2401–2409.

26. Hunter, D. J. (2016). Uncertainty in the era of precision medicine. *N. Engl. J. Med.*, **375**, 711–713.

27. Collins, F. S., Varmus, H. (2015). A new initiative on precision medicine. *N. Engl. J. Med.*, **372**, 793–795.

28. Saußele, S., Krauß, M.-P., Hehlmann, R., Lauseker, M., Proetel, U., Kalmanti, L., Hanfstein, B., Fabarius, A., Kraemer, D., Berdel, W. E., et al. (2015). Impact of comorbidities on overall survival in patients with chronic myeloid leukemia: Results of the randomized CML Study IV. *Blood*, **126**, 42–49.

29. Slamon, D. J., Leyland-Jones, B., Shak, S., Fuchs, H., Paton, V., Bajamonde, A., Fleming, T., Eiermann, W., Wolter, J., Pegram, M., et al. (2001). Use of chemotherapy plus a monoclonal antibody against HER2 for metastatic breast cancer that overexpresses HER2. *N. Engl. J. Med.*, **344**, 783–792.

30. Le Tourneau, C., Delord, J.-P., Gonçalves, A., Gavoille, C., Dubot, C., Isambert, N., Campone, M., Trédan, O., Massiani, M.-A., Mauborgne, C., et al. (2015). Molecularly targeted therapy based on tumour molecular profiling versus conventional therapy for advanced cancer (SHIVA): A multicentre, open-label, proof-of-concept, randomised, controlled phase 2 trial. *Lancet Oncol.*, **16**, 1324–1334.

31. Zacharakis, N., Chinnasamy, H., Black, M., Xu, H., Lu, Y. C., Zheng, Z., Pasetto, A., Langhan, M., Shelton, T., Prickett, T., et al. (2018). Immune recognition of somatic mutations leading to complete durable regression in metastatic breast cancer. *Nat. Med.*, **24**, 724–730.

32. McLaughlin, J., Han, G., Schalper, K. A., Carvajal-Hausdorf, D., Pelekanou, V., Rehman, J., Velcheti, V., Herbst, R., LoRusso, P., Rimm, D. L. (2016). Quantitative assessment of the heterogeneity of PD-L1 expression in non–small-cell lung cancer. *JAMA Oncol.*, **2**, 46–54.

33. Lee, S. M., Falzon, M., Blackhall, F., Spicer, J., Nicolson, M., Chaudhuri, A., Middleton, G., Ahmed, S., Hicks, J., Crosse, B., et al. (2016). Randomized prospective biomarker trial of ERCC1 for comparing platinum and nonplatinum therapy in advanced non-small-cell lung cancer: ERCC1 trial (ET). *J. Clin. Oncol.*, **35**, 402–411.

34. Demetri, G. D., Von Mehren, M., Blanke, C. D., Van den Abbeele, A. D., Eisenberg, B., Roberts, P. J., Heinrich, M. C., Tuveson, D. A., Singer, S., Janicek, M., et al. (2002). Efficacy and safety of imatinib mesylate in advanced gastrointestinal stromal tumors. *N. Engl. J. Med.*, **347**, 472–480.

35. Ascierto, P. A., Kirkwood, J. M., Grob, J.-J., Simeone, E., Grimaldi, A. M., Maio, M., Palmieri, G., Testori, A., Marincola, F. M., Mozzillo, N., et al. (2012). The role of BRAF V600 mutation in melanoma. *J. Transl. Med.*, **10**, 85.

36. Desai, A., Menon, S. P., Dy, G. K. (2016). Alterations in genes other than EGFR/ALK/ROS1 in non-small cell lung cancer: Trials and treatment options. *Cancer Biol. Med.*, **13**, 77.

37. Do, K., O'Sullivan Coyne, G., Chen, A. P. (2015). An overview of the NCI precision medicine trials—NCI MATCH and MPACT. *Chin. Clin. Oncol.*, **4**.

38. Bhatt, J. R., Klotz, L. (2016). *Overtreatment in Cancer–Is It a Problem?* Taylor & Francis, Didcot, UK.

39. Klotz, L. (2012). Cancer overdiagnosis and overtreatment. *Curr. Opin. Urol.*, **22**, 203–209.

40. Esserman, L. J., Thompson, I. M., Reid, B., Nelson, P., Ransohoff, D. F., Welch, H. G., Hwang, S., Berry, D. A., Kinzler, K. W., Black, W. C., et al. (2014). Addressing overdiagnosis and overtreatment in cancer: A prescription for change. *Lancet Oncol.*, **15**, e234–e242.

41. Nogueira, R. G., Jadhav, A. P., Haussen, D. C., Bonafe, A., Budzik, R. F., Bhuva, P., Yavagal, D. R., Ribo, M., Cognard, C., Hanel, R. A., et al. (2018). Thrombectomy 6 to 24 hours after stroke with a mismatch between deficit and infarct. *N. Engl. J. Med.*, **378**, 11–21.

42. Biswas, R. User driven health care. Available at: http://userdrivenhealthcare.blogspot.com/2016/08/role-of-patient-information.html?m=0 (accessed on January 2, 2019).

43. Andritsos, P., Jurisica, I., Glasgow, J. I. (2014). Case-based reasoning for biomedical informatics and medicine. In: *Springer Handbook of Bio-/Neuroinformatics*, Springer, New York, NY, USA, pp. 207–221.

44. Pantazi, S. V., Arocha, J. F., Moehr, J. R. (2004). Case-based medical informatics. *BMC Med. Inform. Decis. Mak.*, **4**, 19.

45. Hay, M. C., Weisner, T. S., Subramanian, S., Duan, N., Niedzinski, E. J., Kravitz, R. L. (2008). Harnessing experience: Exploring the gap between evidence-based medicine and clinical practice. *J. Eval. Clin. Pract.*, **14**, 707–713.

46. Dussart, C., Pommier, P., Siranyan, V., Grelaud, G., Dussart, S. (2008). Optimizing clinical practice with case-based reasoning approach. *J. Eval. Clin. Prac.*, **14**, 718–720.

47. BMJ. Elective Case Log. Available at: bmjcaselogvivek.blogspot.com/ (accessed on January 2, 2019).

Supplementary Readings

Basu, A., Steen, O. B. Allan, M. (2011). Integrating medical education with medical practice: Role of Web 2.0 tools. In: *User-Driven Healthcare and Narrative Medicine: Utilizing Collaborative Social Networks and Technologies*, IGI Global, Hershey, PA, USA, pp. 433–445.

Bjerg, K. (2011). Dimensions of the patient journey: Charting and sharing the patient journey with long term user-driven support systems. In: *User-Driven Healthcare and Narrative Medicine: Utilizing Collaborative Social Networks and Technologies*, IGI Global, Hershey, PA, USA, pp. 410–432.

Fitzpatrick, J., Ako, W. (2011). Developing community ontologies in user driven healthcare. In: *User-Driven Healthcare and Narrative Medicine: Utilizing Collaborative Social Networks and Technologies*, IGI Global, Hershey, PA, USA, pp. 310–327.

Gavgani, V. Z. (2011). Ubiquitous information therapy service through social networking libraries: An operational Web 2.0 service model. In: *User-Driven Healthcare and Narrative Medicine: Utilizing Collaborative Social Networks and Technologies*, IGI Global, Hershey, PA, USA, pp. 673–688.

James, R. (2011). Practical pointers in medicine over seven decades: Reflections of an individual physician. In: *User-Driven Healthcare and Narrative Medicine: Utilizing Collaborative Social Networks and Technologies*, IGI Global, Hershey, PA, USA, pp. 173–183.

Martin, C. M., Biswas, R., Joshi, A., Sturmberg, J. P. (2011). Patient journey record systems (PaJR): The development of a conceptual framework for a patient journey system. In: *User-Driven Healthcare and Narrative Medicine: Utilizing Collaborative Social Networks and Technologies*, IGI Global, Hershey, PA, USA, pp. 75–92.

Norman, C. (2011). Bridging online and offline social networks to promote health innovation: The CoNEKTR model. In: *User-Driven Healthcare and Narrative Medicine: Utilizing Collaborative Social Networks and Technologies*, IGI Global, Hershey, PA, USA, pp. 462–480.

Sarbadhikari, S. N. (2011). Unlearning and relearning in online health education. In: *User-Driven Healthcare and Narrative Medicine: Utilizing Collaborative Social Networks and Technologies*, IGI Global, Hershey, PA, USA, pp. 1648–1363.

Swennen, M. H. (2011). The gap between what is knowable and what we do in clinical practice. In: *User-Driven Healthcare and Narrative Medicine: Utilizing Collaborative Social Networks and Technologies*, IGI Global, Hershey, PA, USA, pp. 335–356.

Young, J. W. (2011). A healing journey with a thousand echoes. In: *User-Driven Healthcare and Narrative Medicine: Utilizing Collaborative Social Networks and Technologies*, IGI Global, Hershey, PA, USA, pp. 01–15.

Chapter 22

Views from the FDA: Precision Medicine, Genetic Variant Databases and Companion Diagnostics

U.S. Food and Drug Administration, Center for Drug Evaluation and Research, Silver Spring, Maryland, USA

Keywords: personalized medicine, precision medicine, next generation sequencing, genetic variants, diagnosis, consensus standards, crowd-sourced data, open-source computing, bioinformatics, genetic testing, *in vitro* diagnostics, standards developing organizations, precision FDA, genotype, Public Human Genetic Variate Databases, standard operating procedures, databases, tailored treatment, companion diagnostics, oncology, clinical trials, metastatic cancer, colon cancer, therascreen, KRAS gene, Vectibix, Erbitux, Herceptin, breast cancer, HER2 gene, medical device, treatment decisions, treatment safety and effectiveness

22.1 Precision Medicine and Personalized Medicine

22.1.1 Introduction

Most medical treatments are designed for the "average patient" as a one-size-fits-all-approach, which may be successful for some patients but not for others. Precision medicine, sometimes known as "personalized medicine" is an innovative approach to tailoring disease prevention and treatment that takes into account

The Road from Nanomedicine to Precision Medicine
Edited by Shaker A. Mousa, Raj Bawa, and Gerald F. Audette
Copyright © 2020 Jenny Stanford Publishing Pte. Ltd.
ISBN 978-981-4800-59-4 (Hardcover), 978-0-429-29501-0 (eBook)
www.jennystanford.com

differences in people's genes, environments, and lifestyles. The goal of precision medicine is to target the right treatments to the right patients at the right time.

Advances in precision medicine have already led to powerful new discoveries and FDA-approved treatments that are tailored to specific characteristics of individuals, such as a person's genetic makeup, or the genetic profile of an individual's tumor. Patients with a variety of cancers routinely undergo molecular testing as part of patient care, enabling physicians to select treatments that improve chances of survival and reduce exposure to adverse effects.

22.1.2   Next Generation Sequencing (NGS) Tests

Precision care will only be as good as the tests that guide diagnosis and treatment. Next Generation Sequencing (NGS) tests (Fig. 22.1)

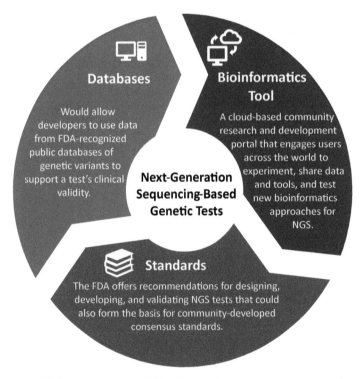

Figure 22.1 FDA's recommendation for next-generation sequencing-based genetic testing.

are capable of rapidly identifying or 'sequencing' large sections of a person's genome and are important advances in the clinical applications of precision medicine. Patients, physicians and researchers can use these tests to find genetic variants that help them diagnose, treat, and understand more about human disease.

22.1.3 The FDA's Role in Advancing Precision Medicine

The FDA is working to ensure the accuracy of NGS tests, so that patients and clinicians can receive accurate and clinically meaningful test results.

The vast amount of information generated through NGS poses novel regulatory issues for the FDA. While current regulatory approaches are appropriate for conventional diagnostics that detect a single disease or condition (such as blood glucose or cholesterol levels), these new sequencing techniques contain the equivalent of millions of tests in one. Because of this, the FDA has worked with stakeholders in industry, laboratories, academia, and patient and professional societies to develop a flexible regulatory approach to accommodate this rapidly evolving technology that leverages consensus standards, crowd-sourced data, and state-of-the-art open-source computing technology to support NGS test development. This approach will enable innovation in testing and research, and will speed access to accurate, reliable genetic tests.

22.1.4 Streamlining FDA's Regulatory Oversight of NGS Tests

In April 2018, the FDA issued two final guidances that recommend approaches to streamline the submission and review of data supporting the clinical and analytical validity of NGS-based tests. These recommendations are intended to provide an efficient and flexible regulatory oversight approach: as technology advances, standards can rapidly evolve and be used to set appropriate metrics for fast growing fields such as NGS. Similarly, as clinical evidence improves, new assertions could be supported. This adaptive approach would ultimately foster innovation among test developers and improve patients' access to these new technologies.

22.1.5   Clinical Databases Guidance

The final guidance "Use of Public Human Genetic Variant Databases to Support Clinical Validity for Genetic and Genomic-Based *In Vitro* Diagnostics" allows developers to use data from FDA-recognized public databases of genetic variants to help support a test's clinical validity and outlines how database administrators can seek recognition for their databases if they meet certain quality recommendations. This approach incentivizes data sharing and provides a more efficient path to market.

22.1.6   Analytical Validation Guidance

The final guidance "Considerations for Design, Development, and Analytical Validation of Next Generation Sequencing (NGS)–Based *In Vitro* Diagnostics (IVDs) Intended to Aid in the Diagnosis of Suspected Germline Diseases" offers recommendations for designing, developing and validating NGS tests. The guidance also encourages community engagement in developing NGS-related standards by standards developing organizations (SDOs) since standards can more rapidly evolve with changes in technology and knowledge and can therefore be used to set appropriate metrics such as specific performance thresholds for fast growing fields such as NGS.

22.1.7   FDA's Bioinformatics Platform

The FDA created precision FDA, a cloud-based community research and development portal that engages users across the world to share data and tools to test, pilot, and validate existing and new bioinformatics approaches to NGS processing. Individuals and organizations in the genomics community can find more information and sign up to participate at http://precision.fda.gov.

22.1.8   Reports and Discussion Papers

- Optimizing FDA's Regulatory Oversight of Next Generation Sequencing Diagnostic Tests
- Report: Paving the Way for Personalized Medicine (also see, Chapter 3 in this volume).

22.1.9   Past Meetings

- 07/27/16: Webinar: Next Generation Sequencing (NGS) Draft Guidances: Implications for Patients and Providers
- 07/27/16: Webinar: Next Generation Sequencing (NGS) Draft Guidances: Technical and Regulatory Aspects
- 11/13/2015: Public Workshop: Use of Databases for Establishing the Clinical Relevance of Human Genetic Variants
- 11/12/2015: Standards Based Approach to Analytical Performance Evaluation of Next Generation Sequencing *In Vitro* Diagnostic Tests
- 02/20/2015: Optimizing FDA's Regulatory Oversight of Next Generation Sequencing Diagnostic Tests Public Workshop

22.2   FDA Recognition of Public Human Genetic Variant Databases

A genetic variant database contains information about genetic differences (also called genetic variants). Researchers submit data to these databases, which collect, organize, and publicly document the evidence supporting links between a human genetic variant and a disease or condition. These databases may make assertions about genetic variants, which are informed assessments of the correlation (or lack thereof) between a disease or condition and a genetic variant based on the current state of knowledge. Better understanding the relationships between genotypes (the genetic code of an organism) and diseases or conditions can aid in the diagnosis and treatment of individuals with genetic conditions.

The FDA's April 2018 final guidance "Use of Public Human Genetic Variant Databases to Support Clinical Validity for Genetic and Genomic-Based *In Vitro* Diagnostics" provides a mechanism for test developers to leverage publicly accessible databases of human genetic variants to support the FDA's regulatory review of genetic and genomic tests. The FDA hopes that this program will encourage database administrators to submit genetic variant information to publicly accessible databases, which could help reduce regulatory burdens on test developers and spur advancements in the evaluation and implementation of precision medicine.

FDA recognition of a database indicates that the FDA believes the data and assertions contained in the database can be considered valid scientific evidence. This program will allow sponsors to use the assertions within FDA-recognized databases to support the clinical validity of their tests.

22.2.1 How to Participate in the FDA Database Recognition Program

Database administrators who wish to seek recognition for their databases should follow the recommendations outlined in the final guidance.

Participation in the FDA database recognition process is voluntary, and participation does not subject the database to FDA oversight, beyond what is needed to retain the recognition. There are no user fees associated with the FDA Database Recognition program.

A request for recognition can be for the entire genetic variant database or for a subset of the database as decided by the database administrator. Submissions should demonstrate that the recommendations in the final guidance have been followed or should explain why an alternative approach was taken.

(1) Prior to submitting an application for database recognition, database administrators should contact FDA staff at OIRPMGroup@fda.hhs.gov to determine how to address their recognition request. Database administrators should then submit an application as an eCopy. The eCopy should be in the form of a DVD, CD, or flash drive and should be sent to the Document Mail Center at the following address:

U.S. Food and Drug Administration
Center for Devices and Radiological Health
Document Mail Center – WO66-G609
10903 New Hampshire Avenue
Silver Spring, MD 20993-0002

The eCopy should meet the technical standards outlined in Attachment 1 of the FDA guidance "eCopy Program for Medical Device Submissions."

(2) The cover letter of this application should identify the appropriate division that will undertake the review of the

application (this information is obtained by contacting FDA staff prior to submission at OIRPMGroup@fda.hhs.gov). The application should be submitted as an "informational meeting" Q-submission based on the FDA guidance document "Requests for Feedback on Medical Device Submissions: The Pre-Submission Program and Meetings with Food and Drug Administration Staff." The following types of documents, which show that the recommendations in the final guidance have been met, should be submitted as part of an application for recognition:

- Standard Operating Procedures (SOPs), policies or other documents related to the recommendations in the final guidance, such as:
  - General operation of the genetic variant database
  - Personally, identifiable information and protected health information confidentiality and privacy
  - Data integrity and security
  - Curation, variant evaluation, and reevaluation
  - Training for curation, evaluation, privacy and security, and other relevant activities
- Validation studies for evaluation SOPs
- Documentation of the qualifications of the individuals evaluating variants and policies for approving those individuals
- Data preservation plan
- Conflict of interest policies and disclosures of conflicts of interest
- A commitment to make all recommended documents publicly accessible via weblinks

(3) Applications should be accompanied by a cover letter, which should include the following information:

- Identification of the submission as an "Informational Meeting" Q-Submission
- Identification of the submission as a request for Database Recognition
- Statement of the types of variants the genetic variant database assertions address (e.g., germline)

- Scope or portion of the database for which recognition is being sought
- Point of contact
- Entity name
- Statement that the submitter believes, to the best of his or her knowledge, that all information submitted are truthful and accurate and that no material fact has been omitted

If you have any questions, please contact us at OIRPMGroup@fda.hhs.gov

22.2.2 Additional Resources

- Webinar: Final Guidances on Next Generation Sequencing-based Tests: May 24, 2018
- Final Guidance: Use of Public Human Genetic Variant Databases to Support Clinical Validity for Genetic and Genomic-Based *In Vitro* Diagnostics

22.3 Personalized Medicine and Companion Diagnostics Go Hand-in-Hand

Personalized medicine is an evolving field of medicine in which treatments are tailored to the individual patient. You may have a condition, for example, that is caused by a mutation in your genes. With advances in personalized medicine, you might be prescribed a medication that targets that specific mutation.

To learn which patients would benefit from a particular drug therapy or, conversely, which patients should not receive the medication, the Food and Drug Administration works with drug and device manufacturers that are developing certain tests called companion diagnostics.

Companion diagnostics (Fig. 22.2) are medical devices that help doctors decide which treatments to offer patients and which dosage to give, tailored specifically to the patient, says Elizabeth A. Mansfield, Ph.D., Deputy Office Director for Personalized Medicine in FDA's Office of *In Vitro* Diagnostics and Radiological Health. The companion diagnostic is essential to the safe and effective use of the drug. They go together.

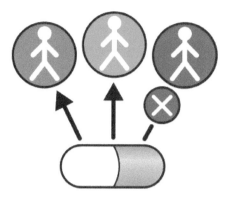

Figure 22.2 Companion diagnostic tests show which patients could be helped by a drug and which patients would not benefit, and could even be harmed.

Because the companion diagnostic test is designed to be paired with a specific drug, the development of both products requires close collaboration between experts in both FDA's device center, which evaluates the test to determine whether it may be cleared or approved, and FDA's drug center, which evaluates the drug to determine whether it may be approved.

Patricia Keegan, M.D., an oncologist and supervisory medical officer in FDA's Division of Oncology Products II, part of FDA's drug center, explains that the agency requires a companion diagnostic test if a new drug works on a specific genetic or biological target that is present in some, but not all, patients with a certain cancer or disease. The companion diagnostic test is used to identify who would benefit from the treatment and sometimes to determine if there are patients who not only would not benefit but could be harmed by use of a certain drug for treatment of their disease.

The process works best when development of the test begins before the drug enters clinical trials, Keegan explains, increasing the likelihood that the participants in the trials are the patients most likely to benefit from the treatment.

FDA's device center is issuing a final guidance on the development, review and approval or clearance of companion diagnostics to help companies identify the need for these tests earlier in the drug development process and to plan for co-development of the drug and companion diagnostic test. The goal is to stimulate early collaborations that will result in faster access to promising new treatments for patients with serious and life-threatening diseases.

"This will give health care providers more confidence in these tests to direct the therapies because the tests and therapies have been developed and evaluated together," Mansfield says.

Most recently, FDA approved a companion diagnostic genetic test to select patients with metastatic colorectal cancer for treatment with the drug Vectibix. The test detects seven mutations in the KRAS gene in colorectal tumor tissue. When the KRAS gene is mutated, those mutations could render Vectibix ineffective in treating the colon cancer.

The patient could be a good candidate for this treatment if he or she does not have a mutation. So, in this case, the approved companion test will be used to identify people who would not be helped by the drug.

This is the second time that FDA has approved this test—called the QIAGEN *therascreen* KRAS RGQ PCR Kit—to help colorectal cancer patients and their health care providers determine the potential effectiveness of a drug. In July 2012, FDA approved the test for use with the drug Erbitux, which was also found to be ineffective in patients with a mutated KRAS gene.

22.3.1 It Began with Herceptin

The road to companion diagnostics began in 1998 with FDA approval of the cancer drug Herceptin, which shuts off a protein present in abnormally high amounts in about one-quarter to one-third of breast cancers. These breast cancers are typically very aggressive. The companion diagnostic test looks for excessive levels of a particular protein, called HER2, in a patient's tumor or for extra copies of the HER2 gene in a patient's tumor, indicating that Herceptin could be an effective treatment for that patient's breast cancer.

Multiple companion diagnostics can be approved for a drug—as they have been for Herceptin—as scientific knowledge evolves with practical application of the therapy, enabling the tests to become more sophisticated.

With the advent of more drugs that target particular genetic mutations, there has been increasing acceptance from drug manufacturers that these diagnostic tests can greatly increase the clinical success of certain medications by carefully identifying patients' tumors where the drug may work, says Keegan. As of December 2018, there are over 30 cleared/approved companion diagnostic tests for selection of drugs to treat various diseases and conditions and are available at https://www. fda.gov/MedicalDevices/ProductsandMedicalProcedures/ InVitroDiagnostics/ucm301431.htm. Most drugs with a companion diagnostic test have been cancer treatments that target specific mutations.

22.3.2 Companion Diagnostics

A companion diagnostic (briefly discussed above) is a medical device, often an *in vitro* device, which provides information that is essential for the safe and effective use of a corresponding drug or biological product. The test helps a health care professional determine whether a particular therapeutic product's benefits to patients will outweigh any potential serious side effects or risks.

Companion diagnostics can:

- identify patients who are most likely to benefit from a particular therapeutic product;
- identify patients likely to be at increased risk for serious side effects as a result of treatment with a particular therapeutic product; or
- monitor response to treatment with a particular therapeutic product for the purpose of adjusting treatment to achieve improved safety or effectiveness.

If the diagnostic test is inaccurate, then the treatment decision based on that test may not be optimal.

On July 31, 2014, the FDA issued "Guidance for Industry: *In Vitro* Companion Diagnostic Devices" to help companies identify the need for companion diagnostics at an earlier stage in the drug development process and to plan for co-development of the drug and companion diagnostic test. The ultimate goal of the guidance is to stimulate early collaborations that will result in faster access to promising new treatments for patients living with serious and life-threatening diseases.

On July 15, 2016, FDA released the draft guidance "Principles for Codevelopment of an *In Vitro* Companion Diagnostic Device with a Therapeutic Product." This guidance document is intended to be a practical guide to assist therapeutic product sponsors and IVD sponsors in developing a therapeutic product and an accompanying IVD companion diagnostic. Electronic comments should be submitted through www.regulations.gov. Written comments may be submitted to the Division of Dockets Management (HFA-305), Food and Drug Administration, 5630 Fishers Lane, Rm. 1061, Rockville, MD 20852. Identify comments with Docket No. FDA-2016-D-1703.

22.3.3   Additional Information

- Webinar Presentation: Draft Guidance on Principles for Codevelopment of an *In Vitro* Companion Diagnostic Device with a Therapeutic Product
- Webinar Slides: Draft Guidance on Principles for Codevelopment of an *In Vitro* Companion Diagnostic Device with a Therapeutic Product
- Webinar Transcript: Draft Guidance on Principles for Codevelopment of an *In Vitro* Companion Diagnostic Device with a Therapeutic Product
- Principles for Codevelopment of an *In Vitro* Companion Diagnostic Device with a Therapeutic Product: Draft Guidance for Industry and Food and Drug Administration Staff (July 15, 2016)
- Guidance for Industry: *In Vitro* Companion Diagnostic Devices (July 31, 2014)
- List of Cleared or Approved Companion Diagnostic Devices (*In Vitro* and Imaging Tools)
- Consumer Update: Personalized Medicine and Companion Diagnostics Go Hand-in-Hand

Chapter 23

Protein-Based Nanoparticle Materials for Medical Applications

**Kelsey DeFrates, BSc,[a,b,] Theodore Markiewicz, BSc,[b]
Pamela Gallo, BSc,[c] Aaron Rack, BSc,[c] Aubrie Weyhmiller, BSc,[a]
Brandon Jarmusik, BSc,[a] and Xiao Hu, PhD[a,b,c]**

[a]Department of Physics and Astronomy, Rowan University,
Glassboro, New Jersey, USA
[b]Department of Biomedical Engineering, Rowan University,
Glassboro, New Jersey, USA
[c]Department of Molecular and Cellular Biosciences, Rowan University,
Glassboro, New Jersey, USA

Keywords: nanoparticle, microsphere, biomaterials fabrication, nanomedicine, bioimaging, drug delivery, silk, keratin, collagen, gelatin, elastin, soy, corn zein, biopolymer, protein, pH variation, spray drying, rapid laminar jet, phase separation, milling, polymer chain collapse, particle composite, tissue engineering

23.1 Introduction

Drug delivery systems are a valuable means of disease treatment and prevention in today's medicine. Prior to the introduction of

The Road from Nanomedicine to Precision Medicine
Edited by Shaker A. Mousa, Raj Bawa, and Gerald F. Audette
Copyright © 2020 Jenny Stanford Publishing Pte. Ltd.
ISBN 978-981-4800-59-4 (Hardcover), 978-0-429-29501-0 (eBook)
www.jennystanford.com

drug particle microencapsulation in the 1950s, drug delivery was based on rudimentary practices such as applying poultices or consuming herbal ingredients [1]. These methods, while moderately effective at the time, are inefficient and pose unnecessary health risks. However, progress in the understanding of pharmacokinetics has led to the development of sophisticated and novel methods for administering a variety of therapeutics throughout the body. Now, drug delivery methods allow for controllable drug-release and targeting to improve the safety and efficacy of treatment. To further enhance drug delivery, nanotechnology has begun to be implemented in the field. Specifically, the use of nanoparticles as carriers is an effective strategy to deploy medications to specifically targeted parts of the body [2, 3].

Nanoparticles, or microspheres, are ideal drug delivery systems for both controlled and targeted drug delivery. Their sizes typically range between 1–100 nm in diameter and can extend to more than 1000 nm [4a]. However, particle sizes smaller than 200 nm are more preferable for use in nanomedicine due to their ability to traverse micro-capillaries [4b]. There are several aspects to consider, such as size and surface charge as examples, before selecting an appropriate nanoparticle material [5, 6]. While nanoparticles can be fabricated from synthetic materials, protein-based nanoparticles have received considerable more attention due to their biodegradability and tunable properties [6, 7]. Advances in medical technology have also brought about techniques to synthesize protein-based materials that offer improved efficacy and reduced costs compared to synthetic materials [8]. Protein polymers are natural macromolecules derived from plants and animals which makes them an easily obtainable, renewable resource. In addition to their biodegradability and tunable properties, nanoparticles fabricated from protein-based materials are often biocompatible and can be easily processed [9, 10]. There are a variety of different protein polymers suitable for nanoparticle-based drug delivery each with their own unique structure-function relationships. In this review, the structure and property relationships of these natural protein-based polymers will be discussed, as well as their methods of preparation. The use of these nanoparticles in medicine will then be reviewed with a focus on their application for nanoparticle-based drug delivery.

23.2   Categories of Protein Materials

Due to the wide range of applications for protein nanoparticles, there are many types of proteins that are used to create protein nanoparticles. The type of protein polymer required may vary depending on the application. In this review, silk fibroin [11], keratin [12], collagen [13], gelatin [14], elastin [15], corn zein [16], and soy protein [17] will be given particular attention due to their popularity in biomaterials research (Fig. 23.1). However, additional protein polymers such as casein [18], fibrinogen [18], hemoglobin [19], bovine serum albumin [20], gluten [20] have also been used to create nanoparticles.

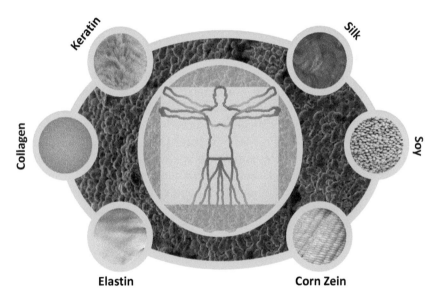

Figure 23.1 Nanoparticle materials can be fabricated from a variety of protein sources, including silk, keratin, collagen, elastin, soy, and corn zein etc. These proteins can then be processed into particles with unique properties for biomedical applications.

23.2.1   Silk Fibroin

Silk fibroin protein is among the most popular natural polymers used for the creation of biomaterials due to its acceptance by the US Food and Drug Administration (FDA), low cost, and abundance.

Commonly extracted from silk produced by the *Bombyx mori* silkworm, fibroin can be easily isolated after removal of the external sericin protein coating through treatment with sodium carbonate. The resulting fibroin protein is made of semi-crystalline structures comprised of a light and heavy chain [21]. An isoelectric point (IEP) below pH 7 and molecular weight of 83 kDa have been reported for regenerated silk fibroin, but the latter value may vary depending on the extraction procedure and duration of treatment [22, 23]. The repetition of amino acids in the pattern $(Gly\text{-}Ser\text{-}Gly\text{-}Ala\text{-}Gly\text{-}Ala)_n$ leads to crystalline beta-sheets that are then stacked in an antiparallel configuration [24–26]. This structure gives silk fibroin robust mechanical properties and high tensile strength. The crystallinity and conformation of silk fibroin can also be modulated to allow for high encapsulation of drugs while preserving their pharmaceutical activity [27].

Silk-based nanoparticles have proven effective in the delivery of both hydrophobic and hydrophilic drugs such as indomethacin and aspirin [28] of varying molecular weights, as well as anti-cancer therapeutics such as doxorubicin [11], bioactive molecules, including growth factors VEGF (vascular endothelial growth factor) and BMP-2 (bone morphogenetic protein 2) and Horseradish peroxidase and glucose oxidase enzymes [29], as well as plasmid DNA [30]. Silk-composite nanoparticles have also been fabricated by combining the protein with other biopolymers such as insulin [31], chitosan [32], and albumin [33] and synthetic polymers such as polyvinyl alcohol [34], polylactic acid [35], and polycaprolactone [36]. These approaches allow for a greater degree of tunability that can potentially increase the efficacy of drug delivery.

23.2.2 Keratin

The use of keratin as a biomaterial has been rapidly expanding over the past 40 years because of its abundance, low cost, biocompatibility, and its ability to biodegrade safely [37]. Keratin is a fibrous structural protein with molecular weight of up to 63 kDa and IEP between pH 4.5 and 5 that is derived from the human or animal epidermis and epidermal appendages, such as hair, scales, feathers, and quills in mammals, reptiles, and birds [38–40]. The keratin protein is most commonly found in epithelial cells. It is a structural protein that provides the framework for cell-cell adhesion to form a protective layer. Keratin structure is a left-handed alpha-helix which can be coiled together with other keratin proteins to form a polymerized complex. There are three different forms of keratin: α-, β-, and γ-keratins. α-keratins contain intermediate filaments, which are involved in the cytoskeleton, and are mainly found in soft tissues. β-keratins also contain intermediate filaments, but are found in hard tissues, such as scales and nails. γ-keratin is not involved in the structural elements of the cytoskeleton [37].

According to recent studies, keratin-based nanoparticles are effective anticancer drug carriers possessing a degree of tumor targeting ability and controlled drug release [12, 41]. Disulfide bonds from cysteine residues and hydrogen bonds from amine groups grant keratin nanoparticles the durability to deliver drugs with high molecular weight to their target location. In addition, keratin is negatively charged allowing positively charged molecules to better adhere to the nanoparticle for more effective transport. The targeting ability of keratin-based nanoparticles is attributed to their pH sensitivity [12, 41–43]. Keratin-based nanoparticles can respond to changes in pH to release their drug contents accordingly in a controlled release. Due to its intrinsic water stability, keratin is also a desirable support polymer for synthetic nanoparticle composites [43]. Silver nanoparticles coated with keratin are shown to have improved stability in aqueous environments [44]. Keratin is also advantageous for supporting cell adhesion and promoting cellular proliferation [41, 45]. Gold nanoparticles coated with keratin are shown to exhibit biocompatibility with improved antibacterial activity [46]. Keratin appears to be an ideal drug carrier which should be investigated further for drug delivery purposes.

23.2.3 Collagen and Gelatin

Collagen is the most abundant biopolymer in the human body [47]. This fibrous protein is a major component of the extracellular matrix and is responsible for maintaining its structure. The majority of collagen is located in connective tissues such as the skin, tendons, and ligaments [48]. Collagen can be divided into two different groups: non-fibrillar and fibrillar, which can be further divided depending on the structure and use. Type 1 of fibrillar collagen is the most common type found in the human body and has a molecular weight in the 100 kDa range. Long, triple helical structures are responsible for strength and flexibility in collagen. This helical structure has high mechanical strength due to a repeating amino acid sequence Gly-X-Y, where "X" and "Y" are commonly proline, hydroxyproline, leucine, or lysine. The individual helical structures, known as tropocollagens, will bind together and form a fibril structure. These fibril structures can then be cross linked together to form suitable cell scaffolds for use in tissue engineering [49].

Due to collagen's biocompatibility and low antigenicity, collagen-based nanoparticles have been used for the delivery of pharmaceuticals such as theophylline, retinol, tretinoin, and lidocaine [13, 50, 51]. Collagen is capable of resembling the microenvironment of some tumors allowing collagen nanoparticles to effectively infiltrate the areas and deliver anticancer therapeutics [52]. Physical properties of collagen nanoparticles such as size, surface area, and absorption capacity, are easy to configure [53]. This makes collagen nanoparticles a prime candidate for controlled drug release strategies.

In comparison, gelatin is a biopolymer derived primarily from insoluble Type I collagen through thermal denaturation or disintegration [54]. Like collagen, gelatin has received much attention in the biomedical field due to its biocompatibility and high abundance. Gelatin contains a triple helical structure, similar to collagen, made of repeating amino acids: alanine, glycine, and proline [55]. Depending on the production process, gelatin can be classified as type A or type B and consist of varying molecular weights. Type A gelatin is extracted through an acidic process, while type B is process under alkaline conditions [56]. Type A gelatin is positively charged and has an IEP of approximately pH 9. Conversely, type B gelatin is negatively charged and has an IEP of pH 5 [56]. Tissue engineering scaffolds have thus been made from gelatin as well. Alternatively, gelatin can also be formed into a gel which can be used in the place of thermoplastic polymers [57].

Gelatin is also a favorable nanoparticle material due to its relatively low antigenicity and non-carcinogenic nature [14, 58]. Gelatin nanoparticles are extensively used as successful anticancer drug carriers [59] and gene delivery vehicles [60]. Gelatin nanoparticles are able to deliver drugs across the blood brain barrier, which is a semipermeable barrier that is highly studied for drug delivery systems [55]. Gelatin nanoparticles have also safely and efficiently carried *NS2*, a recombinant gene from the hepatitis C virus, without negatively impacting the function of the gene [61]. In addition, gelatin can be blended with other natural polymers to enhance their therapeutic behavior. An alginate-gelatin composite nanoparticle benefitted from an electrostatic bond formed between the two polymers and allowed for a more controlled release of the encapsulated drug, doxorubicin [62]. Utilizing gelatin is both a promising and convenient approach for nanoparticle-based delivery of genes, vaccines, and drugs.

23.2.4   Elastin

Elastin is an important protein found in elastic fibers, specifically in the extracellular matrix. It provides support and elasticity to many structures such as the heart, lungs, skin, and blood vessels with high molecular weight species weighing 130 to 140 kDa [63]. It is insoluble and therefore can retain its shape and insolubility after stretching [8]. However, insoluble proteins are often not biocompatible and are difficult to alter. Through the use of recombinant proteins and peptide synthesis, soluble proteins that have elastin-like properties called elastin-like-peptides (ELP) are able to be produced with tunable molecular weights [64]. These polypeptides are derived from tropoelastin, the building block of elastin. This precursor molecule is vital in the exploitation of ELP materials. ELP materials are able to react to stimuli due to their temperature sensitivity, which induces a phase transition [65]. They can then self-assemble by the process of coacervation into a more ordered structure such as beta-spiral structure [66]. This property, along with their biocompatibility, makes them excellent prospects for biomedical applications. Elastin-based proteins also have the ability to communicate with cells through naturally occurring cellular receptors such as elastin binding protein (EBP) [67]. This receptor can be exploited by using tropoelastin-based polymers to induce or inhibit various cell functions [66].

Elastin, or ELP, nanoparticles have proven effective in delivery of cytokines such as BMP-2 and -14 [68], anticancer therapeutics such as doxorubicin [69], and genes [70]. Their ability to self-assembly when exposed to certain temperatures serves as a mechanism to entrap active substances [15] and achieve controlled drug release [69]. The polymer functionality of ELP nanoparticles can be controlled by using a recombinant fabrication technique [69, 70]. This means that variables pertaining to drug release, such as composition and molecular weight, can be tailored for a variety of applications in drug delivery.

23.2.5 Corn Zein

Zein is low-molecular-weight protein (20 kDa), found within the cytoplasm of corn cell endosperm and is insoluble in water except in the presence of alcohol, urea, alkali, and anionic detergents [71]. The protein has an IEP of pH 6.2 and is a mixture of two different peptides: α zein and β zein [72]. α zein is the most widely used variety due to its abundance [73, 74]. Zein has a helical wheel shaped structure with nine homologous units arranged in a non-parallel way with hydrogen bonds stabilizing it. This helical shape gives zein a globular structure similar to insulin and ribonuclease [73]. Zein can be extracted using primary, secondary, and ternary solvents. Primary solvents consist of a compound that dissolves zein in a concentration greater than 10%. Secondary solvents are organic compounds. Ternary solvents are a combination of solvent, water, and alcohol. Zein is commonly used in fibers, adhesives, plastics, ink, chewing gum, and as a preservative coating for some food and pharmaceuticals [74].

Zein nanoparticles are successful drug carriers for encapsulation and controlled release of fat soluble compounds such as α-tocopherol [75–77], other proteins [16], vaccines [16, 78], and vitamins such as D3 [79, 80]. Due to the protein's hydrophobicity, zein nanoparticles are promising oral drug delivery vehicles able to protect encapsulated contents from harsh acidic environments such as in stomach acid [79]. Zein nanoparticles can also have their properties improved by combining the natural polymer with other substances. For example, sodium caseinate was incorporated with zein nanoparticles to improve particle stability in water [81]. Zein nanoparticles are an attractive drug delivery system due to their high stability in a variety of environments and tunable properties in combination with certain molecules.

23.2.6 Soy

Soy protein is a globular protein isolated from soybeans, known as soy protein isolate, and is one of the most abundant types of plant proteins. The globular structure is comprised of two major subunits, conglycinin and glycinin, which contain all amino acids particularly glutamate, aspartate, and leucine [82]. This structure composition gives soy protein relative stability for long storage life [83] and biocompatibility [84]. When the globular protein is treated with enzymes, soy protein hydrolysates below 1 kDa and between 1 and 5 kDa can be obtained and further processed [85]. In addition, soy protein is biodegradable as it can be digested if consumed. For example, soy protein-based edible films are often used as a wax coating for fruits to preserve their quality [86–88]. Soy protein films, scaffolds, and hydrogels have also been applied in tissue engineering for wound healing and transdermal drug delivery [89]. With every amino acid available, soy protein is effective in supporting cellular communication and cell proliferation. The amino acid composition may also attribute to soy protein being used as protection against bacterial infection [83, 90].

Soy protein nanoparticles are becoming more popular due to the high abundance and low cost of the protein, as well as its biodegradability and low immunogenicity. The amino acid composition gives soy protein nanoparticles an advantage in encapsulation of highly hydrophobic drugs [17]. Unlike zein, soy protein nanoparticles are soluble in aqueous environments which can be used in different oral drug delivery scenarios. Soy protein isolates are used as a coating in conjunction with other materials either for protection [91] or for physical or chemical surface modification [17, 92]. For example, magnetic nanoparticles prepared with soy protein isolate benefit from enhanced functional surface area increasing the loading of enzymes [92]. The protein coating also offers a degree of bioinert behavior to otherwise non-immunogenic nanoparticle materials.

23.2.7 Other Proteins: Casein, Fibrinogen, Hemoglobin, Bovine Serum Albumin, Gluten

Along with the many proteins mentioned above, there are some that will be excluded from this review but are worth mentioning.

Casein, fibrinogen, hemoglobin, bovine serum albumin, and gluten are just a few of many. Similar to those previously explained, the use of these proteins depends on their properties and the application's demands. Casein is very useful in hydrophilic environments since casein is a hydrophilic protein in itself. It is useful for water-based environments since as a microsphere they disperse instead of aggregate [18]. As a micro-/nanosphere, fibrinogen polymerizes when used in conjunction with a serine protease and forms a protein mesh that can be used to cover and treat open wounds or used in vitro for more in depth biomedical applications [93]. Hemoglobin as a micro-/nanoparticle can be used as an oxygen deposit to make oxygen releasing biomaterials [19]. Bovine serum albumin can be used to pack prepared protein particles to aid in protein and drug delivery [94]. Gluten as a microsphere can be used as a drug delivery vehicle that is very effective compared to other widely used proteins [20]. While these proteins are not described in further detail in this review, each protein possesses their own unique advantages when applied in nanoparticle-based drug delivery.

23.3 Fabrication Methods

Due to the necessity of obtaining particles of different sizes, shapes and weights, there are many fabrication methods that are available for the creation of nanoparticles. Fabrication methods will also vary depending on the properties of the individual polymers, such as temperature dependence. Fabrication methods that will be discussed in this review include pH variation, spray-drying, phase separation, milling, rapid laminar jet, and polymer chain collapse. The synthesis of blended protein-based nanoparticles will also be discussed. The advantages and disadvantages of these fabrication methods are summarized in Table 23.1.

23.3.1 pH Variation

The drug delivery properties of silk fibroin can be modified by changing many factors during nanoparticle synthesis.

Table 23.1 Advantages and disadvantages of the common protein-based nanoparticle fabrication methods

Method	Advantages	Disadvantages
pH Variation [95]	• Control for particle size • Control secondary structure of protein • Control for zeta potential • Produces chemically and physically stable particles Experimentally simple	• Post-fabrication drug loading is required • Limited to small scale production
Spray-drying [96]	• Cost effective • Experimentally simple • Easily encapsulate hydrophilic drugs • Useful for heat-sensitive samples • Control for particle size	• Limited to small scale production • Challenging to incorporate hydrophobic drugs
Rapid Laminar Jet [97]	• Control for particle size • Production of uniform particles • Production of strong, stable particles	• Possibility of coalescence • Many parameters must be controlled for
Phase Separation [98]	• Specialized equipment is not required • Particle size can be controlled by adjusting protein concentration • Uniform particles are produced	• Particle sizes are limited to 50–500 nm in diameter • Organic solvents are required • Limited to small scale production
Milling [99]	• Cost effective • Large-scale production is possible • Control of nanoparticle size • Experimentally simple	• Heat is released during the process requiring chamber to be cooled • Little control over nanoparticle shape • Nanoparticles must be coarse
Polymer Chain Collapse [100]	• Properties of the nanoparticle can be easily controlled by selection of the precursor chain • Production of particles with high stability • Particles with improved spherical shape are produced	• Particle size is limited to 5–20 nm in diameter • Side reaction may be difficult to control

One of these factors is the pH of the silk fibroin [95]. Particles are made by salting out a fibroin solution with potassium phosphate. The pH of the particles can be controlled depending on what type of potassium phosphate is used in the salting out. Mono potassium phosphate has a pH of 4 and dibasic potassium phosphate has a pH of 9. Silk fibroin particles with a pH of 4 develop silk II rich secondary structures while silk fibroin particles with a pH of 9 developed a silk I rich secondary structure. Particles with the silk II structure or the lower pH are less chemically stable than the particles with a higher pH and the silk I structure. When a positively charged drug is loaded into a negatively charged silk fibroin particle there is a difference in the release depending on the pH of the particle. Particle with the silk II structure and low pH have an increased initial release, whereas the high pH particles have a low release. However, particles at a neutral pH of 7 had an overall increased release over the entire time not just initially.

23.3.2 Spray-Drying

Spray-drying is a technique that is used to fabricate nanoparticles from a liquid sample. The liquid sample is sprayed out of a nozzle into a chamber where heated nitrogen and carbon dioxide gas flow in the direction of the spray (Fig. 23.2). In the bottom of the chamber, there are electrodes which are used to collect the nanoparticles. As the sprayed droplets move towards the bottom of the chamber, they become electrostatically charged due to these electrodes. This is a one-step process that is a quick, cost effective method for small scale protein particle production. One application of spray-drying is for use in drug delivery systems due to the ability of hydrophilic drugs to be encapsulated in these spray-dried nanoparticles [96]. This nanoparticle fabrication technique is useful for samples that are heat-sensitive since the solvent evaporation helps maintain the temperature of the nanoparticle droplets. This method of nanoparticle synthesis also gives the user the ability to control the size of the particle that is produced by changing parameters, such as the size of the nozzle and speed at which they are sprayed out [101].

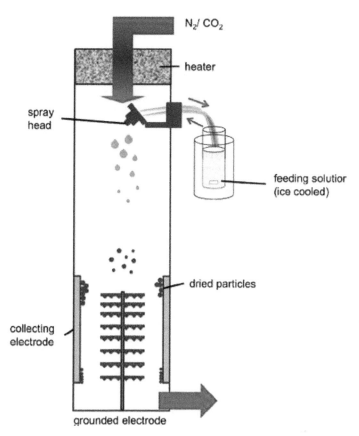

N_2/CO_2

heater

spray head

feeding solutior (ice cooled)

dried particles

collecting electrode

grounded electrode

Figure 23.2 A schematic of a nanoparticle spray-drying system in which the liquid polymer sample is sprayed alongside of heated gas in a chamber that leads to electrodes which are used to collect the charged sprayed nanoparticles. Reproduced with permission from [102], Copyright Springer Nature, 2015.

23.3.3 Rapid Laminar Jet

Particles can be also made using a rapid laminar jet method. The feed liquid will contain a certain number of compounds from which the particle can be made. Spherical drops form when a liquid jet discharges from a small opening at laminar flow conditions. This formation behavior of the drops is resulted because of the surface energy and tension of the jet. The liquid spheres will be dispersed in some type of fluid or gas/air depending on the mechanism. Drop size is-based on the length of the jet. For best results the jet length between breakpoints should be five times the diameter of the stream which gives particle sizes of about twice that of the jet. This laminar breakup of the jet is a result of small disturbances. These disturbances must be controlled to preserve uniformity in drop sizes. To control these disturbances a controlled uniform vibration is applied to the jet. Ideally the frequency of the vibration is close to the naturally occurring frequency for laminar breakup. This leads to a clean controlled breakup and uniform drop sizes [97].

23.3.4 Phase Separation

Out of the various methods of protein nanoparticle fabrication, emulsion-solvent evaporation is the most popular. This technique was the first to form polymer nanoparticles [103]. Organic and aqueous phase separations are the backbone of this method. Prepared polymers are placed in an organic solvent. A surfactant is added to the aqueous phase in order to prevent the fusion of emulsion particles [104]. The solution is then subjected to a mixing method such as ultrasonification. Mini-emulsion droplets of polymer are formed. Finally, the solvent is separated. This is often completed by evaporation of the organic phase. The remaining solution contains polymer nanoparticles which can be collected through a centrifuge. This method produces particles in the 50–500 nm size range. Particle size could be controlled by altering the concentration of polymer solution [105]. This technique is extremely popular due to the availability of conjugated polymers [98].

Another method based on separations is the coacervation method. This is commonly referred to simply as phase separation but for the purposes of this paper it is included in this section. This method requires the separation of two liquid solutions. One will contain the protein polymer and the other is a solvent. Through some means of disrupting equilibrium such as the addition of a salt, coacervation is induced. The charges create electrostatic forces which induce the formation of nanoparticles [98].

23.3.5   Milling

Milling is a fabrication technique for nanoparticles that requires mechanical energy to break down larger particles into fine nanoparticles (Fig. 23.3). This fabrication technique is commonly used for nanoparticles that are to be used in drug delivery [106]. Milling is a cost-effective way to produce nanoparticles in a large-scale production. High energy ball milling involves the subjection of coarse nanoparticles to high energy collisions from the milling balls. Coarse nanoparticle powder is placed in a chamber that contains milling balls and mechanical movement is applied to the cylindrical chamber to accelerate the milling balls which can roll over and collide with the powder. These collisions and other mechanical force from the milling balls causes the coarse nanoparticles to break down into fine nanoparticles. The chamber must be cooled due to the heat energy released from the mechanical energy exerted on the nanoparticles. This fabrication technique allows the user of the system to control the size of the nanoparticles by altering the speed of the rotation of the cylindrical chamber [99].

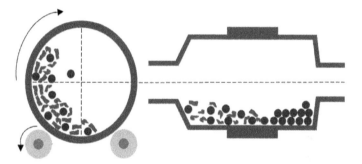

Figure 23.3 The basic mechanism used in high-energy ball milling. As the cylinder rotates, the milling balls are accelerated and through physical force fracture the polymer material that is placed in the chamber.

23.3.6   Polymer Chain Collapse

Single-chain collapse of polymers is a method to produce individual single-chain polymer nanoparticles (SCNP). This method can produce particles in the range of 5–20 nm. In addition, intrachain folding produces particles that have great stability compared to other techniques [100]. Control of the precursor chain can also dictate the properties of the nanoparticle, allowing for the production of distinct molecules [107].

There are different varieties of the SCNP method and the type of reaction is dependent on the functional groups involved. However, all the methods benefit more from intramolecular cross-linking rather than intermolecular cross-linking [108]. Homofunctional chain-collapse involves placing a functional group that is likely to bind with itself on the precursor chain and then performing a reaction that couples the functional group [107]. This method often produces particles that are not globular in shape. Instead, heterofunctional coupling is being looked to for improved results. This requires two functional groups which are orthogonally cross-linked. There are many ways to perform the cross-linking in both hetero and homofunctional chain collapses. Data has shown that this method produces nanoparticles with improved spherical shape [109].

23.3.7 Protein Particle Composite

Protein particle composites contain more than one protein or polymer, the addition of which can be used to tune the mechanical and physical properties of a drug delivery vehicle. The method in which they are fabricated depends on the type of particle desired and the differing properties of the additional component. The properties can be a variety of different things such as mechanical properties, electric properties, electromagnetic properties, elasticity, crystallinity, moldability, and many more. When choosing the different type of additional polymer to add, it usually contains an additional characteristic that the main protein does not. Where one protein's structure may be dependent on pH, adding another material to form a stable complex between the two to withstand a lower pH could make more fabrication methods possible.

For fabrication of these particles, it depends what end product is desired. Any previous or following fabrication method that is described can be used to make composite protein-based particle. The only difference between this method and the others is the fact that a protein composite must be made before or after fabrication. For example, if the particles are going to be fabricated using spray drying, a liquid protein mixture can be made before spray drying is done or individual nanoparticles can be formed and then mixed together to create the same product. If a certain percentage of protein is desired in the final product, it is important that an appropriate fabrication method is selected. Fabrication methods can affect the resulting nature of the particle. Particle behavior depends on their surface composition, geometry, and size among other characteristics. There are many more methods that can be utilized for fabricating protein particles apart from what was described in this review [110].

23.4   Factors to Control Particle Formation

In nanoparticle formation, there are many factors that can be controlled to modify drug delivery, such as size, molecular weight, and shape. These factors are mainly determined by the fabrication technique applied but can also be due to the properties of the polymers themselves.

23.4.1   Size

Nanoparticle size can vary depending on the molecular weight of the protein polymer used. Typically, nanoparticle size ranges from 1–100 nm but they can extend to 1000 nanometers in diameter [4a, 111]. One way to control nanoparticle size is to prevent aggregation of the nanoparticles, which can be done by introducing chemicals that help prevent this aggregation by reducing disulfide bonds or by altering the charge state of the polymers [112]. Other factors related to controlling the size of the nanoparticles vary by the technique used to produce them. With the spray drying nanoparticle manufacturing technique, the size of the particles can be altered by changing the size of the nozzle used to spray the polymeric nanoparticle solution into the drying chamber; the size can also be altered by the speed at which the solution is sprayed [101].

23.4.2 Shape

There are many different forms of nanoparticles. The two fundamental types are nanospheres and nanocapsules. The main difference between these types are that nanospheres contain a polymer matrix inside, whereas nanocapsules have a shell that separates the encapsulated polymer from the outside environment (Fig. 23.4). Solid lipid nanospheres are being studied as potential drug carriers due to their matrix morphology. This allows for controlled release and protection of the drug [113]. These particles can be formed by subjecting alkyl cyanoacrylates to polymerization in emulsion [114]. An additional method is precipitating polymers that have already been altered [113]. Solid nanospheres may also be formed using microfluidics method. This method is extremely cost efficient and allows for more control of particle features. The solvent volatility can be altered to shape the surface. Variances in flow rate and the architecture of the devices can create different geometries [114].

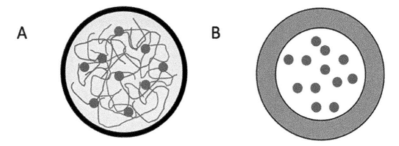

Figure 23.4 (A) A model protein nanosphere. The drug (red) is within a protein matrix (blue); (B) A model of protein nanocapsule. The drug is suspended and encapsulated by a thick protein polymer shell.

Nanocapsules are somewhat the opposite of solid nanoparticles. This is based on their hydrophobic and hydrophilic interactions. The counter methods can be applied to form nanocapsules. Adding an oil to the emulsion polymerization results in a core-shell formation [113]. Essentially, the presence or absence of oil dictates which type of nanoparticle will form. Another type of nanoparticle is the Janus nanoparticle. These particles consist of two different sides, each with their own properties. These properties can include hydrophobicity and hydrophilicity. The combination of functionalities allows for stimuli response and unique assemblies. Janus particles can be formed by masking. This process involves protecting one region of the particle while the other is functionalized. The mask is then removed, and the final product is a particle with a dual nature particle [115]. Another process is self-assembly. First block copolymers undergo phase separation. Then specific cross-linking must occur, followed by disassembly of larger structures [116].

23.4.3   Properties of the Protein

When designing a protein-based nanoparticle, it is important to consider how the protein will interact with the encapsulated drug and physiological environment. Ultimately, a protein with the appropriate molecular weight and IEP must be chosen. The molecular weight of the protein used to create the nanoparticle is important to consider since it can affect how much drug can be effectively stored and the particle targeting mechanism in the body. In some instances, a nanoparticle made from a very high- or low-molecular-weight protein can result in lower encapsulation efficiency. A moderate molecular weight protein is often more appropriate and can help to achieve higher encapsulation efficiency [117]. The molecular weight can also contribute to the pathing of the nanoparticle through the body. In addition to molecular weight, the IEP of the protein will affect the stability of the nanoparticle in different environments. At pH near the IEP, nanoparticles may begin to aggregate and decrease in stability [118]. This can inhibit their circulation throughout the body as well as their drug release. Therefore, a protein with the appropriate IEP and molecular weight must be chosen to ensure that nanoparticles withstand certain environments.

23.5   Novel Applications of Protein-Based Nanoparticles

Protein nanoparticles offer a wide range of uses in medicine as both drug delivery vehicles and bioimaging aids.

23.5.1 Bioimaging

Polymer nanoparticles are gaining traction as contenders to replace typical fluorescent dyes. These are used in non-specific and targeted microscopic imaging [119]. In non-specific imaging, these nanoparticles can be used to dye cells. Data demonstrates that phospholipid encapsulated polymer nanoparticles are successful in providing quality fluorescent imaging of cancer cells. These cells displayed no symptoms of toxicity. In addition, it is possible to tune the wavelength emitted by altering the conjugated protein polymer [120]. These properties, along with an increased circulation period, could lead to applications in vivo. In addition, protein nanoparticles have a bright future in targeted cellular imaging. These particles have an increased uptake due to the enhanced permeability and retention of advanced tumors. Near IR light can provide excellent imaging quality when paired with a polymer nanoparticle-based probe due to the previously mentioned properties [121]. Overall, the applications of these particles in the biomedical imaging field are rapidly growing.

To enhance the biocompatibility and cellular uptake of nanodiamonds (ND), Khalid et al. encapsulated the material in silk fibroin nanospheres using a co-flow technique. Due to silk fibroin's transparency and low background signal, the photoluminescence of the NDs was not diminished. In fact, NDs encapsulated in silk fibroin spheres fluoresced 2–4 times brighter than NDs alone. The 400 to 600 nm spheres were also found to be highly stabile in an aqueous environment, but began to degrade after one week of incubation at 37°C. When introduced to fibroblast cells in vitro, the intracellular mobility and diffusion of NDs was improved [122]. Li et al. also used silk fibroin to create nanoparticles for bioimaging, through hydrothermal treatment that simply involved heating the protein at 200°C for 72 h. This procedure produced water-soluble, nitrogen-doped, photoluminescent-polymer-like carbonaceous nanospheres (CNSs) that measured approximately 70 nm in diameter and could easily be isolated through filtration. These nanoparticles exhibited low cytotoxicity when incubated with HeLa cells and fluoresced in the perinuclear regions once ingested. CNSs could also be used to image tissue at a depth of 60 to 120 μm with no blinking and low photobleaching

[123]. These studies illustrate the improvements that may result from the incorporation of protein into nanoparticles for bioimaging.

In addition to silk fibroin, gelatin nanoparticles have also been utilized as bioimaging platforms. Liu et al. created gelatin nanocapsules containing gold nanoparticles by denaturing gelatin polypeptides that then absorbed onto citrate-stabilized gold nanoparticles. A thin layer of silica was then used to stabilize these particles that measured approximately 50 nm in diameter and hold promise in Raman-active bioimaging [124]. Gelatin has also been used to coat cadmium telluride (CdTe) quantum dots (QDs), leading to an improvement in their cytotoxicity and biocompatibility. In this study, Byrne et al. introduced gelatin single- or multi-stranded polypeptides during QD synthesis, to control their growth and nucleation. Functional groups present in the glycine, proline, and 4-hydroxy proline residues of gelatin were then able to interact with the surface of the CdTe QDs, allowing for their coating. When incubated with macrophages, these "jelly dots" were successfully engulfed by the cells, resulting in the illumination of their membranes. When compared to QDs alone, cells exposed to QDs treated with gelatin showed a lower lysosomal pH and cellular permeability, suggesting decreased toxicity of the particles [125]. In both studies, the ability to further functionalize the surface of these particles due to their natural polymer coating, may further enhance their efficacy in bioimaging.

23.5.2 Drug Delivery Vehicle

Protein-based nanoparticles have also found new use as drug delivery vehicles. In addition to their biocompatibility and biodegradability, the surface of protein nanoparticles can be easily functionalized due to their defined primary structure, while charged proteins can facilitate drug loading through electrostatic interactions [6, 126]. Such particles can also often be fabricated under mild, aqueous conditions, making them easier and safer to process than ones based on synthetic polymers [95]. The use of natural proteins has also been shown to increase cell retention and reduce the effects of toxic byproducts produced during degradation [127].

One such protein used to create nanoparticles for drug delivery is corn zein. Due to its hydrophobic nature, this protein is especially suited for the prolonged, controlled release of pharmaceuticals. Lai et al. noted this effect when they used the protein to create nanoparticles loaded with the chemotherapeutic agent, 5-fluorouracil (5-FU). These particles were synthesized using a standard phase separation procedure and measured approximately 115 nm in diameter. The corn zein particles were able to encapsulate 5-FU at an efficiency of up to 56.7% which was then released after an initial burst of 22.4%. When injected into mice, the nanoparticles remained in circulation for 24 h before localizing to the liver due to their high molecular weight [127]. Corn zein nanoparticles have also been used for the controlled release of vitamin D3 [80], therapeutic proteins such as catalase and superoxide dismutase [128], and anti-diabetic drugs [126]. In the latter study, Xu et al. developed hollow zein-based nanoparticles through a two-step procedure. This fabrication technique began with the mixing of corn zein and sodium citrate ethanol-based solutions. The zein polymer than aggregated around the sodium citrate crystals, resulting in the formation of particles with a sodium citrate core and zein shell. To create hollow particles, the core-shell particles were added to water, leading to the dissolution of the sodium citrate core. The resulting hollow nanoparticles measured less than 100 nm in diameter and were able to encapsulate 30% more drug compared to solid zein particles. This drug was then released in a more sustained, prolonged manner over 200 h. When incubated with 3T3 fibroblast cells, the particles were also successfully internalized by cells without effecting their viability [126].

Other plant-based proteins such as soy protein have also been used to create nanoparticles for the controlled release of nutrients and pharmaceuticals. Due to soy's balanced composition of nonpolar and polar residues, it can act as a versatile carrier by storing drugs with various functional groups. Using a desolvation method and a glutaraldehyde crosslinker, particles measuring between 200 and 300 nm in diameter were fabricated and loaded with curcumin. Curcumin was then released with an initial burst of over 50% within the first 1.5 h, but continued to be released in a more controlled manner over the next 8 h [17]. Although cell studies were not conducted, the established biocompatibility of soy suggests that the particles were act as suitable drug delivery vehicles [84].

Negatively charged proteins such as keratin have also been used to fabricate nanoparticles that are able to incorporate drugs through electrostatic absorption. When prepared through ionic gelation, keratin particles allowed for the long-term and controlled release of the model drug chlorhexidine (CHX). This fabrication technique involved the dropwise addition of a CHX solution to one containing keratin. Negatively charged carboxylate groups on the outside of keratin aggregates attracted the drug, allowing for its retention. CHX was then gradually released over 140 h in a pH-sensitive manner, with greater release occurring at acidic and neutral pH [12]. Cheng et al. also created keratin-based nanoparticles consisting of varying ratios of the oxidized (keratose or KOS) and reduced (kerateine or KTN) forms of the protein. These particles were created using an ultrasonic dispersion technique and measured between 345 and 400 nm, with decreasing diameter upon the addition of more KOS. The addition of KOS also resulted in an increased release rate of the model drug, Amoxicillin (AMO). The nanoparticles were found to be mucoadhesive due to the electrostatic interaction and disulfide bonding between gastric mucin and KTN, and hydrogen binding with KOS. These results suggest that keratin-based nanoparticles may be an ideal carrier for mucoadhesive drug delivery [42].

23.6 Conclusions

Protein polymer and protein composite materials are becoming more accepted in the realm of nanoparticle drug delivery. Their properties are ideal for drug delivery systems and show promise in improving controlled release or targeting delivery mechanisms. Natural protein polymer is relatively cheap, easy to process, and renewable which makes it an attractive material from an economic perspective. The primary advantages that protein-based nanoparticles have over synthetic materials is their biodegradability and biocompatibility. Minimizing the host immune response is an important aspect determining the success of a drug delivery operation. The natural degradation of these protein polymers reduces accumulation of particle byproduct which is also better for the human health.

This review focused on the properties of protein materials, such as silk fibroin, keratin, and elastin, and their usage in nanoparticle drug delivery and biomedical applications. There are a variety of processing methods for protein-based nanoparticles which can tune their resulting properties for more specific applications. While there are still challenges to overcome, there is an increasing demand for biocompatible protein nanoparticles in the medical field. To overcome these challenges, future work involving protein-based nanoparticles must focus on the development of large-scale production techniques that allow these particles to be manufactured in a commercially attractive manner. Functionalized particles capable of targeting specific areas of the body are also likely to be developed to limit off-target effects. With the development of new pharmaceuticals, the fabrication and characteristics of protein nanoparticles must also adapt to provide ideal vehicles for drug delivery. As these new studies emerge and the functionality of these protein materials is improved, the more opportunities there will be for effective disease treatment in the future.

Abbreviations

5-FU:	5-fluorouracil
AMO:	amoxicillin
BMP-2:	bone morphogenetic protein 2
CdTe:	cadmium telluride
CHX:	chlorhexidine
CNSs:	carbonaceous nanospheres
EBP:	elastin binding protein
ELP:	elastin-like-peptides
FDA:	US Food and Drug Administration
IEP:	isoelectric point
KOS:	keratose
KTN:	kerateine
ND:	nanodiamonds

QDs: quantum dots
SCNP: single-chain polymer nanoparticles
VEGF: vascular endothelial growth factor

Disclosures and Conflict of Interest

This chapter was originally published as: DeFrates, K., Markiewicz, T., Gallo, P., Rack, A., Weyhmiller, A., Jarmusik, B., and Hu, X. (2018). Protein Polymer-Based Nanoparticles: Fabrication and Medical Applications, *Int. J. Mol. Sci.*, **19**, 1717, doi:10.3390/ijms19061717, under the Creative Commons Attribution license (http://creativecommons.org/licenses/by/4.0/). It appears here, with edits and updates, by kind permission of the copyright holders.

Author Contributions: K.D., T.M., P.G., A.R., A.W., B.J., and X.H. wrote the paper together; K.D., T.M., P.G., A.R., A.W., and B.J. contributed equally to this chapter. Writing, review, and editing: K.D. and X.H.

Acknowledgments: This study was supported by Rowan University Seed Research Grants, and the New Jersey Health Foundation Research Grants Program (PC111-17).

Conflicts of Interest: The authors declare no conflict of interest.

Corresponding Author

Dr. Xiao Hu
Department of Physics and Astronomy
Rowan University
201 Mullica Hill Road
Glassboro, NJ 08028, USA
Email: hu@rowan.edu

References

1. Rosen, H. (2005). The rise and rise of drug delivery. *Nat. Rev. Drug Discov.*, **4**, 381–385.

2. Ischakov, R., Adler-Abramovich, L., Buzhansky, L., Shekhter, T., Gazit, E. (2013). Peptide-based hydrogel nanoparticles as effective drug delivery agents. *Bioorg. Med. Chem.*, **21**, 3517–3522.

3. Vogelson, C. T. (2001). Advances in drug delivery systems. *Mod. Drug Discov.*, **4**, 49–50.

4a. Bawa, R. (2016). What's in a name? Defining "nano" in the context of drug delivery. In Bawa, R., Audette, G., Rubinstein, I., eds. *Handbook of Clinical Nanomedicine: Nanoparticles, Imaging, Therapy and Clinical Applications*, Pan Stanford Publishing, Singapore, chapter 6, pp. 127–169.

4b. Singh, R., Lillard, J. W. (2009). Nanoparticle-based targeted drug delivery. *Exp. Mol. Pathol.*, **86**, 215–223.

5. Mohanraj, V. J. (2006). Nanoparticles: A review. *Trop. J. Pharm. Res.*, **5**, 561–573.

6. Lohcharoenkal, W., Wang, L., Chen, Y. C., Rojanasakul, Y. (2014). Protein nanoparticles as drug delivery carriers for cancer therapy. *BioMed Res. Int.*, **2014**, 180549.

7. Tarhini, M., Greige-Gerges, H., Elaissari, A. (2017). Protein-based nanoparticles: From preparation to encapsulation of active molecules. *Int. J. Pharm.*, **522**, 172–197.

8. Jao, D., Xue, Y., Medina, J., Hu, X. (2017). Protein-based drug-delivery materials. *Materials*, **10**, 517.

9. Mahmoudi, M., Lynch, I., Ejtehadi, M. R., Monopoli, M. P., Bombelli, F. B., Laurent, S. (2011). Protein–nanoparticle interactions: Opportunities and challenges. *Chem. Rev.*, **111**, 5610–5637.

10. Weber, C., Coester, C., Kreuter, J., Langer, K. (2000). Desolvation process and surface characterisation of protein nanoparticles. *Int. J. Pharm.*, **194**, 91–102.

11. Wongpinyochit, T., Uhlmann, P., Urquhart, A. J., Seib, F. P. (2015). PEGylated silk nanoparticles for anticancer drug delivery. *Biomacromolecules*, **16**, 3712–3722.

12. Zhi, X., Wang, Y., Li, P., Yuan, J., Shen, J. (2015). Preparation of keratin/chlorhexidine complex nanoparticles for long-term and dual stimuli-responsive release. *RSC Adv.*, **5**, 82334–82341.

13. Posadas, I., Monteagudo, S., Ceña, V. (2016). Nanoparticles for brain-specific drug and genetic material delivery, imaging and diagnosis. *Nanomedicine*, **11**, 833–849.

14. Bajpai, A., Choubey, J. (2006). Design of gelatin nanoparticles as swelling controlled delivery system for chloroquine phosphate. *J. Mater. Sci. Mater. Med.*, **17**, 345–358.

15. Herrero-Vanrell, R., Rincon, A., Alonso, M., Reboto, V., Molina-Martinez, I., Rodriguez-Cabello, J. (2005). Self-assembled particles

of an elastin-like polymer as vehicles for controlled drug release. *J. Control. Release*, **102**, 113–122.

16. Hurtado-López, P., Murdan, S. (2006). Zein microspheres as drug/antigen carriers: A study of their degradation and erosion, in the presence and absence of enzymes. *J. Microencapsul.*, **23**, 303–314.

17. Teng, Z., Luo, Y., Wang, Q. (2012). Nanoparticles synthesized from soy protein: Preparation, characterization, and application for nutraceutical encapsulation. *J. Agric. Food Chem.*, **60**, 2712–2720.

18. Saralidze, K., Koole, L. H., Knetsch, M. L. (2010). Polymeric microspheres for medical applications. *Materials*, **3**, 3537–3564.

19. Paciello, A., Amalfitano, G., Garziano, A., Urciuolo, F., Netti, P. A. (2016). Hemoglobin-conjugated gelatin microsphere as a smart oxygen releasing biomaterial. *Adv. Healthcare Mater.*, **5**, 2655–2666.

20. Chen, X., Lv, G., Zhang, J., Tang, S., Yan, Y., Wu, Z., Su, J., Wei, J. (2014). Preparation and properties of BSA-loaded microspheres based on multi-(amino acid) copolymer for protein delivery. *Int. J. Nanomed.*, **9**, 1957.

21. Qi, Y., Wang, H., Wei, K., Yang, Y., Zheng, R.-Y., Kim, I. S., Zhang, K.-Q. (2017). A review of structure construction of silk fibroin biomaterials from single structures to multi-level structures. *Int. J. Mol. Sci.*, **18**, 237.

22. Jetbumpenkul, P., Amornsudthiwat, P., Kanokpanont, S., Damrongsakkul, S. (2012). Balanced electrostatic blending approach: An alternative to chemical crosslinking of Thai silk fibroin/gelatin scaffold. *Int. J. Biol. Macromol.*, **50**, 7–13.

23. Wei, W., Zhang, Y., Shao, H., Hu, X. (2010). Determination of molecular weight of silk fibroin by non-gel sieving capillary electrophoresis. *J. AOAC Int.*, **93**, 1143–1147.

24. Tsuchiya, K., Masunaga, H., Numata, K. (2017). Tensile reinforcement of silk films by the addition of telechelic-type polyalanine. *Biomacromolecules*, **18**, 1002–1009.

25. Im, D. S., Kim, M. H., Yoon, Y. I., Park, W. H. (2016). Gelation behaviors and mechanism of silk fibroin according to the addition of nitrate salts. *Int. J. Mol. Sci.*, **17**, 1697.

26. Wang, F., Yu, H.-Y., Gu, Z.-G., Si, L., Liu, Q.-C., Hu, X. (2017). Impact of calcium chloride concentration on structure and thermal property of Thai silk fibroin films. *J. Therm. Anal. Calorim.*, **130**, 851–859.

27. Cao, Y., Liu, F., Chen, Y., Yu, T., Lou, D., Guo, Y., Li, P., Wang, Z., Ran, H. (2017). Drug release from core-shell PVA/silk fibroin nanoparticles fabricated by one-step electrospraying. *Sci. Rep.*, **7**, 11913.

28. Zhong, T., Jiang, Z., Wang, P., Bie, S., Zhang, F., Zuo, B. (2015). Silk fibroin/copolymer composite hydrogels for the controlled and sustained release of hydrophobic/hydrophilic drugs. *Int. J. Pharm.*, **494**, 264–270.

29. Numata, K., Kaplan, D. L. (2010). Silk-based delivery systems of bioactive molecules. *Adv. Drug Deliv. Rev.*, **62**, 1497–1508.

30. Numata, K., Mieszawska-Czajkowska, A. J., Kvenvold, L. A., Kaplan, D. L. (2012). Silk-based nanocomplexes with tumor-homing peptides for tumor-specific gene delivery. *Macromol. Biosci.*, **12**, 75–82.

31. Yan, H.-B., Zhang, Y.-Q., Ma, Y.-L., Zhou, L.-X. (2000). Biosynthesis of insulin-silk fibroin nanoparticles conjugates and in vitro evaluation of a drug delivery system. *J. Nanopart. Res.*, **11**, 1937.

32. Gupta, V., Aseh, A., Ríos, C. N., Aggarwal, B. B., Mathur, A. B. (2009). Fabrication and characterization of silk fibroin-derived curcumin nanoparticles for cancer therapy. *Int. J. Nanomed.*, **4**, 115.

33. Subia, B., Kundu, S. (2012). Drug loading and release on tumor cells using silk fibroin–albumin nanoparticles as carriers. *Nanotechnology*, **24**, 035103.

34. Wang, X., Yucel, T., Lu, Q., Hu, X., Kaplan, D. L. (2010). Silk nanospheres and microspheres from silk/pva blend films for drug delivery. *Biomaterials*, **31**, 1025–1035.

35. Luz, C. M., Boyles, M. S. P., Falagan-Lotsch, P., Pereira, M. R., Tutumi, H. R., Santos, E. O., Martins, N. B., Himly, M., Sommer, A., Foissner, I. (2017). Poly-lactic acid nanoparticles (PLA-NP) promote physiological modifications in lung epithelial cells and are internalized by clathrin-coated pits and lipid rafts. *J. Nanobiotechnol.*, **15**, 11.

36. de Jesus Sousa-Batista, A., Cerqueira-Coutinho, C., do Carmo, F. S., de Souza Albernaz, M., Santos-Oliveira, R. (2019). Polycaprolactone antimony nanoparticles as drug delivery system for leishmaniasis. *Am. J. Ther.*, **26**(1), e12–e17.

37. Sharma, S., Gupta, A. (2016). Sustainable management of keratin waste biomass: Applications and future perspectives. *Braz. Arch. Biol. Technol.*, **59**, e16150684.

38. Rouse, J. G., Van Dyke, M. E. (2010). A review of keratin-based biomaterials for biomedical applications. *Materials*, **3**, 999–1014.

39. Thomas, P., Said, J. W., Nash, G., Banks-Schlegel, S. (1984). Profiles of keratin proteins in basal and squamous cell carcinomas of the skin. An immunohistochemical study. *Lab. Investig. J. Tech. Methods Pathol.*, **50**, 36–41.

40. Ananthapadmanabhan, K. P., Lips, A., Vincent, C., Meyer, F., Caso, S., Johnson, A., Subramanyan, K., Vethamuthu, M., Rattinger, G., Moore, D. J. (2003). pH-induced alterations in stratum corneum properties. *Int. J. Cosmet. Sci.*, **25**, 103–112.

41. Li, Y., Zhi, X., Lin, J., You, X., Yuan, J. (2017). Preparation and characterization of DOX loaded keratin nanoparticles for pH/GSH dual responsive release. *Mater. Sci. Eng. C*, **73**, 189–197.

42. Cheng, Z., Chen, X., Zhai, D., Gao, F., Guo, T., Li, W., Hao, S., Ji, J., Wang, B. (2018). Development of keratin nanoparticles for controlled gastric mucoadhesion and drug release. *J. Nanobiotechnol.*, **16**, 24.

43. Xu, H., Shi, Z., Reddy, N., Yang, Y. (2014). Intrinsically water-stable keratin nanoparticles and their in vivo biodistribution for targeted delivery. *J. Agric. Food Chem.*, **62**, 9145–9150.

44. Reichl, S. (2009). Films based on human hair keratin as substrates for cell culture and tissue engineering. *Biomaterials*, **30**, 6854–6866.

45. Moll, R., Divo, M., Langbein, L. (2008). The human keratins: Biology and pathology. *Histochem. Cell Biol.*, **129**, 705–733.

46. Tran, C. D., Prosenc, F., Franko, M. (2018). Facile synthesis, structure, biocompatibility and antimicrobial property of gold nanoparticle composites from cellulose and keratin. *J. Colloid Interface Sci.*, **510**, 237–245.

47. Nidhin, M., Vedhanayagam, M., Sangeetha, S., Kiran, M. S., Nazeer, S. S., Jayasree, R. S., Sreeram, K. J., Nair, B. U. (2014). Fluorescent nanonetworks: A novel bioalley for collagen scaffolds and tissue engineering. *Sci. Rep.*, **4**, 5968.

48. Liu, C., Czernuszka, J. (2007). Development of biodegradable scaffolds for tissue engineering: A perspective on emerging technology. *Mater. Sci. Technol.*, **23**, 379–391.

49. Shoulders, M. D., Raines, R. T. (2009). Collagen structure and stability. *Annu. Rev. Biochem.*, **78**, 929–958.

50. Nagarajan, U., Kawakami, K., Zhang, S., Chandrasekaran, B., Unni Nair, B. (2014). Fabrication of solid collagen nanoparticles using electrospray deposition. *Chem. Pharm. Bull.*, **62**, 422–428.

51. Rossler, B., Kreuter, J., Scherer, D. (1995). Collagen microparticles: Preparation and properties. *J. Microencapsul.*, **12**, 49–57.

52. Le, V.-M., Lang, M.-D., Shi, W.-B., Liu, J.-W. (2016). A collagen-based multicellular tumor spheroid model for evaluation of the efficiency of nanoparticle drug delivery. *Artif. Cells Nanomed. Biotechnol.*, **44**, 540–544.

53. Nitta, S. K., Numata, K. (2013). Biopolymer-based nanoparticles for drug/gene delivery and tissue engineering. *Int. J. Mol. Sci.*, **14**, 1629–1654.

54. Gorgieva, S., Kokol, V. (2011). Collagen- vs. gelatine-based biomaterials and their biocompatibility: Review and perspectives. In: Rosario Pignatello, ed., *Biomaterials Applications for Nanomedicine*, IntechOpen, DOI: 10.5772/24118.

55. Yasmin, R., Shah, M., Khan, S. A., Ali, R. (2017). Gelatin nanoparticles: A potential candidate for medical applications. *Nanotechnol. Rev.*, **6**, 191–207.

56. Ratanavaraporn, J., Rangkupan, R., Jeeratawatchai, H., Kanokpanont, S., Damrongsakkul, S. (2010). Influences of physical and chemical crosslinking techniques on electrospun type A and B gelatin fiber mats. *Int. J. Biol. Macromol.*, **47**, 431–438.

57. Touny, A. H., Laurencin, C., Nair, L., Allcock, H., Brown, P. W. (2008). Formation of composites comprised of calcium deficient HAp and cross-linked gelatin. *J. Mater. Sci. Mater. Med.*, **19**, 3193.

58. Stevens, K. R., Einerson, N. J., Burmania, J. A., Kao, W. J. (2002). In vivo biocompatibility of gelatin-based hydrogels and interpenetrating networks. *J. Biomater. Sci. Polym. Ed.*, **13**, 1353–1366.

59. Fuchs, S., Klier, J., May, A., Winter, G., Coester, C., Gehlen, H. (2012). Towards an inhalative in vivo application of immunomodulating gelatin nanoparticles in horse-related preformulation studies. *J. Microencapsul.*, **29**, 615–625.

60. Zwiorek, K., Kloeckner, J., Wagner, E., Coester, C. (2004). Gelatin nanoparticles as a new and simple gene delivery system. *J. Pharm. Pharm. Sci.*, **7**, 22–28.

61. Sabet, S., George, M. A., El-Shorbagy, H. M., Bassiony, H., Farroh, K. Y., Youssef, T., Salaheldin, T. A. (2017). Gelatin nanoparticles enhance delivery of hepatitis C virus recombinant NS2 gene. *PLoS One*, **12**, e0181723.

62. Lee, E. M., Singh, D., Singh, D., Choi, S. M., Zo, S. M., Park, S. J., Han, S. S. (2014). Novel alginate-gelatin hybrid nanoparticle for drug delivery and tissue engineering applications. *J. Nanomater.*, **2014**, 147.

63. Foster, J. A., Mecham, R. P., Franzblau, C. (1976). A high molecular weight species of soluble elastin. *Biochem. Biophys. Res. Commun.*, **72**, 1399–1406.

64. Almine, J. F., Bax, D. V., Mithieux, S. M., Nivison-Smith, L., Rnjak, J., Waterhouse, A., Wise, S. G., Weiss, A. S. (2010). Elastin-based materials. *Chem. Soc. Rev.*, **39**, 3371–3379.

65. Nettles, D. L., Chilkoti, A., Setton, L. A. (2010). Applications of elastin-like polypeptides in tissue engineering. *Adv. Drug Deliv. Rev.*, **62**, 1479–1485.

66. Daamen, W. F., Veerkamp, J., Van Hest, J., Van Kuppevelt, T. (2007). Elastin as a biomaterial for tissue engineering. *Biomaterials*, **28**, 4378–4398.

67. Tarakanova, A., Huang, W., Weiss, A. S., Kaplan, D. L., Buehler, M. J. (2017). Computational smart polymer design based on elastin protein mutability. *Biomaterials*, **127**, 49–60.

68. Bessa, P. C., Machado, R., Nürnberger, S., Dopler, D., Banerjee, A., Cunha, A. M., Rodríguez-Cabello, J. C., Redl, H., van Griensven, M., Reis, R. L. (2010). Thermoresponsive self-assembled elastin-based nanoparticles for delivery of BMPs. *J. Control. Release*, **142**, 312–318.

69. Wu, Y., MacKay, J. A., McDaniel, J. R., Chilkoti, A., Clark, R. L. (2008). Fabrication of elastin-like polypeptide nanoparticles for drug delivery by electrospraying. *Biomacromolecules*, **10**, 19–24.

70. Pina, M. J., Alex, S. M., Arias, F. J., Santos, M., Rodriguez-Cabello, J. C., Ramesan, R. M., Sharma, C. P. (2015). Elastin-like recombinamers with acquired functionalities for gene-delivery applications. *J. Biomed. Mater. Res. Part A*, **103**, 3166–3178.

71. Liu, X., Sun, Q., Wang, H., Zhang, L., Wang, J. Y. (2005). Microspheres of corn protein, zein, for an ivermectin drug delivery system. *Biomaterials*, **26**, 109–115.

72. Majoni, S., Wang, T., Johnson, L. A. (2010). Physical and chemical processes to enhance oil recovery from condensed corn distillers solubles. *J. Am. Oil Chem. Soc.*, **88**, 425–434.

73. Shukla, R., Cheryan, M. (2001). Zein: The industrial protein from corn. *Ind. Crops Prod.*, **13**, 171–192.

74. Anderson, T. J., Lamsal, B. P. (2015). Zein extraction from corn, corn products, and coproducts and modifications for various applications: A review. *Cereal Chem.*, **88**, 159–173.

75. Dhanya, A., Haridas, K., Divia, N., Sudheesh, S. (2012). Development of Zein-Pectin nanoparticle as drug carrier. *Int. J. Drug Deliv.*, **4**, 147.

76. Müller, V., Piai, J. F., Fajardo, A. R., Fávaro, S. L., Rubira, A. F., Muniz, E. C. (2011). Preparation and characterization of zein and zein-chitosan microspheres with great prospective of application in controlled drug release. *J. Nanomater.*, **2011**, 10.

77. Luo, Y., Zhang, B., Whent, M., Yu, L. L., Wang, Q. (2011). Preparation and characterization of zein/chitosan complex for encapsulation

of alpha-tocopherol, and its in vitro controlled release study. *Colloids Surf. B Biointerfaces*, **85**, 145–152.

78. Paliwal, R., Palakurthi, S. (2014). Zein in controlled drug delivery and tissue engineering. *J. Control. Release*, **189**, 108–122.

79. Peñalva, R., Esparza, I., González-Navarro, C. J., Quincoces, G., Peñuelas, I., Irache, J. M. (2015). Zein nanoparticles for oral folic acid delivery. *J. Drug Deliv. Sci. Technol.*, **30**, 450–457.

80. Luo, Y., Teng, Z., Wang, Q. (2012). Development of zein nanoparticles coated with carboxymethyl chitosan for encapsulation and controlled release of vitamin D3. *J. Agric. Food Chem.*, **60**, 836–843.

81. Luo, Y., Teng, Z., Wang, T. T., Wang, Q. (2013). Cellular uptake and transport of zein nanoparticles: Effects of sodium caseinate. *J. Agric. Food Chem.*, **61**, 7621–7629.

82. Hong, C. K., Wool, R. P. (2005). Development of a bio-based composite material from soybean oil and keratin fibers. *J. Appl. Polym. Sci.*, **95**, 1524–1538.

83. Peles, Z., Binderman, I., Berdicevsky, I., Zilberman, M. (2013). Soy protein films for wound-healing applications: Antibiotic release, bacterial inhibition and cellular response. *J. Tissue Eng. Regen. Med.*, **7**, 401–412.

84. Chien, K. B., Shah, R. N. (2012). Novel soy protein scaffolds for tissue regeneration: Material characterization and interaction with human mesenchymal stem cells. *Acta Biomater.*, **8**, 694–703.

85. Song, N., Tan, C., Huang, M., Liu, P., Eric, K., Zhang, X., Xia, S., Jia, C. (2013). Transglutaminase cross-linking effect on sensory characteristics and antioxidant activities of Maillard reaction products from soybean protein hydrolysates. *Food Chem.*, **136**, 144–151.

86. Nandane, A. S., Jain, R. (2015). Study of mechanical properties of soy protein based edible film as affected by its composition and process parameters by using RSM. *J. Food Sci. Technol.*, **52**, 3645–3650.

87. Cho, S. Y., Park, J.-W., Batt, H. P., Thomas, R. L. (2007). Edible films made from membrane processed soy protein concentrates. *LWT-Food Sci. Technol.*, **40**, 418–423.

88. Brandenburg, A., Weller, C., Testin, R. (1993). Edible films and coatings from soy protein. *J. Food Sci.*, **58**, 1086–1089.

89. Tansaz, S., Boccaccini, A. R. (2016). Biomedical applications of soy protein: A brief overview. *J. Biomed. Mater. Res. Part A*, **104**, 553–569.

90. Cappelletti, M., Perazzolli, M., Nesler, A., Giovannini, O., Pertot, I. (2017). The effect of hydrolysis and protein source on the efficacy of protein hydrolysates as plant resistance inducers against powdery mildew. *J. Bioprocess. Biotech.*, **7**, 2.

91. Hadzieva, J., Mladenovska, K., Simonoska Crcarevska, M., Glavaš Dodov, M., Dimchevska, S., Geškovski, N., Grozdanov, A., Popovski, E., Petruševski, G., Chachorovska, M. (2017). Lactobacillus casei encapsulated in soy protein isolate and alginate microparticles prepared by spray drying. *Food Technol. Biotechnol.*, **55**, 173–186.

92. Torabizadeh, H., Mikani, M. (2017). Inulinase immobilization on functionalized magnetic nanoparticles prepared with soy protein isolate conjugated bovine serum albumin for high fructose syrup production. *World Acad. Sci. Eng. Technol. Int. J. Biol. Biomol. Agric. Food Biotechnol. Eng.*, **11**, 546–553.

93. Rejinold, N. S., Muthunarayanan, M., Deepa, N., Chennazhi, K. P., Nair, S. V., Jayakumar, R. (2010). Development of novel fibrinogen nanoparticles by two-step co-acervation method. *Int. J. Biol. Macromol.*, **47**, 37–43.

94. Yu, Z., Yu, M., Zhang, Z., Hong, G., Xiong, Q. (2014). Bovine serum albumin nanoparticles as controlled release carrier for local drug delivery to the inner ear. *Nanoscale Res. Lett.*, **9**, 343.

95. Lammel, A. S., Hu, X., Park, S.-H., Kaplan, D. L., Scheibel, T. R. (2010). Controlling silk fibroin particle features for drug delivery. *Biomaterials*, **31**, 4583–4591.

96. Oliveira, A., Guimarães, K., Cerize, N., Tunussi, A., Poço, J. (2013). Nano spray drying as an innovative technology for encapsulating hydrophilic active pharmaceutical ingredients (API). *J. Nanomed. Nanotechnol.*, **4**, 2.

97. Haas, P. A. (1992). Formation of uniform liquid drops by application of vibration to laminar jets. *Ind. Eng. Chem. Res.*, **31**, 959–967.

98. Yoon, J., Kwag, J., Shin, T. J., Park, J., Lee, Y. M., Lee, Y., Park, J., Heo, J., Joo, C., Park, T. J. (2014). Nanoparticles of conjugated polymers prepared from phase-separated films of phospholipids and polymers for biomedical applications. *Adv. Mater.*, **26**, 4559–4564.

99. Yadav, T. P., Yadav, R. M., Singh, D. P. (2012). Mechanical milling: A top down approach for the synthesis of nanomaterials and nanocomposites. *Nanosci. Nanotechnol.*, **2**, 22–48.

100. Prasher, A., Loynd, C. M., Tuten, B. T., Frank, P. G., Chao, D., Berda, E. B. (2016). Efficient fabrication of polymer nanoparticles via

sonogashira cross-linking of linear polymers in dilute solution. *J. Polym. Sci. Part A Polym. Chem.*, **54**, 209–217.

101. Faheem, A., Haggag, Y. (2015). Evaluation of nano spray drying as a method for drying and formulation of therapeutic peptides and proteins. *Front. Pharmacol.*, **6**, 140.

102. Draheim, C., de Crécy, F., Hansen, S., Collnot, E.-M., Lehr, C.-M. (2015). A design of experiment study of nanoprecipitation and nano spray drying as processes to prepare PLGA nano-and microparticles with defined sizes and size distributions. *Pharm. Res.*, **32**, 2609–2624.

103. Rao, J. P., Geckeler, K. E. (2011). Polymer nanoparticles: Preparation techniques and size-control parameters. *Prog. Polym. Sci.*, **36**, 887–913.

104. Feng, L., Zhu, C., Yuan, H., Liu, L., Lv, F., Wang, S. (2013). Conjugated polymer nanoparticles: Preparation, properties, functionalization and biological applications. *Chem. Soc. Rev.*, **42**, 6620–6633.

105. Tuncel, D., Demir, H. V. (2010). Conjugated polymer nanoparticles. *Nanoscale*, **2**, 484–494.

106. Loh, Z. H., Samanta, A. K., Heng, P. W. S. (2015). Overview of milling techniques for improving the solubility of poorly water-soluble drugs. *Asian J. Pharm. Sci.*, **10**, 255–274.

107. Aiertza, M. K., Odriozola, I., Cabañero, G., Grande, H.-J., Loinaz, I. (2012). Single-chain polymer nanoparticles. *Cell. Mol. Life Sci.*, **69**, 337–346.

108. Hanlon, A. M., Chen, R., Rodriguez, K. J., Willis, C., Dickinson, J. G., Cashman, M., Berda, E. B. (2017). Scalable synthesis of single-chain nanoparticles under mild conditions. *Macromolecules*, **50**, 2996–3003.

109. Verso, F. L., Pomposo, J. A., Colmenero, J., Moreno, A. J. (2014). Multi-orthogonal folding of single polymer chains into soft nanoparticles. *Soft Matter*, **10**, 4813–4821.

110. Balazs, A. C., Emrick, T., Russell, T. P. (2006). Nanoparticle polymer composites: Where two small worlds meet. *Science*, **314**, 1107–1110.

111. Mohanraj, V., Chen, Y. (2006). Nanoparticles: A review. *Trop. J. Pharm. Res.*, **5**, 561–573.

112. Vasicek, T. W., Jenkins, S. V., Vaz, L., Chen, J., Stenken, J. A. (2017). Thermoresponsive nanoparticle agglomeration/aggregation in salt solutions: Dependence on graft density. *J. Colloid Interface Sci.*, **506**, 338–345.

113. Guterres, S. S., Alves, M. P., Pohlmann, A. R. (2007). Polymeric nanoparticles, nanospheres and nanocapsules, for cutaneous applications. *Drug Target Insights*, **2**, 147–157.

114. Kucuk, I., Edirisinghe, M. (2014). Microfluidic preparation of polymer nanospheres. *J. Nanopart. Res.*, **16**, 2626.

115. Lattuada, M., Hatton, T. A. (2011). Synthesis, properties and applications of Janus nanoparticles. *Nano Today*, **6**, 286–308.

116. Deng, R., Liang, F., Qu, X., Wang, Q., Zhu, J., Yang, Z. (2015). Diblock copolymer based Janus nanoparticles. *Macromolecules*, **48**, 750–755.

117. Kouchak, M., Avadi, M., Abbaspour, M., Jahangiri, A., Boldaji, S. K. (2012). Effect of different molecular weights of chitosan on preparation and characterization of insulin loaded nanoparticles by ion gelation method. *Int. J. Drug Dev. Res.*, **4**, 271–277.

118. He, Y. T., Wan, J., Tokunaga, T. (2007). Kinetic stability of hematite nanoparticles: The effect of particle sizes. *J. Nanopart. Res.*, **10**, 321–332.

119. Li, K., Liu, B. (2012). Polymer encapsulated conjugated polymer nanoparticles for fluorescence bioimaging. *J. Mater. Chem.*, **22**, 1257–1264.

120. Howes, P., Green, M., Levitt, J., Suhling, K., Hughes, M. (2010). Phospholipid encapsulated semiconducting polymer nanoparticles: Their use in cell imaging and protein attachment. *J. Am. Chem. Soc.*, **132**, 3989–3996.

121. Yoon, S. M., Myung, S.-J., Kim, I.-W., Do, E.-J., Ye, B. D., Ryu, J. H., Park, K., Kim, K., Kwon, I. C., Kim, M. J. (2011). Application of near-infrared fluorescence imaging using a polymeric nanoparticle-based probe for the diagnosis and therapeutic monitoring of colon cancer. *Dig. Dis. Sci.*, **56**, 3005–3013.

122. Khalid, A., Mitropoulos, A. N., Marelli, B., Simpson, D. A., Tran, P. A., Omenetto, F. G., Tomljenovic-Hanic, S. (2015). Fluorescent nanodiamond silk fibroin spheres: Advanced nanoscale bioimaging tool. *ACS Biomater. Sci. Eng.*, **1**, 1104–1113.

123. Wei, L., Zehui, Z., Biao, K., Shanshan, F., Jinxiu, W., Lingzhi, W., Jianping, Y., Fan, Z., Peiyi, W., Dongyuan, Z. (2013). Simple and green synthesis of nitrogen-doped photoluminescent carbonaceous nanospheres for bioimaging. *Angew. Chem. Int. Ed.*, **52**, 8151–8155.

124. Liu, S. (2005). Nanometer-sized gold-loaded gelatin/silica nanocapsules. *Adv. Mater. (Weinheim)*, **17**, 1862–1866.

125. Byrne, S. J., Williams, Y., Davies, A., Corr, S. A., Rakovich, A., Gun'ko, Y. K., Rakovich, Y. P., Donegan, J. F., Volkov, Y. (2007). "Jelly dots": Synthesis and cytotoxicity studies of CdTe quantum dot–gelatin nanocomposites. *Small*, **3**, 1152–1156.

126. Xu, H., Jiang, Q., Reddy, N., Yang, Y. (2011). Hollow nanoparticles from zein for potential medical applications. *J. Mater. Chem.*, **21**, 18227–18235.

127. Lai, L. F., Guo, H. X. (2011). Preparation of new 5-fluorouracil-loaded zein nanoparticles for liver targeting. *Int. J. Pharm.*, **404**, 317–323.

128. Lee, S., Alwahab, N. S. A., Moazzam, Z. M. (2013). Zein-based oral drug delivery system targeting activated macrophages. *Int. J. Pharm.*, **454**, 388–393.

Chapter 24

Nanomedicine and Phage Capsids

Philip Serwer, PhD, and Elena T. Wright

Department of Biochemistry and Structural Biology,
The University of Texas Health Science Center,
San Antonio, Texas 78229-3900, USA

Keywords: alpha-sheet, cancerous tumors, capsid dynamics, drug delivery vehicles, native gel electrophoresis, neurodegenerative disease, pathogenic viruses, nanomedicine principles, phage therapy, anti-viral therapy, innate immunity

24.1   Introduction

24.1.1   A Principle

The following principle (basics-focus principle) is proposed here as a foundation for uses of phages in nanomedicine. The curing of currently intractable, biochemically complex diseases requires understanding of the disease basics, but it does not require understanding of the disease details. Historical justifications for the basics-focus principle include (1) the development of

The Road from Nanomedicine to Precision Medicine
Edited by Shaker A. Mousa, Raj Bawa, and Gerald F. Audette
Copyright © 2020 Jenny Stanford Publishing Pte. Ltd.
ISBN 978-981-4800-59-4 (Hardcover), 978-0-429-29501-0 (eBook)
www.jennystanford.com

smallpox [1] and rabies [2] vaccines before any details about virus composition and structure were known; (2) the discovery of bacterial cell wall-active [3, 4] and ribosome-active [5] antibiotics before the composition and structure of bacterial cell walls and ribosomes were known and (3) the use of phage therapy for infectious disease in the absence of knowledge of the composition and structure of phages [6, 7].

Applying the basics-focus principle does not mean neglecting the rest of the science. Louis Pasteur 's basics-oriented work on improving wine and beer fermentation was a major part of the foundation for the fields of biochemistry and microbiology [2, 8]. Antibiotics became major tools in investigating the composition, structure and dynamics of bacterial cell walls [4] and ribosomes [5, 9].

A proposed corollary is the following. Basics-focus on practical aspects of neurodegenerative and other diseases will not compromise the remaining science. Indeed, as we will describe in Section 24.6, we think that the remaining science will also be promoted with this approach.

However, to get started, one has to know enough basics. We think likely that, at least in the case of neurodegenerative diseases, some basics will have to be assumed, without rigorous proof. In the discussion below, we will present both key assumptions and the phage-based evidence in the case of neurodegenerative disease. We will also present a basics-oriented strategy for malignant tumor therapy. This strategy includes use of a phage capsid-based drug delivery vehicle (DDV).

24.1.2 The Scientific Environment

Articulation of the basics-focus principle is motivated, in part, by a current environment in which progress is politely described as slow for the curing of both neurodegenerative diseases and metastatic cancer. Less politely, but probably more accurately, most (not all) research on neurodegenerative diseases appears to be mired in its focus on complex biochemical details, with only limited symptomatic relief achieved [10–13]. Focus on these details is the opposite of (1) what has historically been the most successful focus (previous section) and (2) what some fundamentals project to be the optimal focus for developing future therapies for pathogenic virus infections (Section 24.2.5), neurodegenerative diseases (Section 24.3.2) and cancerous tumors (Section 24.4.2). Lack of basics-focus is one possible explanation of why neurodegenerative diseases are incurable at this stage in history.

Two recent books appear to be warning signs that public patience is beginning to exhaust. A recent, cancer-oriented book comes very close to laying the slow-progress blame at the feet of science (really, scientists) [14]. A recent, polio vaccine-oriented book suggests the following. Without the intervention of the law partner of an American President, polio vaccine, as we know it, would not have existed as early as it did [15]. A logical rendition of the current state of phage therapy is likely to exhaust public patience to a new level because, in this case, successful application of the basics has already been achieved ([16], reviewed in [17, 18]). However, apparently, in the US, one can receive phage therapy only on an ad hoc basis [16, 19].

Our entire research history is in the area of phages, with a primary focus on phage assembly. Historically speaking, the Caltech phage group (the home of PS for four years) was, in its early years, supported by the foundation that previously supported the basics focus-oriented work on polio vaccines. This foundation, The March of Dimes (also called the National Foundation), also supported the Pauling-associated work on structure discussed below (see the credits in references [20–23]). That is to say, the basic philosophy in the current article appears to have an indirect linkage to the distant past.

24.1.3 Phage Assembly Basics

The composition and structure of phage T3 and its relative, T7, are illustrated in Fig. 24.1d. The phage particle consists of (1) a DNA-encapsulating shell of the protein product of *gene 10*, called gp10; (2) an external structure (tail) adapted for specific binding to host cells and (3) an internal core stack [24–26]. The various phage proteins are labeled by gp, followed by the number of the encoding gene. A T3 gene is given the same number as its T7 counterpart [27].

The DNA-containing, protein capsid of all well-studied, double-stranded DNA phages begins its existence as a DNA-free capsid, called a procapsid. The T3/T7 procapsid (also called capsid I) is illustrated in Fig. 24.1a [24]. The procapsid subsequently packages a DNA genome and, while so doing, changes its structure to form a more phage-like capsid, called capsid II for T3/T7 (Fig. 24.1b). At the end of packaging, the T3/T7 tail is attached to a ring (called portal or connector) that is on a DNA-filled capsid (Fig. 24.1c) called a head [26]. The connector (1) is the site of DNA entry; (2) occupies a 5-fold vertex of an icosahedral, DNA-containing gp10 shell and (3) forms the base of the core stack [24, 25].

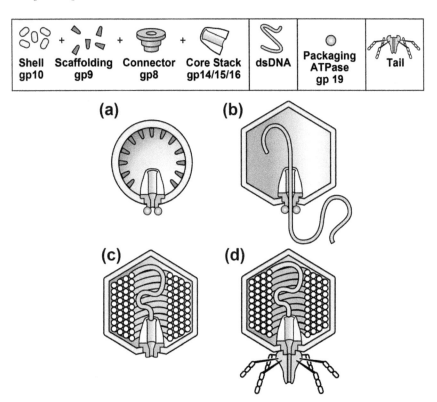

Figure 24.1 The progression of capsids during the *in vivo* assembly of phages T3 and T7. (a) capsid I; (b) capsid II packaging DNA; (c) head; (d) mature phage.

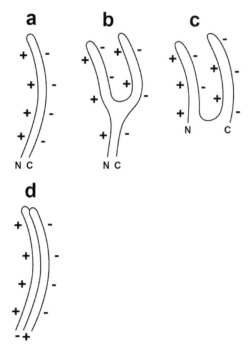

Figure 24.2 A line drawing of the proposed α-sheet-generating polypeptide backbone of the gp10 subunits of (a) hyper-expanded T3/T7 capsid II; (b) an intermediate converting from its state in hyperexpanded to its state in contracted capsid II and (c) contracted capsid II. N and C indicate the N- and C-terminals of gp10; (d) Assembly of two gp10 subunits is shown with the proposed radial staggering. The staggering improves electrical charge-charge-derived energetics. The + symbols indicate the + electrical charge of the α-amino edge; the – symbols indicate the – electrical charge of the α-carboxyl edge.

24.2 Phage Assembly and Dynamic States

24.2.1 The Stability of the Icosahedral Shell of the Related Phages, T3 and T7

One's first impression of the mature T3/T7 capsid is that the DNA-enclosing gp10 shell is extremely stable and unlikely to change states. The shell is resistant to both ionic detergent [28] and the proteases, trypsin and subtilisin [29]. In a mature phage

T3/T7 capsid, the major shell protein has a conformation [25] in common among the various double-stranded DNA phages. This conformation is called the HK97 fold, named after the first phage found to have this fold [30].

The mature gp10 shell of T7 capsid II also has a surprisingly high stability to elevated temperature, as discussed in Section 24.4.3. This characteristic suggests resistance to damage during possible use as a capsid-drug delivery vehicle (DDV).

Nonetheless, dynamism of phage shells is suggested by cryo-electron microscopy (cryo-EM) analysis. After expulsion of DNA from phage HK97, the shell "bows out" more than it does before expulsion [31]. Similarly, T7 capsid II is found 1.4% larger than the mature capsid when purified [25] and in lysates [32]. If one makes the assumption that HK97-type shell proteins have incompressible components, without capacity for storing and releasing energy, the above result with HK97 is impossible. The reason is that DNA expulsion releases pressure from packaged DNA. This would cause contraction, not expansion, if one makes this assumption. Therefore, the shell of HK97 and probably T3/T7 can move internally to store and release energy.

Less direct evidence for shell dynamics arises from analysis of the leakage of DNA from (tail-free) phage T3 heads. The heads are obtained from a T3 mutant; almost no T3 and T7 heads are without a tail in a wild type infection [26]. The DNA leaks from heads in quantized amounts. This phenomenon is seen via the formation of sharp bands during agarose gel electrophoresis of the DNA remaining packaged after 1-hit restriction endonuclease digestion. The DNA remaining packaged is obtained by (1) DNase I-digestion of external DNA; and (2) expulsion from the capsid of DNA remaining packaged [26]. This leakage-quantization phenomenon is best explained by quantized gp10 shell contraction that evolved via selection for control of the rate of infection-initiating DNA injection [26, 33].

In addition, some multi-site T3 mutants undergo enhanced *in vivo* production of the following gp10-shell variants of capsid II: hyper-expanded and contracted. Identification of hyper-expanded capsid II is made by both (1) electron microscopy and (2) calculation of hydration from the density during buoyant density centrifugation in Nycodenz density gradients of a

Nycodenz-impermeable (low density, high hydration) capsid II [34, 35]. A DNA-free version of these capsids was initially investigated [34].

Next, a DNA-containing version of hyper-expanded capsid II was isolated via its Nycodenz impermeability (sealing) and accompanying low density [35]. This particle underwent contraction in the presence of magnesium ATP, but not magnesium ADP. Binding of ATP to gp10 was proposed to be the source of energy [35]. The following was evidence that the sealing had been evolutionarily selected. For wild type capsid II, cryo-EM revealed the complexity of inter-gp10 subunit interactions to be so high [25] that accidental sealing was improbable. Thus, these capsids were proposed [35] to be in states selected for function during DNA packaging (details [36]). The hyper-expansion required shell thinning to the point that β-sheet was proposed as the likely dominant conformation of gp10 major shell protein [35].

24.2.2 α-Sheet, Rather than β-Sheet, in Size-Altered Capsid II?

α-Sheets resemble parallel β-sheets in having a parallel and extended conformation. However, if the amino acids in an α-sheet all have the same chirality (as they do in almost all current proteins), then amino acids alternate side chain positions. That is to say, the conformation is not a helical array of amino acids. If constituent amino acids either alternate in chirality or are all glycine, then an α-sheet can be a helical array of amino acids [21–23, 37, 38]. The above has suggested the possibility that α-sheets began existence abiotically, when proteins were glycine-rich and other amino acids involved were not chiral [37, 38].

An α-sheet-like peptide of 3–6 amino acids is called a nest. Nests are typically glycine-rich. α-Sheets and nests have α-amino groups segregated on one edge and α-carboxyl groups segregated on the opposing edge [21, 37, 38]. Nest-associated α-amino groups are known to bind anions, such as phosphate, via the α-amino group edge. P-loop ATP-binding sites typically have a phosphate-binding nest [37, 38]. Thus, one projects that a more extensive α-sheet is also likely to bind phosphates, including those part of ATP.

α-sheets were discovered via model building in 1951 [21] and were originally called parallel pleated sheets before parallel β-pleated sheet was known to be the more frequent structure. However, extensive α-sheets were found, by the discoverers, to be unlikely for the real world of left-handed amino acids. The reason was "steric hindrance between adjacent side chains" [23] for left-handed, non-glycine amino acids. Bending of α-sheets can reduce steric hindrance enough to make extensive α-sheet possible [22].

Additional unfavorable energetics are expected from the charge separation of α-sheets in the absence of multiple sheet layers. The α-carboxyl edge is negatively charged; the α-amino edge is positively charged at physiological pH. Indeed, stable proteins have only a small percentage of α-sheet-like nests. Among the proteins stable enough to be characterized, α-sheet-like nests are found primarily in ATP binding sites and in the lining of transmembrane pores [37, 38].

The expected increase of α-sheet content for abiotically generated peptides suggests that nests are imprints from times before the existence of living organisms. The idea is that this structure was not completely replaced when increased diversity and chirality arrived for biological amino acids [37, 38].

In theory, evolutionary retention of α-sheet structure would be increased if (1) the α-sheet is curved; (2) cooperativity is symmetry-promoted by incorporation of the protein in a symmetrical structure and (3) the protein binds a nest-stabilizing ATP molecule, thereby initiating a cooperative transition to α-sheet structure. Thus, we propose the following hypothesis. During DNA packaging, the observed T3 capsid II shell dynamics (hyper-expansion and contraction) are caused by the adopting by gp10 of ATP-responsive, dynamic α-sheet conformations.

Specifically, single-layered α-sheet is the proposed structure for gp10 in the shell of the most hyper-expanded version of T3 capsid II (Fig. 24.2a; orientation in the shell is discussed in the next paragraph). The thickness is 0.6–0.9 nm [21, 37], which is thin enough to make possible covering of the entire surface of the observed hyper-expanded T3 capsid II [35]. In addition, the proposed structure for the contracted versions of T3 capsid II is multi-layer α-sheet (Fig. 24.2c), generated by an event approximating the folding of the single-layered α-sheet (Fig. 24.2b), without loss of gp10 subunits. This latter conversion would be assisted by favorable electrical charge-charge interactions (Fig. 24.2b).

Assembly of multiple Fig. 24.2a-like subunits will be inhibited by charge-charge interactions unless the radial positions of neighboring subunits vary so that the negatively and positively charged edges of neighboring subunits are juxtaposed (Fig. 24.2d). This "staggering" of radial position will cause increase in apparent shell thickness when a shell is visualized in a two-dimensional projection of the three-dimensional structure.

Given the polar nature of the two edges of alpha-sheet, one edge is predicted to be at the outer surface of the capsid's shell; the other is predicted to be at the inner surface of the shell. Most likely, the negatively charged edge will be at the outer surface to minimize interaction with other intracellular proteins, most of which are negatively charged at neutral pH [39].

24.2.3   Test of a Prediction: Surface Charge

We have tested the prediction of a relatively high negative surface charge for the gp10 shell of hyper-expanded T3 capsid II. This was done for capsid II that had incompletely packaged DNA, abbreviated ipDNA; an ipDNA-containing capsid is called an ipDNA-capsid. The test was performed by native agarose gel electrophoresis in two dimensions (2d-AGE) (recent reference [26]): 0.30% agarose gel in the first dimension; 2.0% agarose gel in the second dimension. A band of hyper-expanded ipDNA-capsid II was seen (labeled HE-CII in Fig. 24.3a) after GelStar staining, nucleic acid-specific. This band was also seen after Coomassie staining, protein-specific. The effective origin of electrophoresis is indicated by the letter o. For comparison, the position of wild type T3 capsid II was also determined by 2d-AGE (dot labeled WT-CII in Fig. 24.3a). The latter 2d-AGE was performed in a separate first and second dimension gel embedded in the same agarose frame as the gel in Fig. 24.3a. The position of capsid I was determined from a separate analysis (dot labeled CI).

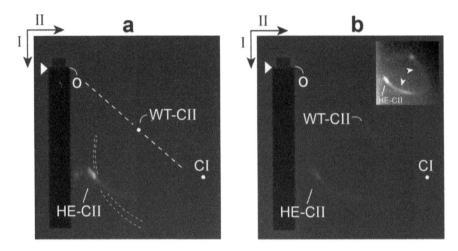

Figure 24.3 Analysis by 2d-AGE of hyper-expanded T3 ipDNA-capsid II. The ipDNA-capsid II is from the Nycodenz gradient-isolation in Fig. 4b of reference [35]. Two fractions of the Nycodenz density gradient were analyzed by 2d-AGE, the (**a**) 1.073 g/mL and (**b**) 1.099 g/mL fractions. The procedure of 2d-AGE is described in reference [26]. The first dimension was run in a 0.30% agarose gel at 2.0 V/cm for 5.0 h. The second dimension was run in a surrounding 2.0% agarose gel at 1.8 V/cm V/cm for 16.0 h. The electrophoresis buffer was 0.09 M Tris-acetate, pH 8.3, 0.001 M $MgCl_2$. The temperature was 25±0.3°C. Seakem LE agarose was used (Lonza, Basel, Switzerland). The arrowheads indicate the leading edges of sample wells. The arrows indicate the directions of the first (I) and second (II) dimension electrophoresis. The curved dashed lines indicate the profile of variable length DNAs (no protein attached) from the DNA fraction of the same Nycodenz gradient. The DNA profile was obtained in a separate quadrant embedded in the same agarose frame as the gels of (a,b).

The effective radius of a particle (R_E), together with the radius of the gel's effective pore (P_E), uniquely determine the straight line drawn from the effective origin to the position of a particle in the gel. This line (dashed) is indicated for the WT-CII position in Fig. 24.3a (R_E = 28.6 nm by small angle X-ray scattering [40]). The angle, θ, between dashed line and first dimension gel decreases as R_E increases [26]. Thus, the particles of the HE-CII band are confirmed in Fig. 24.3a to be relatively large. The shape and orientation of HE-CII band indicate that these particles are also heterogeneous in R_E.

The average particle of the HE-CII band also had an average electrical surface charge density, σ, that was negative and was increased in absolute value. The reasoning is the following. The value of σ has, in general, been found to be proportional to the electrophoretic mobility in the absence of a gel. This mobility has been found to be independent of internal contents, such as ipDNA ([26] and included references). The ratio of σ value for HE-CII to σ value for WT-CII was 1.9. This ratio was determined from the ratio of average distances migrated in the first (low sieving) dimension, corrected for an estimated 5% greater effect of sieving on HE-CII in the first dimension. When CI was substituted for HE-CII, this ratio was also 1.9.

The source of the relatively high negative σ of hyper-expanded ipDNA-capsid II had to be either (1) the σ at the surface of shell-associated gp10 or (2) leakage from the capsid of a segment of (negatively charged) ipDNA, without dissociation of the ipDNA. We concluded that the former possibility was correct because electron micrographs did not reveal any leaked DNA [35]. Thus, qualitatively, the above prediction was confirmed.

Finally, the relatedness of HE-CII and WT-CII particles was confirmed in the 2d-AGE profile of a higher density fraction from the same Nycodenz density gradient. In this case, the HE-CII band was connected to a WT-CII-like band by an arc formed by capsid II particles with intermediate R_E and σ values (Fig. 24.3b and inset). Contrast enhancement was used in the inset of Fig. 24.3b to make the arc more easily seen. Presumably, the arc-forming particles were also intermediate in structure.

24.2.4 Electron Microscopy

In previous electron micrographs of specimens negatively stained with sodium phosphotungstate, hyper-expanded ipDNA-capsids appeared full [35]. This appearance was caused by impermeability to the negative stain, not by filling with DNA [35]. The low-electron density interior was occupied by dried buffer components. The appearance was similar after negative staining with uranyl acetate (Fig. 24.4). However, the following feature was exaggerated in some particles of hyper-expanded NLD capsid II. A thin, dark layer of negative stain separated the light interior from the light gp10 shell (Fig. 24.4). Apparently, after drying of most of the interior, external negative stain leaked through the gp10 shell and then dried in a thin layer. In contrast, a contaminating wild type capsid II-like particle had the traditional negative stain-filled appearance (arrow in Fig. 24.4). The shell of the latter capsids is 2 nm thick.

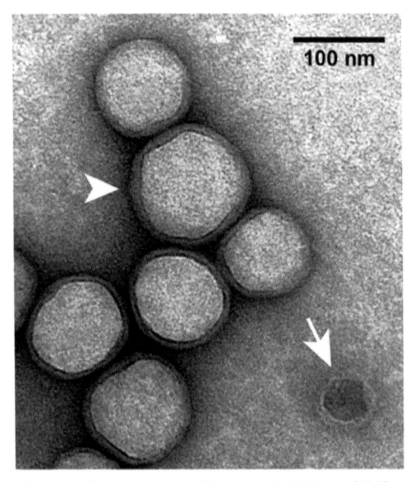

Figure 24.4 Electron microscopy of hyper-expanded NLD capsid II. The sample was the same as the sample used for Fig. 24.3a. Particles in the sample were negatively stained with 1% uranyl acetate. The procedures of specimen preparation and EM were the same as used in reference [35].

In the shell-revealing regions of hyper-expanded ipDNA-capsid II particles, the shell usually appears 3–5 nm thick. However, the thickness of the shell is likely to be significantly less than that

(Section 2.2). This difference is enough so that staining alone is unlikely to be the cause. Thus, the cause of the above difference resides in changes to the gp10 shell that occur in the latter stages of negative staining.

Details were suggested by the observation that, in some regions, the gp10 shell appeared doubled (arrowhead in Fig. 24.4). The radial position staggering proposed in Fig. 24.2d explained the observed doubling via an increase in the staggering distance at the latter stages of negative staining. By this explanation, embedding in the negative stain limited shell disruption to sub-observable levels. The regions of apparent shell thickening, without doubling, could be explained by smaller increase in the staggering distance and superposition of images from multiple planes perpendicular to the direction of observation. In summary, the electron microscopy produced images that were explained by a gp10 structure similar to the one in Fig. 24.2d. Higher resolution, direct determination of structure is needed to test more directly for α-sheet and other structures.

24.2.5 Possible Pathway to Anti-Viral Therapeutics

The above analysis by 2d-AGE has been performed only for phage T3. This analysis is also a relatively simple, inexpensive, sensitive way to determine whether other viruses produce size-altered capsids like those of phage T3. For now, we make the following extrapolation. Pathogenic viruses do make extensive α-sheet-containing shell intermediates, at least in the case of the capsids of viruses with a double-stranded DNA genome, such as herpes viruses.

If, indeed, this is true, then the following is a possible strategy for anti-viral therapy. Find or design therapeutic compounds that target the backbone of extensive α-sheets. The low frequency of extensive α-sheets suggests low therapy toxicity. This strategy has the projected advantage of being insensitive to bypass via single mutations. In addition, a relatively broad spectrum of viruses is likely to be susceptible. Two major factors in reducing the success of anti-viral drug therapies are evolution of resistant viral mutants and presence of narrow drug specificity (examples [41]).

24.3 Dynamic States and Neurodegenerative Disease

24.3.1 Some Details

Neurodegenerative diseases are all characterized by the presence of protein aggregates collectively called amyloid. This name is derived from starch-like texture. The aggregate-forming protein varies with the neurodegenerative disease (reviews: ALS [42], Alzheimer [43], Huntington [44], Parkinson [45], prion-associated [46]). When various proteins form amyloid, they convert from an original mixed-element structure to a predominantly β-sheet structure.

Although β-sheet is the predominant conformation of accumulating amyloid protein, computer simulations reveal that α-sheet is (1) accessible to amyloid-forming proteins [47, 48] and (2) selectively accessible to them at low pH, especially in the protein region thought to initiate the transition to amyloid [48]. The proposal has been made that charge-charge interactions assist assembly to form an α-sheet intermediate that subsequently converts to β-sheet during the formation of amyloid. Molecular dynamics simulation has shown the possibility of conversion of α-sheet to β-sheet [47–49].

Given the association of nest-like structures with membrane pores, the theory is that extensive α-sheet-containing amyloid proteins, which are minority species, generate toxicity by making pores in cell membranes [37, 38, 47, 48]. Mature amyloids are complex enough so that the amount of α-sheet associated with the β-sheet structure is, to our knowledge, not known.

Returning to the biology, the hypothesis has previously been proposed [50] (and expanded [51]) that one of the normal (non-toxic) functions of amyloid-forming proteins is innate immunity. This innate immunity occurs via neutralizing of products of both pathogen-generated and physical insults. By this hypothesis, neurodegenerative disease occurs when control of this insult-neutralizing activity is lost and, in some cases, the innate immune proteins start neutralizing themselves. A result is accumulation of toxic aggregates. Indeed, more recent studies have shown anti-bacterial and anti-viral effects of the amyloid for Alzheimer disease [52, 53].

The expanded proposal [51] originally was that the neutralizing interaction occurred via the extension by innate immune proteins of β-sheet structure generated by the insult. However, the above considerations suggest substitution of α-sheet structure for β-sheet structure.

24.3.2 A Proposal for Reduction to Practice

We propose the following working assumptions to implement the basics-focus principle for neurodegenerative diseases. (1) Pathogenic viral infection and other insults trigger activation of amyloid-forming host innate immune proteins. These proteins convert to extensive α-sheet structure and, then, co-assemble with and inactivate extensive α-sheet-rich structures generated by the insults. These activities are currently obscure. (2) Extensive α-sheets are rare to non-existent in endogenous proteins of the virus host, other than activated innate immune system proteins. (3) Neurodegenerative diseases are caused by α-sheet-dependent damage caused by hyper-produced, activated innate immune proteins. These proteins eventually form β-sheet-rich amyloid.

Other virus-induced innate immunity activities (cytokine-induced neuroinflammation, for example) are known to be associated with the onset of neurodegenerative disease [54]. By the above hypothesis, these activities are secondary.

The proposal of (1)–(3) includes the fundamental idea that higher organisms have retained an immune system that has anciently derived targets and mode of action. This system is based on an abiotically derived protein structure. This system is difficult to analyze because proteins with more recently evolved composition and structure force both the innate immune system and its targets into intermittent status.

A way to reduce these ideas to practice is the following, as similarly suggested [51] before the introduction of the above α-sheet proposal. (1) Select for low-pathogenicity viruses that yield unstable, α-sheet-rich complexes when innate immune system proteins co-assemble with intracellular intermediates of this virus. (2) Infect patients with these viruses. To manage adaptive immunity, several such viruses would be used in succession. To isolate the needed viruses, one could begin by trying to find cases of spontaneous (presumably limited) disease-remission that have a correlation with a previous virus infection. The correlation would presumably not be dramatic and finding it would require determination. The reason is that most infections are expected to have the opposite effect, i.e., to trigger disease [54, 55]. Therefore, the desired post-infection remissions are likely to be rare.

Of course, selecting for any immune system-avoiding virus has an obvious potential danger of ending with a problem of virus virulence. Presumably, virus virulence would be carefully monitored during selection. In any case, the selection would include screening for low virulence. Low virulence might, indeed, be necessarily co-selected during selection for an immune complex-destabilizing phenotype.

24.4 Phage Assembly and Development of Gated, Targeted Drug Delivery Vehicles

24.4.1 Avoiding Immune Systems

Here, we change direction to discuss use of phage capsids as DDVs for the therapy of cancerous tumors. Administration of phage capsids is accompanied by the problem of phage capsid removal/neutralization by host immune systems. Both adaptive and innate immune systems will act to remove or neutralize foreign nanoparticles, including phages [56]. Optimization of a DDV-based strategy must include a response to this problem. One response to adaptive immunity is to serially change the phage-source of the DDV, in analogy to what is proposed above for low-virulence, therapeutic, eukaryotic viruses. In addition, phages propagate rapidly enough so that directed evolution can be used to reduce removal of a phage-derived DDV by any immune system.

However, probably the most significant potential advantage of a phage capsid-DDV is gating via the connector (formed by gp8 for T3/T7; Fig. 24.1). The connector of phages T3 and T7 also exists for all other studied double-stranded DNA phages (reviews [57, 58]). Thus, if connectors can, in general, be used as gates to capsid-DDVs, nature will provide many immunologically unrelated DDVs.

Before adaptive immunity becomes limiting (which takes 1–2 weeks in the case of phage K [56]), innate immune systems reduce effectiveness of any unprotected nanoparticle. Apparently the most active system is the mononuclear phagocyte system (MPS), formerly called the reticuloendothelial system. The MPS moves nanoparticles primarily to the liver and spleen [59, 60]. Published tests of the efficiency of the MPS have been made with phages lambda, P22 and K. Phages lambda and P22 are lysogenic. When in a lysogenic state, lambda and P22 presumably could not have experienced and adapted to their environments. In contrast, phage K is lytic and had continuously experienced and adapted to its environment. Therefore, one expects that past genetic adaptation would cause phage K to have a longer lifetime in mouse circulation. This is the case. Complete removal occurs in over 24 hr for phage K [56] and 3–4 h for phages lambda and P22 [61, 62].

That the above difference in lifetimes can be adaptive for phage lambda was shown by 10-cycles of the following: mutagenic growth, followed by mouse passage [61]. The adapted lambda was removed from mouse circulation more slowly than wild type lambda. Removal was by approximately a factor of 10 per day, several orders of magnitude more slowly than before adaptation [61]. A comparable removal rate was found for lytic phage T3 without any laboratory adaptation [63]. Apparently, the needed adaptation had occurred for T3 in the wild. Systematic attempts to make lytic phages more long-lived in circulation have not been made to our knowledge.

24.4.2 Targeting Tumors

Perhaps the most dramatic potential application of the basics-focus principle is to the therapy of cancerous tumors, including metastatic tumors. Focus on details (rather than basics) leads to targeting of either tumor-concentrated biochemistry or a patient's immune systems. However, the biochemistry of tumors overlaps the biochemistry of normal cells. Thus, essentially all drug and radiation therapies are toxic, which limits the extent of the therapy and makes patients sick [64–66]. Multi-faceted toxicity of immunotherapy is emerging as a problem [67]. In addition, when a cancer becomes metastatic, none of the above therapies can systematically block bypass of the therapy via evolution of cancer cells.

However, tumors do have one targetable, basic characteristic that they apparently cannot evolve to eliminate. This characteristic is the presence of 10–200 nm-sized pores in the blood vessels that feed tumors. Healthy blood vessels do not have these pores [68, 69].

Thus, one should be able to achieve specific drug delivery to tumors with the following strategy: use of a DDV small enough to fit in these pores, but too large to pass through healthy blood vessels. In addition, tumors have relatively poor lymphatic drainage. Thus, when a DDV enters a tumor, it is slow to leave. The combined effect of the pores and the poor drainage is called the enhanced permeability and retention (EPR) effect [68, 69].

Knowledge of the EPR effect is not new. The EPR effect has been the foundation for improved drug delivery via liposomal DDVs. Several products are FDA approved [70], the earliest being Doxil [71]. However, results have been disappointing. A recent review indicates the following limitations [72]: (1) "rapid loss of the drug cargo...often immediately after...systemic administration"; (2) "in many cases, less than 1% of the administered nanoparticle dose reaches the malignant tissue"; (3) "lack of release of drug in tumors"; (4) safety of the DDV; (5) safety of impurities (bacterial endotoxin which can generate a cytokine storm); (6) immunotoxicity (complement-generated, for example) and (7) standardization during manufacturing scale-up. A major uncertainty is the extent of the EPR effect in early metastases.

In evaluating the use of a phage capsid-DDV, we initially note that phages have never been found to be toxic to humans. Phages T3 and T7 were presumably isolated from sewage [73] and have passed through humans many times with the associated likelihood that they and their capsids already have evolved to avoid human immune systems. Furthermore, phage capsids with DDV potential are assembled *in vivo* and easily purified with structure uniform enough to obtain a 3–4 Å cryo-EM structure of the shell in the case of T7 [25]. If gated, a phage capsid is well on its way to removing the above limitations.

Indeed, one form of T3 capsid II is sufficiently gated so that it does not allow Nycodenz (molecular weight = 821) in its internal cavity until the temperature is raised. That is to say, the elevated temperature opens a gate. The gate is closed by lowering temperature, with the loaded Nycodenz not leaking detectably. The connector is the likely location of the gate [74].

24.4.3 Adequate Loading of a Phage Capsid-DDV

Nonetheless, a major barrier still exists to implementation of a gated T3 phage capsid-DDV. When we used elevated temperature to increase permeability (open the gate) in the presence of 10 mg/mL doxorubicin, the T3 gp10 shell was damaged. Native gel electrophoresis suggested disassembly to small aggregates, possibly monomers of gp10 (unpublished data). This limitation was possibly caused by detergent characteristics of doxorubicin, which has a positively charged and a hydrophobic region. These two regions are also present in most anti-cancer drugs.

Although this limitation has not yet been bypassed, we have found that the T7 counterpart of the loadable T3 capsid II is much more stable (80–83°C). Future work will focus on the T7 version of this capsid II.

24.5 Prospects for the Future

Hypothesis-derived theory and practice can differ, possibly because of variables not in the theory. Also, hypotheses can be incorrect. The above proposals for treating neurodegenerative disease and cancerous tumors are not exempt from this pattern. Nonetheless, when basic considerations point in a simple direction, attempts to go in that direction should, it seems to us, be made. However, in the absence of a change in priorities, the probability is very close to zero that the above strategies will be tested experimentally within the next few years.

As support for this pessimism, exhibit A is the current status of phage therapy of infectious disease. For phage therapy, the basics point in a clear direction (details [17]). In addition, going in this direction is supported by (1) historical [6, 7, 73] and present-day [16] examples of dramatic success of doing that and (2) recent development of technologies that should dramatically improve results, if deployed. These include technologies of computerized database use/phage storage/phage retrieval, in addition to updated procedures of phage isolation and characterization. Yet, few, if any, systematic attempts at implementation are being made. Phage therapy is performed on an ad hoc basis [16, 75], which slows implementation, sometimes to the point that it is too late.

24.6 Relationship to Scientific Details

In Section 24.1.1, we gave examples of scientific output triggered by de facto past clinically oriented use of the basics-focus principle. The finding of a curative, low-virulence virus, as proposed above, should do the same in the case of neurodegenerative diseases. Existence of such a virus would likely trigger experiments that use the virus to determine the interaction of virus metabolism and assembly with cellular events. Cellular events would be better understood. One finding might be that the above assumptions need to be modified or abandoned. However, the overall direction embodied in the assumptions could still be accurate enough to obtain a cure.

A gated, uniform-size, uniform-structure DDV can also be used to track the progression of tumors via the loading of the DDV with a trackable compound. Nycodenz, for example, is x-ray opaque and can be tracked by use of micro-CT. Such tracking would be used both clinically and also to analyze pathways of receptor mediated endocytosis and lysosome targeting of endocytic vesicles.

In other words, the division between basic science and clinical practice is not sharp.

Abbreviations

cryo-EM: cryo-electron microscopy
DDV: drug delivery vehicle
EPR: enhanced permeability and retention
MPS: mononuclear phagocyte system

Disclosures and Conflict of Interest

This chapter was originally published as: Serwer, P., and Wright, E. T. (2018). Nanomedicine and phage capsids, *Viruses*, **10**, 307, doi:10.3390/v10060307, under the Creative Commons Attribution license (http://creativecommons.org/licenses/by/4.0/). It appears here, with edits and updates, by kind permission of the copyright holders.

Author Contributions: P.S. wrote the manuscript, designed the experiment for Fig. 24.3 and performed the electron microscopy; E.T.W. contributed procedural suggestions, performed the 2D-PAGE and prepared specimens for electron microscopy.

Acknowledgments: We thank Martin Adamo and Richard Ludueña for comments on drafts of this manuscript. This work was supported by the Welch Foundation (AQ-764) and the San Antonio Area Foundation.

Conflicts of Interest: The authors declare no conflict of interest. The founding sponsors had no role in the design of the study; in the collection, analyses, or interpretation of data; in the writing of the manuscript, and in the decision to publish the results.

Corresponding Author

Dr. Philip Serwer
Department of Biochemistry and Structural Biology
The University of Texas Health Science Center
7703 Floyd Curl Drive
San Antonio, Texas 78229-3900, USA
Email: serwer@uthscsa.edu

References

1. Riedel, S. (2005). Edward Jenner and the history of smallpox and vaccination. *Proc. (Baylor Univ. Med. Cent.)*, **18**, 21–25. Available at: https://www.ncbi.nlm.nih.gov/pmc/articles/PMC1200696/ (accessed on July 12, 2019).

2. Schwartz, M. (2001). The life and works of Louis Pasteur. *J. Appl. Microbiol.*, **91**, 597–601.

3. Tan, S. Y., Tatsumura, Y. (2015). Alexander Fleming (1881–1955): Discoverer of penicillin. *Singap. Med. J.*, **56**, 366–367.

4. Kong, K.-F., Schneper, L., Mathee, K. (2010). Beta-lactam antibiotics: From antibiosis to resistance and bacteriology. *APMIS*, **118**, 1–36.

5. Jelić, D., Antolović, R. (2016). From erythromycin to azithromycin and new potential ribosome-binding antimicrobials. *Antibiotics*, **5**, 29.

6. Summers, W. C. (1999). *Felix d'Herelle and the Origins of Molecular Biology*, Yale Univ. Press: New Haven, CT, USA.

7. Kutter, E., De Vos, D., Gvasalia, G., Alavidze, Z., Gogokhia, L., Kuhl, S., Abedon, S. T. (2010). Phage therapy in clinical practice: Treatment of human infections. *Curr. Pharm. Biotechnol.*, **11**, 69–86.

8. Dubos, R. J. (1976). *Louis Pasteur: Free Lance of Science*, Scribner: New York, NY, USA.

9. Yonath, A. (2005). Antibiotics targeting ribosomes: Resistance, selectivity, synergism, and cellular regulation. *Annu. Rev. Biochem.*, **74**, 649–679.

10. Graham, W. V., Bonito-Oliva, A., Sakmar, T. P. (2017). Update on Alzheimer's disease therapy and prevention strategies. *Annu. Rev. Med.*, **68**, 413–430.

11. Piemontese, L. (2017). New approaches for prevention and treatment of Alzheimer's disease: A fascinating challenge. *Neural. Regen. Res.*, **12**, 405–406.

12. Gitler, A. D., Dhillon, P., Shorter, J. (2017). Neurodegenerative disease: Models, mechanisms, and a new hope. *Dis. Models Mech.*, **10**, 499–502.

13. Santiago, J. A., Potashkin, J. A. (2014). A network approach to clinical intervention in neurodegenerative diseases. *Trends Mol. Med.*, **20**, 694–703.

14. Leaf, C. (2013). *The Truth in Small Doses*, Simon and Schuster: New York, NY, USA.

15. Rose, D. W. (2016). *Friends and Partners: The Legacy of Franklin D. Roosevelt and Basil O'Connor in the History of Polio*, Elsevier: Amsterdam, The Netherlands.

16. Schooley, R. T., Biswas, B., Gill, J. J., Hernandez-Morales, A., Lancaster, J., Lessor, L., Barr, J. J., Reed, S. L., Rohwer, F., Benler, S., et al. (2017). Development and use of personalized bacteriophage-based therapeutic cocktails to treat a patient with a disseminated resistant *Acinetobacter baumannii* infection. *Antimicrob. Agents Chemother.*, **61**, e00954-17.

17. Serwer, P. (2017). Restoring logic and data to phage-cures for infectious disease. *AIMS Microbiol.*, **3**, 706–712.

18. Sankar, A., Merril, C. R., Biswas, B. (2014). Therapeutic and prophylactic applications of bacteriophage components in modern medicine. *Cold Spring Harb. Perspect. Med.*, **4**, a012518.

19. Weber, L. (2017). Sewage saved this man's life. Someday it could save yours. Bacteriophages—viruses found in soil, water and human waste—may be the cure in a post—antibiotic world. *HuffPost*, US Edition. Available at: http://www.huffingtonpost.com/entry/antibioticresistant-superbugs-phage-therapy_us_5913414de4b05e1ca203f7d4 (accessed on July 12, 2019).

20. Edgar, R. S., Feynman, R. P., Klein, S., Lielausis, I., Steinberg, C. M. (1962). Mapping experiments with R mutants of bacteriophage T4D. *Genetics*, **47**, 179–186. Available at: https://www.ncbi.nlm.nih.gov/pmc/articles/PMC1210321/ (accessed on July 12, 2019).

21. Pauling, L., Corey, R. B. (1951). The pleated sheet, a new layer configuration of polypeptide chains. *Proc. Natl. Acad. Sci. USA*, **37**, 251–256.

22. Pauling, L., Corey, R. B. (1951). The structure of feather rachis keratin. *Proc. Natl. Acad. Sci. USA*, **37**, 256–261.

23. Pauling, L., Corey, R. B. (1951). Configurations of polypeptide chains with favored orientations around single bonds: Two new pleated sheets. *Proc. Natl. Acad. Sci. USA*, **37**, 729–740.

24. Guo, F., Liu, Z., Vago, F., Ren, Y., Wu, W., Wright, E. T., Serwer, P., Jiang, W. (2013). Visualization of uncorrelated, tandem symmetry mismatches in the internal genome packaging apparatus of bacteriophage T7. *Proc. Natl. Acad. Sci. USA*, **110**, 6811–6816.

25. Guo, F., Liu, Z., Fang, P.-A., Zhang, Q., Wright, E. T., Wu, W., Zhang, C., Vago, F., Ren, Y., Jakana, J., et al. (2014). Capsid expansion mechanism of bacteriophage T7 revealed by multistate atomic models derived from cryo-EM reconstructions. *Proc. Natl. Acad. Sci. USA*, **111**, E4606–E4614.

26. Serwer, P., Wright, E. T., Liu, Z., Jiang, W. (2014). Length quantization of DNA partially expelled from heads of a bacteriophage T3 mutant. *Virology*, **456–457**, 157–170.

27. Pajunen, M. I., Elizondo, M. R., Skurnik, M., Kieleczawa, J., Molineux, I. J. (2002). Complete nucleotide sequence and likely recombinatorial origin of bacteriophage T3. *J. Mol. Biol.*, **319**, 1115–1132.

28. Serwer, P., Pichler, M. E. (1978). Electrophoresis of bacteriophage T7 and T7 capsids in agarose gels. *J. Virol.*, **28**, 917–928.

29. Serwer, P., Hayes, S. J., Watson, R. H. (1982). The structure of a bacteriophage T7 procapsid and its *in vivo* conversion product probed by digestion with trypsin. *Virology*, **122**, 392–401.

30. Wikoff, W. R., Liljas, L., Duda, R. L., Tsuruta, H., Hendrix, R. W., Johnson, J. E. (2000). Topologically linked protein rings in the bacteriophage HK97 capsid. *Science*, **289**, 2129–2133.

31. Duda, R. L., Ross, P. D., Cheng, N., Firek, B. A., Hendrix, R. W., Conway, J. F., Steven, A. C. (2009). Structure and energetics of encapsidated DNA in bacteriophage HK97 studied by scanning calorimetry and cryoelectron microscopy. *J. Mol. Biol.*, **391**, 471–483.

32. Yu, G., Vago, F., Zhang, D., Snyder, J. E., Yan, R., Zhang, C., Benjamin, C., Jiang, X., Kuhn, R. J., Serwer, P., et al. (2014). Single-step antibody-based affinity cryo-electron microscopy for imaging and structural analysis of macromolecular assemblies. *J. Struct. Biol.*, **187**, 1–9.

33. Serwer, P., Wright, E. T., Demeler, B., Jiang, W. (2018). States of phage T3/T7 capsids: Buoyant density centrifugation and cryo-EM. *Biophys. Rev.*, **10**, 583–596.

34. Serwer, P., Wright, E. (2016). Testing a proposed paradigm shift in analysis of phage DNA packaging. *Bacteriophage*, **6**, e1268664.

35. Serwer, P., Wright, E. T. (2017). ATP-driven contraction of phage T3 capsids with DNA incompletely packaged *in vivo*. *Viruses*, **9**, 119.

36. Serwer, P. (2011). Proposed ancestors of phage nucleic acid packaging motors (and cells). *Viruses*, **3**, 1249–1280.

37. Milner-White, E. J., Russell, M. J. (2008). Predicting the conformations of peptides and proteins in early evolution. A review article submitted to Biology Direct. *Biol. Direct*, **3**, 3.

38. Milner-White, E. J., Russell, M. J. (2011). Functional capabilities of the earliest peptides and the emergence of life. *Genes*, **2**, 671–678.

39. Naryzhny, S. (2016). Towards the full realization of 2DE power. *Proteomes*, **4**, 33.

40. Stroud, R. M., Serwer, P., Ross, M. J. (1981). Assembly of bacteriophage t7. Dimensions of the bacteriophage and its capsids. *Biophys. J.*, **36**, 743–757.

41. Van der Linden, L., Wolthers, K. C., van Kuppeveld, F. J. M. (2015). Replication and inhibitors of enteroviruses and parechoviruses. *Viruses*, **7**, 4529–4562.

42. Taylor, P. J., Brown, R. H., Jr., Cleveland, D. W. (2016). Decoding ALS: From genes to mechanism. *Nature*, **539**, 197–206.

43. Van Dam, D., Vermeiren, Y., Dekker, A. D., Naudé, P. J. W., De Deyn, P. P. (2016). Neuropsychiatric disturbances in Alzheimer's disease: What have we learned from neuropathological studies? *Curr. Alzheimer Res.*, **13**, 1145–1164.

44. Bates, G. P., Dorsey, R., Gusella, J. F., Hayden, M. R., Kay, C., Leavitt, B. R., Nance, M., Ross, C. A., Scahill, R. I., Wetzel, R., et al. (2015). Huntington disease. *Nat. Rev. Dis. Primers*, **1**, 15005. Available at: https://www.nature.com/articles/nrdp20155 (accessed on July 12, 2019).

45. Kalia, L. V., Lang, A. E. (2016). Parkinson disease in 2015: Evolving basic, pathological and clinical concepts in PD. *Nat. Rev. Neurol.*, **12**, 65–66.

46. Collinge, J. (2016). Mammalian prions and their wider relevance in neurodegenerative diseases. *Nature*, **539**, 217–226.

47. Daggett, V. (2006). Alpha-sheet: The toxic conformer in amyloid diseases? *Acc. Chem. Res.*, **39**, 594–602.

48. Armen, R. S., Alonso, D. O. V., Daggett, V. (2004). Anatomy of an amyloidogenic intermediate: Conversion of β-sheet to α-sheet in transthyretin. *Structure*, **12**, 1847–1863.

49. Kellock, J., Hopping, G., Caughey, B., Daggett, V. (2016). Peptides composed of alternating L- and D-amino acids inhibit amyloidogenesis in three distinct amyloid systems independent of sequence. *J. Mol. Biol.*, **428**, 2317–2328.

50. Bandea, C. I. (2013). Aβ, tau, α-synuclein, huntingtin, TDP-43, PrP and AA are members of the innate immune system: A unifying hypothesis on the etiology of AD, PD, HD, ALS, CJD and RSA as innate immunity disorders. *bioRχiv.*

51. Serwer, P. (2018). Hypothesis for the cause and therapy of neurodegenerative diseases. *Med. Hypotheses*, **110**, 60–63.

52. Vijaya Kumar, D. K., Choi, S. H., Washicosky, K. J., Eimer, W. A., Tucker, S., Ghofrani, J., Lefkowitz, A., McColl, G., Goldstein, L. E., Tanzi, R. E., et al. (2016). Amyloid-β peptide protects against microbial infection in mouse and worm models of Alzheimer's disease. *Sci. Transl. Med.*, **8**, 340ra72.

53. White, M. R., Kandel, R., Hsieh, I.-N., De Luna, X., Hartshorn, K. L. (2018). Critical role of C-terminal residues of the Alzheimer's associated β-amyloid protein in mediating antiviral activity and modulating viral and bacterial interactions with neutrophils. *PLoS ONE*, **13**, e0194001.

54. Deleidi, M., Isacson, O. (2012). Viral and inflammatory triggers of neurodegenerative diseases. *Sci. Transl. Med.*, **4**, 121ps3.

55. Itzhaki, R. F., Lathe, R., Balin, B. J., Ball, M. J., Bearer, E. L., Braak, H., Bullido, M. J., Carter, C., Clerici, M., Cosby, S. L., et al. (2016). Microbes and Alzheimer's disease. *J. Alzheimers Dis.*, **51**, 979–984.

56. Hodyra-Stefaniak, K., Miernikiewicz, P., Drapała, J., Drab, M., Jończyk-Matysiak, E., Lecion, D., Kaz´ mierczak, Z., Beta, W., Majewska, J., Harhala, M., et al. (2015). Mammalian host-versus-phage immune response determines phage fate *in vivo*. *Sci. Rep.*, **5**, 14802.

57. Aksyuk, A. A., Rossmann, M. G. (2011). Bacteriophage assembly. *Viruses*, **3**, 172–203.

58. Fokine, A., Rossmann, R. G. (2014). Molecular architecture of tailed double-stranded DNA phages. *Bacteriophage*, **4**, e28281.

59. Hume, D. A. (2006). The mononuclear phagocyte system. *Curr. Opin. Immunol.*, **18**, 49–53.

60. Yona, S., Gordon, S. (2015). From the reticuloendothelial to mononuclear phagocyte system—The unaccounted years. *Front. Immunol.*, **6**, 328.

61. Merril, C. R., Biswas, B., Carlton, R., Jensen, N. C., Creed, G. J., Zullo, S., Adhya, S. (1996). Long-circulating bacteriophage as antibacterial agents. *Proc. Natl. Acad. Sci. USA*, **93**, 3188–3192.

62. Merril, C. R., Scholl, D., Adhya, S. L. (2003). The prospect for bacteriophage therapy in Western medicine. *Nat. Rev. Drug Discov.*, **2**, 489–497.

63. Serwer, P., Wright, E. T., Williams, T. L., Demeler, B., Lee, J. C. (2017). Phage-based therapies: What is the "real thing"? In *Proceedings of the XXV Biennial Conference on Phage/Virus Assembly*, Ellicott City, MD, USA, 20–25 August 2017.

64. Schaue, D., McBride, W. H. (2015). Opportunities and challenges of radiotherapy for treating cancer. *Nat. Rev. Clin. Oncol.*, **12**, 527–540.

65. Chabner, B. A., Roberts, T. G. (2005). Chemotherapy and the war on cancer. *Nat. Rev. Cancer*, **5**, 65–72.

66. Mehlen, P., Puisieux, A. (2006). Metastasis: A question of life or death. *Nat. Rev. Cancer*, **6**, 449–458.

67. Kottschade, L. A. (2018). Incidence and management of immune-related adverse events in patients undergoing treatment with immune checkpoint inhibitors. *Curr. Oncol. Rep.*, **20**, 24.

68. Kobayashi, H., Watanabe, R., Choyke, P. L. (2014). Improving conventional enhanced permeability and retention (EPR) effects, what is the appropriate target? *Theranostics*, **4**, 81–89.

69. Nakamura, H., Fang, J., Maeda, H. (2015). Development of next-generation macromolecular drugs based on the EPR effect: Challenges and pitfalls. *Expert Opin. Drug Deliv.*, **12**, 53–64.

70. Weissig, V., Pettinger, T. K., Murdock, N. (2014). Nanopharmaceuticals (Part 1): Products on the market. *Int. J. Nanomed.*, **9**, 4357–4373.

71. Barenholz, Y. (2012). Doxil®—The first FDA-approved nano-drug: Lessons learned. *J. Control. Release*, **160**, 117–134.

72. Anchordoquy, T. J., Barenholz, Y., Boraschi, D., Chorny, M., Decuzzi, P., Dobrovolskaia, M., Farhangrazi, Z. S., Farrell, D., Gabizon, A., Ghandehari, H., et al. (2017). Mechanisms and barriers in cancer nanomedicine: Addressing challenges, looking for solutions. *ACS Nano*, **11**, 12–18.

73. Summers, W. C. (2001). Bacteriophage therapy. *Annu. Rev. Microbiol.*, **55**, 437–451.

74. Serwer, P., Wright, E. T., Chang, J. T., Liu, X. (2014). Enhancing and initiating phage-based therapies. *Bacteriophage*, **4**, e961869.

75. Boodman, E. To save a young woman besieged by superbugs, scientists hunt a killer virus. Available at: https://www.statnews.com/2017/11/10/superbug-phage-mallory-smith/ (accessed on July 12, 2019).

Chapter 25

Pharmacogenomics: Ethical, Social, and Public Policy Issues

Yann Joly, PhD, and Denise Avard, PhD

Centre of Genomics and Policy, Faculty of Medicine, McGill University, Quebec, Canada

Keywords: pharmacogenomics, genomics, genetic testing, precision medicine, personalized medicine, regulation, bioethics, biotechnology, informed consent, research, professional liability, direct to consumer genetic tests, privacy protection, return of results, ethnic stratification, clinical trial

25.1   Introduction

Pharmacogenomics has been described as one of the most promising research areas to stem from the genetic revolution [28]. Proponents proclaim it has the potential to transform health care and pharmaceutical development, leading the way toward

The Road from Nanomedicine to Precision Medicine
Edited by Shaker A. Mousa, Raj Bawa, and Gerald F. Audette
Copyright © 2020 Jenny Stanford Publishing Pte. Ltd.
ISBN 978-981-4800-59-4 (Hardcover), 978-0-429-29501-0 (eBook)
www.jennystanford.com

a new era of personalized medicine whereby the right amount of the right kind of drug would be provided to the right patient. Pharmacogenomic stratification is becoming more frequent in clinical drug trials, and already a handful of pharmacogenomic tests have been approved by regulators and are now included as part of drug labels in many developed countries. Perhaps because of the necessity for researchers to convince industry and policy makers of the pharmaceutical potential of pharmacogenomics, this type of research has been described in a particularly favorable light in both scientific and lay publications. For example, in recent scientific publications, words such as "revolution," "rebirth," "renewal," "revitalization," "hope," "new age," "new century," "new era," "new paradigm," "transformation," and "illumination" have been used by authors to describe pharmacogenomics or its probable impact on medical practice [39]. Pharmacogenomic research has also often been described as the field of genomic research that will have a substantial impact on medical practice in a relatively short time span (usually stated as 5 to 10 years) [66]. Coupled with this attractive picture, the popular expression "personalized medicine" is often employed as a synonym of pharmacogenomics, conferring a stronger and more positive appeal to policy makers and the general population alike [23, 67].

However, issues linked to research ethics—a multidisciplinary field of inquiry interested in ensuring that research is conducted in a way that serves the needs of participants and society—have often been associated with genetic research. Could these issues impede the development of new pharmacogenomic drugs and diagnostic tests? Indeed, both pharmaceutical research and genomic research have generated their share of ethical concerns in the past. Since it integrates important aspects of both disciplines, pharmacogenomics could exacerbate well-known problems, as well as create completely novel ones.

This chapter will introduce the reader to the ethics of pharmacogenomic research. Following a general introduction to research ethics, we will present a critical overview of pharmaco-genomic research that penetrates the idyllic vision of this discipline pictured at times in both scientific and popular media. Genomic exceptionalism and its implications for pharmacogenomics will also be explored in this section. Following this discussion, we will present the main ethical issues raised by pharmacogenomics,

Definitions of Key Terms

Ancestry informative markers (AIMs): A subset of genetic markers (single-nucleotide polymorphisms [SNPs]) used to estimate the biogeographical origins of an individual and ascertain what proportion of ancestry is derived from each geographical region.

Anonymization: A process of irrevocably stripping data of direct identifiers so as to prevent subject reidentification. The risk of reidentification of individuals from remaining indirect identifiers is low or very low.

Broad consent: Consent to a wide (broadly delimited) range of options.

Casuistry: A branch of applied ethics casuistry that takes a practical approach to morality by examining cases, rather than theories, to resolve moral problems by applying general principles of morality to particular instances.

Consequentialism: A system of ethics that bases the morality of an action upon the consequences of the outcome.

Deontology: A system of ethics that judges the morality of an action on the basis of the action's adherence to a rule or rules. Deontology focuses on inputs, the actions and will of individuals, rather than outputs, and the goals achieved.

Genetic exceptionalism: A belief, popularized in the 1990s during the development of the Human Genome Project, that genetic information requires greater protection than other personal or health information because of its special, more sensitive nature.

Incidental findings: Unanticipated or unintended research findings or results concerning a participant that are outside the research objectives provided in the informed consent process.

Laboratory services (or lab-developed tests [LDTs]): Assays developed in a particular laboratory for internal use, or research use only, and therefore not distributed or sold to any other labs, health care facilities, or consumers.

Orphan populations: Populations that either have a genotype leading to a condition for which there is currently no effective therapy or that have been defined by the pharmaceutical industry as too small to be attractive as a drug market.

Personalized medicine: A medical model that customizes and tailor's health care decisions and practices to individual patients on the basis of genetic and environmental profiles.

Principlism: A system of ethics based on the four moral principles of autonomy, beneficence, nonmaleficence, and justice. Viewed as a practical and pluralistic approach to ethics that bridges various ethical, theological, and social approaches toward moral decision making, principlism operates by specifying how the principles are to be used in specific situations and then balancing the principles with other competing moral principles.

Specific consent: Specific and fully informed consent provided by research participants that is confined to one particular research purpose and time period.

Test kit (or in vitro diagnostic test kit): A pharmacogenomic or genetic testing medical device intended for use in the collection, preparation, and examination of specimens taken from the human body to diagnose a disease or other condition in order to cure, treat, or prevent such disease or condition. These test kits may be sold by manufacturers directly to consumers, testing laboratories, clinicians, or other approved recipients, depending on the device.

Virtue ethics: An ethical theory based on the intentions of an individual. As opposed to deontological or consequentialist ethics, which derives rightness or wrongness from the outcome of the act itself, virtue ethics focuses on the character of an individual as a driving force for ethical behavior.

Vulnerable population: Individuals or groups who have a greater probability than the population as a whole of being harmed, coerced, or exploited because of financial circumstances or place of residence; health, age, education, functional, or developmental status; the ability to communicate effectively; and personal characteristics, such as race, ethnicity, and sex.

including data privacy, direct-to-consumer (DTC) sale of pharmacogenomic tests, and race-based stratification. We hope that our chapter will provide readers with the methodological tools and academic interest to start their own personal ethical inquiry into this research field.

25.2 The Ethics of Pharmacogenomics

Ethics is a broad academic discipline interested in moral options in particular domains of inquiry [12]. Bioethics is the field of ethics concerned with the impact of advances in the biomedical sciences (e.g., stem cell research, gene therapy, pharmacogenomics, population genetics) on humanity and its environment. It is a constantly evolving multidisciplinary field of reflection closely related, yet complementary, to medical ethics [44]. In North American biotechnology research, a movement has emerged in favor of aggregating ethical reflection with other social science fields in order to enrich the debate. Hence in the United States, we now refer to the ethical, legal, and social issues (ELSI) of biotechnology research [103], while in Canada the adopted acronym is genomics and its ethical, economic, environmental, legal, and social aspects (GE$_3$ LS) [29].

Ethical reflection can be made at a theoretical level (e.g., what constitutes beneficial health innovation in a contemporary society?) or at a more applied level (e.g., should this particular pharmacogenomic project receive approbation from a local ethics committee to start recruiting research participants?). At the applied level, practitioners will usually resort to core ethical theories (principlism, consequentialism, deontological ethics, casuistry, and virtue ethics) and norms to guide them in their decision process. The rights-centered approach to bioethics based on principlism remains the most popular approach used in the ethical literature. It has been criticized by several authors, however, for simplifying and somewhat limiting the depth of ethical reflection [26].

Pharmacogenomics, situated at the crossroads of pharmacology and genomics, could be expected to, at minimum, generate similar ethical issues to those raised in both of these research fields. It also seems logical to assume that it may create a few novel issues of its own. Yet it is somewhat surprising to discover that pharmacogenomics has been treated in a relatively conservative, quiescent vein in the ethics literature [33].

Early pharmacogenomics proponents generally posited that the potential ethical issues would be less numerous and easier to handle in this context than in other types of genetic research (see Fig. 25.1 for an example) [83]. According to these proponents, because pharmacogenomics mainly focuses on the reactions to drugs of already-sick patients, rather than on a more controversial research topic such as disease genes, gene therapy, or population genetics, it should raise comparatively minor ethical issues [31].

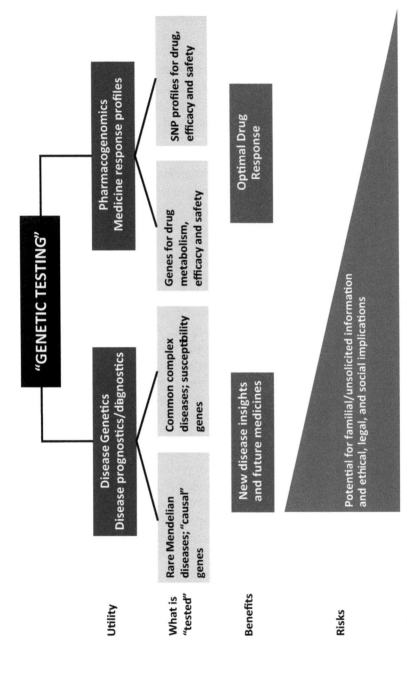

Figure 25.1 Early model for the ethics of pharmacogenomics (see Ref. [83]).

Could a somewhat complacent vision of pharmacogenomic research explain why there are only few ethical guidelines in existence that specifically address the ethical issues raised by pharmacogenomic research and development [81]? Recent research on pharmacogenomic ethics suggests that pharmacogenomics is not readily distinguishable from other types of biotechnology research and could raise similar ethical issues like informed consent, data privacy, genetic discrimination, DTC sale of genetic tests, and incidental findings [97].

Moreover, a case can be made for pharmacogenomics-specific concerns (see Fig. 25.2). Pharmacogenomics can raise ethical issues usually associated with pharmaceutical drug research and development (e.g., conflicts of interest, orphan drugs/orphan patient, access to medicine) and ethical issues more specifically associated with pharmacogenomic research, including genotyping and drug development based on racial and ethnic categories, the redefinition of the traditional model of drug development, and assessment of the cost and benefit of new pharmacogenomic tests and medicines.

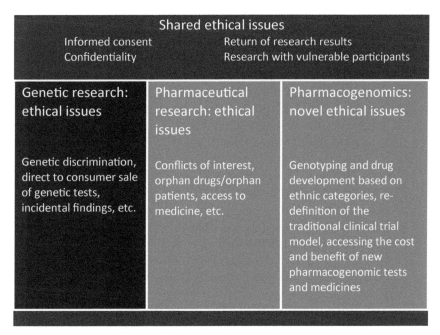

Figure 25.2 The ethics of pharmacogenomic research.

Lastly, the ongoing debate on the relevance of the concept of "genetic exceptionalism" [69] is also pertinent to the field of pharmacogenomic research. There is some consensus, though, that setting apart genetic data for special protective treatment can be both unpractical (e.g., because of the genetic nature of most medical data) and counterproductive (e.g., because treating genetic data differently from other data contributes to the marginalization of this type of information), making exceptionalism an imperfect solution at best [37].

25.2.1 Hopes and Accomplishments

As discussed by some authors [99], to conduct a proper ethical assessment of pharmacogenomics, it is necessary to reach beyond the rhetoric and hype that have been associated with this discipline to critically survey the nature and concrete realization of pharmacogenomics since it first made regular scientific headlines in the mid-1990s.

At the outset, it must be stated that "pharmacogenomics" does not equate to "personalized medicine." The latter expression refers to a much broader world than that of pharmacogenomic tests and drugs, including the general customization of health care, the tailoring of medical decisions, and the provision of recommendations and practices to individual patients [67]. Moreover, given that pharmacogenomics tends to propose treatment options applicable to genetically defined groups, it is more accurate to describe its methods as "group-tailored medicine" rather than as "personalized medicine" [89].

In addition, the cost of pharmacogenomic drugs and tests remains high and their availability—and utility—can be limited [94]. While this is not a surprise in such a nascent industry, it is a reason to remain level headed about the rate of progress and discovery. In the United States, where pharmacogenomics has been progressing most rapidly, pharmaceutical companies are routinely using pharmacogenomic techniques to identify new therapeutic markers. But the validation of these numerous markers is proving to be a challenging task. Pharmacogenomic tests are required, recommended, or mentioned for around 82 drugs currently in use. Most of these drugs (63%) are used for cancer treatment [80]. The first test using deoxyribonucleic acid (DNA) chips technology to determine the way patients will metabolize and respond to a variety of drugs received marketing approval in the United States in 2006 [9]. The number of pharmacogenomics-related drugs (such as Gleevec and Herceptin) remains small but growing, and their cost is high. Several companies, such as 23andMe and Navigenics, have started providing genetic testing services (including some pharmacogenetic tests) to consumers over the Internet. The clinical utility of many of these tests remains questionable [23, 82]. The biggest immediate challenge for pharmacogenomics is probably the magnitude of genes and polymorphisms involved in drug responses to most known drugs, as well as the substantial role played by environmental factors in drug responses [70, 71].

In light of this brief overview of the hopes and accomplishments of pharmacogenomic research, one can say that the clinical impact of pharmacogenomic research at the current time is modest but growing. It is difficult to determine with great precision if or when pharmacogenomics will have a major impact on the development of new medicines, clinical practice, and health care management. Thus, there remains great potential for this genetic discipline, but concrete realizations are emerging at a somewhat slower pace than was originally anticipated—or at least publically proclaimed—by the media [54, 57] and research community.

25.3 Pharmacogenomics: A Selective Review of Ethical Issues

This section presents a number of ethical issues currently confronted by pharmacogenomics. It is not meant as a comprehensive list but as a good illustration of the ethics of pharmacogenomics "here" and "now." Critical prospective analysis remains necessary to anticipate emerging issues.

25.3.1 Informed Consent

In pharmacogenomics, the breadth of permissible research for a given study, also known as the scope of consent, remains controversial [85]. At the root of the controversy is whether it is necessary to disclose to participants all the specific research objectives for which their samples will be used. Traditionally, both legal and ethical norms relating to consent in biomedical research require that, as part of the consent process, researchers disclose to participants the specific objectives for which their samples will be used [95]. However, this process of "specific consent" might be outdated in the modern era of large-scale biobanks and longitudinal studies. In this context, it is perceived by some critics as impractical, since samples have to be destroyed following completion of the specified research. Moreover, participants will have to be recontacted every time to give new samples, while they might have preferred being left alone by the research team [16].

Some argue that if there are no restrictions on the use of collected samples for future, as of yet, unspecified projects, pharmacogenomic research would proceed more efficiently and the value of samples would be maximized [24]. Moreover, the high costs of creating and maintaining a biobank should be considered. But shouldn't these important considerations be balanced with the ethical duty of respecting a research participant's autonomy by disclosing to him or her how his or her samples will be used in the research? Many policy documents and recommendations in pharmacogenomic research advocate a compromise between both interests. For example, regarding genetic studies, some authors suggest obtaining "broad consent" for a range of related studies over a defined period of time, with specific consent for research projects that could pose particular problematic issues to the research participant [81]. Another important area of concern is the issue of consent to pharmacogenomic "add on" studies. These studies are conducted in conjunction with clinical trials, but research subjects are not necessarily obligated to participate in them in order to be part of the clinical trial. There are two main types of add-on studies, (1) a specific add-on study, where a request is made to study specific genetic variants that are thought to be associated with the performance of the drug being investigated in the clinical trial, and (2) a general add-on study, where a request is made to store participants' DNA samples in a biobank for a certain period of time following the completion of the clinical trial for future, as of yet undefined research. These two types may also be combined in a clinical study, where the researcher desires to study specific genetic variants but also wants to store samples for a long term for future research [33].

Add-on studies are not troublesome per se; in fact, their addition to the main clinical trial could be a potential source of individual benefit for some participants (e.g., by discovering the genetic cause of certain adverse effects). This being said, for obvious reasons, fundamental research should still be clearly distinguished from clinical trials and medical care. Although pharmacogenomic add-on studies might become commonplace in the future, they are as of now still a relatively novel area of study, especially for participants who may be unfamiliar with medical research in general. Investigators and researchers must therefore ensure that participants are aware during the initial screening or recruitment process that these are generally optional studies that are distinct from the main clinical trial [33].

25.3.2 Confidentiality

Privacy protection is an ethical and legal issue often associated with genomic research. It is a particularly important issue for pharmacogenomics as there will be not only genomic information included in a participants' research file but also medical information and other types of personal data (e.g., lifestyle, familial health data). On the one hand, some industry researchers have argued that pharmacogenomic information is not particularly sensitive health information [88]. Scholars, on the other hand, counter that pharmacogenomic research, like other types of genetic research, can produce incidental findings (some of those related to susceptibility to disease, paternity, etc.) and that databases used in pharmacogenomics are vulnerable to third-party misuse (i.e., potential misuse by governmental law enforcement agencies, insurers, employers, and drug companies) [27, 72]. In pharmacogenomics, the privacy concerns are complex because private pharmaceutical companies often control sample collections. This raises concerns regarding the long-term governance of the samples. For example, what will happen to the samples if a private company becomes bankrupt or is sold [33, 100]? Another scenario to consider is the case where law enforcement officials, in the context of a criminal investigation, request access to the database.

Usually, pharmacogenomic researchers require access to a large number of samples from various geographical locations to validate new biomarkers and translate interesting preliminary findings into concrete clinical applications [96]. This entails sharing of samples and data between research projects, making it that much harder to adequately protect personal information. Sharing can be done by depositing the samples and data in a public or controlled access repository (e.g., the National Institutes of Health dbGaP and the Wellcome Trust Case Control Consortium) or through bilateral or multilateral agreements between researchers. Samples are usually anonymized or coded to protect the identity of research participants. However, studies by Lin et al. [58], Malin and Sweeney [61], Homer [76], and Elrich et al. [14], have shown the limits of sample anonymization to fully protect the identity of research participants and the possibility of reidentifying individuals participating in genetic research. Adding additional means of privacy protection, such as firewalls, access codes, confidentiality agreements, and oversight committees, will become increasingly important.

Privacy is a serious concern, but one must temper it with the reminder that the risk of identifying an individual and using their information for illicit purposes remains low. Very few people have the combination of skills, interest, information, equipment and ethical disregard to reidentify participants from their genetic data [59]. It is therefore not surprising that pharmacogenomic research confidentiality breaches resulting in adverse consequences for research participants remain rare. For example, results from research on genetic discrimination in the field of personal insurance have yet to convincingly demonstrate misappropriation or substantial use of genetic research data by insurers (with the exception of rare monogenic life-threatening diseases) [76]. Moreover, cryptography and anonymization techniques are constantly improving to better protect privacy. Nevertheless, the recognized limits of privacy protection mechanisms should be disclosed clearly to research participants on the informed consent form. The 2008 adoption of sample coding terminology standards by the International Conference on Harmonisation (ICH) should facilitate this necessary disclosure [34].

25.3.3 Clinical Trials and Genotyping

Pharmacogenomics is increasingly used to streamline clinical trials. By genotyping potential research participants and by only including those that, according to their genetic profile, are more likely to respond well to the experimental drug or to avoid the risk of adverse events, clinical trials should be safer and less expensive [80, 36]. In theory, this would lead to smaller and more targeted clinical trials and better protection of research participants. The disadvantage of using genetic information to select research participants in clinical trials is that the drug is no longer fully tested on a full spectrum of participants, representative of the general population, and therefore there is no guarantee that a genetically heterogeneous group can use the drug. Considering the widespread medical practice of off-label prescribing, the implications of genotyping clinical trial participants should be carefully reviewed [89]. Drug development strategies that incorporate pharma- cogenomic genotyping are based on the assumption that certain polymorphisms will be identified that can predict the response to a specific drug. This leads to the stratification of clinical trial participants into subgroups on the basis of genotype. Genotyping as either an inclusion or exclusion criteria to stratify research participants might lead to a loss of the benefits of research participation or to unfair representation in the clinical trial. In other words, this genotype-induced stratification could lead to subject selection biases [35]. Marketing considerations could also lead pharmaceutical companies to apply pharmacogenomic stratification to improve the attractiveness of a particular drug. For example, a company that has struggled to get a given drug onto hospital formularies and public systems reimbursement lists might carry out further trials on only a genotypically selected subpopulation of patients and use the results of that study to portray the given drug in a more positive light. Conversely, if clinical trials can be designed to accommodate smaller but better-defined populations, this could lead to an improved analysis of risk–benefit ratios [35].

Another challenge of genotyping is that small groups of participants with rare genetic profiles could find themselves excluded from the clinical trial process that would lead to the development of promising new drugs. The resulting drugs would not be authorized for use on this excluded "orphan population." However, it should be recognized that pharmacogenomics does not so much *create* orphan populations as *reveal* already existing ones to researchers and clinicians through the use of genotyping. Used in a scientific and ethical way, genotyping can avoid subjecting research participants to serious adverse effects for a drug that is unlikely to yield them any tangible benefits. Further, it is still too early to tell if there will be many people with orphan genotypes or if, as many pharmacogenomic researchers claim, this will remain a purely theoretical issue.

25.3.4 Race and Ethnic Stratification

It is important to note that although race and ethnicity are often used interchangeably in population research literature, the two terms describe related but distinct concepts. The term "race" refers to biological differences between groups that are assumed to have different biogeographical ancestries or genetic makeup [65]. On the other hand, ethnicity refers to an intricate multidimensional construct of various biological and geographical factors, as well as historical and social influences among populations that may or may not share a common genetic origin [65].

Race-genetics debates have subsisted for more than a century, and since the 1950s, stratification by ethnic origin has been common practice in pharmacogenetics [77]. Despite the longevity of the debates the issue is current and controversial. The story of BiDil is a telling example. In 2005, the US Food and Drug Administration (FDA) approved the sale of this antihypertensive drug produced by the company Nitromed. However, the approval was for use only in African Americans, as this was the only group on whom researchers tested the drug [45]. Although BiDil was shown to be a very beneficial drug for its target population, by exclusively testing BiDil in African American patients, the researchers did not demonstrate greater efficacy of the drug in this population than in other groups. In addition, the BiDil approval raised concerns over how pharmaceutical companies could deploy—or exploit—race as a marketing tool through the patent system [55, 77]. Although medical history is replete with the usage of race as a proxy in medical, research, diagnosis and treatments, this type of medical practice is controversial and has been refuted by part of the medical community [2]. Thus, it stands as a concern to see race frequently used as a variable to stratify participants in current pharmacogenomic research [55, 77].

One of the principal concerns lies with defining race in terms of pharmacogenomics and other omics research. Not only is this a complex endeavor, categorizing DNA samples by ethnic categories is also highly inconsistent. Divergent and often conflicting methodologies are used to define and interpret race in clinical practice and in academic disciplines such as genetic epidemiology, molecular biology, and biological anthropology. Often implicitly, biomedical researchers and clinicians invoke surrogate concepts to define and identify race, including skin color, hair type, national origin, geographic location, personal experience and citizenship. Further complicating matters is the common practice of relying on self-reports, which can be based on nonbiological factors [55]. Indeed, it is not uncommon for the same individual to report their racial identity differently in different contexts and at different points in their lives [9]. Scholars warn that self-identified race and ethnicity should not be used exclusively to predict a patient's genotype or drug response as it often lacks precision and scientific validity [8].

More recently, the term "ethnicity," which draws more on cultural, socioeconomic, religious, and political aspects, has surfaced as an alternate way of characterizing individual and population differences. Yet, ethnic definition is as fraught with difficulty as a racial definition when used as a basis for pharmacogenomic stratification. Both suffer from the twin problems of ambiguity and malleability. Because biological, cultural, political, religious, and socioeconomic systems change over time, it is axiomatic that racial and ethnic definitions change as well.

But since clinicians and researchers continue to use these imprecise concepts, it can seriously impact health outcome predictions [8]. In the case of genomics research, imprecise use of race and ethnicity can be misguiding in that it may validate the notion that discrete genetic groups exist, and may miscommunicate the complex relationships among social identity, ancestry, socioeconomics, and health [8]. Uncertainty can also lead to greater risk of potential harm to patients who are prescribed drugs on the basis of these determinants [76]. The use of race as a proxy can further have a considerable negative social impact on population groups. Even in the case of carefully and conservatively publicized results, the media and fringe, racist organizations could misrepresent or exaggerate the "racial" element of the research to fit personal agendas. Hence, use of racial and ethnic categories in pharmacogenomics could increase stigmatization and discrimination, while decreasing access to information, surveillance, and valuable treatment [13].

Despite the pitfalls associated with using race and ethnicity in genomic research, an increasing number of researchers demonstrate the significance of considering human genetic diversity (in the form of ethnicity or ancestry) for the success of personalized precision medicine [60]. Nevertheless, it must be noted that from a clinical genomic and population genomic point of view, race should be understood as a mere contextual, instrumental value [60]. For example, in the case of pharmacogenomics, the statistical value of race is a meagre 0.3% [46]. Nonetheless, the frequency of certain health-related phenotypes do indeed show variation by ethnicity, such as breast cancer [60]. This therefore begs the question of how researchers should use ethnicity to maximize possible benefits while ensuring minimal risk.

Furthermore, given the profit-oriented nature of the private pharmaceutical industry, could population groups in developing countries be vulnerable to inappropriate uses of ethnicity as a proxy for genotyping in clinical trials? For all its flaws, racial or ethnic stratification remains widely used in pharmacogenomic research today. This is because this type of stratification can simplify the drug development process and has proven to be an interesting proxy for genotyping in several pharmacogenomic studies [19]. Considering the time and cost involved in the drug development process and the oft-urgent need to bring new drugs to the market, wholly discarding race and ethnicity as a stratification marker at this stage could seriously inhibit valuable pharmacogenomic research.

However, as genome sequencing and SNP scans become increasingly inexpensive and as biotechnology evolves into the personal genome era, the use of racial categorization will hopefully become moot. Indeed, the use of genotyping to ascertain population stratification removes one of the primary analytic arguments for the utility of race and other social identities in genetic research [73]. In the meantime, it has been suggested that the scientific community develop and adopt consensus practices for the use of ethnicity and ancestry in determinants of health and medical care, study design, and interpretation of results [8]. It has also been suggested that researchers should use AIMs instead of relying on self-reports or arbitrarily defined physical criteria to assign racial or ethnic categories to research participants. Researchers could also describe and justify the group denomination used in the research protocol and consult with the population group that will be involved in the research [9, 65].

25.3.5   Return of Results

The return of research results has become an increasingly popular topic in the literature. The importance of reporting general research results (results that concern a group of participants) back to the scientific and lay communities is now well recognized. At the very least, researchers should provide general information on the progress of their research as a sign of gratitude and respect to the research participants who contributed to their project [50].

In the case of personal results associated directly with an identified individual, the appropriate course of action is more complex. The early guidelines on pharmacogenomic research tended to recommend nondisclosure because this type of research was judged too fundamental to generate findings that would be sufficiently robust or relevant to be of interest for research participants [75]. Communicating fundamental or non-validated results is generally associated with the following ethical issues: participants could misunderstand results, results could be an unjustified source of stress and anxiety, results could be refuted by future research, information could be misused by third parties, and participants could seek to have additional information on their results from their family doctor, thus unnecessarily burdening the health care system [7, 64].

However, the landscape has changed in the last 20 or so years, and it is now possible, through an analysis of the ethical guidelines and literature applicable to both genetic and pharmacogenomic research, to identify a general trend toward the reporting of results on the condition that they meet set criteria such as analytical validity, clinical validity, clinical utility, and an absence of outstanding ethical issues linked with the disclosure [43, 51]. Yet, it should be recognized that these criteria are expressed in a wide variety of terms and expressions, creating a certain degree of uncertainty as to the exact threshold required for disclosure.

The situation regarding the disclosure of incidental findings is even less clear than that of personal results. An incidental finding can be described as a research finding that has potential health or reproductive importance and is discovered in the course of conducting research but is beyond the aims of the study [68]. An example in pharmacogenomics would be a researcher studying the implication of certain polymorphisms on drug responses who finds out that these same polymorphisms are associated with a greater predisposition to arthritis. Until recently, very little was made of the question of incidental findings outside the context of clinical research. Incidental findings were, by default, treated similarly to research results. But with the advent of research fields like personal genomics (via high-throughput technologies such as genome-wide association studies or exome sequencing) [62] and neuroimaging, the quantity of medical data available to researchers has grown exponentially, putting the question of incidental findings at the forefront. There is still little consensus within the international community as to when and how incidental findings should be addressed by researchers [7]. The 2008 recommendations by Wolf et al. are an interesting first attempt to introduce a flexible grid comprising proposed options on how to handle different types of incidental findings [101].

In 2013, the ACMG published a "minimum list" of gene and gene variant incidental findings, subsequently updated in 2016, to be reported from clinical sequencing [21, 47]. The basis for this list is the consideration that certain findings may be of clinical utility, and if reported, would therefore likely have medical benefit for the patients and their families [21]. Additionally, the ACMG is considering the incorporation of pharmacogenomic variants onto this "minimum list", with a preliminary version already being developed [47]. The ACMG's attempt to define cases where there would be sufficient reasons to report incidental findings raises important questions as to researchers' professional and ethical obligations, and the possibility of legal liability in the case of unreported incidental findings [86]. Concerns were also raised in relation to uncertainties regarding accuracy of genotypic predictions, technological gaps in sequence coverage, and the potential harms of reporting false positives [86]. While such professional guidance can prove helpful, the ACMG approach has generally not been followed outside of the U.S. and remains controversial in this country as well [52]. In contrast to the proposition of the ACMG, EuroGentest and the European Society of Human Genetics recommend that the analysis of diagnostic laboratories should filter results to avoid the chances of incidental findings [61]. Ideally, more empirical data on return of incidental findings and the outcome of such initiatives should be carefully studied before implementing a systemic, variant specific, approach, such as that proposed by the ACMG.

Nevertheless, making a decision to return research results or incidental findings is only the first step in the construction of a more complete return of results framework that could also consider the following questions:

(1) Who should be returning the results?
(2) To whom should the results be returned?
(3) Is genetic counseling appropriate/necessary?
(4) How should the results be communicated?
(5) Should the results be included in the medical file?

Answers to these complex questions are slowly emerging from the literature and guidelines. It is becoming apparent that a comprehensive return of results framework should address these questions prospectively, at the outset of the research project. Research participants can then be offered from the start the option to receive available research results and incidental findings rather, than being presented with distressing news at an unexpected future moment in time.

25.3.6 Pediatric Research

It is consistently recognized that despite being considered a vulnerable population, children should be able to benefit from having access to research [5, 40]. Clinical research is needed to establish the value and validity of new pharmacogenomic tests and drugs in children. However, carrying out research in pediatric pharmacogenomics presents considerable challenges. Children are not miniature adults; neonates, infants, and adolescents are set apart by their rapid development and differ in their responses to drugs. This developmental issue demonstrates the complexity of associating genotypic and phenotypic information in pediatric populations [46].

Foremost on the list of ethical issues associated with pediatric pharmacogenomics could be the need to better define the acceptable risk level in the case of invasive sampling (e.g., a biopsy of an organ involved in drug metabolism). Pediatric pharmacogenomics also raises challenges that are methodological in nature that impact the informed consent and return-of-results processes. Informed consent in pediatric pharmacogenomics is complicated by, amongst other factors, the need to adapt complex pharmacogenomic information to a child's level of understanding. Moreover, very young children might not remember their participation in a pharmacogenomic research study after a few years. Even if they do, upon reaching adulthood, they might have a different opinion from that of their parent on the question of the return of results, making this issue particularly challenging to address prospectively in a pediatric population [40].

Children are often considered an orphan population from a drug development standpoint since so little research has been done on them historically [40]. Including children in pharmacogenomic drug trials may both positively and negatively influence their therapeutic orphan status. Pharmacogenomics could foster investment in orphan markets as it may reduce the cost of drug trials. In fact, it has been suggested that treatments that are not commercially viable at present may well become viable with reduced clinical trial costs through pharmacogenomics. However, it may also reveal specific genetic characteristics of nonresponders, thereby identifying new orphan groups for whom it may be even more economically disadvantageous to develop treatment. This raises the concern that already-orphaned children may be further stratified on the basis of genotype, making them even less desirable targets for the development of new treatments [17].

There is a statistics-related issue regarding whether there will be enough children with matching diseases and genotypes to conduct pediatric pharmacogenomic clinical trials. Pediatric clinical trials need to be performed with many healthy and sick research participants at different stages of development in order to have appropriate statistical power. Obtaining a sufficient sample size to generate robust study results could be difficult because of the difficulties of recruiting children as research participants [6].

25.3.7 Regulatory Approval Issues

This chapter has posited that the clinical impact of pharmacogenomics remains modest so far. Not unlike other "-omics" fields of research (genomics, proteomics, metabolomics), pharmacogenomic researchers have encountered difficulties in translating promising pharmacogenomic findings into valuable clinical applications. This situation is known as the "translation problem" in genomic research [49]. The transfer problem is likely the result of a combination of scientific, technical, and regulatory hurdles. In pharmacogenomics, one of the regulatory hurdles that have been identified relates to the regulatory approval framework for new tests [57].

The challenge for government and policy makers is to develop a regulatory framework sufficiently stringent to ensure that only tests that are sufficiently robust and useful are offered to patients and refunded by health care payers. At the same time, the entry of valuable new tests that could potentially save lives should not be delayed needlessly because of cumbersome requirements linked with regulatory approval. An ideal system would be simple, efficient, and inexpensive for researchers and drug companies yet sufficiently structured and thorough to ensure the safety of the users of the technology [42].

In Europe, Canada, and the United States, pharmacogenomic tests have to respond to different requirements, depending on whether they are sold as test kits to be used directly by the health care provider/consumer or as laboratory-offered services. Tests marketed as kits usually have to meet more stringent requirements, while in-lab developed tests are not scrutinized as closely if the laboratory that developed them met specified regulatory requirements. This difference in regulatory treatment could result in unnecessary qualitative dissimilarities and could also explain, in part, the low clinical utility of a number of pharmacogenomic tests sold directly to the consumer over the Internet [42, 48]. Regulatory requirements may also be partly responsible for differences in availability of new genetic tests between national jurisdictions in both developed and developing countries. For example, there are currently more pharmacogenomic tests approved and available to consumers in the United States than in most other countries [41]. Given the current weight, in terms of time and money, of obtaining regulatory approval, private companies are naturally focusing their energies on obtaining approval in countries possessing a combination of friendly regulatory approval mechanisms and attractive potential consumer markets. Jurisdictional harmonization of regulatory requirements and processes could alleviate this problem. It should be noted that this "availability problem" is not unique to pharmacogenomics and is also well known in the entire field of pharmaceutical development. Nevertheless, the high cost linked with the development of pharmacogenomic tests and drugs could substantially amplify this particular issue [42, 43].

On a positive note, policy makers in a number of countries are aware of the issues linked with the approval of new pharmaco-genomic tests and drugs and are currently in the process of refining their national systems [42]. The introduction of initiatives such as codevelopment policies aimed at simultaneously obtaining the approval of new combinations of tests and drugs and of encouraging signs for future regulatory coordination and technology transfer [42].

25.3.8 Professional Liability

The intersection of professional liability and pharmacogenomic research is currently in emergence. It is beginning to define new areas of responsibility and liability for various actors. Pharmacogenomic research has already generated a considerable amount of new knowledge on the way genetics can impact individual responses to drugs. This new medical knowledge, once sufficiently validated, will impact the duties and the professional liability of a number of stakeholders in the medication chain, namely, researchers, pharmaceutical companies, doctors, genetic counselors, nurses, and pharmacists [38]. The fundamental question—who is liable for what?—is beginning to be addressed by the literature and case law. Given the growing scientific knowledge, drug manufacturers could eventually be expected to know the genomic risks particular to the drug they are developing and marketing [84]. Thus, a manufacturer may face liability if clinical data was to show that certain genotypes are more susceptible to adverse side effects to a drug that is subsequently marketed without adequate genetic warnings [88]. For example, a class action suit involving the manufacturer of LYMErix, a Lyme disease vaccine, alleged that the drug manufacturer should have known the particular risks presented by the drug for only 30% of the population with a specific genotype. As is often the case with novel types of liability suits, the manufacturer agreed to a small settlement without acknowledging any liability and withdrew the vaccine from the market [11]. Researchers involved in clinical trials who do not use available and reliable pharmacogenetic information in clinical trials to reduce risks to research participants, by excluding those who are unlikely to respond, or for whom the drug would be harmful, could also eventually face liability. It could also become a legal duty in the future to disclose to both research participants and at-risk relatives clinical research results in pharmacogenomics that are associated with serious adverse effects [88]. These scenarios demonstrate why it is imperative that pharmacogenomics be integrated into clinical standards in a timely manner if these stakeholders hope to minimize professional liability.

Pharmacogenomics can also identify genetic markers that reduce the efficacy of a given drug in particular individuals. A number of product liability claims have been made against drug manufacturers based on the argument that the drug was ineffective due to genetic variation, and that the plaintiff was owed a "duty to warn" of such information [51]. In face of such claims, the courts predominantly concluded that drug manufacturers are only required to provide a warning in relation to any known potential risks associated with the drug, and that the duty to warn does not extend to the drug's efficacy [51, 91].

Physicians could see their professional duties expanded by the integration of pharmacogenomic drugs and tests into clinical practice. Pharmacogenomics may expand physicians' duty of care, that is, the legal obligation to improve and safeguard the health of a patient using reasonable means. A first pharmacogenomics-related liability suit against a family physician involving the death of child due to fluoxetine toxicity was discussed in the US journals *Forbes* and *Fortune* in 2000 [92, 93]. Although it was settled out of court in 2001 and little publicized, this case could be a harbinger of things to come for the medical profession. New or expanded duties for physicians could include a duty to offer pharmacogenetic testing prior to prescribing a new medication, a duty to prescribe in accordance with patients' pharmacogenetic test results, a duty to refer patients to genetic counselors in specific cases, and a duty to recontact at-risk patients and known relatives following a novel pharmacogenomic breakthrough [1, 6]. Doctors and pharmaceutical companies might also be expected to use validated pharmacogenomic tests to investigate the genomic source of drug side effects on patients, thus improving the pharmaco-vigilance process. Nurses, pharmacists, and genetic counselors could also see their traditional roles affected if a considerable quantity of pharmacogenomic tests and drugs was to become part of clinical practice [88]. It is however, difficult to clearly determine which new duties could befall each of these professionals at this point, given that the impact of pharmaco-genomics on clinical practice remains relatively modest.

25.3.9 Pharmacogenomic Tests Sold Directly to Consumers over the Internet

DTC genetic testing (DTC-GT) claims to have placed genomics directly into the hands of consumers by allowing patients to easily obtain information about their genetic makeup, sometimes without having to even see a doctor, genetic counselor, or pharmacist. The type of DTC genetic tests being offered to the public varies from testing for lifestyle factors, nutrigenomics, ancestry, rare diseases, and susceptibility for common chronic diseases and includes pharmacogenetic testing for drug responses [18].

Proponents of DTG-GT argue that the benefits of such services include providing a greater choice for consumers, empowering people to become more involved in making decisions about their medication, and at the same time avoiding charges of paternalistic notions of consumers [25].

However, the use of DTC-GT is controversial and could raise ethical, legal, and social issues such as false advertising and the risk of misleading patients/consumers about their health status; regulatory issues; and concerns about the possibility that the samples collected by the DTC company are also being used for research purposes [20, 30]. Furthermore, there are questions as to the actual preparedness of health professionals and the health care system to deal with patients who have DTC pharmacogenomic tests in hand [102]. These ethical, legal, and social issues will likely become more prominent (or urgent) as full-genome scans become widely available and affordable for consumers. In the case of pharmacogenomics, DTC problems may arise when a patient actually acts on the information without medical input. Pharmacogenomic research is still in its infancy and has only a limited number of clinical applications. This reality is often not disclosed to consumers by DTC-GT companies. Even when the pharmacogenomics-based tests will be adequately validated, the consumer may still misinterpret or misuse this information when deciding which medicines to purchase over the counter or independently adjust his or her drug dosage. For example, the pharmacogenetic testing for cytochrome P450 (CYP450) polymorphisms to assess how individuals metabolize an antidepressant drug is currently being offered by some DTC-GT companies to advise drug choice and dosage. Patients on antidepressants, depending on their test results, may change the dose of antidepressant medication after receiving information on, for instance, their CYP450 profile. In the absence of professional guidance, this practice is not without risks and might have adverse health outcomes [48].

The lack of regulatory guidance about what can be posted and sold on DTC-GT websites is concerning. This issue was propelled into the headlines in June 2010 when the FDA challenged the health information role of companies providing DTC-GT. The US Government Accountability Office (GAO) was asked to investigate DTC companies due to concerns that the advertisement of these companies has expanded rapidly despite the lack of evidence about the validity and clinical utility of genetic tests. This is all the more troubling, given the impact these tests could have on the medical decisions of consumers [40, 96].

Seemingly, progress has been made since. In October 2018, The FDA approved the 23andMe Personal Genome Service Pharmacogenetic Reports test as a direct-to-consumer test to detect 33 variants for multiple genes. Specifically, these genetic variants may be associated with a patient's ability to metabolize certain medications. The concerns expressed by the FDA in 2010 are evidently still salient. For example, the FDA's news release concerning the 23andMe test states very clearly that the test is not intended to provide information as to a patient's ability to respond to a given medication, and that health care providers should not use the test as an aid to make treatment decisions [15]. The FDA even indicated that any results from the 23andMe test should be confirmed with "independent pharmacogenetic testing before making any medical decisions" [15].

Furthermore, knowledge about how health professionals should use pharmacogenomics is still developing, and we do not have sufficient guidance about how health professionals could help patients make decisions based on information obtained from DTC-GT. The American College of Clinical Pharmacology (ACCP) aims to build awareness amongst its members to prepare them for the day a patient will walk in with a genome report card in hand to ask for medical advice. Their position statement recommends the following:

- Consult with knowledgeable and trusted colleagues and pay attention to the population at risk and the weight of the genetic factor.
- Appreciate that many professionals do not have expertise in genetics and pharmacology.
- Recognize the scientific and clinical limitations of each test and ensure that the genetic tests are validated and supported by evidence [3].

Finally, it has been noted that some DTC-GT companies have built "in-house" biobanks, in which they store all data they provide. This raises a number of ethical issues. There are concerns that consumers are not always aware that the data generated by the DTC company, to fulfill the terms of their service, may be used later for research purposes, or used for other pharmaceutical purposes. This contravenes the principle of informed consent, an essential requisite of participating in a research study. A commentary by Howard et al. notes that in the DTC-GT context there seems to be a blurred line between individuals as consumers and as research participants. It is felt that "requesting additional information could still be understood by consumers as an additional service that they purchased and not an explicit invitation to take part in the research" [32]. Also, there is some concern about the possibility that privacy will not be sufficiently protected, since, often, on company websites it is not made clear whether the data used for their research will be anonymized or not [32].

These social, ethical, and legal issues related to DTC-GT testing will need to be addressed to facilitate the eventual use of pharmacogenomic information and to protect the consumer.

25.4 Conclusion

Pharmacogenomic research is a promising discipline that could have a substantially positive impact on pharmaceutical development and population health in the future. Like research in most other biomedical fields, it raises its share of ethical, legal, and social issues. Without falling into the excesses of genetic exceptionalism, these issues should not be obscured in an attempt to circumvent ethical research and oversight. Instead, all stakeholders would greatly benefit from proactively engaging in a multidisciplinary reflection on the ethics of pharmacogenomics [78]. Not only will this type of proactive engagement promote trust in scientific research and in the pharmaceutical industry, it will also ensure that research is not derailed at the most inopportune moment by unexpected ethical developments. More importantly, stakeholders should be interested in the ethics of pharmacogenomics because it is beneficial for research participants and the population in general. Pharmaceutical companies, academic researchers, and ethics committees have a common goal in promoting safe and efficient research and development on promising new drugs that could one day save lives. Ensuring that research practices are ethically sound is an integral part of this process.

Beyond the research environment, stakeholders will also need to reflect on the impact that new pharmacogenomic drugs and tests will have on health care delivery once they enter clinical practice. This sort of foresight exercise will promote safe and effective use of new pharmacogenomic products and facilitate a fair distribution of the benefits of pharmacogenomic research.

This chapter aimed to assist the reader in taking the important first steps toward a reflective process on the ethics of pharmacogenomics. The ethical issues presented therein constitute a necessarily incomplete overview of the major ethical dilemmas raised by pharmacogenomics. Some issues will likely be easily resolved and disappear over time. Others may persist, and new ones will probably emerge within the next few years. Consequently, it is critical to continuously revisit the practices and impacts of this promising research discipline.

Abbreviations

ACCP: American College of Clinical Pharmacology

ACMG: American College of Medical Genetics and Genomics

AIMs: ancestry informative markers

CYP450: cytochrome P450

DNA: deoxyribonucleic acid

DTC: direct-to-consumer

DTC-GT: DTC genetic testing

ELSI: ethical, legal, and social issues

FDA: US Food and Drug Administration

GAO: US Government Accountability Office

GE$_3$ LS: genomics and its ethical, economic, environmental, legal, and social aspects

ICH: International Conference on Harmonisation

LDTs: lab-developed tests

SNPs: single-nucleotide polymorphisms

Disclosures and Conflict of Interest

This chapter is a completely revised and updated version of the original published as Joly, Y., Avard, D. (2014). Pharmacogenomics: Ethical, Legal, and Social Issues. In: Vizirianakis, I. S., ed. *Handbook of Personalized Medicine: Advances in Nanotechnology, Drug Delivery, and Therapy*, Pan Stanford Publishing Pte. Ltd., chapter 19, pp. 813–844, and appears here with kind permission of the publisher. The authors would like to acknowledge the editorial assistance of Francis Hemmings and Edward Dove and the financial support of the Canadian Pharmacogenomics Network for Drug Safety (Genome Canada/Genome BC) and the Montreal University Pharmacogenomics Centre. The authors gratefully acknowledge the editorial and research assistance of Miriam Pinkesz.

Corresponding Author

Dr. Yann Joly
Centre of Genomics and Policy
Faculty of Medicine, McGill University
740 Dr. Penfield Avenue, #5101
Montréal H3A0G1, Quebec, Canada
Email: yann.joly@ mcgill.ca

References

1. Abdul-Rahman (Zawati), M. H. (2010). La pharmacogénomique et la responsabilité civile des médecins. In: Duguet, A. M., ed. *Droit de la santé publique dans un contexte translational: IVe forum des jeunes chercheurs*, Les Études Hospitalières, Bordeaux Centre, chapter 9.

2. Acquaviva, K. D., Mintz, M. (2010). Perspective: Are we teaching racial profiling? The dangers of subjective determinations of race and ethnicity in case presentations. *Acad. Med.*, **85**(4), 702–705.

3. Ameer, B., Krivoy, N. (2009). Direct-to-consumer/patient advertising of genetic testing: A position statement of the American College of Clinical Pharmacology. *J. Clin. Pharmacol.*, **49**(8), 886–888.

4. Avard, D., Joly, Y. (2008). Improved understanding of genetic and genomic influences on drug disposition and action: Implications for children. *Pediatr. Drugs*, **10**(5), 275–278.

5. Avard, D., Silverstein, T., Sillon, G., Joly, Y. (2009). Researchers' perceptions of the ethical implications of pharmacogenomics research with children. *Public Health Genomics*, **12**(3), 191–201.

6. Bird, S. (2003). A GP's duty to follow up test results. *Aust. Fam. Physician*, **32**(1–2), 45–46.

7. Bredenoord, A. L., Kroes, H. Y., Cuppen, E., Parker, M., van Delden, J. J. M. (2011). Disclosure of individual genetic data to research participants: The debate reconsidered. *Trends Genet.*, **27**(2), 41–47.

8. Bonham, V. L., Green, E. D., Pérez-Stable, E. J. (2018). Examining how race, ethnicity, and ancestry data are used in biomedical research. *JAMA*, **320**(15), 1533–1534.

9. Caulfield, T., et al. (2009). Race and ancestry in biomedical research: Exploring the challenges. *Genome Med.*, **1**, 1–8.

10. De Leon, J. (2006). AmpliChip CYP450 test: Personalized medicine has arrived in psychiatry. *Expert Rev. Mol. Genet.*, **6**(3), 277–286.

11. Donovan, M. J. (2010). Legal issues stemming from the advancement of pharmacogenomics. *UCLA J. Law Technol.*, **14**(1), 31–65.

12. Driver, J. (2007). *Ethics the Fundamentals*, Blackwell Publishing Ltd, Malden.

13. Égalité, N., Özdemir, V., Godard, B. (2007). Pharmacogenomics research involving racial classification: Qualitative research findings on researchers' views, perceptions and attitudes towards socioethical responsibilities. *Pharmacogenomics*, **8**(9), 1115–1126.

14. Elrich, Y., Shor, T., Pe'er, I., Carmi, S. (2018) Identity inference of genomic data using long-range familial searches. *Science,* **362**(6415), 690–94.

15. FDA (2018). FDA authorizes first direct-to-consumer test for detecting genetic variants that may be associated with medication metabolism. Available at: https://www.fda.gov/NewsEvents/ Newsroom/PressAnnouncements/ucm624753.htm (accessed on February 6, 2019).

16. Fond de la recherche en santé Québec (2006). Final report, Advisory Group on a Governance Framework for Data Banks and Biobanks used for Health Research. Available at: https://www.bibliotheque. assnat.qc.ca/DepotNumerique_v2/AffichageFichier.aspx?idf=67934 (accessed on February 14, 2019).

17. Freund, C., Clayton, E. W. (2003). Pharmacogenomics and children: Meeting the ethical challenges. *Am. J. Pharmaco. Genomics*, **3**(6), 399–404.

18. Goddard, K. A. B., et al. (2009). Health-related direct-to-consumer genetic tests: A public health assessment and analysis of practices related to Internet-based tests for risk of thrombosis. *Public Health Genomics*, **12**(2), 92–104.

19. Gonzalez Burchard E., et al. (2003). The importance of race and ethnic background in biomedical research and clinical practice. *N. Engl. J. Med.*, **348**(12), 1170–1175.

20. Gray, S., Olopade, O. I. (2003). Direct-to-consumer marketing of genetic tests for cancer: Buyer beware. *J. Clin. Oncol.*, **21**(17), 3191–3193.

21. Green, R. C., et al. (2013). ACMG recommendations for reporting of incidental findings in clinical exome and genome sequencing. *Genet. Med.*, **15**(7), 565–574.

22. Gurwitz, D., Bregman-Eschet, Y. (2009). Personal genomics services: Whose genomes? *Eur. J. Hum. Genet.*, **17**(7), 883–889.

23. Haga, S. B., Warner, L. R., O'Daniel, J. (2009). The potential of a placebo/ nocebo effect in pharmacogenetics. *Public Health Genomics*, **12**(3), 158–162.

24. Hansson, M. G. (2009). Ethics and biobanks. *British J. Cancer*, **100**(1), 8–12.

25. Harvey, A. (2010). Genetic risks and healthy choices: Creating citizen-consumers of genetic services through empowerment and facilitation. *Sociol. Health Illn.*, **32**(3), 365–381.

26. Have, H. T., Gordijn, B. eds (2001). *Bioethics in a European Perspective*, Kluwer, Netherlands.

27. Henrikson, N. B., Burke, W., Veenstra, D. L. (2008). Ancillary risk information and pharmacogenetic tests: Social and policy implications. *Pharmacogenomics J.*, **8**(2), 85–89.

28. Hines, R. N., McCarver, D. G. (2006). Pharmacogenomics and the future of drug therapy. *Pediatric Clin. N. Am.*, **53**(4), 591–619.

29. Hogarth, S. (2008). GE3LS capacity issues: An environmental scan for CIHR Institute of Genetics. Available at: https://www. genomecanada.ca/sites/default/files/pdf/en/GELS_environmental-final-reportJan08.pdf (accessed on February 14, 2019).

30. Hogarth, S., Javitt, G., Melzer, D. (2008). The current landscape for direct-to-consumer genetic testing: Legal, ethical, and policy issues. *Ann. Rev. Genomics Hum. Genet.*, **9**, 161–182.

31. Hood, E. (2003). Pharmacogenomics the promise of personalized medicine. *Environ. Health Perspect.*, **111**(11), A581–A589.

32. Howard, H., Knoppers, B. M., Borry, P. (2010). Blurring lines. The research activities of direct-to-consumer genetic testing companies raise questions about consumers as research subjects. *EMBO Rep.*, **11**(8), 579–582.

33. Howard, H. C., Joly, Y., Avard, D., Laplante, N., Phillips, M., Tardif, J. C. (2011). Informed consent in the context of pharmacogenomic 2 research: Ethical considerations. *Pharmacogenetics J.*, **11**(3), 155–161.

34. International Conference on Harmonisation (2007). International Conference on Harmonisation of Technical Requirements for Registration of Pharmaceuticals for Human Use. Available at: http:// www.ich.org/fileadmin/Public_Web_Site/ICH_Products/Guidelines/ Efficacy/E15/Step4/E15_Guideline.pdf (accessed on February 14, 2019).

35. Issa, A. (2000). Ethical considerations in clinical pharmacogenomics research. *Trends Pharmacol. Sci.*, **21**(7), 247–249.

36. Issa, A. M., Thorogood, A., Joly, Y., Knoppers, B. M. (2018) Accelerating evidence gathering and approval of precision medicine therapies: The FDA takes aim at rare mutations. *Genet. Medicine*, 1–3.

37. Joly, Y. (2004). La pharmacogénomique: Perspectives et enjeux éthico-juridiques. *Lex Electronica*, **9**(3), 1–19.

38. Joly, Y. (2006). Life insurers' access to genetic information: A way out of the stalemate. *Health Law Rev.*, **14**(3), 14–21.

39. Joly, Y. (2005). Biotechnologies et brevets: Le cas de la pharmaco-génomique. *Lex Electronica*, **10**(2), 1–134.

40. Joly, Y., Sillon, G., Silverstein, T., Krajinovic, M., Avard, D. (2008). Pharmacogenomics: Don't forget the children. *Curr. Pharm. Pers. Med.*, **6**(2), 77–84.

41. Joly, Y., Ramos-Paque, E. (2010). Approval of new pharmacogenomic tests: Is the Canadian regulatory process adequate? *Can. J. Law Technol.*, **8**(2), 215–241.

42. Joly, Y., et al. (2011a). Regulatory approval for new pharmaco-genomics tests: A comparative overview. *Food Drug Law J.*, **66**(1), 1–24.

43. Joly, Y., et al. (2011b). Diagnostic testing for vaccinomics: Is the regulatory approval framework adequate? *OMICS*, **15**(9), 597–605.

44. Jonsen, A. R. (2007). A history of bioethics as discipline and discourse. In: Jecker, N. S., Jonsen, A. R., Pearlman, R. A., eds. *Bioethics: An Introduction to the History, Methods and Practices*, Jones and Bartlett Publishers, Sudbury, chapter 1.

45. Kahn, J. (2008). Exploiting race in drug development: BiDil's interim model of pharmacogenomics. *Social Stud. Sci.*, **38**(5), 737–758.

46. Kahn, J. (2013). *Race in a bottle: The Story of BiDil and Racialized Medicine in a Post-Genomic Age*, Columbia University Press, New York, NY.

47. Kalia, S. S., et al. (2017). Recommendations for reporting of secondary findings in clinical exome and genome sequencing, 2016 Update (ACMG SF V2.0): A policy statement of the American College of Medical Genetics and Genomics. *Genet. Med.*, **19**(2), 249–255.

48. Katsanis, S. H., Javitt, G., Hudson, K. (2008). A case study of personalized medicine. *Science*, **320**(5872), 53–54.

49. Khoury, M. J., et al. (2007). The continuum of translation research in genomic medicine: How can we accelerate the appropriate

integration of human genome discoveries into healthcare and disease prevention. *Genet. Med.*, **9**(10), 665–674.

50. Knoppers, B. M., Joly, Y., Simard, J., Durocher, F. (2006). The emergence of an ethical duty to disclose genetic research results: International perspectives. *Eur. J. Hum. Genet.*, **14**(11), 1170–1178.

51. Knoppers, B. M., Zawati, M. H., Seìneìcal, K. (2015). Return of genetic testing results in the era of whole-genome sequencing. *Nat. Rev. Genet.*, **16**(9), 553–559.

52. Knoppers, B. M., Nguyen M. T., Sénécal K., Tassé A. M., Zawati M. H. (2016). Next-generation sequencing and the return of results. *Cold Spring Harb. Perspect. Med.*, **6**(10), 1–10.

53. LaBarre v. Bristol Myers Squibb Co., 544 F. Appx. 120, 125 (3d Cir. 2013).

54. Langreth, R., Waldholz, M. (1999). New era of personalized medicine. *Wall Street J.*, April 16, 1999.

55. Lee, S. S.-J. (2007). The ethical implications of stratifying by race in pharmacogenomics. *Clin. Pharmacol. Ther.*, **81**(1), 122–125.

56. Lemonick, M. D., Cray, D., Park, A., Thomas, C. B., Thompson, D. (2001). Brave new pharmacy, *Time*, Jan. 15, 2001.

57. Lesko, L. J., Woodcock, J. (2004). Translation of pharmacogenomics and pharmacogenetics: A regulatory perspective. *Nat. Rev. Drug Discov.*, **3**(9), 763–769.

58. Lin, Z., Owen, A. B., Altman, R. B. (2004). Genomic research and human subject privacy. *Science*, **305**(5681), 183.

59. Lowrance, W. W., Collin, F. S. (2007). Identifiability in genomic research. *Science*, **317**(5838), 600–602.

60. Maglo, K. N., Mersha, T.B., Martin, L.J. (2016) Population genomics and the statistical values of race: An interdisciplinary perspective on the biological classification of human populations and implications for clinical genetic epidemiological research. *Front. Genet.*, **7**(22), 1–13.

61. Malin, B., Sweeney, L. (2004). How (not) to protect genomic data privacy in a distributed network: Using trail re-identification to evaluate and design anonymity protection systems. *J. Biomed. Inf.*, **37**(3), 179.

62. Matthijs, G., Souche, E., Alders, M., Corveleyn, A., Eck, S., Feenstra, I., Race, V., Sistermans, E., Sturm, M., Weiss, M., Yntema, H., Bakker, E., Scheffer, H., Bauer, P., EuroGentest; European Society of Human Genetics

(2015). Guidelines for diagnostic next-generation sequencing. *Eur. J. Hum. Genet.*, **24**(1), 2–5.

63. McGuire, A. L., Caulfield, T., Cho, M. K. (2008). Research ethics and the challenge of whole-genome sequencing. *Nat. Rev. Genet.*, **9**(2), 152–156.

64. McGuire, A. L., James, R., Lupski, J. R. (2010). Personal genome research: What should the participant be told? *Trends Genet.*, **26**(5), 199–201.

65. Mersha, T., Abebe, T. (2015) Self-reported race/ethnicity in the age of genomic research: Its potential impact on understanding health disparities. *Hum. Genomics*, **9**(1), 1–15.

66. Middleton, E. G., Freeman, L., Brewster, A. S., Foster, C., Roses, A. (2000). From gene-specific tests to pharmacogenetics. *Community Genet.*, **3**(4), 198–203.

67. Millenson, M. L. (2006). The promise of personalized medicine: A conversation with Michael Svinte. *Health Affairs*, **25**(2), w54–w60.

68. Miller, F. G., Mello, M. M., Joffe, S. (2008). Incidental findings in human subjects research: What do investigators owe research participants? *J. Law Med. Ethics*, **36**(2), 271–279.

69. Murray, T. H. (1997). Genetic exceptionalism and 'future diaries': Is genetic information different from other medical information. In: Rothstein, M. A., ed. *Genetic Secrets: Protecting Privacy and Confidentiality in the Genetic Era*, Yale University Press, New Haven, chapter 3.

70. Nebert, D. W., Jorge-Nebert, L., Vesell, E. S. (2003). Pharmaco-genomics and "individualized drug therapy": High expectations and disappointing achievements. *Am. J. Pharm.*, **3**(6), 361–370.

71. Nebert, D. W., Zhang, G., Vesell, E. S. (2008). From human genetics and genomics to pharmacogenetics and pharmacogenomics: Past lessons, future directions. *Drug Metab. Rev.*, **40**(2), 187–224.

72. Netzer, C., Biller-Andorno, N. (2004). Pharmacogenetic testing, informed consent and the problem of secondary information. *Bioethics*, **18**(4), 344–360.

73. Ng, P. C., Zhao, Q., Levy, S., Strausberg, R. L., Venter, J. C. (2008). Individual genomes instead of race for personalized medicine. *Clin. Pharmacol. Ther.*, **84**(3), 306–309.

74. Nils, H., et al. (2008). Resolving individuals contributing trace amounts of DNA to highly complex mixtures using high-density SNP genotyping microarrays. *PLoS Genet.*, **4**(8), e1000167.

75. Nuffield Council on Bioethics (2003). Pharmacogenetics: Ethical issues, Nuffield Council on Bioethics, London. Available at: http://nuffieldbioethics.org/wpcontent/uploads/2014/07/Pharmacogenetics-Report.pdf (accessed on February 15, 2019).

76. Otlowski, M. F. A., Stranger, M. J. A., Taylor, S., Barlow-Stewart, K.,Treloar, S. (2007). The use of legal remedies in Australia for pursuing allegations of genetic discrimination: Findings of an empirical study. *Int. J. Discriminat. Law*, **9**(1), 3–35.

77. Ozdemir, V., Graham, J. E., Godard, B. (2008). Race as a variable in pharmacogenomics science: From empirical ethics to publication standards. *Pharmacogenet. Genom.*, **18**(10), 837–841.

78. Özdemir, V., Joly, Y., Kirby, E., Arvard, D., Knoppers, B. M. (2013). Beyond ELSI: Where to from here? From "regulating" to anticipating and shaping the innovation trajectory in personalized medicine. In: Lam, Y. M. F., Cavallari, L. H., eds. *Pharmacogenomics: Challenges and Opportunities in Therapeutic Implementation*, Elsevier, Amsterdam, chapter 11.

79. Peterson-Iyer, K. (2008). Pharmacogenomics, ethics, and public policy. *Kennedy Inst. Ethics J.*, **18**(1), 35–56.

80. Pharmacogenomics Knowledge Base Drug label annotations. Available at: https://www.pharmgkb.org/labelAnnotations (accessed on February 15, 2019).

81. Phillips, M. S., Joly, Y., Silverstein, T., Avard, D. (2007). Consent in pharmacogenomic research. *GenEdit*, **V-2**, 1–10.

82. Prainsack, B., Reardon, J., Hindmarsh, R., Gottweis, H., Naue, U., Lunshof, J. E. (2008). Personal genomes: Misdirected precaution. *Nature*, **456**(7218), 34–35.

83. Roses, A. D. (2000). Pharmacogenetics and the practice of medicine. *Nature*, **405**(6788), 857–865.

84. Rothstein, M. A. (2005). Liability issues in pharmacogenomics. *Louisiana Law Rev.*, **66**(5), 117–124.

85. Salvaterra, E., et al. (2008). Banking together. *EMBO Rep.*, **9**(4), 307–313.

86. Roche, M. I., Berg, J. S. (2015). Incidental findings with genomic testing: Implications for genetic counseling practice. *Curr. Genet. Med. Rep.*, **3**(4), 166–176.

87. Sass, H. M. (2004). Bioethical issues in genetic screening and patient information. *T. Klin. J. Med. Ethics Law Hist.*, **12**(1), 1–9.

88. Sillon, S., Joly, Y., Feldman, S., Avard, D. (2008). An ethical and legal overview of pharmacogenomics: Perspectives and issues. *Med. Law*, **27**(4), 843.

89. Smart, A., Martin, P., Parker, M. (2004). Tailored medicine: Whom will it fit? The ethics of patient and disease stratification. *Bioethics*, **18**(4), 1467.

90. Smith, C. (2003). Drug target validation: Hitting the target. *Nature*, **422**(6929), 341–347.

91. Solomon V. Bristol Myers Squibb Co., 916 F. Supp.2d 556 (D.N.J. 2013).

92. Stipp, D. (2000a). A DNA tragedy. *Forbes Mag.*, October 30, 2000.

93. Stipp, D. (2000b). A DNA tragedy. *Fortune Mag.*, October 30, 2000.

94. Tucker, L. (2008). Pharmacogenomics: A primer for policymakers, National Health Policy Forum, Washington, D.C. Available at: https://www.nhpf.org/library/background-papers/BP_Pharmacogenomics_01-28-08.pdf (accessed on December 14, 2018).

95. US Government (1949). Nuremberg Code, *Trials of War Criminals before the Nuremberg Military Tribunals under Control Council Law No. 10*, U.S. Government Printing Office, Washington, D.C., vol. 2, pp. 181–182.

96. US Government Accountability Office (2010). Direct-to-consumer genetic tests: Misleading test results are further complicated by deceptive marketing and other questionable practices. Available at: www.gao.gov/new.items/d10847t.pdf (accessed on December 14, 2018).

97. Van Delden, J., Bolt, I., Kalis, A. M., Derijks, J., Leufkens, H. (2004). Tailormade pharmacotherapy: Future developments and ethical challenges in the field of pharmacogenomics. *Bioethics*, **18**(4), 303–321.

98. Warner, A., et al. (2011). Challenges in obtaining adequate genetic sample sets in clinical trials: The perspective of the industry pharmacogenomics working group. *Clin. Pharmacol. Ther.*, **89**(4), 529–534.

99. Williams-Jones, B., Corrigan, O. P. (2003). Rhetoric and hype: Where's the "ethics." *Pharmacogenomics*, 3(6), 375–383.

100. Winickoff, D. E., Winickoff, R. N. (2003). The charitable trust as a model for genomic biobanks. *N. Engl. J. Med.*, **349**(12), 1180–1184.

101. Wolf, S. M., et al. (2008). Managing incidental findings in human subjects research: Analysis and recommendations. *J. Law Med. Ethics*, **36**(2), 219–248.

102. Wu, A. H. B., Babic, N., Yeo, K. J. (2009). Implementation of pharmacogenomics into the clinical practice of therapeutics: Issues for the clinician and the laboratorian. *Per. Med.*, **6**(3), 315–327.

103. Yesley, M. (2008). What's ELSI got to do with it? Bioethics and the human genome project. *New Genetics Soc.*, **27**(1), 1–6.

Chapter 26

From Single Level Analysis to Multi-Omics Integrative Approaches: A Powerful Strategy towards the Precision Oncology

Maria Eugenia Gallo Cantafio, PhD,[a,*] Katia Grillone, PhD,[a,*] Daniele Caracciolo, MD,[a] Francesca Scionti, PhD,[a] Mariamena Arbitrio, MD,[b] Vito Barbieri, MD,[c] Licia Pensabene, MD,[d] Pietro Hiram Guzzi, PhD,[e] and Maria Teresa Di Martino, PhD[a]

[a]*Department of Experimental and Clinical Medicine, Magna Graecia University, Salvatore Venuta University Campus, Catanzaro, Italy*
[b]*CNR-Institute of Neurological Sciences, UOS of Pharmacology, Catanzaro, Italy*
[c]*Medical Oncology Unit, Mater Domini Hospital, Salvatore Venuta University Campus, Catanzaro, Italy*
[d]*Department of Medical and Surgical Sciences Pediatric Unit, Magna Graecia University, Catanzaro, Italy*
[e]*Department of Medical and Surgical Sciences, Magna Graecia University, Catanzaro, Italy*

Keywords: multi-omics data, biomarkers, precision medicine, integromics, cancer, genomics, epigenomics, transcriptomics, proteomics, metabolomics, tailored therapy, personalized medicine, integrative data, molecular profile, high-throughput platform, single nucleotide polymorphism, copy number variation, next generation sequencing, microarray

---

*These authors contributed equally to this work.

The Road from Nanomedicine to Precision Medicine
Edited by Shaker A. Mousa, Raj Bawa, and Gerald F. Audette
Copyright © 2020 Jenny Stanford Publishing Pte. Ltd.
ISBN 978-981-4800-59-4 (Hardcover), 978-0-429-29501-0 (eBook)
www.jennystanford.com

26.1   Introduction

Omics technologies offer a new view of biological function and organization at level of different molecular systems. High-throughput studies generate genomic and transcriptomic data which finally lead to dissection of the cancer molecular profile. However, because human genomes are complex and regulated at multiple levels, a new challenge could be the integration of information coming from different biological layers. The integration of "omics" including genomics, epigenomics, transcriptomics, proteomics, and metabolomics, into physiological and clinical studies will provide new clues on the mechanism of tumor initiation, progression, and metastatic spread, as well as the discovering of novel targets for therapeutic intervention. Personalized medicine aims to ensure customized healthcare, which proposes disease prevention, medical decision, and tailored therapy for each condition/patient, taking information from integrative studies. However, omics integration is still in its infancy. Currently, single-omics analyses on cancer patient samples have provided valuable data available to the scientific community, mostly in the field of genomics, although matched clinical annotation is still limited. In this review, we first discuss general concepts on the generation of omics data, along with currently available strategies for multi-omics data analysis and integration in the context of clinical oncology. Finally, we discuss emerging challenges to translate information arising from this new knowledge into personalized anticancer therapies.

26.2   Omics Data Production

In the last decades, the improvement or development of new omics technologies notably ameliorated personalized medicine, in the prevention or treatment settings, by providing a broad range of information from genetic to metabolic tumor-specific features. Here, we will describe several exemplary studies reporting various applications of the main technologies, summarized in Table 26.1, adopted in cancer research to profile each biological layer.

Table 26.1 Summary of the technologies which have provided relevant information in cancer research to investigate different aspects of each biological level and are described in this chapter

Data type	Main platforms		Applications
Genomic	Microarray	Array-CGH SNP-array Array-CGH + SNP	Identification of CNVs Identification of CNVs, copy neutral of LOH, SNPs genotyping in defined sequences
	DNA-seq	WES WGS Targeted exon-seq	Identification of DNA mutations and CNVs
Epigenomic	Affinity enrichment-based methods	MeDip-Seq MBD-Seq	DNA-methylation profiling
	Bisulfite conversion-based methods	BS-Seq OxBS-Seq	
	Capture-based methods		
	Restriction enzymes-based methods		
	ChIP-Seq		Identification of chromatin-associated proteins
	MNase-Seq ATAC-Seq DNase II-Sseq		Investigation of chromatin accessibility
	4C-Seq HiC-Seq		Investigation of the 3D structure of the genome

(Continued)

Table 26.1 (*Continued*)

Data type	Main platforms	Applications
Transcriptomic	Microarray	Quantification of a wide set of defined sequences simultaneously
	RNA-Seq	Detection and quantification of theoretically all RNA sequences including lncRNAs and microRNAs
Proteomic	LC–MS/MS MALDI-TOF/TOF MS	Analysis of complex protein mixtures with high sensitivity
	ICAT SILAC iTRAQ	Labeled proteins quantification
	X-ray crystallography NMR	Identification of the 3D structure of proteins
	RPPA	Quantification of either total proteins or post-translationally modified proteins
Metabolomic	NMR	Discrimination of metabolic markers
	MS	Analysis of complex metabolite mixtures with high sensitivity

Abbreviations: CGH, Comparative genomic hybridization; CNV, copy number variation; SNP, single-nucleotide polymorphism; LOH, loss of heterozygosity; WES, whole exome sequencing; WGS, whole genome sequencing; MeDip, methylated DNA immunoprecipitation; MBD, methyl-CpG-binding domain; BS, bisulfite; OxBS, oxidative bisulfite; ChIP, chromatin immunoprecipitation; MNase, micrococcal nuclease; ATAC, assay for transposase-accessible chromatin; 4C, chromosome conformation capture-on-chip; HiC, high-throughput chromosome conformation capture; LC–MS/MS, liquid chromatography-tandem mass spectrometry; MALDI-TOF/TOF MS, matrix assisted laser desorption ionization time-of-flight; ICAT, isotope-coded affinity tag; SILAC, stable isotope labeling by/with amino acids in cell culture; iTRAQ, isobaric tags for relative and absolute quantitation; NMR, nuclear magnetic resonance; RPPA, reverse phase protein array; lncRNA, long non-coding RNA.

26.2.1 Genomic Profile

The cancer genome carries several somatic changes that include single nucleotide substitutions, small insertions and deletions (indels), structural rearrangements, and copy number variations (CNVs). Some of these somatic alterations are targetable driver mutations, which contribute to cancer development and disease progression. The remaining ones are passengers, which have no "fitness" effects and thus do not contribute to cancer. However, the prevalence of somatic mutations shows inter- and intratumor heterogeneity ranging from approximately 0.001 per megabase (Mb) to more than 400 per Mb [1]. This variability is a consequence of clonal tumor evolution in which selection pressure in different contexts (microenvironment, host immune system, therapy) lead to a mosaic of different clones with varying degrees of somatic truncal and branched mutations in a Darwinian evolution process. In particular, tumors with increased genomic instability, such as melanomas, lung cancers, and cancers with DNA repair defects, are prone to develop branched aberrations [2]. Although many cancer drugs have been developed to specifically target clonal mutations in known truncal driver genes, which occurred early in cancer evolution, later branched abnormalities may confer resistance to therapies and thus may not account for the whole tumor evolution. The accurate detection of clonal and branched subclonal cancer mutations is a critical step in personalized cancer care as mutational profiling is used in clinical practice to classify patients in subgroups and associate the subtypes with clinical outcomes for better prognosis and treatment. McGranahan and colleagues identified approximately 15% of branched druggable mutations in *IDH1* in glioblastomas and approximately 20% in PI3K-AKT pathway effectors in all cancer types [3] Until now, a large amount of oncogenetic alterations from various tumor types have been collected in curated databases such as the Cancer Genome Atlas (TCGA), the International Cancer Genome Consortium, and the Cancer Cell Line Encyclopedia [4].

DNA sequencing technologies now allow targeted or whole exome sequencing (WES) and whole genome sequencing (WGS) of multiple tumors to identify genetic alterations. WES investigates the coding regions of the genome, whereas WGS focuses on the entire DNA sequence. In both, the tumor genome is compared with a patient's germline sequence or a reference genome, thus parallel data capture and analysis is needed to classify variants as germline and somatic. Many reports demonstrate the power of massively parallel sequencing. For example, Wedge and colleagues [5] sequenced the whole genomes of 112 primary and metastatic prostate cancer samples. They identified 28 genes with an excess of coding driver mutations, five of which (*TBL1XR1*, *ZMYM3*, *IL6ST*, *CASZ1*, and *TBX3*) were previously unknown drivers in prostate cancer. They also reported loss of *CHD1* and *BRCA2* as early events in cancer development of ETS fusion-negative cancers and losses of *CDH12* and *ANTXR2* were associated with poorer recurrence-free survival. In addition, Miao and colleagues [6] analyzed the WES of 249 tumors and matched normal tissues from patients with clinically annotated outcomes to immune checkpoint therapy. They found that clonal driver alterations in *PIK3CA* and *KRAS* were enriched in patients with complete or partial response to treatment, while clonal driver mutations in *EGFR* were enriched in patients with progressive disease.

Complementary to WGS/WES, oligonucleotide array-based methods have been widely used to identify copy number (CN) changes as well as CN neutral differences associated with loss of heterozygosity (LOH). Currently, there are three types of chromosomal microarray analysis (CMA) platforms that differ in technology, resolution, and detection: array comparative genomic hybridization (array-CGH), single nucleotides polymorphism array (SNP-array), and array-CGH supplemented with SNP probes. Array-CGH platforms are designed for the detection of CN, while SNP-array and array-CGH plus SNP probes combine classic CN analysis with SNP genotyping. However, SNP density on CN+SNP platforms is typically lower than on the traditional SNP-array. For instance, Yeung et al. [7], in a retrospective analysis of 68 patients with myelodysplastic syndromes (MDS), found that 73% of patients had abnormal CMA, carrying either copy number LOH (32%) or CNV (41%). Patients harboring chromosomal abnormalities showed a lower overall survival ($p = 0.04$). A new and promising area of research is the use of liquid biopsies to genotype cell-free DNA (cfDNA) and circulating tumor cells (CTCs). Circulating cell-free DNA (cfDNA) is released from apoptotic or necrotic cancer cells. Circulating cell-free DNA harbors mutations, which are representative of the genetic background of the cell of origin and their levels are related to tumor stage and prognosis [8]. For example, Zill and colleagues [9] analyzed the somatic mutation landscape of 70 cancer genes from cfDNA deep-sequencing analysis of 21,807 patients with treated, late-stage cancers. They found that patterns and frequencies of driver alterations in advanced cancers reflect patterns found in early-stage disease. Circulating tumor cells derive from primary and metastatic tumor cells and are present into the bloodstream. Enumeration of CTCs has been proven as a prognostic marker for metastatic cancer and response to therapy [10]. Also, CTCs have similar clonal and subclonal structures with matched primary tumor. Comparative genomic analyses of CTCs, bone marrow (BM) tumor cells, and peripheral blood germline DNA from multiple myeloma (MM) patients showed that 100% of clonal mutations in patient BM were detected in CTCs and that 99% of clonal mutations in CTCs were present in BM MM [11]. Inherited germline variants were also associated with somatic mutations in known cancer genes, suggesting that a specific germline background may contribute to tumor development. Carter and colleagues [12] found that a germline haplotype at locus 19p13.3 increases likelihood of somatic mutations in *PTEN*. This locus includes two genes, *GNA11* and *STK11*, known to be involved in the PIK3CA/mTOR signaling pathway in which *PTEN* plays a major repressive role. In particular, GNA11 activates mTOR signaling whereas STK11 inhibits mTOR activity. The authors suggested a model in which the minor allele of 19p13.3 impairs mTOR signaling, conferring selective pressure to later somatic mutation of *PTEN*. In terms of cancer prevention, genome-wide association studies (GWAS) have identified many common low-penetrance germline variants that are linked with cancer predisposition, providing direct evidence of polygenic susceptibility. An example is the SNP rs149574 in *N*-acetyltransferase 2 (*NAT2*) gene, which is functional and has been reported to modify the effect of smoking in bladder cancer [13].

26.2.2 Epigenomic Profile

Epigenomic profiling provides the possibility to discover significant associations between chromatin features and genomic function at gene expression level. The epigenetic modifications, such as DNA methylation at the 5′ position of cytosine, or post-translational modifications of histone tails, are heritable changes that regulate gene expression pattern by altering DNA accessibility and chromatin structure without altering the genotype. As a consequence of these modifications, several biological processes are differentially modulated, sometimes producing pathologic conditions. In particular, defects in chromatin modifiers and remodelers have been associated with the etiology of both solid and hematological tumors [14]. Approximately thirty percent of all the reported cancer driver genes have been related to chromatin remodeling [15]. For example, alterations in epigenetic enzymes such as the methylcytosine dioxygenase (TET2) [16], the DNA methyltransferase 3α (DNMT3A) [17], and the histone-lysine *N*-methyltransferase (EZH2) [18], as well as the aberrant promoter DNA methylation causing the silencing of genes such as *CDKN2A* or *MLH1* [19], have been associated with tumor development or progression. Considering the important role of the epigenetic enzymes in tumorigenesis, different targeted drugs have been developed. Among them, HDAC and DNMT inhibitors have been approved for treatment of hematological cancers and are under clinical evaluation for solid tumors and for combination protocols with non-epigenetic drugs, while many others drugs targeting other players in nucleosome remodeling are currently undergoing preclinical and clinical studies with very promising findings [20, 21]. The idea to identify disease-related epigenetic biomarkers and to exploit the epigenetic reprogramming to revert a pathological state to a healthy one, led to deeper investigation of the human epigenome. For this purpose, next-generation technologies, combined with methylation microarrays, are being increasingly adopted to improve our understanding. In particular a strong interest is devoted to the DNA methylome whose characterization is essential to disclose cell-specific transcriptomes that could not be explained by limiting the analyses to a gene level. For DNA methylation profiling we can distinguish different genome-wide approaches such as (i) affinity enrichment-based methods [22, 23], (ii) restriction rnzymes-based methods [24], (iii) bisulfite conversion-based methods [25, 26], (iv) capture-based methods [27], and (v) third-generation sequencing based-methods [28, 29]. All these techniques share the preservation of the methylation signature that would be otherwise lost during polymerase chain reaction (PCR) amplification needed for DNA sequencing. The readout of these approaches is entrusted to microarrays, such as the Infinium Human Methylation 450 Bead Chip array (Illumina, Inc., San Diego, CA, USA) that cover 450.000 CpGs, or sequencing [30]. Apart from the technologies adopted to reveal the DNA methylation profiling, it is worth mentioning other methods used to investigate the epigenome such as (i) chromatin immunoprecipitation sequencing (ChIP-Seq) [31, 32], which is performed to identify chromatin-associated proteins; (ii) DNase I sequencing, assay for transposase-accessible chromatin sequencing (ATAC-Seq), and micrococcal nuclease sequencing (MNase-seq) [33, 34] enabling to reveal chromatin accessibility; and (iii) chromosome conformation capture-on-chip sequencing (4C-seq) and high-throughput chromosome conformation capture sequencing (HiC-seq) which provide information about the global three-dimensional (3D) structure of the nucleus [35].

Many of the data produced through the technologies mentioned above are collected in several databases such as MethyCancer [36–38] including DNA methylation data; Histone Database [39] and HIstome [40] including histone modification data; and Cistrome cancer including chromatin remodelers data combined with tumor molecular profiling data. Several studies that focused on the epigenomic profiling of cancer patients led to the characterization of epigenetic players in tumor initiation or progression and to the identification of diagnostic or prognostic biomarkers. Guo and colleagues identified a signature of differentially methylated genes (*AGTR1, GALR1, SLC5A8, ZMYND10,* and *NTSR1*) as biomarkers for non-small cell lung cancer (NSCLC) diagnosis by integrating three high-throughput DNA methylation microarray datasets including 458 samples (352 NSCLC and 106 normal tissues) from GEO (Gene Expression Omnibus) and TCGA [41]. Exner et al. performed a microarray-based methylation analysis of 360 promoters of genes previously described as hypermethylated in several tumors on 22 rectal cancer DNA samples and eight control DNA samples identifying two novel three gene-based signatures comprising *TFPI2-DCC-PTGS2* and *TMEFF2-TWIST1-PITX2* which are able to label tumor samples respect to adjacent tissues or blood, respectively. Moreover, the authors found the methylation of *CDKN2A* as a negative prognostic factor for overall survival of rectal cancer patients [42]. Legendre and colleagues reported a differential methylation analysis performed through a paired-end whole-genome bisulfite sequencing on cell-free DNA from plasma of 40 metastatic breast cancer (MBC) patients that led to the identification of hypermethylation hotspots within CpG islands of 21 genes that are unique for MBC patients compared with a pool of 40 disease-free survivors or 40 healthy individuals. Their findings suggested that a DNA hypermethylation signature might be of prognostic relevance [43]. Qu and colleagues analyzed 111 Cutaneous T cell lymphomas (CTCLs) by ATAC-seq by identifying chromatin signatures which enable the discrimination between leukemic, host, and normal CD4$^+$ T cells and by revealing that clinical response to HDAC inhibitors is associated with specific changes in chromatin accessibility. The regulome profiling could then be used as prognostic factor due to the patient-specific chromatin landscape in response to epigenetic drugs [44]. Cai and colleagues, through the 4C-capture method followed by next-generation sequencing, revealed genome-wide interacting partners in correspondence of the 8q24 locus by proposing an approach able to explain how genetic variants affect this well-known hotspot increasing prostate cancer risk [45].

26.2.3 Transcriptomic Profile

Transcriptomic technologies allow for the production of information on the total transcripts of a genome or a specific cell by the use of two high-throughput methods: (i) microarrays, which allow the simultaneous detection and quantification of thousands of previously identified transcripts by hybridization of targets on high-density array containing complementary probes; (ii) RNA sequencing (RNA-Seq), which uses high-throughput massive parallel sequencing combined with computational methods to detect and quantify the complete set of RNA transcripts. Comparison of transcriptomes in different tissues, conditions, time points, or even at single cell level gives information on how genes are regulated and differentially expressed disclosing details about the biology of the system. Moreover, expression profiles can also help to infer the functions of previously unannotated genes. Thereby, the lowering of the technology costs and increased sensitivity allowed a large amount of studies. Many consortium efforts have produced transcriptomic data sets of (i) cancer cell lines, such as the Encyclopedia of DNA Elements (ENCODE) [46], the Cancer Cell Line Encyclopedia (CCLE) [47], and Genentech [48]; (ii) normal tissues, such as the Genotype-Tissue Expression (GTEx) project [49] and the Human Protein Atlas (HPA) [50]; and (iii) tumor tissues such as TCGA [51] and the Stand Up To Cancer-Prostate Cancer Foundation (SU2C-PCF) project [52]. RNA-seq has become the most robust and comprehensive transcriptome profiling technology, virtually replacing all expression microarrays. An example of the clinical utility of RNA-seq has been demonstrated by several studies disclosing a large number of new actionable genetic events [53] or the real-time management of pediatric tumors [54] as well as the characterization of metastatic tumors [55]. Moreover, the advance in RNA-Seq library preparation methods, resulted in enhanced sensitivity and effectiveness of single-cell *in situ* RNA-Seq also performed in fixed tissues [56].

The applications of the transcriptome analysis span in a broad range of biomedical research, including diagnosis, disease classification, and monitoring, or response to, treatments. For example, Huet and colleagues [57] developed a predictor of progression-free survival based on a gene expression signature, including 23 genes, in follicular lymphoma. Using this strategy the authors were able to identify, at diagnosis, patients with an increased risk of progression when initially treated with rituximab and chemotherapy.

An additional example was reported by Boyault et al., who established one of the first transcriptomic molecular classification systems for hepatocellular carcinoma (HCC), composed of six groups, G1-G6 and G1-G3 groups were associated with a high rate of chromosomal instability and overexpression of proliferation genes. Specifically, the G3 subtype was reported as having the worst prognosis, bearing increased allelic loss, including chromosome 17 deletion and *TP53* mutation, *CDKN2A* hypermethylation, and increased expression of cyclins [58]. Further studies established other transcriptomic HCC signatures correlated with adverse biologic features and clinicopathologic observations [59]. Apart from gene expression profiling, transcriptomics has also been applied to non-coding RNAs (ncRNA), which are untranslated transcripts with several biological functions [60]. Many of the ncRNAs affect disease states, including cancer, cardiovascular, and neurological diseases. A class of ncRNAs, the microRNAs (miRNAs), have been widely analyzed by expression to generate miRNA signatures able to stratify human tumors in different subtypes correlating with clinical features or to be used as biomarkers or candidate therapeutic targets [61, 62]. miRNA alterations were identified during breast cancer transition from ductal carcinoma *in situ* to invasive ductal carcinoma by Volinia and colleagues. A nine-microRNA signature was identified and specifically, let-7d, miR-210, and miR-221 were downregulated in the *in situ* and upregulated in the invasive transition, thus featuring an expression reversal along the cancer progression path [63].

Based on miRNA expression profiles, Namkung and colleagues identified three pancreatic ductal adenocarcinoma tumor subtypes associated with prognosis. These subtypes showed significantly different survival for patients with similar clinical features, demonstrating that the prognostic molecular subgroup has independent prognostic utility [64]. Li and colleagues reported novel diagnostic tools for deeper prognostic substratification in five intrinsic subtypes of primary glioblastoma (GBM) in TCGA dataset based upon miRNA expression profiles. miRNA signatures revealed that high-risk scores strongly correlated with poor overall survival as compared with patients who had low-risk scores. This evidence suggests that transcriptomic technologies allow the identification of genes and pathways that can be associated with oncodriver signatures. Moreover, the nontargeted nature of transcriptomics let us define novel transcriptional networks in complex systems such is the cancer disease.

26.2.4 Proteomic Profile

Proteomics is the study of the entire set of proteins in any given cell, including the set of all protein isoforms and modifications, the interactions between them, the structural description of proteins, and their higher-order complexes [65, 66]. Therefore, proteomics is the next step to study biological systems because proteins are responsible for most cellular processes; their analysis would more accurately reflect cellular status to determine the mechanism that underlies disease initiation, progression, and dissemination [67, 68]. To analyze the complex protein mixtures with higher sensitivity, the main technology presently in use is mass spectrometry (MS) [69], combined with liquid chromatography or matrix-assisted laser desorption ionization (MALDI-TOF/TOF) [68, 70, 71]. However, new methods were recently developed for quantitative proteomic such as isotope-coded affinity tag (ICAT) labeling, stable isotope labeling with amino acids in cell culture (SILAC), and isobaric tag for relative and absolute quantitation (iTRAQ) [72–74]. X-ray crystallography and nuclear magnetic resonance (NMR) spectroscopy are two major high-throughput techniques that provide 3D structure of proteins [75]. These technologies are adopted in cancer research to generate data on (i) differential protein expression levels, (ii) protein–gene expression correlation, (iii) differential protein expression comparisons between different cancer phenotypes and subtypes, and (iv) the associations of protein expression with survival prognosis in cancer patients. Proteomic data are collected in several databases including PRIDE (proteomics identification database) and Global Proteome Machine [76]. Other databases such as KEGG, IPA (Ingenuity Pathway Analysis), Pathway Knowledge Base Reactome, or BioCarta include comprehensive data regarding metabolism, signaling, and protein interactions [77–79]. Global proteomic profile is increasingly being carried out in both cancer cell lines and patient-derived samples to provide information that are useful for cancer type classification [80–83] and drug sensitivity/resistance prediction [84, 85].

Tyanova and colleagues used quantitative proteomics to examine the functional difference between breast cancer subtypes, related to energy metabolism, clearly selecting four proteins (HER2, Grb7, FOXA1, and MLPH) which may represent novel potential therapeutic targets [80]. Moreover, in ovarian cancer, the application of proteomic analysis highlighted the differential expression of five proteins (serotransferrin, amyloid A1, hemopexin, C-reactive protein, and albumin) which improved the diagnostic performance of the model discriminating between benign and malignant tumors. The identification of these proteins shed light on the molecular signaling pathways that are associated with ovarian cancer development [84]. Similarly, by using an iTRAQ approach combined with high-resolution MS analysis, several proteins were found dysregulated during gastric cancer progression, demonstrating their potential use as specific biomarkers and/or therapeutic targets [73]. Moreover, by using MS-based proteomic analysis on 130 clinical breast cancer samples, Yanovich and colleagues demonstrated intertumor heterogeneity across three breast cancer subtypes and healthy tissues, identifying four proteomic clusters. One of them represents a novel luminal subtype, characterized by increased PI3K signaling, demonstrating the importance of deep proteomic analysis for clinical decision-making [86]. The application of a quantitative targeted proteomic approach was also applied to identify four candidate biomarkers of drug resistance (GRP75, APOA1, PRDX2, and ANXA) in ovarian cancer cell lines and patient biopsies, after carboplatin and paclitaxel treatments [87].

In the last decades, a promising new array application, named reverse phase protein array (RPPA), has been developed to measure either total or post-translationally modified proteins [88, 89]. This technology allows the investigation of protein–protein interactions or biochemical reactions revealing information on the cellular processes driving tumor growth and response to treatments in cancer patients. An example of the application of RPPA has been reported by Masuda and colleagues, which demonstrated that the level of ribosomal protein S6 phosphorylated at serine residue 235/236 (p-RPS6 S235/236) was most significantly correlated with the resistance of HCC cells to sorafenib. The high expression of p-RPS6 S235/236 was confirmed immunohistochemically in HCC biopsies from patients who responded poorly to sorafenib, suggesting a novel biomarker for drug resistance [90]. In recent years, additional RPPA data generation is improving, for identification of cancer subtypes and targeted therapy through integration with other data platforms.

26.2.5   Metabolomic Profile

Another way to understand the amount of endogenous proteins within a biological system is to measure the metabolome—the endpoint of the omics cascade. Metabolites are small molecules interacting with proteins helping various biological functions. Detection of metabolites is carried out in cells, tissues, or biofluids by either NMR spectroscopy or MS [91]. Metabolomics have important potential in oncology, including the early detection and diagnosis of cancer and as both a predictive and pharmacodynamic marker of drug effect for therapeutic evaluation [92]. Moreover, metabolomics can provide a link between laboratory and clinics, particularly because metabolic and molecular imaging technologies enable a non-invasive discrimination of metabolic markers *in vivo* [93]. Despite this, the knowledge about metabolomics in clinical cancer research is poor. However, metabolomics allows for a complete global assessment of a cellular state, taking into account genetic regulation, altered kinetic activity of enzymes, and changes in metabolic reactions. Thus, compared with genomics or proteomics, metabolomics reflects changes in phenotype, and therefore, functions [91]. Metabolomic data obtained from NMR or MS methods were analyzed to provide information about metabolic profile of samples, including quantitation and association of putative biomarkers with respect to a particular characteristic or outcome, such as tumor grade or response to therapy. To accomplish this, bioinformatic databases were created to store metabolomic data of endogenous metabolites of biological samples simultaneously, such as the Human Metabolome Database, Genome Alberta, and Genome Canada [94]. An example of a metabolomics approach has been reported in several studies such as in the analysis of patients with acute myeloid leukemia (AML), in which significant changes in carbohydrate metabolism in blood were assessed [95]. Specifically, high levels of the well-known oncometabolite 2-hydroxyglutarate (a product of IDH1/IDH2 mutations) were detected in AML patients with poor prognosis, suggesting its potential role as a prognostic marker [96]. Other studies reported an association between fatty acids metabolites and carcinogenesis. Serum levels of unsaturated free fatty acids were revealed to be diagnostic indicators of early-stage colorectal cancer [97]. Similarly, amino acids also play important role in tumor development. As a matter of fact, cancer cell studies indicated serine and glycine metabolism as necessary resources for cancer cell metastatization and malignant potential [98]. Moreover, decreased citrate and elevated spermine levels were detected in prostatic fluid from men with prostate cancer, compared with noncancer patients, identifying new prostate cancer biomarkers for clinical diagnostic [99]. Branched-chain amino acids, including leucine, isoleucine, and valine, were found at high levels in blood from patients with human pancreatic adenocarcinoma compared with healthy controls [100].

Therefore, metabolomics data can contribute, with other omics, to an integrative approach to decipher cancer disease facilitating and accelerating the clinical practice.

26.3 Integrative Analysis Tools

As discussed before, the application of high-throughput technologies for the monitoring of almost all the key players within cells (DNA, RNA, ncRNAs, proteins, or metabolites) leads to the possibility to investigate cancer samples at different levels. The ultimate goal of these efforts is to discriminate among cancers samples to deliver precision and personalized treatments. From a computer science point of view, the key challenge to be faced is the integration of such heterogeneous data, since many different works have demonstrated that integromic analysis provides the opportunity to better understand molecular phenomena [101]. For instance, Hofree and colleagues developed a strategy to cluster patients integrating clinical data and functional relationships among a set of genes [102]. In parallel the integration of transcriptome analysis with proteomic data (protein–protein interactions) has been used in ovarian cancer [103]. More recently, Singh and colleagues introduced a novel approach analysis able to integrate even the mutation analysis [104]. Consequently, the need for the introduction of novel frameworks, algorithms, and tools for the integrated analysis arose. Here we focus on available tools for the integrated analysis to offer to the reader a synergistic point of view [105].

(a) dChip-GemiNi [106] is a web server able to integrate and to analyze miRNA and mRNA expression data. It is based on the analysis of time-series data, i.e., a set of temporally sorted observations, in which for each time there exist both an mRNA and a miRNA observation. dChip-GemiNi may be used through a web interface and it may be also downloaded for running it in a local environment. In dchip-GemiNi experimental data are compared with respect to an experimental model derived from publicly available data. This model has been built using a workflow composed of four steps. Initially, existing databases have been mined to derive known associations among miRNA–mRNA (e.g., TargetScan [107]) and transcription factor (TF) sites. Such associations have been integrated with data derived from expression databases. These two steps produce an initial network of associations among TF, miRNA, and mRNA. Then significant motifs have been mined and ranked. Each motif is a small subgraph representing the association among TF, miRNA, and mRNA.

(b) MAGIA [108] is a web server for the integrated analysis of mRNA, TFs, and miRNA. It includes miRNA–mRNA associations derived from eight different databases, and it also includes experimentally validated TF–miRNA interactions and TF–gene interactions. The user that would analyze its own data using MAGIA needs to upload into the web server experimental miRNA and gene/transcripts expression data. MAGIA is able to mine time series experiments in which for each sample there exists a pair miRNA/mRNA experiment (referred to as matched data), and two-class experiment (referred to as unmatched data). Results of MAGIA are small networks of association among miRNA, mRNA and TFs ranked by score.

(c) mirConnX [109] is based on a genome-wide approach, i.e., the associations are analyzed on a genome scale, but only referring to human and mouse data. The approach of analysis is based on the comparison of a first network, derived from data extracted from public databases, and a second one derived from experimental data provided by the user. It considers both miRNA–mRNA associations derived from many existing databases and associations among TF and genes.

Experimental data provided by the user are mined to build a network obtained by analyzing all the possible pairwise interactions between TFs, miRNAs, and genes across the samples/replicates. Networks are finally merged into a single model using a simple weighted sum function (S) producing a novel network in which edges, which are found in both networks, have a higher weight.

(d) IntegraMiR [110] analyzes experimental data whose samples belong to two classes (e.g., healthy vs. disease). Initially, it searches and ranks differentially expressed miRNAs and mRNAs. Then it focuses on the functional comparison of transcripts by performing a gene set enrichment analysis (GSEA). Finally, the associations among miRNA and mRNA are derived on the basis of expression levels and biological consideration derived from GSEA. Finally, association among genes and TFs are derived from existing databases. Resulting networks of association are mined to reconstruct motifs (i.e., association among miRNA–mRNA and TF) considering only differentially expressed genes. These FFLs are then organized considering the kind of deregulation and ranked by using a statistical approach and visualized to the user.

(e) XCMS Online is a new platform available through an online web interface. It enables the analysis of metabolomics data by integrating both genomic and proteomic data, such as LC–MS data. It is based on a cloud infrastructure, therefore, it enables the sharing of data easily among collaborator [111]. Ruffalo and colleagues [112] developed a model for the integration of expression data, mutational data, and protein interaction networks. The model is based on the use of protein interaction networks to integrate experimental data into a single model and then to derive knowledge by mining this network. Authors used the model to analyze cancer related genes. They used public available data from TCGA to predict a gene association with cancer, shows improved predictive power in recovering cancer-related genes in known pathways.

(f) The MR4Cancer [113] is a web server to prioritize key genes involved in cancer controls, also known as master regulator

(MR) genes. Authors extracted all the cancer specific regulator genes for 26 cancer types by analyzing TCGA data and then they extracted regulators that are not directly related to cancer from public databases. The list of these regulators has been used to build a reference model. User may upload into the web server its own experimental data (e.g., expression data) and MR4Cancer outputs ranked MRs by enrichment testing against the predefined.

(g) The Cancer Systems Biology Database (CancerSysDB) is a database for the analysis of cancer-related data that integrate multiple data types and multiple studies [114].

Regarding the integration of the epigenomic data together with RNA expression profiling and clinical data, we can mention databases such as:

(a) MENT, which is a database containing integrated data of DNA methylation and gene expression of normal and tumor tissues together with clinical data from GEO and TCGA [115].

(b) MethHC, which includes a systematic integration of DNA methylation and mRNA/microRNA expression data from human cancers [109].

(c) Wanderer, which is a web tool allowing user-friendly access to gene expression and DNA methylation data from TCGA [116].

(d) MethCNA, a database in which raw array data obtained by Infinium HumanMethylation450 bead chip and deposited in TCGA and GEO databases are collected and re-analyzed through a pipeline that includes multiple computational tools and resources for omics data integration. In this database DNA methylation and copy number alteration data refer to exactly the same genetic loci from the same DNA specimen, providing an important advantage respect than other databases that instead integrate data deriving from different patients and platforms [117].

26.4 Integrative Analysis Approaches in Cancer Research

Several studies focused on the integration of multi-omics data in order to provide a global view of the tumor landscape, to accurately stratify patients that may or not benefit from current treatment options, as well as identify new potential targets and new diagnostic and prognostic biomarkers, often by taking advantage of the access to wide databases, such as those previously mentioned, collecting data from thousands of patients.

Among these datasets, TCGA is the most extensive, and includes multi-omics data deposited from many centers involved in the TCGA research network, as well as patient's clinical metadata prospectively collected. TCGA data currently refers to 33 cancer types from more than 11,000 patients that have been obtained through different high-throughput technologies such as DNA-seq, SNP-based platforms, array-based DNA methylation-seq, microRNA-seq, RNA-seq, and RPPA by providing a comprehensive view of the human cancer molecular bases. Each platform produces data that are informative about DNA mutational status, SNP, methylation, loss of heterozygosity (LOH), copy number variation, miRNA expression, gene expression, and protein expression.

An example of the TGCA-based multi-omics data integration is represented by the study published from the Cancer Genome Atlas Network on the breast cancer molecular landscape. This study analyzed different sets of breast tumors, through the six platforms mentioned above, to provide an extensive characterization enabling the identification of many subtype-specific alterations. As matter of fact, thanks to the multi-omics cross-integration analysis, it has been found that the four main breast cancer subtypes (luminal A, luminal B, basal-like, and HER2$^+$) have specific genomic, proteomic, and clinical features. This study highlighted druggable targets for each group such as many players of PI3K and RAS-RAF-MEK pathways in basal-like tumors in which enhanced activity of the HIF1α/ARNT pathway was also found, suggesting the possibility to use bioreductive drugs or angiogenesis inhibitors in this orphan disease. In HER$^+$ tumors, mutations of PIK3CA, PTEN, INPP4B, IK3R1, and within HER-family members were identified instead, while in luminal/ER$^+$ subtypes, there were many mutated genes including p53 and RB1 [118]. Moreover, the huge amount of information included in TGCA offers the possibility to perform sophisticated analyses with an enhanced statistical power by integrating multi-omics data, not only from different patients with the same type of cancers, but also from different tumor types (TCGA Pan-Cancer Project). In fact, Weinstein and colleagues demonstrated that analysis of the molecular aberrations and their functional roles across tumor types could extend therapies effective in one cancer type to others with a similar genomic profile. Thus, the Pan Cancer TCGA data set provides a major opportunity to develop an integrated picture of commonalities, differences, and emergent themes across tumor species [119]. Multiplatform integrated analysis of different cancer types revealed molecular classification within and across tissues of origin [120, 121].

An application of integromics includes the development of models that combine proteomic and genomic data with the aim to accurately predict patient survival. Interestingly, Zhu and colleagues generated RPPA data from tumors for which genomic, transcriptomic, and clinical features have been previously collected in the TCGA database, in order to provide the possibility to integrate multi-omics data from different layers to globally characterize these tumors [122].

Using the same cell line, Akbani and colleagues integrated RPPA together with genomic and transcriptomic data from TCGA to identify similarities and differences in pathways and network biology within and across tumor lineages, as well as biomarker and target discovery spanning multiple tumor lineages by providing a framework for determining the prognostic, predictive, and therapeutic relevance of the functional proteome [120].

Apart from TCGA-based analyses, several studies focused on the integration of different platforms to answer specific questions about various cancer types. Koplev and colleagues performed an integrative analysis on 726 pan-cancer cell lines profiled for gene expression, protein expression, and phosphorylation by explaining key regulators mechanisms involved in tumorigenesis. In particular, they revealed enrichment for HDAC inhibitors as inducers of epithelial–mesenchymal transition and kinase inhibitors as mesenchymal-to-epithelial transition (MET) promoters [123]. Snyder and colleagues instead performed an integrated analysis of multi-omics data generated through WES, RNA-seq, and T cell receptor sequencing from 29 patients with locally advanced or metastatic urothelial carcinoma which were treated with the checkpoint-blockade atezolizumab. With this approach, they evaluated the role of somatic, immune, and clinical patient-specific features in response to atezolizumab [124].

Mancikrnaova and colleagues focused on the study of the medullary thyroid carcinoma by combining DNA methylation data with mRNA/miRNA expression data. They identified specific genes involved in tumor progression that were negatively regulated through promoter methylation and indicated JAK/Stat pathway as potential target in RETM918T medullary thyroid carcinomas in enhancing the antitumor activity of RET inhibitor vandetanib [125]. Piccolo et al. summarized genomic data in tracking pathways to explain how germline, genetic, and epigenetic variations regulate gene expression changes in normal cells, in order to identify mechanisms that underlie breast cancer susceptibility [126]. Robles et al. found biomarkers to allow early detection and prognostic assessment of lung cancer by combining data regarding genomics (DNA methylation), transcriptomics (miRNA and mRNA expression), and metabolomics data (pro-inflammatory cytokines and metabolites from urine) in order to provide additional information for therapeutic tailoring [127]. With a similar approach, Li and colleagues combined genomic, transcriptomic, proteomic, and metabolomic data obtained by profiling three cell lines that were representative of HCC, each with a different metastatic potential in order to evaluate the influence of metabolism in metastatization process. They revealed 12 altered genes at different levels of specific pathways involved in cell metabolism such as sucrose and glutathione metabolism and glycolysis. Moreover, they reported an association between uridinediphosphate (UDP)-glucose pyrophosphorylase 2 (UGP2) and cell migration and invasion *in vitro* and *in vivo* [128]. Recently, our group used an integromics approach to shed new light within the molecular architecture of multiple myeloma (MM) hyperdiploid (HD-MM) and the non-hyperdiploid (nHD-MM) subtypes. By integrating annotated MM patient mRNA/miRNA dataset information, a specific gene and miRNA expression profile for HD-MM was found. Indeed, from this analysis a significant role of the STAT3 pathway, as well as the Transforming Growth Factor β (TGFβ) and the transcription regulator nuclear protein-1 (NUPR1), was demonstrated, thus defining novel molecular features of HD-MM that may translate in novel relevant therapeutic targets characterization [129].

26.5   Discussion

Molecular approaches, including mutational analysis, RNA and miRNA expression profiling, and epigenetic characterization, greatly enhance the understanding of pathogenesis and allows prognostic stratification for many cancers, thus driving the rational design of novel targeted therapies. In addition, to gain a broader perspective of the molecular aberrations that contribute to tumor development, it is valuable to consider data at the pathway more than at single gene level. Furthermore, by tracking pathway activities using multiple types of omics data, including both genomic alterations and gene expression changes, it is possible to delineate a landscape view, which might include tumor cells and the micro-environment, providing a novel approach to define personalized therapeutics. Accordingly, the expression of cDNAs harboring mutations identified in human cancer if combined with protein functional assays may address whether these mutations are crucial for disease progression (driver mutations) or have been generated as the consequence of genomic instability, without biologic relevance (passenger mutations). High-throughput data may provide inputs for preclinical and clinical studies towards the precision medicine starting from the exploration of a wide amount of patient-specific omics information (see a possible workflow represented in Fig. 26.1).

Even if the theoretical value of precision medicine approaches as integrative genomics is now well-established, it has to be demonstrated its value in providing a paradigm shift in the real life. This is a crucial point that needs to be approached in terms of feasibility, effectiveness, and equity. It is clear that a major point is that all validated approaches must be offered in light of clinical utility, which means they provide useful information for clinical decision-making and not provide only redundant data. In this light, a biomarker-driven approach can allow patient selection and, therefore, reduce toxicity and costs, with a major benefit for patients, the health system, and other stakeholders. In terms of equity the major point is not indeed the cost of molecular analysis, which is lowering but a true access for all patients and a full access to all molecular targeted drugs which might be identified, that can be highly expensive.

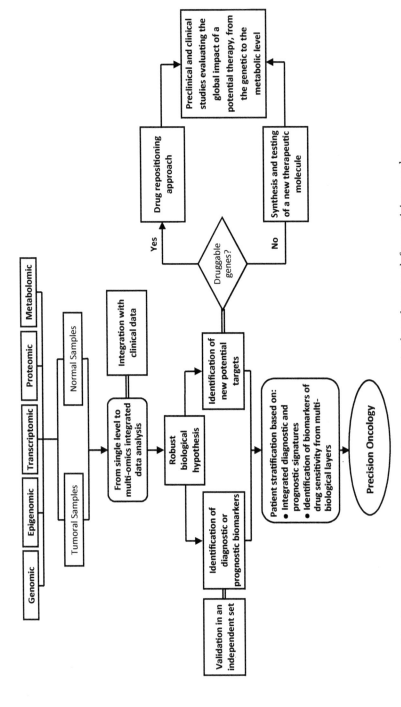

Figure 26.1 Representative flowchart of a multi-omics integrative-based approach for precision oncology.

At this aim is crucial the setting up of laboratory networks with free sharing of molecular data and programs to negotiate access to innovative drugs for nationwide coverage. These considerations resound more for those patients with a rare disease, in which more specific and personalized treatments are eagerly awaited. In cancer treatment, precision medicine offers the opportunity to choose the drugs not only based on the type of cancer, but also on the genomic driver aberrations that can be tissue-agnostic but shared by different malignancies In drug development, precision medicine means finding new drugs that act as "keys" to certain "locks" in the body. In the case of rare disease, molecular diagnosis opens the door for new treatment options that cannot be explored by conventional pivotal trials.

All together these points show that the empowering of integrative genomics as a novel tool for precision medicine indeed is not only a technical but also a social and health policy challenge.

The integration of omics profiles with clinical variables lead to improved prognostic performance over the use of clinical variables alone. In fact, the integrated portrait of omics architecture provides a comprehensive view which likely outperforms the predictive capability of single gene changes. Developing models that accurately can predict patient survival using prognostic and predictive biomarkers obtained from aggregation of multi-dimensional omics data, is a challenge in the era of precision oncology. A promising new direction for enhancing all the omics techniques is the integration of data-driven network models with prior biological knowledge. This strategy lead to a significantly improved biomarkers identification compared for example to the top genes obtained by conventional differential gene expression analysis. A great step forward in precision medicine could derive from a pan-cancer analysis of multiple omics profiles on a genome-wide scale, in order to understand the shared patterns across cancer types and identify shared actionable targets at a multilayer level. It therefore appears evident that the integration of omics data represents a powerful tool to allow clinical translation of this integrated dissection of cancer biology (Fig. 27.2).

However, major efforts will be also necessary for the adoption in the clinics of raw data obtained from the described technology platforms, in order to translate wet biologic information to an improved therapeutic approach for cancer patients.

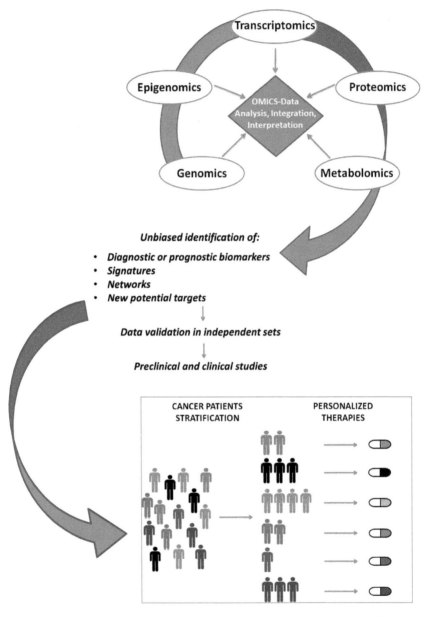

Figure 26.2 Schematic workflow summarizing major steps occurring from omics data production to personalized clinical decision-making.

Abbreviations

4C:	chromosome conformation capture-on-chip
array-CGH:	array comparative genomic hybridization
AML:	acute myeloid leukemia
ATAC:	assay for transposase-accessible chromatin
ATAC-Seq:	assay for transposase-accessible chromatin sequencing
BM:	bone marrow
BS:	bisulfate
CancerSysDB:	Cancer Systems Biology Database
CCLE:	Cancer Cell Line Encyclopedia
cfDNA:	cell-free DNA
CGH:	comparative genomic hybridization
ChIP:	chromatin immunoprecipitation
ChIP-Seq:	chromatin immunoprecipitation sequencing
CMA:	chromosomal microarray analysis
CN:	copy number
CNV:	copy number variation
CTC:	circulating tumor cell
CTCLs:	cutaneous T cell lymphomas
DNMT:	DNA methyltransferase
DNMT3A:	DNA methyltransferase 3 α
ENCODE:	Encyclopedia of DNA Elements
EZH2:	histone-lysine *N*-methyltransferase
GBM:	glioblastoma
GEO:	Gene Expression Omnibus
GSEA:	gene set enrichment analysis
GTEx:	Genotype-Tissue Expression
GWAS:	genome-wide association studies
HCC:	hepatocellular carcinoma

HDAC:	histone deacetylase
HiC:	high-throughput chromosome conformation capture
HPA:	Human Protein Atlas
ICAT:	isotope-coded affinity tag
IPA:	ingenuity pathway analysis
iTRAQ:	isobaric tags for relative and absolute quantitation
LC–MS/MS:	liquid chromatography-tandem mass spectrometry
lncRNA:	long non-coding RNA
LOH:	loss of heterozygosity
MALDI-TOF/ TOF MS:	matrix assisted laser desorption ionization time-of-flight
MBC:	metastatic breast cancer
MBD:	methyl-CpG-binding domain
MDS:	myelodysplastic syndromes
MeDip:	methylated DNA immunoprecipitation
MET:	mesenchymal-to-epithelial transition
miRNAs:	microRNAs
MM:	multiple myeloma
MNase:	micrococcal nuclease
MNase-seq:	micrococcal nuclease sequencing
MR:	master regulator
MS:	mass spectrometry
NAT2:	*N*-acetyltransferase 2
ncRNA:	non-coding RNAs
NMR:	nuclear magnetic resonance
NSCLC:	non-small cell lung cancer
NUPR1:	nuclear protein-1
OxBS:	oxidative bisulfate
PCR:	polymerase chain reaction
PRIDE:	proteomics identification database
RPPA:	reverse phase protein array
RNA-Seq:	RNA sequencing

SILAC:	stable isotope labeling by/with amino acids in cell culture
SNP:	single-nucleotide polymorphism
SNP-array:	single nucleotides polymorphism array
TCGA:	Cancer Genome Atlas
TET2:	methylcytosine dioxygenase
TF:	transcription factor
TGFβ:	transforming growth factor β
UDP:	uridinediphosphate
UGP2:	uridinediphosphate-glucose pyrophosphorylase 2
WES:	whole exome sequencing
WGS:	whole genome sequencing

Disclosures and Conflict of Interests

This chapter was originally published as: Cantafio, M.E.G, Grillone, K., Caracciolo, D., Scionti, F., Arbitrio, M., Barberi, V., Pensabene, L. Guzzi, P.H. and Di Martino, M.T. "From Single Level Analysis to Multi-Omics Integrative Approaches: A Powerful Strategy towards the Precision Oncology," *High-Throughput* **2018**, 7, 33, doi:10.3390/ht7040033 under the Creative Commons Attribution license (http://creativecommons.org/licenses/by/4.0/). It appears here, with edits and updates, by kind permission of the authors and publisher, MDPI (Basel). The authors declare that they have no conflict of interest and have no affiliations or financial involvement with any organization or entity discussed in this chapter. No writing assistance was utilized in the production of this chapter and the authors have received no payment for it.

Corresponding Author

Dr. Maria Teresa Di Martino
Department of Experimental and Clinical Medicine
Magna Graecia University
Salvatore Venuta University Campus
Viale Europa, 88100 Catanzaro, Italy
Email: teresadm@unicz.it

References

1. Alexandrov, L. B., Nik-Zainal, S., Wedge, D. C., Aparicio, S. A., Behjati, S., Biankin, A. V., Bignell, G. R., Bolli, N., Borg, A., Borresen-Dale, A. L., et al. (2013). Signatures of mutational processes in human cancer. *Nature*, **500**, 415–421.

2. Burrell, R. A., McGranahan, N., Bartek, J., Swanton, C. (2013). The causes and consequences of genetic heterogeneity in cancer evolution. *Nature*, **501**, 338–345.

3. McGranahan, N., Favero, F., de Bruin, E. C., Birkbak, N. J., Szallasi, Z., Swanton, C. (2015). Clonal status of actionable driver events and the timing of mutational processes in cancer evolution. *Sci. Transl. Med.*, **7**, 283ra254.

4. Collins, D. C., Sundar, R., Lim, J. S. J., Yap, T. A. (2017). Towards Precision Medicine in the Clinic: From Biomarker Discovery to Novel Therapeutics. *Trends Pharmacol. Sci.*, **38**, 25–40.

5. Wedge, D. C., Gundem, G., Mitchell, T., Woodcock, D. J., Martincorena, I., Ghori, M., Zamora, J., Butler, A., Whitaker, H., Kote-Jarai, Z., et al. (2018). Sequencing of prostate cancers identifies new cancer genes, routes of progression and drug targets. *Nat. Genet.*, **50**, 682–692.

6. Miao, D., Margolis, C. A., Vokes, N. I., Liu, D., Taylor-Weiner, A., Wankowicz, S. M., Adeegbe, D., Keliher, D., Schilling, B., Tracy, A., et al. (2018). Genomic correlates of response to immune checkpoint blockade in microsatellite-stable solid tumors. *Nat. Genet.*, **50**, 1271–1281.

7. Yeung, C. C. S., McElhone, S., Chen, X. Y., Ng, D., Storer, B. E., Deeg, H. J., Fang, M. (2018). Impact of copy neutral loss of heterozygosity and total genome aberrations on survival in myelodysplastic syndrome. *Mod. Pathol.*, **31**, 569–580.

8. Diaz, L. A., Jr., Bardelli, A. (2014). Liquid biopsies: Genotyping circulating tumor DNA. *J. Clin. Oncol.*, **32**, 579–586.

9. Zill, O. A., Banks, K. C., Fairclough, S. R., Mortimer, S. A., Vowles, J. V., Mokhtari, R., Gandara, D. R., Mack, P. C., Odegaard, J. I., Nagy, R. J., et al. (2018). The landscape of actionable genomic alterations in cell-free circulating tumor DNA from 21,807 advanced cancer patients. *Clin. Cancer Res.*, **24**, 3528–3538.

10. Krebs, M. G., Hou, J. M., Ward, T. H., Blackhall, F. H., Dive, C. (2010). Circulating tumour cells: Their utility in cancer management and predicting outcomes. *Ther. Adv. Med. Oncol.*, **2**, 351–365.

11. Mishima, Y., Paiva, B., Shi, J., Park, J., Manier, S., Takagi, S., Massoud, M., Perilla-Glen, A., Aljawai, Y., Huynh, D., et al. (2017). The mutational landscape of circulating tumor cells in multiple myeloma. *Cell Rep.*, **19**, 218–224.

12. Carter, H., Marty, R., Hofree, M., Gross, A. M., Jensen, J., Fisch, K. M., Wu, X., DeBoever, C., Van Nostrand, E. L., Song, Y., et al. (2017). Interaction landscape of inherited polymorphisms with somatic events in cancer. *Cancer Discov.*, **7**, 410–423.

13. Garcia-Closas, M., Rothman, N., Figueroa, J. D., Prokunina-Olsson, L., Han, S. S., Baris, D., Jacobs, E. J., Malats, N., De Vivo, I., Albanes, D., et al. (2013). Common genetic polymorphisms modify the effect of smoking on absolute risk of bladder cancer. *Cancer Res.*, **73**, 2211–2220.

14. Nebbioso, A., Tambaro, F. P., Dell'Aversana, C., Altucci, L. (2018). Cancer epigenetics: Moving forward. *PLoS Genet.*, **14**, e1007362.

15. Vogelstein, B., Papadopoulos, N., Velculescu, V. E., Zhou, S., Diaz, L. A., Jr., Kinzler, K. W. (2013). Cancer genome landscapes. *Science*, **339**, 1546–1558.

16. Delhommeau, F., Dupont, S., Della Valle, V., James, C., Trannoy, S., Masse, A., Kosmider, O., Le Couedic, J. P., Robert, F., Alberdi, A., et al. (2009). Mutation in TET2 in myeloid cancers. *N. Engl. J. Med.*, **360**, 2289–2301.

17. Ley, T. J., Ding, L., Walter, M. J., McLellan, M. D., Lamprecht, T., Larson, D. E., Kandoth, C., Payton, J. E., Baty, J., Welch, J., et al. (2010). DNMT3A mutations in acute myeloid leukemia. *N. Engl. J. Med.*, **363**, 2424–2433.

18. Morin, R. D., Johnson, N. A., Severson, T. M., Mungall, A. J., An, J., Goya, R., Paul, J. E., Boyle, M., Woolcock, B. W., Kuchenbauer, F., et al. (2010). Somatic mutations altering EZH2 (Tyr641) in follicular and diffuse large B-cell lymphomas of germinal-center origin. *Nat. Genet.*, **42**, 181–185.

19. Beggs, A. D., Jones, A., El-Bahrawy, M., Abulafi, M., Hodgson, S. V., Tomlinson, I. P. (2013). Whole-genome methylation analysis of benign and malignant colorectal tumours. *J. Pathol.*, **229**, 697–704.

20. Ganesan, A. (2018). Epigenetics: The first 25 centuries. *Philos. Trans. R. Soc. Lond. B Biol. Sci.*, **373**.

21. Amodio, N., D'Aquila, P., Passarino, G., Tassone, P., Bellizzi, D. (2017). Epigenetic modifications in multiple myeloma: Recent advances on the role of DNA and histone methylation. *Expert Opin. Ther. Targets*, **21**, 91–101.

22. Serre, D., Lee, B. H., Ting, A. H. (2010). MBD-isolated Genome Sequencing provides a high-throughput and comprehensive survey of DNA methylation in the human genome. *Nucleic Acids Res.*, **38**, 391–399.

23. Brinkman, A. B., Simmer, F., Ma, K., Kaan, A., Zhu, J., Stunnenberg, H. G. (2010). Whole-genome DNA methylation profiling using MethylCap-seq. *Methods*, **52**, 232–236.

24. Maunakea, A. K., Nagarajan, R. P., Bilenky, M., Ballinger, T. J., D'Souza, C., Fouse, S. D., Johnson, B. E., Hong, C., Nielsen, C., Zhao, Y., et al. (2010). Conserved role of intragenic DNA methylation in regulating alternative promoters. *Nature*, **466**, 253–257.

25. Hayatsu, H. (2008). Discovery of bisulfite-mediated cytosine conversion to uracil, the key reaction for DNA methylation analysis: A personal account. *Proc. Jpn. Acad. Ser. B Phys. Biol. Sci.*, **84**, 321–330.

26. Shi, D. Q., Ali, I., Tang, J., Yang, W. C. (2017). New insights into 5hmC DNA modification: Generation, distribution and function. *Front. Genet.*, **8**, 100.

27. Dapprich, J., Ferriola, D., Mackiewicz, K., Clark, P. M., Rappaport, E., D'Arcy, M., Sasson, A., Gai, X., Schug, J., Kaestner, K. H., et al. (2016). The next generation of target capture technologies–large DNA fragment enrichment and sequencing determines regional genomic variation of high complexity. *BMC Genom.*, **17**, 486.

28. Clark, T. A., Spittle, K. E., Turner, S. W., Korlach, J. (2011). Direct detection and sequencing of damaged DNA bases. *Genome Integr.*, **2**, 10.

29. Schatz, M. C. (2017). Nanopore sequencing meets epigenetics. *Nat. Methods*, **14**, 347–348.

30. Barros-Silva, D., Marques, C. J., Henrique, R., Jeronimo, C. (2018). Profiling DNA methylation based on next-generation sequencing approaches: New insights and clinical applications. *Genes (Basel)*, **9**.

31. Barski, A., Cuddapah, S., Cui, K., Roh, T. Y., Schones, D. E., Wang, Z., Wei, G., Chepelev, I., Zhao, K. (2007). High-resolution profiling of histone methylations in the human genome. *Cell*, **129**, 823–837.

32. Johnson, D. S., Mortazavi, A., Myers, R. M., Wold, B. (2007). Genome-wide mapping of *in vivo* protein-DNA interactions. *Science*, **316**, 1497–1502.

33. Neph, S., Vierstra, J., Stergachis, A. B., Reynolds, A. P., Haugen, E., Vernot, B., Thurman, R. E., John, S., Sandstrom, R., Johnson, A. K., et al. (2012). An expansive human regulatory lexicon encoded in transcription factor footprints. *Nature*, **489**, 83–90.

34. Cui, K., Zhao, K. (2012). Genome-wide approaches to determining nucleosome occupancy in metazoans using MNase-Seq. *Methods Mol. Biol.*, **833**, 413–419.

35. Splinter, E., de Wit, E., Nora, E. P., Klous, P., van de Werken, H. J., Zhu, Y., Kaaij, L. J., van Ijcken, W., Gribnau, J., Heard, E., et al. (2011). The inactive X chromosome adopts a unique three-dimensional conformation that is dependent on Xist RNA. *Genes Dev.*, **25**, 1371–1383.

36. He, X., Chang, S., Zhang, J., Zhao, Q., Xiang, H., Kusonmano, K., Yang, L., Sun, Z. S., Yang, H., Wang, J. (2008). MethyCancer: The database of human DNA methylation and cancer. *Nucleic Acids Res.*, **36**, D836–D841.

37. Cancer Genome Atlas Research Network (2008). Comprehensive genomic characterization defines human glioblastoma genes and core pathways. *Nature*, **455**, 1061–1068.

38. Mei, S., Meyer, C. A., Zheng, R., Qin, Q., Wu, Q., Jiang, P., Li, B., Shi, X., Wang, B., Fan, J., et al. (2017). Cistrome cancer: A web resource for integrative gene regulation modeling in cancer. *Cancer Res.*, **77**, e19–e22.

39. Marino-Ramirez, L., Levine, K. M., Morales, M., Zhang, S., Moreland, R. T., Baxevanis, A. D., Landsman, D. (2011). The histone database: An integrated resource for histones and histone fold-containing proteins. *Database (Oxford)*, **2011**, bar048.

40. Khare, S. P., Habib, F., Sharma, R., Gadewal, N., Gupta, S., Galande, S. (2012). HIstome—A relational knowledgebase of human histone proteins and histone modifying enzymes. *Nucleic Acids Res.*, **40**, D337–D342.

41. Guo, S., Yan, F., Xu, J., Bao, Y., Zhu, J., Wang, X., Wu, J., Li, Y., Pu, W., Liu, Y., et al. (2015). Identification and validation of the methylation biomarkers of non-small cell lung cancer (NSCLC). *Clin. Epigenetics*, **7**, 3.

42. Exner, R., Pulverer, W., Diem, M., Spaller, L., Woltering, L., Schreiber, M., Wolf, B., Sonntagbauer, M., Schroder, F., Stift, J., et al. (2015). Potential of DNA methylation in rectal cancer as diagnostic and prognostic biomarkers. *Br. J. Cancer*, **113**, 1035–1045.

43. Legendre, C., Gooden, G. C., Johnson, K., Martinez, R. A., Liang, W. S., Salhia, B. (2015). Whole-genome bisulfite sequencing of cell-free DNA identifies signature associated with metastatic breast cancer. *Clin. Epigenetics*, **7**, 100.

44. Qu, K., Zaba, L. C., Satpathy, A. T., Giresi, P. G., Li, R., Jin, Y., Armstrong, R., Jin, C., Schmitt, N., Rahbar, Z., et al. (2017). Chromatin accessibility landscape of cutaneous T cell lymphoma and dynamic response to HDAC inhibitors. *Cancer Cell*, **32**, 27–41.

45. Cai, M., Kim, S., Wang, K., Farnham, P. J., Coetzee, G. A., Lu, W. (2016). 4C-seq revealed long-range interactions of a functional enhancer at the 8q24 prostate cancer risk locus. *Sci. Rep.*, **6**, 22462.

46. Consortium, E. P. (2012). An integrated encyclopedia of DNA elements in the human genome. *Nature*, **489**, 57–74.

47. Barretina, J., Caponigro, G., Stransky, N., Venkatesan, K., Margolin, A. A., Kim, S., Wilson, C. J., Lehar, J., Kryukov, G. V., Sonkin, D., et al. (2012). The Cancer Cell Line Encyclopedia enables predictive modelling of anticancer drug sensitivity. *Nature*, **483**, 603–607.

48. Klijn, C., Durinck, S., Stawiski, E. W., Haverty, P. M., Jiang, Z., Liu, H., Degenhardt, J., Mayba, O., Gnad, F., Liu, J., et al. (2015). A comprehensive transcriptional portrait of human cancer cell lines. *Nat. Biotechnol.*, **33**, 306–312.

49. Consortium, G. T. (2013). The Genotype-Tissue Expression (GTEx) project. *Nat. Genet.*, **45**, 580–585.

50. Uhlen, M., Fagerberg, L., Hallstrom, B. M., Lindskog, C., Oksvold, P., Mardinoglu, A., Sivertsson, A., Kampf, C., Sjostedt, E., Asplund, A., et al. (2015). Proteomics. Tissue-based map of the human proteome. *Science*, **347**, 1260419.

51. Cancer Genome Atlas Research Network, Weinstein, J. N., Collisson, E. A., Mills, G. B., Shaw, K. R., Ozenberger, B. A., Ellrott, K., Shmulevich, I., Sander, C., Stuart, J. M. (2013). The Cancer Genome Atlas Pan-Cancer analysis project. *Nat. Genet.*, **45**, 1113–1120.

52. Robinson, D., Van Allen, E. M., Wu, Y. M., Schultz, N., Lonigro, R. J., Mosquera, J. M., Montgomery, B., Taplin, M. E., Pritchard, C. C., Attard, G., et al. (2015). Integrative clinical genomics of advanced prostate cancer. *Cell*, **162**, 454.

53. Mody, R. J., Wu, Y. M., Lonigro, R. J., Cao, X., Roychowdhury, S., Vats, P., Frank, K. M., Prensner, J. R., Asangani, I., Palanisamy, N., et al. (2015). Integrative Clinical sequencing in the management of refractory or relapsed cancer in youth. *JAMA*, **314**, 913–925.

54. Oberg, J. A., Glade Bender, J. L., Sulis, M. L., Pendrick, D., Sireci, A. N., Hsiao, S. J., Turk, A. T., Dela Cruz, F. S., Hibshoosh, H., Remotti, H., et al. (2016). Implementation of next generation sequencing into pediatric hematology-oncology practice: Moving beyond actionable alterations. *Genome Med.*, **8**, 133.

55. Robinson, D. R., Wu, Y. M., Lonigro, R. J., Vats, P., Cobain, E., Everett, J., Cao, X., Rabban, E., Kumar-Sinha, C., Raymond, V., et al. (2017). Integrative clinical genomics of metastatic cancer. *Nature*, **548**, 297–303.

56. Kurimoto, K., Yabuta, Y., Ohinata, Y., Ono, Y., Uno, K. D., Yamada, R. G., Ueda, H. R., Saitou, M. (2006). An improved single-cell cDNA amplification method for efficient high-density oligonucleotide microarray analysis. *Nucleic Acids Res.*, **34**, e42.

57. Huet, S., Tesson, B., Salles, G. (2018). Predictive gene-expression score for follicular lymphoma—Authors' reply. *Lancet Oncol.*, **19**, e282.

58. Boyault, S., Rickman, D. S., de Reynies, A., Balabaud, C., Rebouissou, S., Jeannot, E., Herault, A., Saric, J., Belghiti, J., Franco, D., et al. (2007). Transcriptome classification of HCC is related to gene alterations and to new therapeutic targets. *Hepatology*, **45**, 42–52.

59. Hoshida, Y., Nijman, S. M., Kobayashi, M., Chan, J. A., Brunet, J. P., Chiang, D. Y., Villanueva, A., Newell, P., Ikeda, K., Hashimoto, M., et al. (2009). Integrative transcriptome analysis reveals common molecular subclasses of human hepatocellular carcinoma. *Cancer Res.*, **69**, 7385–7392.

60. Amodio, N., Raimondi, L., Juli, G., Stamato, M. A., Caracciolo, D., Tagliaferri, P., Tassone, P. (2018). MALAT1: A druggable long non-coding RNA for targeted anti-cancer approaches. *J. Hematol. Oncol.*, **11**, 63.

61. Ochoa, A. E., Choi, W., Su, X., Siefker-Radtke, A., Czerniak, B., Dinney, C., McConkey, D. J. (2016). Specific micro-RNA expression patterns distinguish the basal and luminal subtypes of muscle-invasive bladder cancer. *Oncotarget*, **7**, 80164–80174.

62. Raimondi, L., De Luca, A., Morelli, E., Giavaresi, G., Tagliaferri, P., Tassone, P., Amodio, N. (2016). MicroRNAs: Novel crossroads between myeloma cells and the bone marrow microenvironment. *Biomed. Res. Int.* **2016**, 6504593.

63. Volinia, S., Galasso, M., Sana, M. E., Wise, T. F., Palatini, J., Huebner, K., Croce, C. M. (2012). Breast cancer signatures for invasiveness and prognosis defined by deep sequencing of microRNA. *Proc. Natl. Acad. Sci. U.S.A.*, **109**, 3024–3029.

64. Namkung, J., Kwon, W., Choi, Y., Yi, S. G., Han, S., Kang, M. J., Kim, S. W., Park, T., Jang, J. Y. (2016). Molecular subtypes of pancreatic cancer based on miRNA expression profiles have independent prognostic value. *J. Gastroenterol. Hepatol.*, **31**, 1160–1167.

65. Karczewski, K. J., Snyder, M. P. (2018). Integrative omics for health and disease. *Nat. Rev. Genet.*, **19**, 299–310.

66. Poste, G. (2012). Biospecimens, biomarkers, and burgeoning data: The imperative for more rigorous research standards. *Trends Mol. Med.*, **18**, 717–722.

67. Hanash, S., Taguchi, A. (2011). Application of proteomics to cancer early detection. *Cancer J.*, **17**, 423–428.

68. Baker, E. S., Liu, T., Petyuk, V. A., Burnum-Johnson, K. E., Ibrahim, Y. M., Anderson, G. A., Smith, R. D. (2012). Mass spectrometry for translational proteomics: Progress and clinical implications. *Genome Med.*, **4**, 63.

69. Yates, J. R., 3rd. (2013). The revolution and evolution of shotgun proteomics for large-scale proteome analysis. *J. Am. Chem. Soc.*, **135**, 1629–1640.

70. Aslam, B., Basit, M., Nisar, M. A., Khurshid, M., Rasool, M. H. (2017). Proteomics: Technologies and their applications. *J. Chromatogr. Sci.*, **55**, 182–196.

71. Padoan, A., Basso, D., Zambon, C. F., Prayer-Galetti, T., Arrigoni, G., Bozzato, D., Moz, S., Zattoni, F., Bellocco, R., Plebani, M. (2018). MALDI-TOF peptidomic analysis of serum and post-prostatic massage urine specimens to identify prostate cancer biomarkers. *Clin. Proteom.*, **15**, 23.

72. Walther, T. C., Mann, M. (2010). Mass spectrometry-based proteomics in cell biology. *J. Cell Biol.*, **190**, 491–500.

73. Fernandez-Coto, D. L., Gil, J., Hernandez, A., Herrera-Goepfert, R., Castro-Romero, I., Hernandez-Marquez, E., Arenas-Linares, A. S., Calderon-Sosa, V. T., Sanchez-Aleman, M. A., Mendez-Tenorio, A., et al. (2018). Quantitative proteomics reveals proteins involved in the progression from non-cancerous lesions to gastric cancer. *J. Proteom.*, **186**, 15–27.

74. Wang, X., He, Y., Ye, Y., Zhao, X., Deng, S., He, G., Zhu, H., Xu, N., Liang, S. (2018). SILAC-based quantitative MS approach for real-time recording protein-mediated cell-cell interactions. *Sci. Rep.*, **8**, 8441.

75. Shin, J., Lee, W., Lee, W. (2008). Structural proteomics by NMR spectroscopy. *Expert Rev. Proteom.*, **5**, 589–601.

76. Vizcaino, J. A., Cote, R. G., Csordas, A., Dianes, J. A., Fabregat, A., Foster, J. M., Griss, J., Alpi, E., Birim, M., Contell, J., et al. (2013). The PRoteomics IDEntifications (PRIDE) database and associated tools: Status in 2013. *Nucleic Acids Res.*, **41**, D1063–D1069.

77. Croft, D., O'Kelly, G., Wu, G., Haw, R., Gillespie, M., Matthews, L., Caudy, M., Garapati, P., Gopinath, G., Jassal, B., et al. (2011). Reactome: A database of reactions, pathways and biological processes. *Nucleic Acids Res.*, **39**, D691–D697.

78. Kanehisa, M., Goto, S., Sato, Y., Furumichi, M., Tanabe, M. (2012). KEGG for integration and interpretation of large-scale molecular data sets. *Nucleic Acids Res.*, **40**, D109–D114.

79. Kanehisa, M., Furumichi, M., Tanabe, M., Sato, Y., Morishima, K. (2017). KEGG: New perspectives on genomes, pathways, diseases and drugs. *Nucleic Acids Res.*, **45**, D353–D361.

80. Tyanova, S., Albrechtsen, R., Kronqvist, P., Cox, J., Mann, M., Geiger, T. (2016). Proteomic maps of breast cancer subtypes. *Nat. Commun.*, **7**, 10259.

81. Lawrence, R. T., Perez, E. M., Hernandez, D., Miller, C. P., Haas, K. M., Irie, H. Y., Lee, S. I., Blau, C. A., Villen, J. (2015). The proteomic landscape of triple-negative breast cancer. *Cell Rep.*, **11**, 990.

82. Bohnenberger, H., Kaderali, L., Strobel, P., Yepes, D., Plessmann, U., Dharia, N. V., Yao, S., Heydt, C., Merkelbach-Bruse, S., Emmert, A., et al. (2018). Comparative proteomics reveals a diagnostic signature for pulmonary head-and-neck cancer metastasis. *EMBO Mol. Med.*, **10**.

83. Wilhelm, M., Schlegl, J., Hahne, H., Gholami, A. M., Lieberenz, M., Savitski, M. M., Ziegler, E., Butzmann, L., Gessulat, S., Marx, H., et al. (2014). Mass-spectrometry-based draft of the human proteome. *Nature*, **509**, 582–587.

84. Swiatly, A., Horala, A., Matysiak, J., Hajduk, J., Nowak-Markwitz, E., Kokot, Z. J. (2018). Understanding ovarian Cancer: ITRAQ-Based proteomics for biomarker discovery. *Int. J. Mol. Sci.*, **19**.

85. Ali, M., Khan, S. A., Wennerberg, K., Aittokallio, T. (2018). Global proteomics profiling improves drug sensitivity prediction: Results from a multi-omics, pan-cancer modeling approach. *Bioinformatics*, **34**, 1353–1362.

86. Yanovich, G., Agmon, H., Harel, M., Sonnenblick, A., Peretz, T., Geiger, T. (2018). Clinical proteomics of breast cancer reveals a novel layer of breast cancer classification. *Cancer Res.*, **78**, 6001–6010.

87. Cruz, I. N., Coley, H. M., Kramer, H. B., Madhuri, T. K., Safuwan, N. A., Angelino, A. R., Yang, M. (2017). Proteomics analysis of ovarian cancer cell lines and tissues reveals drug resistance-associated proteins. *Cancer Genom. Proteom.*, **14**, 35–51.

88. Creighton, C. J., Huang, S. (2015). Reverse phase protein arrays in signaling pathways: A data integration perspective. *Drug Des. Dev. Ther.*, **9**, 3519–3527.

89. Mueller, C., deCarvalho, A. C., Mikkelsen, T., Lehman, N. L., Calvert, V., Espina, V., Liotta, L. A., Petricoin, E. F., 3rd. (2014). Glioblastoma cell enrichment is critical for analysis of phosphorylated drug targets and proteomic-genomic correlations. *Cancer Res.*, **74**, 818–828.

90. Masuda, M., Chen, W. Y., Miyanaga, A., Nakamura, Y., Kawasaki, K., Sakuma, T., Ono, M., Chen, C. L., Honda, K., Yamada, T. (2014). Alternative mammalian target of rapamycin (mTOR) signal activation in sorafenib-resistant hepatocellular carcinoma cells revealed by array-based pathway profiling. *Mol. Cell Proteom.*, **13**, 1429–1438.

91. Patti, G. J., Yanes, O., Siuzdak, G. (2012). Innovation: Metabolomics: The apogee of the omics trilogy. *Nat. Rev. Mol. Cell Biol.*, **13**, 263–269.

92. Zhang, A., Sun, H., Yan, G., Wang, P., Han, Y., Wang, X. (2014). Metabolomics in diagnosis and biomarker discovery of colorectal cancer. *Cancer Lett.*, **345**, 17–20.

93. Brown, D. G., Rao, S., Weir, T. L., O'Malia, J., Bazan, M., Brown, R. J., Ryan, E. P. (2016). Metabolomics and metabolic pathway networks from human colorectal cancers, adjacent mucosa, and stool. *Cancer Metab.*, **4**, 11.

94. Wishart, D. S., Feunang, Y. D., Marcu, A., Guo, A. C., Liang, K., Vazquez-Fresno, R., Sajed, T., Johnson, D., Li, C., Karu, N., et al. (2018). HMDB 4.0: The human metabolome database for 2018. *Nucleic Acids Res.*, **46**, D608–D617.

95. Chen, W. L., Wang, J. H., Zhao, A. H., Xu, X., Wang, Y. H., Chen, T. L., Li, J. M., Mi, J. Q., Zhu, Y. M., Liu, Y. F., et al. (2014). A distinct glucose metabolism signature of acute myeloid leukemia with prognostic value. *Blood*, **124**, 1645–1654.

96. Chaturvedi, A., Araujo Cruz, M. M., Jyotsana, N., Sharma, A., Yun, H., Gorlich, K., Wichmann, M., Schwarzer, A., Preller, M., Thol, F., et al. (2013). Mutant IDH1 promotes leukemogenesis *in vivo* and can be specifically targeted in human AML. *Blood*, **122**, 2877–2887.

97. Zhang, Y., He, C., Qiu, L., Wang, Y., Qin, X., Liu, Y., Li, Z. (2016). Serum unsaturated free fatty acids: A potential biomarker panel for early-stage detection of colorectal cancer. *J. Cancer*, **7**, 477–483.

98. Locasale, J. W., Grassian, A. R., Melman, T., Lyssiotis, C. A., Mattaini, K. R., Bass, A. J., Heffron, G., Metallo, C. M., Muranen, T., Sharfi, H., et al. (2011). Phosphoglycerate dehydrogenase diverts glycolytic flux and contributes to oncogenesis. *Nat. Genet.*, **43**, 869–874.

99. Giskeodegard, G. F., Bertilsson, H., Selnaes, K. M., Wright, A. J., Bathen, T. F., Viset, T., Halgunset, J., Angelsen, A., Gribbestad, I. S., Tessem, M. B. (2013). Spermine and citrate as metabolic biomarkers for assessing prostate cancer aggressiveness. *PLoS One*, **8**, e62375.

100. Mayers, J. R., Wu, C., Clish, C. B., Kraft, P., Torrence, M. E., Fiske, B. P., Yuan, C., Bao, Y., Townsend, M. K., Tworoger, S. S., et al. (2014). Elevation of circulating branched-chain amino acids is an early event in human pancreatic adenocarcinoma development. *Nat. Med.*, **20**, 1193–1198.

101. Guzzi, P. H., Milenkovic, T. (2018). Survey of local and global biological network alignment: The need to reconcile the two sides of the same coin. *Brief. Bioinform.*, **19**, 472–481.

102. Hofree, M., Shen, J. P., Carter, H., Gross, A., Ideker, T. (2013). Network-based stratification of tumor mutations. *Nat. Methods*, **10**, 1108–1115.

103. Zhang, W., Ota, T., Shridhar, V., Chien, J., Wu, B., Kuang, R. (2013). Network-based survival analysis reveals subnetwork signatures for predicting outcomes of ovarian cancer treatment. *PLoS Comput. Biol.*, **9**, e1002975.

104. Przytycki, P. F., Singh, M. (2017). Differential analysis between somatic mutation and germline variation profiles reveals cancer-related genes. *Genome Med.*, **9**, 79.

105. Guzzi, P. H., Di Martino, M. T., Tagliaferri, P., Tassone, P., Cannataro, M. (2015). Analysis of miRNA, mRNA, and TF interactions through network-based methods. *EURASIP J. Bioinform. Syst. Biol.*, **2015**, 4.

106. Yan, Z., Shah, P. K., Amin, S. B., Samur, M. K., Huang, N., Wang, X., Misra, V., Ji, H., Gabuzda, D., Li, C. (2012). Integrative analysis of gene and miRNA expression profiles with transcription factor-miRNA feed-forward loops identifies regulators in human cancers. *Nucleic Acids Res.*, **40**, e135.

107. Agarwal, V., Bell, G. W., Nam, J. W., Bartel, D. P. (2015). Predicting effective microRNA target sites in mammalian mRNAs. *Elife*, **4**.

108. Bisognin, A., Sales, G., Coppe, A., Bortoluzzi, S., Romualdi, C. (2012). MAGIA(2): From miRNA and genes expression data integrative analysis to microRNA-transcription factor mixed regulatory circuits (2012 update). *Nucleic Acids Res.*, **40**, W13–W21.

109. Huang, G. T., Athanassiou, C., Benos, P. V. (2011). mirConnX: Condition-specific mRNA-microRNA network integrator. *Nucleic Acids Res.*, **39**, W416–W423.

110. Afshar, A. S., Xu, J., Goutsias, J. (2014). Integrative identification of deregulated miRNA/TF-mediated gene regulatory loops and networks in prostate cancer. *PLoS One*, **9**, e100806.

111. Forsberg, E. M., Huan, T., Rinehart, D., Benton, H. P., Warth, B., Hilmers, B., Siuzdak, G. (2018). Data processing, multi-omic pathway mapping, and metabolite activity analysis using XCMS online. *Nat. Protoc.*, **13**, 633–651.

112. Ruffalo, M., Koyuturk, M., Sharan, R. (2015). Network-based integration of disparate omic data to identify "silent players" in cancer. *PLoS Comput. Biol.*, **11**, e1004595.

113. Ru, B., Tong, Y., Zhang, J. (2018). MR4Cancer: A web server prioritizing master regulators for cancer. *Bioinformatics*.

114. Krempel, R., Kulkarni, P., Yim, A., Lang, U., Habermann, B., Frommolt, P. (2018). Integrative analysis and machine learning on cancer genomics data using the Cancer Systems Biology Database (CancerSysDB). *BMC Bioinform.*, **19**, 156.

115. Baek, S. J., Yang, S., Kang, T. W., Park, S. M., Kim, Y. S., Kim, S. Y. (2013). MENT: Methylation and expression database of normal and tumor tissues. *Gene*, **518**, 194–200.

116. Diez-Villanueva, A., Mallona, I., Peinado, M. A. (2015). Wanderer, an interactive viewer to explore DNA methylation and gene expression data in human cancer. *Epigenetics Chromatin*, **8**, 22.

117. Deng, G., Yang, J., Zhang, Q., Xiao, Z. X., Cai, H. (2018). MethCNA: A database for integrating genomic and epigenomic data in human cancer. *BMC Genom.*, **19**, 138.

118. Cancer Genome Atlas Network (2012). Comprehensive molecular portraits of human breast tumours. *Nature*, **490**, 61–70.

119. Tomczak, K., Czerwinska, P., Wiznerowicz, M. (2015). The Cancer Genome Atlas (TCGA): An immeasurable source of knowledge. *Contemp. Oncol. (Pozn)*, **19**, A68–A77.

120. Akbani, R., Ng, P. K., Werner, H. M., Shahmoradgoli, M., Zhang, F., Ju, Z., Liu, W., Yang, J. Y., Yoshihara, K., Li, J., et al. (2014). A pan-cancer proteomic perspective on The Cancer Genome Atlas. *Nat. Commun.*, **5**, 3887.

121. Hoadley, K. A., Yau, C., Wolf, D. M., Cherniack, A. D., Tamborero, D., Ng, S., Leiserson, M. D. M., Niu, B., McLellan, M. D., Uzunangelov, V., et al. (2014). Multiplatform analysis of 12 cancer types reveals molecular classification within and across tissues of origin. *Cell*, **158**, 929–944.

122. Zhu, B., Song, N., Shen, R., Arora, A., Machiela, M. J., Song, L., Landi, M. T., Ghosh, D., Chatterjee, N., Baladandayuthapani, V., et al. (2017). Integrating clinical and multiple omics data for prognostic assessment across human cancers. *Sci. Rep.*, **7**, 16954.

123. Koplev, S., Lin, K., Dohlman, A. B., Ma'ayan, A. (2018). Integration of pan-cancer transcriptomics with RPPA proteomics reveals mechanisms of epithelial-mesenchymal transition. *PLoS Comput. Biol.*, **14**, e1005911.

124. Snyder, A., Nathanson, T., Funt, S. A., Ahuja, A., Buros Novik, J., Hellmann, M. D., Chang, E., Aksoy, B. A., Al-Ahmadie, H., Yusko, E., et al. (2017). Contribution of systemic and somatic factors to clinical response and resistance to PD-L1 blockade in urothelial cancer: An exploratory multi-omic analysis. *PLoS Med.*, **14**, e1002309.

125. Mancikova, V., Montero-Conde, C., Perales-Paton, J., Fernandez, A., Santacana, M., Jodkowska, K., Inglada-Perez, L., Castelblanco, E., Borrego, S., Encinas, M., et al. (2017). Multilayer OMIC data in medullary thyroid carcinoma identifies the STAT3 pathway as a potential therapeutic target in RET(M918T) tumors. *Clin. Cancer Res.*, **23**, 1334–1345.

126. Piccolo, S. R., Hoffman, L. M., Conner, T., Shrestha, G., Cohen, A. L., Marks, J. R., Neumayer, L. A., Agarwal, C. A., Beckerle, M. C., Andrulis, I. L., et al. (2016). Integrative analyses reveal signaling pathways underlying familial breast cancer susceptibility. *Mol. Syst. Biol.*, **12**, 860.

127. Robles, A. I., Harris, C. C. (2017). Integration of multiple "OMIC" biomarkers: A precision medicine strategy for lung cancer. *Lung Cancer*, **107**, 50–58.

128. Li, Z., Liu, H., Niu, Z., Zhong, W., Xue, M., Wang, J., Yang, F., Zhou, Y., Zhou, Y., Xu, T., et al. (2018). Temporal proteomic analysis of pancreatic β-cells in response to lipotoxicity and glucolipotoxicity. *Mol. Cell. Proteom.*

129. Di Martino, M. T., Guzzi, P. H., Caracciolo, D., Agnelli, L., Neri, A., Walker, B. A., Morgan, G. J., Cannataro, M., Tassone, P., Tagliaferri, P. (2015). Integrated analysis of microRNAs, transcription factors and target genes expression discloses a specific molecular architecture of hyperdiploid multiple myeloma. *Oncotarget*, **6**, 19132–19147.

Chapter 27

Harmonizing Outcomes for Genomic Medicine: Comparison of eMERGE Outcomes to ClinGen Outcome/ Intervention Pairs

Janet L. Williams, MS,[a] Wendy K. Chung, MD, PhD,[b] Alex Fedotov, PhD,[c] Krzysztof Kiryluk, MD,[d] Chunhua Weng, PhD,[e] John J. Connolly, PhD,[f] Margaret Harr, MS,[f] Hakon Hakonarson, MD, PhD,[f,g] Kathleen A. Leppig, MD,[h] Eric B. Larson, MD,[i] Gail P. Jarvik, MD, PhD,[j] David L. Veenstra, Pharm D, PhD,[k] Christin Hoell, MS,[l] Maureen E. Smith, MS,[l] Ingrid A. Holm, MD, MPH,[m] Josh F. Peterson, MD, MPH,[n] and Marc S. Williams, MD[a]

[a]Genomic Medicine Institute, Geisinger, Danville, Pennsylvania, USA
[b]Departments of Pediatrics and Medicine, Columbia University, New York, USA
[c]Irving Institute for Clinical and Translational Research,
Columbia University, New York, USA
[d]Department of Medicine, Division of Nephrology,
Columbia University, New York, USA
[e]Department of Biomedical Informatics, Columbia University, New York, USA
[f]Children's Hospital of Philadelphia, Philadelphia, USA
[g]Perelman School of Medicine, University of Pennsylvania, Philadelphia, USA
[h]Genetic Services, Kaiser Permanente of Washington, Seattle, Washington, USA
[i]Kaiser Permanente Washington Health Research Institute, Seattle, Washington, USA
[j]Departments of Medicine (Medical Genetics) and Genome Sciences,
University of Washington, Seattle, Washington, USA
[k]Department Pharmacy, University of Washington, Seattle, Washington, USA
[l]Center for Genetic Medicine, Northwestern University, Chicago, Illinois, USA
[m]Division of Genetics and Genomics, Boston Children's Hospital,
and Department of Pediatrics, Harvard Medical School, Boston, Massachusetts, USA
[n]Departments of Biomedical Informatics and Medicine,
School of Medicine, Vanderbilt University, Nashville, Tennessee, USA

The Road from Nanomedicine to Precision Medicine
Edited by Shaker A. Mousa, Raj Bawa, and Gerald F. Audette
Copyright © 2020 Jenny Stanford Publishing Pte. Ltd.
ISBN 978-981-4800-59-4 (Hardcover), 978-0-429-29501-0 (eBook)
www.jennystanford.com

Keywords: genomics, genomic medicine, health outcomes, evidence, standards, Electronic Medical Records and Genomics, Clinical Genome Resource, precision public health, Patient-Reported Outcomes Measurement Information System, patient-reported outcome measures, Patient-Centered Outcomes Research Institute, National Human Genome Research Institute, hereditary breast and ovarian cancer syndrome, Lynch syndrome, familial hypercholesterolemia, actionability, Implementing Genomics in Practice Network, Clinical Sequencing Exploratory Research Consortium, aortopathies, hypertrophic cardiomyopathy, inherited arrhythmias, chromosomal deletion, 22q deletion

27.1 Introduction

Genomic medicine is defined by the National Human Genome Research Institute (NHGRI) as, "an emerging medical discipline that involves using genomic information about an individual as part of their clinical care (e.g., for diagnostic or therapeutic decision-making) and the health outcomes and policy implications of that clinical use" [1]. Prior research has demonstrated that genomic medicine has promise for improving health outcomes. As a result, it is beginning to emerge into the clinical practice for selected indications including pharmacogenomics [2], precision oncology [3], and diagnosis of complex conditions suspected be genetic [4]. Large-scale research programs such as the All of Us program funded by the United States National Institutes of Health (NIH) [5] and smaller private clinical research programs [6, 7] are beginning to explore the integration of genomic information with other health information to assess the impact on patient outcomes that, it is hoped, will ultimately result in more programs in precision public health.

Several barriers to the implementation of genomic medicine have been identified [8]. One of the most important of these is the lack of evidence of the clinical utility of the interventions. Stated another way, while there is strong evidence about the association of genomic variation with genetic disorders, there is, with few exceptions, inadequate information about the impact on outcomes (both positive and negative) of implementing genomic medicine into clinical care [9, 10]. This lack of evidence results in a reluctance of healthcare systems to invest in and payers to reimburse for genomic medicine interventions. There is a general agreement that evidence of the impact of genomic medicine on health outcomes must be generated. There are many barriers to the generation of evidence [9, 10], one of which is the lack of agreed-upon outcomes to measure the impact of conditions of interest.

The NHGRI has funded several large collaborations to study genomic medicine in clinical care. These include, but are not limited to, the Implementing Genomics in Practice (IGNITE) network [11], the Clinical Sequencing Evidence-Generating Research (CSER) consortium [12], and the Electronic Medical Records and Genomics (eMERGE) network [13]. All three of these groups have a workgroup tasked to develop outcomes for site-specific and network projects. While these groups have worked to harmonize outcomes within each project, it was not until 2017 that an effort started to try to harmonize outcomes across these and potentially other NHGRI-funded projects. This was initially accomplished by creating formal liaisons between each of the respective outcomes groups, and by holding joint meetings between the networks/consortium [14]. While this has resulted in some convergence, the differences between the projects and the lack of alignment of the project timelines have hindered the agreement on a standard set of outcomes across the three networks.

eMERGE is in its third phase of funding. The focus of this phase is the return of genomic results to participants [15]. A total of just over 25,000 participants will be sequenced on a next-generation sequencing platform, eMERGEseq, that contains 109 genes and a number of single nucleotide variants, including pharmacogenomic variants that may also be returned to participants [16]. The eMERGE Outcomes Working Group (OWG) was tasked to develop outcome measures for a set of genetic disorders for which the associated genes would be interrogated by sequencing. The OWG identified another NHGRI-funded project, the Clinical Genome Resource (ClinGen) [17] that had a relevant activity that could be used to move outcomes harmonization forward. Herein we report the results of a comparison between the eMERGE-defined outcomes and the ClinGen outcome intervention pairs.

27.2 Materials and Methods

eMERGE network sites represented on the OWG selected a disorder(s) for which their site developed clinical outcome measures. The outcomes were organized into three categories, process outcomes, intermediate outcomes, and health outcomes (Table 27.1). While health outcomes are of the greatest interest, the relatively short project timeline necessitated reliance on the process and intermediate outcomes for which a chain of evidence exists relating them to health outcomes of interest. Sites developed outcomes using their own approach, with the expectation that any proposed outcomes would have evidence of its relevance to clinical care. Emphasis was given to outcomes that were related to published clinical and practice guidelines where available. Once the draft outcomes were developed, they were presented to the OWG for discussion and revisions. The penultimate draft was submitted to the eMERGE coordinating center that, under the direction of one of the OWG co-chairs (JP), was tasked to develop the outcomes into a collection tool that could be created in REDCap [18] using a standard format. The coordinating center worked with the individual sites to create the final version of the outcomes.

The ClinGen Actionability Working Group (AWG) was tasked to assess the relative actionability of returning a genomic variant identified in an asymptomatic patient undergoing next-generation sequencing [19]. This was to be accomplished through four activities:

(1) Develop rigorous and standardized procedures for categorically defining "clinical actionability"; a concept that includes a known ability to intervene and thereby avert a poor outcome due to a previously unsuspected high risk of disease.

(2) Nominate genes and diseases to score for "clinical actionability."

(3) Produce evidence-based reports and semi-quantitative metric scores using a standardized method for nominated gene-disease pairs.

(4) Make these reports and actionability scores publicly available to aid broad efforts for prioritizing those human genes with the greatest relevance for clinical intervention.

Table 27.1 The framework of outcomes for clinical implementation

Outcome type	Description	Examples
Process	The specific steps in a process that lead—either positively or negatively—to a particular health outcome	Lipid profile performed after the return of a pathogenic variant in *LDLR*, a gene associated with familial hypercholesterolemia
Intermediate	A biomarker associated—either positively or negatively—to a particular health outcome	An LDL cholesterol level at or below the target level of 100 mg/dL in response to interventions recommended based on presences of a pathogenic variant in *LDLR*
Health	Change in the health of an individual, group of people or population which is attributable to an intervention or series of interventions	Decrease in myocardial infarction, or cardiac revascularization procedures in response to interventions recommended based on presences of a pathogenic variant in *LDLR*

The AWG has developed a set of outcome intervention pairs [20] that have been scored using a standardized approach informed by evidence-based summaries as described in a methods paper from 2016 [21]. The published outcome intervention pairs' table represents those that have been scored by the AWG. The evidence summary also contains interventions and outcomes that were not formally scored. Both the table and the associated evidence summary were reviewed to completely ascertain the interventions and outcomes that had been reviewed by the AWG.

For the comparison, each site participating in the exercise compared the set of outcomes developed for the disorder in eMERGE to the corresponding outcome intervention pair published on the AWG website. If the eMERGE outcome was represented in the scored AWG outcome intervention pair, it was categorized as concordant. If it was not represented in the scored AWG outcome intervention pair, but was noted in the evidence summary, it was also categorized as concordant with the annotation that it did not cross the threshold for scoring by the AWG. If the outcome was not present in either the scored list or evidence summary, it was categorized as discordant. Conversely, if an outcome intervention was present on the AWG scored list, but not represented as an eMERGE outcome, it was also categorized as discordant. The evidence summaries were not comprehensively reviewed for outcomes to compare to eMERGE outcomes.

The sites' comparisons were compiled and reviewed by one of the authors (MSW) who also independently compared the eMERGE outcomes to the AWG outcome intervention pairs. No differences were noted between the sites' scores and the second review for the AWG outcome intervention pairs. A few outcomes were identified in the evidence summaries that had not been scored by the sites, and these were added to the comparison table. The final comparison table was reviewed and approved by all the authors.

27.3 Results

A total of 12 disorders were scored (Tables 27.2 and 27.3). The full comparison table with all defined eMERGE outcomes for each disorder is provided in the supplemental materials. Three gene/variant disorder pairs with outcomes defined by eMERGE do not have an AWG actionability score or evidence summary. *CFTR*/Cystic Fibrosis is being returned by eMERGE but has not yet been evaluated by the ClinGen AWG. While adult familial hypercholesterolemia (FH associated with the genes *LDLR*, *APOB*, and *PCSK9*) has been evaluated by both the OWG and AWG, FH in the pediatric population has only been evaluated by the OWG. This is because ClinGen initially focused on conditions in the adult population. However, this year, a pediatric AWG is being convened by ClinGen and one of their first conditions to evaluate will be pediatric FH. Finally, eMERGE is studying a large, well-characterized copy number variant (CNV) at chromosome 22q11.2 that encompasses many genes. The AWG is only looking at single gene-disorder associations at present.

Table 27.2 Disorders with equivalent definitions from eMERGE and ClinGen

Disorder	Genes	eMERGE Outcomes	AWG Scored O/I Pair	AWG evidence review
OTC Deficiency	*OTC*	**Process**		
		Metabolic testing	No	Yes
		Metabolic crisis plan in EHR	No	No
		Intermediate		
		Low protein diet	Yes	
		Prescription for nitrogen scavenger	Yes	
		Health		
		Metabolic protocol applied during illness	Yes (Hyperammonemic encephalopathy)	
Tuberous sclerosis	*TSC1, TSC2*	**Process**		
		Imaging studies	Yes	
		Assessment for lymphangioleiomyoma-tosis (LAM)	Yes	
		Intermediate		
		Discontinuation of estrogen containing medications (F)	No	Yes
		Use of inhibitor of renin-aldosterone-angiotensin system as first line therapy for hypertension	No	No
		Avoid ACE inhibitor	No	No
		No	Use of mTOR inhibitor	

(*Continued*)

Table 27.2 (*Continued*)

Disorder	Genes	eMERGE Outcomes	AWG Scored O/I Pair	AWG evidence review
		Health		
		No	Development of SEGA, non-SEGA tumors, LAM	
HBOC (Breast)	*BRCA1, BRCA2*	**Process**		
		Breast self-exam	Yes	
		Breast imaging	Yes	
		Specialty referral	No	Yes
		Intermediate		
		Risk reducing mastectomy	Yes	
		Selective estrogen receptor modulator	No	Yes
		Aromatase inhibitor	No	No
		Discontinuation HRT	No	No
		Health		
		Breast cancer	Yes	
		Vital status	No	Yes
HBOC (Ovarian	*BRCA1, BRCA2*	**Process**		
		Pelvic US	Yes	No
		CA 125	Yes	No
		Specialty referral	Yes	No
		Intermediate		
		Prophylactic BSO or TAH/BSO	Yes	No
		Oral contraceptives	Yes	No
		Health		
		Ovarian, fallopian, peritoneal or endometrial cancer	Yes	
		vital status	No	Yes

Disorder	Genes	eMERGE Outcomes	AWG Scored O/I Pair	AWG evidence review
Adult FH	LDLR, APOB, PCSK9	**Process**		
		Laboratory testing (lipid, CRP)	No	Yes
		Coronary CT angiogram	No	Yes
		Echocardiogram	No	Yes
		ECG	No	No
		Stress test	No	No
		Specialty referral	No	No
		No	No	Cardiac catheterization
		Intermediate		
		Lipid lowering therapy	Yes (statins)	High-intensity statins
		Aspirin	No	Yes
		Coronary revascularization	No	No
		No	High cholesterol	
		Health		

Of the remaining nine gene(s)-disorder pairs defined by eMERGE, five had equivalent definitions from the AWG, while four had some differences which raised interesting issues that impacted the comparison. These two groups will be discussed separately.

The five disorders with equivalent definitions from both groups and the associated genes are presented in Table 27.2. It should be noted that the eMERGE project is only returning results from two genes that are associated with breast and/or ovarian cancer risk (*BRCA1* and *BRCA2*). Three genes with evidence for association with breast cancer are on the eMERGEseq platform (*ATM*, *CHEK2*, *PALB2*), but were not used to develop outcomes. These have been scored by the AWG but had much lower actionability scores than *BRCA1* and *BRCA2*; therefore, they were excluded from the comparison for the purposes of this study.

Table 27.3 Disorders with differing definitions between eMERGE and ClinGen

Disorder	Genes	eMERGE outcomes	ClinGen Actionability Working Group			
			Lynch syndrome (MLH1, MSH2, MSH6, PMS2)		Familial adenomatous polyposis (FAP)	
			Scored 0/1 Pair	Evidence Review	Scored 0/1 Pair	Evidence Review
Colorectal cancer	MLH1, MSH2, MSH6, PMS2, FAP					
		Process				
		Specialist Referral	No	No	No	Yes (gastroenterology)
		Intermediate				
		CRC Screening	Yes		No	No
		Other cancer screening	Yes		No	Yes
		Familial Cascade Testing	No	Yes	No	Yes
		No			Colectomy	
		Health				
		CRC (polyps, hospitalization, death)	Yes		Yes	
		Gynecologic cancer (endometrial, ovarian)	Yes		N/A	N/A

Disorder	Genes	eMERGE outcomes	ClinGen Actionability Working Group	
			Arterial tortuosity syndrome (SLC2A10)	FTAAD (FBN1, TGFBR1/2, SMAD3, ACTA2, MYLK, MYH11)
Aortopathies	FBN1, TGFBR1/2, SMAD3, ACTA2, MYLK, MYH11	**Process**		
		Aortic Imaging	Yes	Yes
		Magnetic Resonance Angiography	Yes	Yes
		High risk pregnancy management	Yes	Yes
		No	Recommendation to avoid contact sports	Yes
		No	Ophthalmologic eval	Yes
		Intermediate		
		Medication (beta-blocker, ARB)	Yes (both)	Yes (beta-blocker)
		Prophylactic surgical intervention	No	Yes

(Continued)

Table 27.3 (*Continued*)

Disorder	Genes	eMERGE outcomes	ClinGen Actionability Working Group					
			Dilated cardiomyopathy (*TNNT2, LMNA, DMD*)		Hypertrophic cardiomyopathy (*ACTC1,CSRP3, MYBPC3, MYH7, MYL2, MYL3,PRKAG2, TNNI3, TNNT2, TPM1*)		Arrhythmogenic right ventricular cardiomyopathy (*DSC2, DSP, DSG2, PKP2, TMEM43*)	
			Scored O/I pair	Evidence review	Scored O/I pair	Evidence review	Scored O/I pair	Evidence review
Cardiomyopathies	*ACTC1, DSC2, DSG2, DSP, LMNA, MYBPC3, MYCH7, MYL2, MYL3, PKP2, TMEM43, TNNI3, TNNT2, TPM1*	**Process**						
		EKG	Yes		No	Yes	No	Yes
		Echocardiogram	Yes		No	Yes	No	Yes
		Holter monitor	No	No	No	Yes	No	Yes
		Loop recorder	No	No	No	Yes	No	No
		Stress test	No	No	No	Yes	No	No
		Electrophysiology study	No	No	No	No	No	Yes
		Cardiac MRI	No	No	No	No	No	Yes

Disorder	Genes	eMERGE outcomes	ClinGen Actionability Working Group		
			Brugada syndrome (SCN5A)	Catecholaminergic polymorphic ventricular tachycardia (RYR2)	Romano-Ward Long QT syndromes (KCNH2, KCNQ1, SCN5A)
Inherited arrhythmias	KCNH2, KCNQ1, RYR2, SCN5A	**Intermediate**			
		Specialty referral	Yes	No	No
		Medications	Yes	No	Yes
		Implantable defibrillator	Yes	Yes	Yes
		Documentation of activity	No	No	Yes
		Health			
		Sudden cardiac death	Yes	Yes	Yes
		Reduce heart failure	Yes	No	No
		Process			
		EKG	Yes	Yes	Yes
		Echocardiogram	No	No	No
		Holter monitor	No	Yes	No
		Loop recorder	Yes	No	No
		Stress test	No	Yes	No

(Continued)

Table 27.3 *(Continued)*

Disorder	Genes	eMERGE outcomes	ClinGen Actionability Working Group				
		Electrophysiology study	No	No	No	No	No	No
		Cardiac MRI	No	No	No	No	No	No
		Trial sodium channel blocker	No	Yes	No	No	No	No
		Personal history of arrhythmias	No	Yes	No	Yes	No	Yes
		Specialty referral	No	Yes	No	No	No	No
		Intermediate						
		Symptoms suggestive of arrhythmia	No	Yes	No	Yes	No	Yes
		Medications	No	Yes (quinidine)	Yes			Yes (beta-blockers are ineffective for LQT3)
		Activity restriction	Yes		No	Yes	No	Yes
		ICD	Yes	Yes	No	No	Yes	Yes
		Health						
		Sudden cardiac death	Yes	Yes		Yes		Yes

Comparing AWG scoring to the eMERGE outcomes list demonstrates significant concordance. Only two of the outcome intervention pairs scored by AWG were not present in the eMERGE outcomes. Both of these represented health outcomes (diagnosis of tumors and/or lymphangioleiomyomatosis (LAM) in the tuberous sclerosis complex (TSC) and high cholesterol in adult FH. For the latter, lipid values will be obtained from EHR review so a determination can be made as to whether a participant who has been tested is at a goal. Thus, while this is not explicitly represented in the eMERGE outcomes, it should be added given the robust association between low-density lipoprotein cholesterol (LDLC) and cardiovascular events [22–24]. For the TSC health outcomes, eMERGE will be capturing information about the prior diagnosis of sub-ependymal giant astrocytoma (SEGA), other TSC-associated non-SEGA tumors, and LAM. It is also possible that the diagnostic evaluation prompted by the genomic result could lead to a diagnosis of one of the conditions. However, given the short time period of the eMERGE project, a long-term longitudinal follow-up is not feasible, in contrast to the AWG score, which is meant to inform interventions over a patient's lifetime.

While most of the eMERGE outcomes are not represented in the AWG scored outcome intervention pairs, most are discussed in the evidence review that accompanies the scored pairs. The AWG methodology does not score all possible outcome intervention pairs, rather it focuses on those interventions that have the strongest impact on the most important health outcomes of interest.

Hereditary breast and ovarian cancer syndrome (HBOC), associated with *BRCA1/2*, illustrates an interesting difference in the OWG and AWG approaches. The eMERGE OWG developed outcomes for HBOC as a whole, while the AWG has organized this around the two primary cancer types, breast, and ovarian and associated gynecologic cancers. This is logical as the outcome intervention pairs for the two types of cancers are quite different. This is not incompatible with the eMERGE outcomes, and Table 27.2 reflects how the outcomes can be separated to allow comparison.

A more important difference in the approach between the two groups is illustrated in Table 27.3. The four disorders represented, cardiomyopathy, inherited arrhythmogenic disorders, aortopathies, and colorectal cancer (CRC) predisposition illustrate the tension between pragmatic decisions to reduce the burden to collect outcomes of interest at the expense of capturing outcomes that are specific to individual disorders lumped within the overarching category of disorders. Some of these differences are clinically significant as discussed below.

27.3.1 Colorectal Cancer Predisposition

The eMERGE outcomes combine two disorders, Lynch syndrome (LS) and the rarer familial adenomatous polyposis (FAP), while these are scored separately by the ClinGen AWG. There is good concordance between eMERGE and the AWG scored intervention outcome pairs. One significant difference is in FAP, for which the AWG does not score CRC surveillance. Review of the evidence summary presents the rationale that the polyp burden reduces the effectiveness of surveillance. The outcome intervention pair scored by the AWG for FAP is colectomy to prevent CRC. This is consistent with the clinical guidelines for FAP [25], although this recommendation may not be as relevant for patients with attenuated FAP, as they have fewer polyps than FAP (hundreds vs. thousands). Colectomy is listed as an option for reducing the risk of CRC in patients with LS, but is generally not indicated due to the effectiveness of routine colonoscopy in prevention. Another difference between FAP and LS is that the non-CRC tumors differ and occur at a higher frequency in LS. This necessitates different screening approaches which are detailed in the AWG evidence reports. Finally, the AWG evidence reports also discuss the use of aspirin (LS) and non-steroidal anti-inflammatory drugs other than aspirin (FAP) to reduce the CRC risk. These should be considered for inclusion in the eMERGE outcomes.

27.3.2 Aortopathies

The OWG developed outcomes to accommodate all disorders that could result in aortic root dilation and other arteriopathies. The AWG divided these into arterial tortuosity syndrome (associated with variants in *SLC2A10*), and Familial Thoracic Aortic Aneurysms and Dissections (FTAAD associated with seven genes; Table 27.3). The AWG scored each of these FTAAD genes separately, although the evidence summary was the same for all seven genes. The actionability scores for the seven gene-disorder pairs were identical. As with CRC, there was very good concordance between the eMERGE outcomes and the AWG scored outcome intervention pairs. Indeed, the only discrepancies were recommendations for avoidance of contact sports and evaluation by an ophthalmologist, both present as a scored recommendation for arterial tortuosity syndrome, present in the evidence summary for FTAAD but not scored, and absent from eMERGE. Given that many of these disorders have associated ophthalmologic findings, this should be considered as an outcome by the eMERGE OWG. Recommendations to avoid activities such as contact sports are difficult to extract from medical records, so they were not considered for practical reasons.

There is one other issue with the aortopathies that complicates outcome development. There are two multiple malformation syndromes that can be seen in patients with variants in some of these genes, the Marfan and Loeys-Dietz syndromes. This complexity was acknowledged by the ClinGen AWG, as both disorders have been scored as separate entities. These syndromes are associated with many other medical issues; however, the scored outcome intervention pairs are concordant with the recommendations for aortic root dilation represented in arterial tortuosity syndrome and FTAAD.

However, the evidence summary goes into much more detail about the other medical issues associated with these syndromes. The eMERGE OWG recognizes this issue and it is anticipated that a targeted clinical evaluation will occur in conjunction with the return of results.

27.3.3 Cardiomyopathies

The eMERGEseq platform has 14 genes associated with three forms of cardiomyopathy: dilated, hypertrophic, and arrhythmogenic right ventricular (ARVC). One form was developed to capture outcomes for all three disorders. The ClinGen AWG scored each of the three disorders separately, and further scored each of the five ARVC genes separately, although as with FTAAD, the scores were identical for each of the five genes. The major risk for all three of these disorders is sudden death, and this health outcome is common across all the conditions. Related to this, an implantable cardiac defibrillator (ICD) is also present across all conditions. Not surprisingly, given the differences in the clinical course of these three conditions, beyond sudden cardiac death and ICD, there is a considerably more difference in the other outcomes. Most of these differences appropriately reflect the clinical differences between the conditions. There is only one AWG recommendation that is not reflected in the OWG outcomes. A creatine kinase determination is recommended for dilated cardiomyopathy associated with variants in *DMD*. However, *DMD* is not included on the eMERGEseq platform, explaining this difference. One gene associated with dilated cardiomyopathy, *LMNA*, is associated with several other disorders. One of them is Emery-Dreifuss Muscular Dystrophy (EDMD), which was scored separately by the AWG. There were other outcome intervention pairs scored for EDMD in addition to those related to cardiomyopathy. The eMERGE network decided that it would only return variants in *LMNA* associated with dilated cardiomyopathy, so outcomes for the other disorders were not considered. One other issue with the cardiomyopathies reviewed by the AWG is that variants in *TNNT2* can cause either dilated or hypertrophic cardiomyopathy. This pleiotropy will be more of an issue in the next group of disorders.

27.3.4   Inherited Arrhythmias

The eMERGEseq platform has four genes associated with three inherited arrhythmogenic disorders: Brugada syndrome, catecholaminergic polymorphic ventricular tachycardia (CPVT), and Romano-Ward Long QT syndromes (LQT). As with the cardiomyopathies, the major risk is for sudden death. This health outcome is represented across all conditions. ICD is an AWG recommendation for two of the three conditions. CPVT is the exception given the effectiveness of the beta-blockade to prevent sudden cardiac death in this disorder. There are numerous differences between the OWG outcomes and the AWG that reflect the differences in the conditions. The most notable absence from the eMERGE outcomes were medications to avoid in each condition. The AWG evidence reports provide detailed lists of medications and other substances to avoid as they can provoke abnormal cardiac rhythms. These are important to document and should be considered in addition to the eMERGE outcomes, as the documentation of medications associated with adverse events are relatively easy to find on the chart review.

As noted with *TNNT2* previously, one gene (*SCN5A*) is associated with two different arrhythmogenic disorders: Brugada syndrome and LQT3. There are several unique aspects to disorders associated with variants in *SCN5A*. For patients with Brugada syndrome, a trial of therapy with sodium channel blockers is indicated. The recommended anti-arrhythmic drug is quinidine. Both recommendations are specific only for the arrhythmogenic disorders associated with variants in *SCN5A*. For LQT3, the treatment with beta-blockers is not indicated as these have been shown to be ineffective in this condition. These findings argue persuasively for outcomes that are not only condition specific but gene and potentially even variant specific when appropriate.

27.4 Discussion

The results of this study show that it is possible to compare outcomes from two projects despite differences in the project objectives and methods. The important finding is that outcomes that are represented across multiple projects can be prioritized to harmonize the outcome definitions and develop guidance for their collection. This will facilitate the collection of prioritized outcomes from a wider set of research projects and clinical implementations, allowing evidence to accumulate at a faster rate to support clinical use. An example of the power of this type of approach for a genetic condition is cystic fibrosis (CF). Certified CF centers that receive funding from the CF Foundation are required to collect and submit many standard outcome measures. The outcomes are compared across sites and opportunities to improve care are identified, followed by implementation at the centers. This approach, which is also being used in other settings, has resulted in a dramatic improvement in multiple outcomes of interest for patients with CF [26]. The hope is that similar improvements in care could be realized across the many conditions for which genomic information can be used to inform care.

While there was generally good agreement for the high-level outcomes across the various conditions, there are some significant differences—the highlighting of which could inform further efforts to harmonize outcomes. eMERGE and ClinGen have very different objectives. The eMERGE network is studying the impact of implementation of genomic information into clinical care. To fully understand this impact, the outcomes are much more granular and detailed to allow chart abstractors to identify relevant information from the EHR. For example, in the cardiomyopathies (Table 27.3), process outcomes include five different interventions that assess the cardiac conduction system and two imaging modalities. The ClinGen scored outcome/intervention pairs only list one assessment of the cardiac conduction system and one imaging modality, and that was only for dilated cardiomyopathy. This is understandable as the scored pairs represent the results of the evidence synthesis that identifies the interventions and outcomes that drive clinical actionability, the key objective for ClinGen—a much different objective compared to eMERGE. Nonetheless, most of the eMERGE outcomes were identified in the ClinGen evidence reviews, although the reviews identified a few outcomes not included in the eMERGE OWG outcomes that are worthy of consideration for inclusion. Additionally, the AWG scored some gene-disorder pairs that, while on the eMERGEseq platform, are not being routinely returned. If the OWG proceeds with outcomes development for these genes, the AWG outcome intervention pairs and evidence summary will be used to inform the process.

A more complex issue is illustrated by the conditions in Tables 27.2 and 27.3, that is, how best to map outcomes for separate but related disorders. While it may be desirable to create outcomes specific for each disorder within a category, the time and effort required to do this are significant. Therefore, the eMERGE OWG opted to develop one outcome form for an overarching disorder category that encompasses multiple conditions. While this reduces the resources needed to create the outcome forms and simplifies the work for the chart abstractor, it will require more effort by the OWG after the abstraction to map the outcomes that are specific to the relevant disorder in order to determine whether appropriate condition-specific management goals were achieved. Challenges with this issue are also evident in the ClinGen

AWG scoring as some conditions lump all genes under one disorder (e.g., familial hypertrophic cardiomyopathy), while others have a separate score for each gene (e.g., FTAAD, ARVC). In these examples the scored outcome intervention pairs are identical across the different genes, raising the question as to the value added from this approach. In contrast, the three LQT disorders have different interventions based on the causal gene, supporting separate scoring of the outcome intervention pair. A further complication involves a pleiotropy of disorders associated with variants in the same gene. The issues with *SCN5A* and *LMNA* described previously illustrate the challenges of developing outcomes for disorders associated with variants in these genes. The most precise solution would be to develop outcomes based on the established genotype-phenotype correlations, but this further increases the complexity. This issue has led to the creation within ClinGen of the Lumping and Splitting Working Group (LSWG) [27]. The goal of the LSWG is to engage with a broad range of stakeholders to gather input "... to coordinate disease classification and categorization in order to harmonize disease categorization and classification for the greater community". The work product from this group will be incorporated into the ongoing efforts for outcomes harmonization.

Chromosome 22q11.2 deletion syndrome (22q11.2DS) is the most common chromosomal microdeletion disorder with approximately 3.0 million base pairs deleted (ranging from 0.7–3.0 Mb) resulting in a loss of ~90 known or predicted genes, including 46 protein-coding genes and 7 microRNAs, 10 non-coding RNAs, and 27 pseudogenes (Figure 1) [28]. The 22q11.2DS results most commonly from de novo non-homologous meiotic recombination events occurring in approximately 1 in every 1000 fetuses and 1 in 2000 live births. About 4% of infants with 22q11.2DS succumb to it, while cardiac defects, hypocalcemia, and airways disease are risk factors for early death, with the median age of death at 3–4 months. However, most individuals with 22q11.2DS survive well into adulthood, at which time approximately 50% of them develop schizophrenia.

While ClinGen (currently) makes no recommendations with respect to 22q11.2DS we note the syndrome has become a model for understanding rare and frequent congenital anomalies such as heart defects, medical conditions including immunodeficiency, allergies, asthma, and psychiatric and developmental differences, which may provide a platform into better understanding these phenotypes, while affording opportunities for translational strategies across the lifespan for both patients with 22q11.2DS and for those with these associated features in the general population. The diverse phenotype and outcomes of nearly every organ system make this population valuable for understanding the variables that impact on the manifestations of the deletion, which is relatively consistent from person to person.

The eMERGESeq panel captures six SNPs (five in the *COMT* gene and one flanking the region), which can be used to capture 22q11.2DS, while existing genotype data can be readily used to detect the syndrome. Current efforts aim at assessing the prevalence of 22q11.2DS in respective eMERGE cohorts, and to determine a health outcome across multiple organ systems and outcome measures as available.

We are using PennCNV and XHMM to derive CNVs from eMERGESeq data, as well as existing array data. Data will be returned to participating sites for outcome evaluation of relevant phenotypes (e.g., heart defects, immunodeficiency, allergy, asthma, psychiatric, and developmental differences) and for additional validation, if required.

This study represents a pilot to assess the feasibility of harmonizing outcomes across two notable research projects. As such the results are descriptive and limited to the two projects assessed. The study did not include the evaluation of outcomes for any clinical genomic medicine implementation projects. However, one eMERGE site reports the genomic results on a large scale in a clinical research setting [7]. Institutional authors (MSW, JLW), in conjunction with the Genetic Screening and Counseling Program at the institution, have aligned the eMERGE and institutional outcomes for the disorders shared in common between the two efforts (data not shown). The availability of the outcomes from eMERGE aided in the prioritization of the institutional outcomes, while input from the authors, both of whom are members of the eMERGE OWG, influenced the outcome definitions for the OWG. This illustrates that the harmonization of outcomes is not only feasible but may represent a generalizable approach. Mapping outcomes to standardized, structured terminologies such as the

International Classifications of Disease-Clinical Modification (ICD-CM) or the Systematized Nomenclature of Medicine-Clinical Terms (SNOMED-CT) would facilitate generalizability and reduce the reliance on manual collection, although it is important to note that many critical outcomes are not currently represented as structured data so some manual review will be required. It is possible that outcome "algorithms" could be developed. These would be similar to phenotyping algorithms that eMERGE has developed, disseminated across multiple healthcare and electronic health record systems and made publicly available through the Phenotype Knowledgebase-PheKB [29]. This could further reduce, although not eliminate, the burden of manual review.

Another limitation of this study was the outcomes and process measures such as cost, reimbursement, institutional visibility, access, etc., which also play a role in decisions about implementation were not assessed. We also did not focus on patient-centered outcomes, which are not always aligned with health or other outcomes. Measuring outcomes from the perspective of the patient has been identified as a deficiency in much medical research as evidenced by the creation of the Patient-centered Outcomes Research Institute (PCORI) in 2010 [30]. The PCORI vision statement ("patients and the public have the information they can use to make decisions that reflect their desired health outcomes") emphasizes that part of precision medicine is understanding what outcomes the patient desires, which will vary from patient to patient. Patient engagement is a key part of the All of Us project [5], therefore, developing and harmonizing patient-centered outcomes for genomic medicine is important. Of interest, the NIH funded the development and harmonization of a large set of patient-centered outcome measures now included in the Patient-Reported Outcomes Measurement Information System (PROMIS®) [31] made available through the Department of Health and Human Services. These measures can be reviewed and revised as necessary to develop patient-reported outcomes for genomic medicine. This also illustrates that a process led by the NIH to collect and harmonize outcome measures across its portfolio of projects is a successful approach and can promote the use of standardized measures going forward.

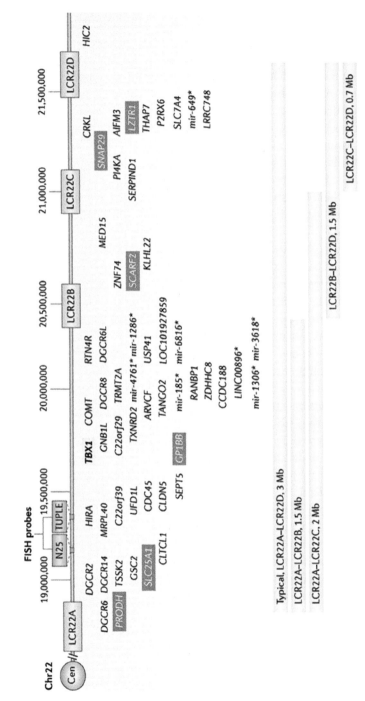

Figure 27.1 The depiction of the chromosome 22q11.2 deletion including the deleted genes and variations of the common deletions reported.

27.5   Conclusions

The definition and harmonization of common outcomes to develop evidence and assess the value of genomic medicine implementation are needed to further the goals embodied in precision public health. The approach proposed in this study will be applied to other NHGRI-funded genomic implementation projects. The resulting outcomes will be made publicly available and their use will be encouraged for outcome measurement, collection, and research to accelerate the implementation of those interventions that demonstrate improved value.

Abbreviations

ARVC:	arrhythmogenic right ventricular
AWG:	ClinGen Actionability Working Group
CF:	cystic fibrosis
ClinGen:	Clinical Genome Resource
CRC:	colorectal cancer
CPVT:	catecholaminergic polymorphic ventricular tachycardia
CSER:	Clinical Sequencing Exploratory Research Consortium
EDMD:	Emery-Dreifuss muscular dystrophy
eMERGE:	Electronic Medical Records and Genomics
FAP:	familial adenomatous polyposis
FH:	familial hypercholesterolemia
FTAAD:	familial thoracic aortic aneurysms and dissections
HBOC:	hereditary breast and ovarian cancer syndrome
ICD-CM:	International Classifications of Disease-Clinical Modification
ICD:	implantable cardiac defibrillator
IGNITE:	Implementing Genomics in Practice
LAM:	lymphangioleiomyomatosis
LDLC:	low-density lipoprotein cholesterol
LQT:	Romano-Ward Long QT syndromes
LS:	Lynch syndrome

LSWG: Lumping and Splitting Working Group
NHGRI: National Human Genome Research Institute
NIH: United States National Institutes of Health
OWG: eMERGE Outcomes Working Group
PCORI: Patient-Centered Outcomes Research Institute
PROMIS®:   Patient-Reported Outcomes Measurement
             Information System
SEGA: sub-ependymal giant astrocytoma
SNOMED-CT: Systematized Nomenclature of Medicine-Clinical
             Terms
TSC: tuberous sclerosis complex

Disclosures and Conflict of Interests

This chapter was originally published as: Williams, J. L.; Chung, W. K.; Fedotov, A.; Kiryluk, K.; Weng, C.; Connolly, J. J.; Harr, M.; Hakonarson, H.; Leppig, K. A.; Larson, E. B.; Jarvik, G. P.; Veenstra, D. L.; Hoell, C.; Smith, M. E.; Holm, I. A.; Peterson, J. F.; Williams, M. S. (2018). Harmonizing outcomes for genomic medicine: comparison of eMERGE outcomes to ClinGen outcome/intervention pairs. *Healthcare*, **6**, 83, doi: 10.3390/healthcare6030083, under the Creative Commons Attribution license (http://creativecommons.org/licenses/by/4.0/). It appears here, with edits and updates, by kind permission of the authors and publisher, MDPI (Basel).

Author Contributions: Conceptualization and Methodology, M.S.W., J.L.W., D.L.V., J.F.P., H.H.; Validation, M.S.W.; Formal Analysis, All authors; Data Curation, M.S.W., G.P.J., J.F.P., W.K.C., I.A.H., C.H., H.H., J.J.C.; Writing—Original Draft Preparation, M.S.W.; Writing—Review & Editing, All authors; Supervision, M.S.W.; Project Administration, M.S.W., G.P.J., M.E.S.; Funding Acquisition, G.P.J., E.B.L., H.H., J.F.P., M.E.S., W.K.C., K.K., C.W., M.S.W.

Funding: This work was supported by U01HG8657 (Group Health Cooperative/University of Washington); U01HG8685 (Brigham and Women's Hospital); U01HG8672 (Vanderbilt University Medical Center); U01HG8679 (Geisinger Clinic); U01HG8680 (Columbia University Health Sciences); U01HG8684 (Children's Hospital of Philadelphia); U01HG8673 (Northwestern University); U01HG8701

(Vanderbilt University Medical Center serving as the Coordinating Center).

Acknowledgments: Northwestern University—Laura Rasmussen-Torvik, Lisa Castillo for creating the cardiomyopathy outcome forms; Vanderbilt—Brittany City for coordinating and editing eMERGE outcomes forms, Department of Medicine, Columbia University—Katherine Crew for creating the breast cancer outcomes form.

Conflicts of Interest: No authors have a conflict of interest to declare.

Corresponding Author

Dr. Marc S. Williams
Genomic Medicine Institute, Geisinger
100 Academy Drive
Danville, PA 17822, USA
Email: mswilliams1@geisinger.edu

References

1. Definition Genomic Medicine NHGRI. Available at: https://www. Genome.Gov/27552451/what-is-genomic-medicine/ (accessed on January 10, 2019).

2. Relling, M. V., Evan, W. E. (2015). Pharmacogenomics in the Clinic. *Nature*, **526**, 343–350.

3. Haslem, D. S., Van Norman, S. B., Fulde, G., Knighton, A. J., Belnap, T., Butler, A. M., Rhagunath, S., Newman, D., Gilbert, H., Tudor, B. P., et al. (2017). A retrospective analysis of precision medicine outcomes in patients with advanced cancer reveals improved progression-free survival without increased health care costs. *J. Oncol. Pract.*, **13**, e108–e119.

4. Stark, Z., Schofield, D., Martyn, M., Rynehart, L., Shrestha, R., Alam, K., Lunke, S., Tan, T. Y., Gaff, C. L., White, S. M. (2018). Does genomic sequencing early in the diagnostic trajectory make a difference? A follow-up study of clinical outcomes and cost-effectiveness. *Genet. Med.*, **15**, (doi: 10.1038/s41436-018-0006-8)

5. All of Us. Available at: https://allofus.nih.gov/ (accessed on January 10, 2019).

6. Inova Translational Medicine Institute. Available at: https://www. inova.org/itmi/home (accessed on January 10, 2019).

7. Williams, M. S., Buchanan, A. H., Davis, F. D., Faucett, W. A., Hallquist, M. L. G., Leader, J. B., Martin, C. L., McCormick, C. Z., Meyer, M. N., Murray, M. F., et al. (2018). Patient-Centered precision health in a learning health care system: Geisinger's Genomic medicine experience. *Health Aff.*, **37**, 757–764.

8. Manolio, T. A., Chisolm, R. L., Ozenberger, B., Roden, D. M., Williams, M. S., Wilson, R., Bick, D., Bottinger, E., Brilliant, M. H., Eng, C., et al. (2013). Implementing genomic medicine in the clinic: The future is here. *Genet. Med.*, **15**, 258–267.

9. Phillips, K. A., Deverka, P. A., Sox, H. C., Khoury, M. J., Sandy, L. G., Ginsburg, G. S., Tunis, S. R., Orlando, L. A., Douglas, M. P. (2017). Making genomic medicine evidence-based and patient-centered: A structured review and landscape analysis of comparative effectiveness research. *Genet. Med.*, **19**, 1081–1091.

10. Horgan, D., Jansen, M., Leyens, L., Lal, J. A., Sudbrak, R., Hackenitz, E., Bußhoff, U., Ballensiefen, W., Brand, A. (2014). An index of barriers for the implementation of personalised medicine and pharmacogenomics in Europe. *Public Health Genom.*, **17**, 287–298.

11. IGNITE Network. Available at: https://ignite-genomics.org/about-ignite/ (accessed on January 10, 2019).

12. CSER Consortium. Available at: https://cser-consortium.org/ (accessed on January 10, 2019).

13. eMERGE Network. Available at: https://emerge.mc.vanderbilt.edu/ (accessed on January 10, 2019).

14. eMERGE and CSER Joint Meeting. Available at: https://www. Genome. Gov/27567557/emerge-cser-the-convergence-of-genomics-and-medicine/ (accessed on January 10, 2019).

15. eMERGE Phase 3. Available at: https://www.Genome. Gov/27540473/ electronic-medical-records-and-genomics-emerge-network/ (accessed on January 10, 2019).

16. eMERGEseq Platform. Available at: https://emerge.mc.vanderbilt. edu/the-emergeseq-platform/ (accessed on January 10, 2019).

17. Clinical Genome Resource. Available at: https://www.Clinicalgenome. Org/ (accessed on January 10, 2019).

18. REDCap. Available at: https://www.project-redcap.org/ (accessed on January 10, 2019).

19. ClinGen Actionability Working Group. Available at: https://www. clinicalgenome.org/working-groups/actionability/ (accessed on January 10, 2019).

20. Actionability Outcome Intervention Pairs. Available at: https://search. clinicalgenome.org/kb/actionability (accessed on January 10, 2019).

21. Hunter, J. E., Irving, S. A., Biesecker, L. G., Buchanan, A., Jensen, B., Lee, K., Martin, C. L., Milko, L., Muessig, K., Niehaus, A. D., et al. (2016). A standardized, evidence-based protocol to assess clinical actionability of genetic disorders associated with genomic variation. *Genet. Med.*, **18**, 1258–1268.

22. Stamler, J., Wentworth, D., Neaton, J. D. (1986). Is relationship between serum cholesterol and risk of premature death from coronary heart disease continuous and graded? Findings in 356,222 primary screenees of the Multiple Risk Factor Intervention Trial (MRFIT). *JAMA*, **256**, 2823–2828.

23. Castelli, W. P., Anderson, K., Wilson, P. W., Levy, D. (1992). Lipids and risk of coronary heart disease. The Framingham Study. *Ann. Epidemiol.*, **2**, 23–28.

24. Levinson, S. S. (2017). Critical review of 2016 ACC guidelines on therapies for cholesterol lowering with reference to laboratory testing. *Clin. Chim. Acta*, pii:S0009-8981(17)30431-X.

25. Kohlmann, W., Gruber, S. B. (2018). Lynch Syndrome. In: Adam, M. P., Ardinger, H. H., Pagon, R. A., eds. *GeneReviews®*, University of Washington, Seattle, WA, USA. Available at: https://www.Ncbi. Nlm. Nih.Gov/books/NBK1211/ (accessed on January 10, 2019).

26. Khan, A. A., Nash, E. F., Whitehouse, J., Rashid, R. (2017). Improving the care of patients with cystic fibrosis (CF). *BMJ Open Qual.*, **6**, e000020.

27. ClinGen Lumping and Splitting Working Group. Available at: https:// www.Clinicalgenome.Org/working-groups/lumping-and-splitting/ (accessed on January 10, 2019).

28. McDonald-McGinn, D. M., Emanuel, B. S., Zackai, E. H. (2018). 22q11.2 Deletion Syndrome. In: Adam, M. P., Ardinger, H. H., Pagon, R. A., eds. *GeneReviews®*, University of Washington, Seattle, WA, USA. Available at: https://www.Ncbi.Nlm.Nih.Gov/books/NBK1523/ (accessed on January 10, 2019).

29. Phenotype Knowledgebase. Available at: https://phekb.Org/ (accessed on January 10, 2019).

30. Patient Centered Outcomes Research Institute. Available at: https:// www. Pcori.Org/about-us (accessed on January 10, 2019).

31. PROMIS®. Available at: http://www.Healthmeasures.Net/explore-measurement-systems/promis (accessed on January 10, 2019).

Chapter 28

Estrogen and Androgen Blockade for Advanced Prostate Cancer in the Era of Precision Medicine

Tetsuya Fujimura, MD, PhD,[a] Kenichi Takayama, MD, PhD,[b]
Satoru Takahashi, MD, PhD,[c] and Satoshi Inoue, MD, PhD[b]

[a]*Department of Urology, Jichi Medical University, Tochigi, Japan*
[b]*Department of Systems Aging Science and Medicine,*
Tokyo Metropolitan Institute of Gerontology, Tokyo, Japan
[c]*Department of Urology, Nihon University School of Medicine, Tokyo, Japan*

Keywords: prostate cancer, androgen receptor, estrogen-related receptor, stem cell, androgen deprivation therapy, selective estrogen receptor modulators, personalized medicine, precision medicine, cancer, cancer specific survival, diethylstilbestrol, estrogen receptor, androgen-responsive element, estrogen response element, dihydrotestosterone, Cytochrome P450, biochemical recurrence, progression-free survival, overall survival, protein–protein interactions, Forkhead Box O1, platelet derived growth factor subunit A, growth differentiation gene, phosphatase and tensin homolog, vascular endothelial growth factor A, Wnt Family Member 5A, transforming growth factor beta 1, ATM serine/threonine kinase, breast cancer 1/2, amyloid precursor protein, Octamer-Binding Transcription Factor 1, tripartite motif containing 36, Forkhead Box P1

The Road from Nanomedicine to Precision Medicine
Edited by Shaker A. Mousa, Raj Bawa, and Gerald F. Audette
Copyright © 2020 Jenny Stanford Publishing Pte. Ltd.
ISBN 978-981-4800-59-4 (Hardcover), 978-0-429-29501-0 (eBook)
www.jennystanford.com

28.1   Introduction

Prostate cancer (PC) is the second-most frequently diagnosed cancer in men worldwide, with 1.1 million new cases estimated to have occurred in 2012 [1]. PC is the fifth leading cause of death due to cancer worldwide, with the highest mortality rates reported in the Caribbean (29.3 per 100,000) and Southern (24.4 per 100,000) and Middle Africa (24.2 per 100,000) [1]. Mortality rates due to PC have decreased in most of the developed countries, including those in the North America, Oceania, and Northern and Western Europe [1]. In contrast, mortality rates have increased in some Asian and Central and Eastern European countries, such as Korea, China (Hong Kong), and Russia. Chemoprevention, prostate-specific antigen (PSA) screening for early detection, and innovative treatments for advanced PC are necessary to reduce the resultant mortality due to PC.

Systematic treatment for advanced or metastatic PC includes androgen deprivation therapy (ADT) or chemotherapy. Conventional ADT involved surgical or medical castration and the administration of anti-androgen agents, such as bicalutamide, flutamide, and nilutamide. Recently, new anti-androgen agents, such as enzalutamide and abiraterone, have been approved for castration-resistant PC (CRPC) [2]. Docetaxel and cabazitaxel are also available for CRPC. However, effects of these new agents are transitory. Moreover, they are relatively expensive, costing $10,759 for 160 mg of enzalutamide; $9817 for 1000 mg of abiraterone per month; $1919 for 120 mg of docetaxel; and $10,639 for 40 mg of cabazitaxel, respectively (https://www.drugs.com/price-guide/). To improve cancer-specific survival (CSS) and reduce the cost incurred on drugs used for advanced PC treatment, durable and economic therapeutic strategies are warranted worldwide. Here, we reviewed the literature pertaining to endocrine therapy and proposed a new therapeutic strategy involving estrogen and androgen signal blockade (EAB) for advanced PC.

28.2   Initiating Endocrine Therapy for PC

Pioneering work by Huggins [3] established diethylstilbestrol (DES) administration as a low-cost but effective treatment for metastatic PC. However, due to adverse side effects associated with DES treatment, including the exacerbation of heart failure, vascular complications, and gynecomastia, DES therapies have been replaced with the application of luteinizing hormone-releasing hormone analogs and anti-androgen agents, such as leuprolide, goserelin, bicalutamide, and flutamide, since the past two decades. The precise functional network of androgen receptors (ARs), co-factors, and micro-RNA (miR) has been clarified using next generation sequencing in these lines [4, 5]. Recently, the use of selective estrogen modulators (SERMs) for advanced PC treatment has reemerged due to the discovery of various nuclear receptors and their functional analyses in context of PC.

28.3 Nuclear Receptors Associated with Endocrine Therapy for PC

Initially, the action of estrogens was believed to be mediated via the blockade of the pituitary–testicular axis, which effectively decreased circulating androgen levels and induced tumor regression. This concept was supported by immunohistochemical and in situ hybridization studies conducted in the 1990s, which could not identify any detectable estrogen receptor (ER) levels in the epithelial compartments of the human prostatic tissue [6]. The classical ER, ERα, is dominantly expressed in the stromal compartment but not in the glandular epithelium of the normal human prostate [6–8]. Estrogens actions on prostatic epithelium have been considered to be exerted via ERα-mediated paracrine mechanism. Conversely, the exposure of humans or rodents to estrogens induced proliferative changes and squamous metaplasia in their prostates [9–11]. Noble strain rats treated with androgen plus estrogens over a long period have been reported to show high PC incidence [12]. The mechanism underlying estrogen action was clarified in the past two decades owing to the successful cloning of nuclear receptors, such as ERα, ERβ, and estrogen-related receptors (ERRs: ERRα, ERRβ, and ERRγ) from 1980 to 1996 [13–16]. Evidence suggests an overlap between ERR and ER biology.

In the 1990s, the status of ERα expression in PC remained controversial [17–23]. ERα protein expression was more frequently observed in higher-grade metastatic cancers than in low-to-moderate grade tumors [17]. Conversely, other investigators noted the presence of ERα expression in well-differentiated adenocarcinoma but not in poorly differentiated tumors and metastatic lesions [24–26]. In our study, over 70% of the malignant epithelium showed no ERα expression, whereas distinct ERα immunoreactivity was identified in 7% of the cancerous loci in 50 patients with localized PC [27]. Previous researchers have reported variable expression patterns of ERα in PC [28].

Novel ERβ cloned from a rat prostate cDNA library was found to be localized in the epithelial compartment of the rat prostate [14]. ERα and ERβ are paralogs and they share 86%, 23%, 17%, and 100% amino acid identity in DNA-binding domains (DBDs), N-termini, C-termini, and ligand-binding domains (LBDs), respectively. However, because of the divergence in their LBDs, the two ER subtypes bind ligands (agonists or antagonists) with different affinities [29]. Although several studies have investigated the expression of the second ER, ERβ, in normal and malignant epithelium of the human prostate, the available data on ERβ protein expression in the human prostate remains controversial [17–23, 30, 31]. While one study reported a decrease in the ERβ protein expression in PC, another reported an increase. This discrepancy may be attributed to the presence of C-terminal truncated splice variant of ERβ, ERβl-ERβ5 [32] and the specificity of ERβ antibody [33]. The expression of ERβ3 is limited to the testis, whereas ERβ1, ERβ2, ERβ4, and ERβ5 are expressed in the prostate [32]. Anderson showed the schematic view of antibody epitopes and ERβ isoforms, and validated 13 commercially available or in-house produced ERβ antibodies [33]. Out of the 13 ERβ-targeting antibodies, PPZ0506 and 14C8 appeared to specifically target ERβ in formalin fixed paraffin embedded-treated cell lines [33].

Next, cDNA for ERRα was isolated by screening cDNA libraries with probes corresponding to the DBD of human ERα [16]. ERRα has a high homology to ERα at DBD and also recognizes estrogen response element (ERE) [34–38]. ERRα mRNA was highly expressed in the heart and skeletal muscle and expressed to a lesser degree in the kidney, pancreas, small intestine, and colon in humans [39]. ERRα is now established to be associated with unfavorable biomarkers in human breast cancer [40, 41].

Moreover, ERRα and ERRγ are associated with unfavorable and favorable biomarkers, respectively, in human PC [42, 43].

The steroid and xenobiotic receptor (SXR), also known as the human pregnane X receptor, constitutes members of the nuclear receptor superfamily of ligand-activated transcription factors [44, 45]. The elimination of dihydrotestosterone (DHT), which is the active androgen in the prostate, is critical for successful endocrine therapy for PC. Testosterone is inactivated in the liver and prostate by Cytochrome P450 (CYP) enzymes [46–48]. SXR regulates CYP3A4 and CYP2B6, which are responsible for the hydroxylation of testosterone in the liver and prostate [48–50]. SXR and its target genes *CYP3A4* and *CYP2B6* participate in the regulation of PC via intra-prostatic testosterone metabolism [49–52].

28.4 Clinicopathological Features of Nuclear Receptors in PC

wtERβ expression was significantly lower in cancers than in benign epithelium and inversely correlated with the Gleason grade. In contrast, ERβcx (ERβ2) was significantly more expressed in high-grade cancers than in low-grade tumors. The CSS of patients with lower wtERβ expression was significantly worse than that of patients with higher wtERβ expression. ERβcx (ERβ2) expression was correlated with the Gleason grade and inversely correlated with EBβ expression [27]. Higher ERβcx (ERβ2) expression was significantly correlated with poor CSS.

This finding was corroborated by the results of the following investigations of ERβ splice variants. Tissue microarrays constructed using samples from 566 men who had undergone radical prostatectomy were analyzed using immunohistochemistry for wtERβ, ERβ2, ERβ5, and G-protein-coupled receptor-30 (GPR 30) [53]. Intense cytoplasmic wtERβ staining was independently associated with time to recurrence (Hazards ratio (HR) 1.7, 95% confidence interval (CI) 1.1–2.6, $p = 0.01$) and PC-specific mortality (HR 3.9, 95% CI 1.8–24.9, $p = 0.01$). Similarly, intense nuclear ERβ2 staining was independently associated with PC-specific mortality (HR 3.9, 95% CI 1.1–13.4, $p = 0.03$) [53]. Samples obtained from 100 patients with cT3N0M0 PC were evaluated to determine the expression of AR, ERα, and wtERβ as well as the clinical outcomes, including biochemical recurrence (BCR), progression-free survival (PFS), and overall survival (OS) [54]. BCR was defined with two consecutive PSA values of ≥0.2 ng/mL. Disease progression was diagnosed by digital rectal examination, computed tomography, or bone scintigraphy. AR expression was not associated with any of the above outcomes; however, patients with high ERα or low wtERβ immunoreactivity scores than those with negative ERα or high wtERβ immunoreactivity scores showed 6.03-, 10.93-, and 10.53-times greater hazards for BCR, PFS, and OS, respectively. Nuclear ERβ2 and cytoplasmic ERβ5 expression served as significant prognostic factors for BCR and PFS in 144 men with localized PC, who had undergone radical prostatectomy [54].

ERRα expression was elevated in PC, particularly in those with a higher Gleason grade, compared with benign epithelium foci, and higher ERRα expression acted as a significant prognostic predictor of CSS [42]. In contrast, ERRγ expression decreased in PC and acted as a preferable prognostic marker of PC [43].

SXR immunoreactivity was significantly lower in cancerous lesions than in the benign foci of specimens obtained from radical prostatectomy. CSS in patients with high SXR expression was significantly increased, and the combined data of SXR and CYP3A4 showed that a higher expression of SXR and CYP3A4 was correlated with better CSS [49]. In addition, CYP2B6 was abundantly localized in the cytoplasm of normal epithelial cells compared with that of PC cells. CYP2B6 immunoreactivity was inversely correlated with higher Gleason grade. Patients with decreased CYP2B6 expression showed poor CSS [51]. CYP2B6 overexpression in LNCaP cells significantly decreased testosterone-induced proliferation. Thus, SXR, CYP3A4, and CYP2B6 may regulate PC progression via testosterone metabolism.

28.5 Functional Analysis of Nuclear Receptors and Associated Factors for Endocrine Therapy

We summarize the pathway of the nuclear receptors associated with endocrine therapy for PC (Fig. 28.1). In the classical pathway, ERα binds to EREs to regulate its gene transcription [55]. ERα also participates in several non-classical pathways, including ERE-independent gene transcription via protein–protein interactions with transcription factors, and rapid non-genotropic pathways. ERα functions have been investigated in two mouse models of aggressive PC: Phosphatase and Tensin Homolog (PTEN)-deficient and Hi-MYC mice [56]. ERα promoted cell proliferation in PTEN-deficient tumors by regulating phosphatidylinositol-3 kinase (PI3K) and mitogen-activated protein kinase (MAPK) signaling pathways and glucose sensitivity.

Various genes of cell proliferation and bone metastasis gens are known to be recruited by ERβ [57]. E2 imparts paradoxical effects in PC because E2 biphasically regulates prostate tumor growth by suppressing *Forkhead Box O1* (*FOXO1*) and *Platelet Derived Growth Factor Subunit A* (*PDGFA*) expression levels through ERβ and Kruppel-like factor (KLF) 5 pathways [58, 59]. E2 treatment decreased KLF5-dependent *FOXO1* transcription in PC cells through ERβ, inhibiting apoptosis and increasing tumor weight in mouse xenograft models [59]. In contrast, when mice were treated with higher doses of E2, prostate tumor growth was suppressed through ERβ and KLF5 pathways. In fact, E2 inhibited *PDGFA* transcription and suppressed angiogenesis through ERβ and KLF5 pathways. PDGFA recovered angiogenesis inhibited by E2. Both *PDFGA* and *FOXO1* expressions were markedly suppressed by higher doses of E2, and angiogenesis was insufficient for prostate tumor growth, leading to the suppression of tumor growth [59]. ERβ and NFκB also regulate PC activating *Interleukin (IL)-12*, *Growth differentiation gene (GDF)-1*, *IL-8*, and *Receptor-like tyrosine kinase (RYK)* [60].

ERβ activation is a target for treating early stage PC to prevent cancer progression [61, 62]. RNA sequencing and immunohistochemistry were conducted to compare gene expression profiles in the ventral prostate of young (2-month-old) and aging (18-month-old) ERβ/ERβ mice and their wild-type littermates. ERβ modulates AR signaling by repressing AR driver RORc and increasing AR co-repressor dachshund family (DACH 1/2). ERβ loss resulted in the upregulation of genes whose expression is associated with poor prognosis in PC, accompanied with the downregulation of tumor-suppressive or tumor-preventive genes. ERβ agonist (LY3201) treatment resulted in the nuclear import of PTEN and repression of AR signaling.

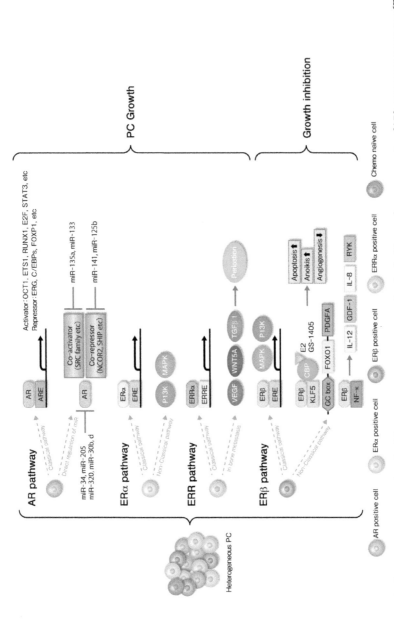

Figure 28.1 Nuclear receptors associated with prostate cancer progression including androgen receptor (AR), estrogen receptor (ER), and estrogen-related receptor (ERR).

ERβ activation is also a treatment option for CRPC [62]. The influence of the ERβ-specific ligand 8β-VE2 was investigated using three kinds of VCaP cells through pretreatment, under ADT, and under maximum ADT. 8β-VE2 treatment reduced the overexpression of AR as well as AR splice variants (ARVs) lacking the ligand binding domain in VCaP cells under maximum ADT.

ERRα has been reported to promote several traits of cancer progression, such as proliferation [63–65], epithelial–mesenchymal transition [66], resistance to hypoxia [67], angiogenesis [68], and cell migration [69] in prostate, breast, or colon cancer. ERRα was also involved in the bone tumor progression of CRPC [70]. Increased ERRα levels in tumor cells led to rapid tumor progression, with both bone destruction and formation accompanied with osteoclasts and osteoblasts. *Vascular Endothelial Growth Factor A (VEGF-A), Wnt Family Member 5A (WNT5A)*, and *Transforming Growth Factor Beta 1 (TGFβ1)* were upregulated by *ERRα* expression in tumor cells. In addition, *ERRα* regulated tumor stromal microenvironment by stimulating pro-metastatic factor *periostin* expression. With no natural ligands of ERRα involved miR-135a, modulated ERRα function [71, 72]. miR-135a downregulated ERRα expression through specific sequences of its 3′-untranslated region (UTR) and also decreased the cell invasive potential of ERRα pathway [72].

28.6 SERMs for PC

SERMs are involved in major therapeutic advancements in clinical practice for breast cancer, osteoporosis, and PC. The first-generation clomiphene, the second-generation toremifene and raloxifene, and the third-generation ospemifene and bazedoxifene have been used in diseases affecting women [73]. SERMs are economical, costing $90 for 50 mg of clomiphene; $1313 for 60 mg of toremifene; and $16 for 60 mg of raloxifene per month, respectively. SERMs are synthetic ligands for ERs that can exhibit either estrogenic or anti-estrogenic effects, depending on the tissue type [73]. SERMs have tissue-specific agonist–antagonist activity [74]. For example, raloxifene exhibited diverse activities via ER depending on ERα or ERβ expression in the target organ [75]. When ERE-luciferase (ERE-LUC) and either ERα or

ERβ were co-transfected into HEK 293 cells, toremifene acted as a potent antagonist of the 17β-estradiol-stimulated transactivation of ERα (at 1 μM) and ERβ (at 5 μM), respectively [74]. In contrast, toremifene (0.1 μM) served as a potent antagonist of ERα (95% inhibition), but not of ERβ (20% inhibition). Thus, toremifene is a more selective antagonist of ERα, the activation of which is implicated in prostate epithelial growth, than of ERβ [74]. Toremifene significantly reduced PC incidence in a transgenic adenocarcinoma of mouse prostate model [76]. Toremifene treatment resulted in a significant reduction in prostate tumor growth and PC3M (a human PC cell line) proliferation, expressing ERα [77].

Details of clinical investigations of SERMs are summarized in Table 28.1 [78–82]. A phase II trial of toremifene was conducted in 15 patients with CRPC [78]. The patients received toremifene at a dose of 300 to 600 mg/m^2, which is a relatively high dosage. The treatment was well-tolerated and toxicity was mild; however, no objective responses were achieved [78]. Another study used toremifene at a dose of 20 to 60 mg to prevent PC incidence in men with high-grade prostatic intraepithelial neoplasia (HGPIN) [79]. A total of 514 patients with HGPIN and no PC evidence on screening biopsy were randomized to 20, 40, or 60 mg toremifene dose or to a placebo daily for 12 months. The patients underwent re-biopsy at 6 and 12 months. The cumulative risk of PC was decreased in patients on 20 mg toremifene compared with that in placebo controls (24.4% vs. 31.2%, $p < 0.05$). The annualized rate of prevention was 6.8 cancers/100 treated men. In patients with no biopsy evidence of cancer at baseline and at 6 months, the 12-month incidence of PC was decreased by 48.2% with 20 mg toremifene as compared with the placebo controls (9.1% vs. 17.4%, $p < 0.05$). Thus, 20-mg dose was found to be most effective, but the cumulative (40 mg, 29.2%; 60 mg, 28.1%) and 12-month incidences of PC (40 mg, 14.3%; 60 mg, 13.0%) were lower for each toremifene dose than those for the placebo. In addition, in a 3-year phase III, double-blinded, multicenter trial, 1590 men with HGPIN and negative PC findings on the prostate biopsy were randomly assigned to a toremifene citrate (20 mg) treatment or placebo group, with equal number of patients in each group [81]. Cancer was detected in 34.7%

Table 28.1 Clinical trials of selective estrogen receptors modulators for prostatic diseases

Years (References)	Subjective	Objective	Design	Treatments	Number	Results
2001 [78]	CRPC	Cancer Control	Experimental	ADT+Toremifene 300–640 mg/m²	15	No cancer inhibitory effect
2006 [79]	HGPIN	Cancer prevention	RCT	Toremifene 20, 40, 60 mg	514	Cancer prevention in 20 mg group
2008 [80]	PC	Osteoporosis prevention	RCT	Toremifene 80 mg	197	Increased bone density
2008 [82]	PC	Lipid profile improvement	RCT	Toremifene 80 mg	1389	Decreased T Cho, LDL, HDL, TG
2013 [81]	HGPIN	Cancer prevention	RCT	Toremifene 20 mg	1467	Not significant cancer prevention
2006 [84]	CRPC	Cancer control	Experimental	Raloxifene 60 mg	13	Partial effect (5 of 13 patients)
2017 [85]	CRPC	Cancer control	Experimental	Raloxifene 60 mg + Bicaltamide 50 mg	18	Partial effect (4 of 18 patients)
2008 [86]	CRPC	Cancer control	Experimental	Fulvestrant 500 mg, 250 mg	20	No patients reduced >50% PSA reduction
2015 [87]	Hormone naïve PC	Testosterone reduction	RCT	ADT, ADT+ GTx-758 1000 mg, or ADT+GTx-758 2000 mg	164	Superior testosterone reduction in GTx-756 group
2015 [88]	Hormone naïve PC	Cancer control	RCT	ADT, ADT+ Toremifene 60 mg, or ADT+Raloxifene 60 mg	15	ADT+ toremifene significantly improved BCR

Abbreviations: PC, prostate cancer; CRPC, castration-resistant prostate cancer; ADT, androgen deprivation therapy; HGPIN, high-grade prostatic intra-epithelial neoplasia; RCT, randomized clinical trial; T Cho, total cholesterol; LDL, low-density lipoprotein; HDL, high-density lipoprotein; TG, total glyceride; BCR, Biochemical recurrence.

and 32.3% of the men in the placebo and treatment groups, respectively, without any difference ($p = 0.39$, log-rank test) in PC-free survival. The 3-year Kaplan–Meier PC-free survival estimate was 54.9% (99% CI, 43.3–66.5%) in the placebo group and 59.5% (99% CI, 48.1–70.9%) in the treatment group. Some previous studies had focused on adverse events associated with ADT, including osteoporosis and lipid metabolism [82, 83]. However, no study has yet demonstrated the inhibitory effect for PC or for serum PSA changes.

Raloxifene, an ER agonist in the bone tissue [73], has been developed for osteoporosis treatment in women, which demonstrated some tumor-inhibitory effects against CRPC in a pilot study [84]. Raloxifene inhibited androgen-independent PC growth in 5 of 13 patients (28%). Another investigation was conducted using a combination of raloxifene and bicalutamide to evaluate treatment toxicity [85], in which 18 men with CRPC were administered a combination of bicalutamide (50 mg) and raloxifene (60 mg) over 28-day cycles. Although none of the patients required dose reduction, the resultant clinical benefit was limited. Only 4 of the 18 patients experienced >50% PSA reduction, with a median PFS of 1.9 months (1.8–2.8 months).

Fulvestrant is a pure estrogen antagonist with no agonist activity [86, 89]. Fulvestrant was administered via intramuscular injection at a dose of 500 mg on day 0, followed by 250 mg on day 14, day 28, and monthly thereafter in 20 patients with CRPC [86]. Median time to progression was 4.3 months, and median OS was 19.4 months. No patient showed >50% PSA reduction or any radiological improvement.

Only one study has been conducted in patients with treatment-naïve PC [87], in which 164 patients were randomized to oral GTx-758, ERα agonist, 1000 mg/day, 2000 mg/day, or leuprolide depot. Although leuprolide reduced the total testosterone level to <50 ng/dL in most patients compared with GTx-758, GTx-758 was superior in lowering the free testosterone level and PSA. GTx-758 reduced the side effects of estrogen deficiency, such as hot flushes, bone loss, and insulin resistance, but increased the incidence of venous thromboembolic events. Therefore, the oncological outcomes of SERMs have not yet been comprehensively investigated for use in patients with treatment-naïve PC.

We thus hypothesized that additional SERMs may prolong the durability of ADT because androgen and estrogen signaling drive PC progression. In our study, we conducted a prospective randomized clinical phase IIA trial to investigate the effects of SERMs (toremifene and raloxifene) in combination with ADT in treatment-naïve bone metastatic PC [88]. Men with treatment-naïve bone metastatic PC were randomly assigned into receive ADT, toremifene (60 mg) plus ADT (TOPADT), or raloxifene (60 mg) plus ADT (RAPADT). A total of 15 men, 5 each, were allocated to one of the three treatment arms. The basal serum PSA level was 198 ng/mL (median; range, 30–8428). Bone metastases were graded as 1 (n = 11), 2 (n = 3), and 3 (n = 1) depending on the extent of the disease. During the median follow-up period of 1370 days (range, 431–1983), BCR occurred in 3, 0, and 2 men in the ADT, TOPADT, and RAPADT groups, respectively. The 5-year BCR-free rate was 30%, 100%, and 53% in the ADT, TOPADT, and RAPADT groups, respectively (p = 0.04, ADT vs. TOPADT; p = 0.48, ADT vs. RAPADT; and p = 0.12, TOPADT vs. RAPADT). Although ERβ agonists were expected to have tumor-inhibitory effect, the study did not prove a distinct tumor-inhibitory effect. Further additional ERβ agonists are warranted with respect to their potential role in the inhibition of human PC. Thus, we for the first time demonstrated clinical benefits of EAB in patients with treatment-naïve bone metastatic PC.

To date, anti-cancer effects of EAB have not yet been completely investigated in treatment-naïve PC patients. We assumed that the concurrent use of SERMs can prolong the duration of ADT efficacy in treatment-naïve men with bone metastatic PC. In the future, additional clinical trials with larger cohorts will be warranted to confirm our promising phase IIA findings. Furthermore, functional investigations are required to clarify the mechanism underlying EAB in treatment-naïve PC.

28.7 Molecular Diagnosis for Precision Medicine

Determination of the therapeutic strategy for breast cancer depends on the expression patterns of ERα, progesterone receptor, and human epidermal growth factor receptor 2 (HER 2) in needle biopsy samples [90]. Therapeutic strategies for non-small cell lung cancer are based on mutation of epidermal growth factor receptor (EGFR), Kirsten murine sarcoma virus (KRAS), or anaplastic lymphoma kinase (ALK) [91]. However, pretreatment diagnosis using the estimated gene expression is not yet prevalent in patients with PC. Molecular diagnosis using immunohistochemistry; fluorescent in situ hybridization; and DNA, RNA, and/or micro-RNA analyses have been discussed in recent articles [92–100]. Studies in which gene sets of SC-like cells, micro-RNA, or cell-cycle progression markers reflect more aggressive diseases are limited to localized PC. A recent study has revealed that of the analysis of mutations, such as *ATM Serine/Threonine Kinase* (*ATM*) and *Breast Cancer 1/2* (*BRCA1/2*), or the expression of specific genes can indicate the risk, metastatic potential, or radio sensitivity of PC [100, 101]; however, investigations regarding metastatic PC through molecular diagnosis are limited.

We have previously defined an AR transcriptional network in PC cells using Chromatin immunoprecipitation on chip (ChIP–chip) and 5′-cap analysis of gene expression (CAGE) analyses [102, 103] as well as functional analysis of AR-related genes, including *Amyloid precursor protein* (*APP*), Octamer-Binding Transcription Factor 1 (*Oct1*), *Tripartite Motif Containing 36* (*TRIM36*), and *Forkhead Box P1* (*FOXP1*), in PC [104–106]. Following experiments revealed the association between BCR and the expression of *AR*, *ERα*, *Sox2*, *CRP*, and *Her2* as well as the association between CSS and the expression of *AR*, *Oct1*, *TRIM36*, *Sox2*, *Klf4*, *c-Myc*, and *ERα* in patients with metastatic PC using laser microdissection technique (LMD) [107].

Various nuclear receptors; their related co-regulators; and miR, including ARs, ERα, ERβ, and ERRs participate in the development or regulation of metastatic PC. Here, we proposed a new therapeutic strategy for metastatic PC (Fig. 28.2). In summary, some patients with homogeneous PC cells sensitive to ADT alone, while others with heterogeneous PC cells, including ERα, ERβ, and ERRα, were eligible for EAB. In conclusion, we assume that an initial precision medicine induces dormancy in PC cells. In the future, there is a scope of precision medicine contributing to a longer and active life in patients with metastatic PC through the use of LMD of biopsy specimens, genomic analysis, or liquid biopsy of blood samples.

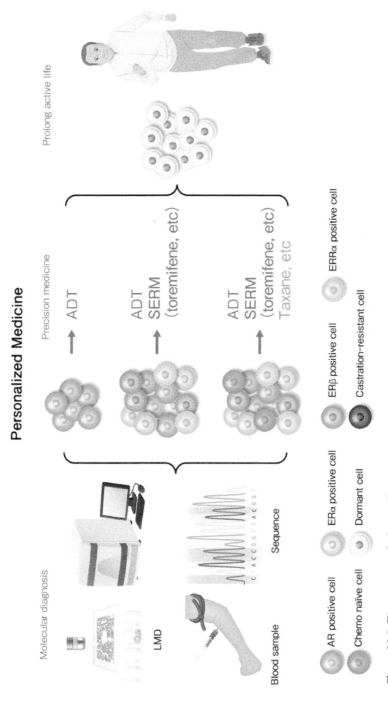

Figure 28.2 Time course of advanced prostate cancer.

In classical pathways down-stream genes of AR, ER, or ERR are activated through binding AR, ER, or ERR with androgen-responsive elements (ARE), estrogen-responsive elements (ERE), or estrogen-related receptor elements (ERRE), respectively. AR is directly regulated by micro-RNAs (miRs). ERα and ERRα contribute prostate cancer (PC) progression via *phosphatidylinositol-3 kinase* (Pl3K), mitogen-activated protein kinase (MAPK) and vascular endothelial growth factor A (VEGF-A), Wnt Family Member 5A (WNT5A), and *Transforming Growth Factor Beta 1* (*TGFβ1*) VEGF/WNT5A/TGF pathways, respectively. Estradiol and GS-1405, ERβ agonist, suppress *Forkhead Box O1* (*FOXO1*) and *platelet derived growth factor subunit A* (*PDGFA*) expression levels through ERβ and Kruppel-like factor (KLF) 5 pathways, and inhibit PC growth via apoptosis, anoikis, and angiogenesis. ERβ and NFκB also regulate PC activating Interleukin *(IL)-12, Growth differentiation gene (GDF)-1, IL-8*, and *Receptor-like tyrosine kinase (RYK)*.

Conventionally most of the patients with advanced PC receive ADT after initial diagnostic prostate biopsy irrespective of the PC cell type. Consequently, PC becomes castration resistant status with distant metastasis.

Precision or personalized medicine can separate patients according to types of PC cells: AR positive, AR, ERα, ERβ, and ERRα positive, or chemo-naïve type. Precision medicine changes advanced PC cells to dormant cells and provides patients with long and active life.

Abbreviations

ADT: androgen deprivation therapy
ALK: anaplastic lymphoma kinase
APP: amyloid precursor protein
AR: androgen receptor
ARE: androgen-responsive element
BCR: biochemical recurrence
BRCA1/2:   breast cancer 1/2
CAGE: 5′-cap analysis of gene expression
CI: confidence interval

CRPC:	castration-resistant PC
CSS:	cancer specific survival
CYP:	Cytochrome P450
DBDs:	DNA-binding domains
DES:	diethylstilbestrol
DHT:	dihydrotestosterone
EAB:	estrogen and androgen signal blockade
EGFR:	epidermal growth factor receptor
ER:	estrogen receptor
ERE:	estrogen response element
ERR:	estrogen-related receptor
ERRE:	estrogen-related receptor elements
FOXO1:	Forkhead Box O1
FOXP1:	Forkhead Box P1
GDF:	growth differentiation gene
GPR 30:	G-protein-coupled receptor-30
HGPIN:	high-grade prostatic intraepithelial neoplasia
HR:	hazards ratio
IL:	interleukin
KLF:	Kruppel-like factor
KRAS:	Kirsten murine sarcoma virus
LBDs:	ligand-binding domains
LMD:	laser microdissection technique
MAPK:	mitogen-activated protein kinase
miR:	micro-RNA
Oct1:	octamer-binding transcription factor 1
OS:	overall survival
PC:	prostate cancer
PDGFA:	platelet derived growth factor subunit A
PFS:	progression-free survival
PI3K:	phosphatidylinositol-3 kinase
PSA:	prostate-specific antigen
PTEN:	phosphatase and tensin homolog
RAPADT:	raloxifene plus ADT

RYK: receptor-like tyrosine kinase
SERMs: selective estrogen receptor modulators
SXR: steroid and xenobiotic receptor
TGFβ1:   transforming growth factor beta 1
TOPADT: toremifene plus ADT
TRIM36:   tripartite motif containing 36
VEGF-A: vascular endothelial growth factor A
WNT5A: Wnt Family Member 5A

Disclosures and Conflict of Interest

This chapter was originally published as: Fujimura, T., Takayama, K., Takahashi, S., Inour, S. (2018). Estrogen and androgen blockade for advanced prostate cancer in the era of precision medicine, *Cancers*, **10**, 29, doi: 10.3390/cancers10020029, under the Creative Commons Attribution license (http://creativecommons. org/licenses/by/4.0/). It appears here, with edits and updates, by kind permission of the authors and publisher, MDPI (Basel).

The work described in this chapter was supported by grants from the Japan Society for Promotion of Science. The founding sponsors had no role in the design of the study; in the collection, analyses, or interpretation of data; in the writing of the manuscript; and in the decision to publish the results. The authors declare that they have no conflict of interest. No writing assistance was utilized in the production of this chapter and the authors have received no payment for it.

Corresponding Author

Dr. Satoshi Inoue
Department of Systems Aging Science and Medicine
Tokyo Metropolitan Institute of Gerontology
35-2 Sakae-cho, Itabashi-ku, Tokyo 173-0015, Japan
Email: sinoue@tmig.or.jp

References

1. Torre, L. A., Bray, F., Siegel, R. L., Ferlay, J., Lortet-Tieulent, J., Jemal, A. (2015). Global cancer statistics, 2012. *CA Cancer J. Clin.*, **65**, 87–108.

2. Gillessen, S., Attard, G., Beer, T. M., Beltran, H., Bossi, A., Bristow, R., Carver, B., Castellano, D., Chung, B. H., Clarke, N., et al. (2017). Management of patients with advanced prostate cancer: The report of the advanced prostate cancer consensus conference APCCC 2017. *Eur. Urol.*, **73**, 178–211.

3. Huggins, C., Hodges, C. V. (1972). Studies on prostatic cancer. I. The effect of castration, of estrogen and androgen injection on serum phosphatases in metastatic carcinoma of the prostate. *CA Cancer J. Clin.*, **22**, 232–240.

4. Takayama, K. I., Misawa, A., Inoue, S. (2017). Significance of microRNAs in androgen signaling and prostate cancer progression. *Cancers*, **9**, 102.

5. Obinata, D., Takayama, K., Takahashi, S., Inoue, S. (2017). Crosstalk of the androgen receptor with transcriptional collaborators: Potential therapeutic targets for castration-resistant prostate cancer. *Cancers*, **9**, 22.

6. Ehara, H., Koji, T., Deguchi, T., Yoshii, A., Nakano, M., Nakane, P. K., Kawada, Y. (1995). Expression of estrogen receptor in diseased human prostate assessed by non-radioactive in situ hybridization and immunohistochemistry. *Prostate*, **27**, 304–313.

7. Kirschenbaum, A., Ren, M., Erenburg, I., Schachter, B., Levine, A. C. (1994). Estrogen receptor messenger RNA expression in human benign prostatic hyperplasia: Detection, localization, and modulation with a long-acting gonadotropin-releasing hormone agonist. *J. Androl.*, **15**, 528–533.

8. Hiramatsu, M., Maehara, I., Orikasa, S., Sasano, H. (1996). Immunolocalization of oestrogen and progesterone receptors in prostatic hyperplasia and carcinoma. *Histopathology*, **28**, 163–168.

9. Sugimura, Y., Cunha, G. R., Yonemura, C. U., Kawamura, J. (1988). Temporal and spatial factors in diethylstilbestrol-induced squamous metaplasia of the developing human prostate. *Hum. Pathol.*, **19**, 133–139.

10. Triche, T. J., Harkin, J. C. (1971). An ultrastructural study of hormonally induced squamous metaplasia in the coagulating gland of the mouse prostate. *Lab. Investig. J. Tech. Methods Pathol.*, **25**, 596–606.

11. Levine, A. C., Kirschenbaum, A., Droller, M., Gabrilove, J. L. (1991). Effect of the addition of estrogen to medical castration on prostatic size, symptoms, histology and serum prostate specific antigen in 4 men with benign prostatic hypertrophy. *J. Urol.*, **146**, 790–793.

12. Noble, R. L. (1980). Production of Nb rat carcinoma of the dorsal prostate and response of estrogen-dependent transplants to sex hormones and tamoxifen. *Cancer Res.*, **40**, 3547–3550.

13. Jensen, E. V., Jacobson, H. I., Smith, S., Jungblut, P. W., De Sombre, E. R. (1972). The use of estrogen antagonists in hormone receptor studies. *Gynecol. Investig.*, **3**, 108–123.

14. Kuiper, G. G., Enmark, E., Pelto-Huikko, M., Nilsson, S., Gustafsson, J. A. (1996). Cloning of a novel receptor expressed in rat prostate and ovary. *Proc. Natl. Acad. Sci. U.S.A.*, **93**, 5925–5930.

15. Mosselman, S., Polman, J., Dijkema, R. (1996). ER β: Identification and characterization of a novel human estrogen receptor. *FEBS Lett.*, **392**, 49–53.

16. Giguere, V., Yang, N., Segui, P., Evans, R. M. (1988). Identification of a new class of steroid hormone receptors. *Nature*, **331**, 91–94.

17. Bonkhoff, H., Fixemer, T., Hunsicker, I., Remberger, K. (1999). Estrogen receptor expression in prostate cancer and premalignant prostatic lesions. *Am. J. Pathol.*, **155**, 641–647.

18. Royuela, M., de Miguel, M. P., Bethencourt, F. R., Sanchez-Chapado, M., Fraile, B., Arenas, M. I., Paniagua, R. (2001). Estrogen receptors α and β in the normal, hyperplastic and carcinomatous human prostate. *J. Endocrinol.*, **168**, 447–454.

19. Leav, I., Lau, K. M., Adams, J. Y., McNeal, J. E., Taplin, M. E., Wang, J., Singh, H., Ho, S. M. (2001). Comparative studies of the estrogen receptors β and α and the androgen receptor in normal human prostate glands, dysplasia, and in primary and metastatic carcinoma. *Am. J. Pathol.*, **159**, 79–92.

20. Pasquali, D., Rossi, V., Esposito, D., Abbondanza, C., Puca, G. A., Bellastella, A., Sinisi, A. A. (2001). Loss of estrogen receptor β expression in malignant human prostate cells in primary cultures and in prostate cancer tissues. *J. Clin. Endocrinol. Metab.*, **86**, 2051–2055.

21. Pasquali, D., Staibano, S., Prezioso, D., Franco, R., Esposito, D., Notaro, A., De Rosa, G., Bellastella, A., Sinisi, A. A. (2001). Estrogen receptor β expression in human prostate tissue. *Mol. Cell. Endocrinol.*, **178**, 47–50.

22. Horvath, L. G., Henshall, S. M., Lee, C. S., Head, D. R., Quinn, D. I., Makela, S., Delprado, W., Golovsky, D., Brenner, P. C., O'Neill, G., et al. (2001). Frequent loss of estrogen receptor-β expression in prostate cancer. *Cancer Res.*, **61**, 5331–5335.

23. Latil, A., Bieche, I., Vidaud, D., Lidereau, R., Berthon, P., Cussenot, O., Vidaud, M. (2001). Evaluation of androgen, estrogen (ER α and ER

β), and progesterone receptor expression in human prostate cancer by real-time quantitative reverse transcription-polymerase chain reaction assays. *Cancer Res.*, **61**, 1919–1926.

24. Castagnetta, L. A., Miceli, M. D., Sorci, C. M., Pfeffer, U., Farruggio, R., Oliveri, G., Calabro, M., Carruba, G. (1995). Growth of LNCaP human prostate cancer cells is stimulated by estradiol via its own receptor. *Endocrinology*, **136**, 2309–2319.

25. Hobisch, A., Hittmair, A., Daxenbichler, G., Wille, S., Radmayr, C., Hobisch-Hagen, P., Bartsch, G., Klocker, H., Culig, Z. (1997). Metastatic lesions from prostate cancer do not express oestrogen and progesterone receptors. *J. Pathol.*, **182**, 356–361.

26. Konishi, N., Nakaoka, S., Hiasa, Y., Kitahori, Y., Ohshima, M., Samma, S., Okajima, E. (1993). Immunohistochemical evaluation of estrogen receptor status in benign prostatic hypertrophy and in prostate carcinoma and the relationship to efficacy of endocrine therapy. *Oncology*, **50**, 259–263.

27. Fujimura, T., Takahashi, S., Urano, T., Ogawa, S., Ouchi, Y., Kitamura, T., Muramatsu, M., Inoue, S. (2001). Differential expression of estrogen receptor β (ERβ) and its C-terminal truncated splice variant ERβcx as prognostic predictors in human prostatic cancer. *Biochem. Biophys. Res. Commun.*, **289**, 692–699.

28. Prins, G. S., Korach, K. S. (2008). The role of estrogens and estrogen receptors in normal prostate growth and disease. *Steroids*, **73**, 233–244.

29. Gustafsson, J. A. (1999). Estrogen receptor β—A new dimension in estrogen mechanism of action. *J. Endocrinol.*, **163**, 379–383.

30. Lau, K. M., LaSpina, M., Long, J., Ho, S. M. (2000). Expression of estrogen receptor (ER)-α and ER-β in normal and malignant prostatic epithelial cells: Regulation by methylation and involvement in growth regulation. *Cancer Res.*, **60**, 3175–3182.

31. Nelson, A. W., Groen, A. J., Miller, J. L., Warren, A. Y., Holmes, K. A., Tarulli, G. A., Tilley, W. D., Katzenellenbogen, B. S., Hawse, J. R., Gnanapragasam, V. J., et al. (2017). Comprehensive assessment of estrogen receptor β antibodies in cancer cell line models and tissue reveals critical limitations in reagent specificity. *Mol. Cell. Endocrinol.*, **440**, 138–150.

32. Moore, J. T., McKee, D. D., Slentz-Kesler, K., Moore, L. B., Jones, S. A., Horne, E. L., Su, J. L., Kliewer, S. A., Lehmann, J. M., Willson, T. M. (1998). Cloning and characterization of human estrogen receptor β isoforms. *Biochem. Biophys. Res. Commun.*, **247**, 75–78.

33. Andersson, S., Sundberg, M., Pristovsek, N., Ibrahim, A., Jonsson, P., Katona, B., Clausson, C. M., Zieba, A., Ramstrom, M., Soderberg, O., et al. (2017). Insufficient antibody validation challenges oestrogen receptor β research. *Nat. Commun.*, **8**, 15840.

34. Yang, N., Shigeta, H., Shi, H., Teng, C. T. (1996). Estrogen-related receptor, hERR1, modulates estrogen receptor-mediated response of human lactoferrin gene promoter. *J. Biol. Chem.*, **271**, 5795–5804.

35. Kraus, R. J., Ariazi, E. A., Farrell, M. L., Mertz, J. E. (2002). Estrogen-related receptor α 1 actively antagonizes estrogen receptor-regulated transcription in MCF-7 mammary cells. *J. Biol. Chem.*, **277**, 24826–24834.

36. Cheung, C. P., Yu, S., Wong, K. B., Chan, L. W., Lai, F. M., Wang, X., Suetsugi, M., Chen, S., Chan, F. L. (2005). Expression and functional study of estrogen receptor-related receptors in human prostatic cells and tissues. *J. Clin. Endocrinol. Metab.*, **90**, 1830–1844.

37. Vanacker, J. M., Bonnelye, E., Chopin-Delannoy, S., Delmarre, C., Cavailles, V., Laudet, V. (1999). Transcriptional activities of the orphan nuclear receptor ERR α (estrogen receptor-related receptor-α). *Mol. Endocrinol.*, **13**, 764–773.

38. Zhang, Z., Teng, C. T. (2001). Estrogen receptor α and estrogen receptor-related receptor α1 compete for binding and coactivator. *Mol. Cell. Endocrinol.*, **172**, 223–233.

39. Shi, H., Shigeta, H., Yang, N., Fu, K., O'Brian, G., Teng, C. T. (1997). Human estrogen receptor-like 1 (ESRL1) gene: Genomic organization, chromosomal localization, and promoter characterization. *Genomics*, **44**, 52–60.

40. Ariazi, E. A., Clark, G. M., Mertz, J. E. (2002). Estrogen-related receptor α and estrogen-related receptor γ associate with unfavorable and favorable biomarkers, respectively, in human breast cancer. *Cancer Res.*, **62**, 6510–6518.

41. Suzuki, T., Miki, Y., Moriya, T., Shimada, N., Ishida, T., Hirakawa, H., Ohuchi, N., Sasano, H. (2004). Estrogen-related receptor α in human breast carcinoma as a potent prognostic factor. *Cancer Res.*, **64**, 4670–4676.

42. Fujimura, T., Takahashi, S., Urano, T., Kumagai, J., Ogushi, T., Horie-Inoue, K., Ouchi, Y., Kitamura, T., Muramatsu, M., Inoue, S. (2007). Increased expression of estrogen-related receptor α (ERRα) is a negative prognostic predictor in human prostate cancer. *Int. J. Cancer*, **120**, 2325–2330.

43. Fujimura, T., Takahashi, S., Urano, T., Ijichi, N., Ikeda, K., Kumagai, J., Murata, T., Takayama, K., Horie-Inoue, K., Ouchi, Y., et al. (2010). Differential expression of estrogen-related receptors β and γ (ERRβ and ERRγ) and their clinical significance in human prostate cancer. *Cancer Sci.*, **101**, 646–651.

44. Di Masi, A., De Marinis, E., Ascenzi, P., Marino, M. (2009). Nuclear receptors CAR and PXR: Molecular, functional, and biomedical aspects. *Mol. Asp. Med.*, **30**, 297–343.

45. Blumberg, B., Sabbagh, W., Jr., Juguilon, H., Bolado, J., Jr., van Meter, C. M., Ong, E. S., Evans, R. M. (1998). SXR, a novel steroid and xenobiotic-sensing nuclear receptor. *Genes Dev.*, **12**, 3195–3205.

46. Isaacs, J. T., McDermott, I. R., Coffey, D. S. (1979). Characterization of two new enzymatic activities of the rat ventral prostate: 5 α-androstane-3 β, 17 β-diol 6 α-hydroxylase and 5 α-androstane-3 β, 17 β-diol 7 α-hydroxylase. *Steroids*, **33**, 675–692.

47. Isaacs, J. T., McDermott, I. R., Coffey, D. S. (1979). The identification and characterization of a new C19O3 steroid metabolite in the rat ventral prostate: 5 α-androstane-3 β, 6 α, 17 β-triol. *Steroids*, **33**, 639–657.

48. Waxman, D. J., Lapenson, D. P., Aoyama, T., Gelboin, H. V., Gonzalez, F. J., Korzekwa, K. (1991). Steroid hormone hydroxylase specificities of eleven cDNA-expressed human cytochrome P450s. *Arch. Biochem. Biophys.*, **290**, 160–166.

49. Fujimura, T., Takahashi, S., Urano, T., Tanaka, T., Zhang, W., Azuma, K., Takayama, K., Obinata, D., Murata, T., Horie-Inoue, K., et al. (2012). Clinical significance of steroid and xenobiotic receptor and its targeted gene CYP3A4 in human prostate cancer. *Cancer Sci.*, **103**, 176–180.

50. Fujimura, T., Takahashi, S., Urano, T., Kumagai, J., Murata, T., Takayama, K., Ogushi, T., Horie-Inoue, K., Ouchi, Y., Kitamura, T., et al. (2009). Expression of cytochrome P450 3A4 and its clinical significance in human prostate cancer. *Urology*, **74**, 391–397.

51. Kumagai, J., Fujimura, T., Takahashi, S., Urano, T., Ogushi, T., Horie-Inoue, K., Ouchi, Y., Kitamura, T., Muramatsu, M., Blumberg, B., et al. (2007). Cytochrome P450 2B6 is a growth-inhibitory and prognostic factor for prostate cancer. *Prostate*, **67**, 1029–1037.

52. Zhang, B., Cheng, Q., Ou, Z., Lee, J. H., Xu, M., Kochhar, U., Ren, S., Huang, M., Pflug, B. R., Xie, W. (2010). Pregnane X receptor as a therapeutic target to inhibit androgen activity. *Endocrinology*, **151**, 5721–5729.

53. Schade, G. R., Holt, S. K., Zhang, X., Song, D., Wright, J. L., Zhao, S., Kolb, S., Lam, H. M., Levin, L., Leung, Y. K., et al. (2016). Prostate

cancer expression profiles of cytoplasmic ERβ1 and nuclear ERβ2 are associated with poor outcomes following radical prostatectomy. *J. Urol.*, **195**, 1760–1766.

54. Megas, G., Chrisofos, M., Anastasiou, I., Tsitlidou, A., Choreftaki, T., Deliveliotis, C. (2015). Estrogen receptor (α and β) but not androgen receptor expression is correlated with recurrence, progression and survival in post prostatectomy T3N0M0 locally advanced prostate cancer in an urban Greek population. *Asian J. Androl.*, **17**, 98–105.

55. McDevitt, M. A., Glidewell-Kenney, C., Jimenez, M. A., Ahearn, P. C., Weiss, J., Jameson, J. L., Levine, J. E. (2008). New insights into the classical and non-classical actions of estrogen: Evidence from estrogen receptor knock-out and knock-in mice. *Mol. Cell. Endocrinol.*, **290**, 24–30.

56. Takizawa, I., Lawrence, M. G., Balanathan, P., Rebello, R., Pearson, H. B., Garg, E., Pedersen, J., Pouliot, N., Nadon, R., Watt, M. J., et al. (2015). Estrogen receptor α drives proliferation in PTEN-deficient prostate carcinoma by stimulating survival signaling, MYC expression and altering glucose sensitivity. *Oncotarget*, **6**, 604–616.

57. Dey, P., Jonsson, P., Hartman, J., Williams, C., Strom, A., Gustafsson, J. A. (2012). Estrogen receptors β1 and β2 have opposing roles in regulating proliferation and bone metastasis genes in the prostate cancer cell line PC3. *Mol. Endocrinol.*, **26**, 1991–2003.

58. Nakajima, Y., Akaogi, K., Suzuki, T., Osakabe, A., Yamaguchi, C., Sunahara, N., Ishida, J., Kako, K., Ogawa, S., Fujimura, T., et al. (2011). Estrogen regulates tumor growth through a nonclassical pathway that includes the transcription factors ERβ and KLF5. *Sci. Signal.*, **4**, ra22.

59. Nakajima, Y., Osakabe, A., Waku, T., Suzuki, T., Akaogi, K., Fujimura, T., Homma, Y., Inoue, S., Yanagisawa, J. (2016). Estrogen exhibits a biphasic effect on prostate tumor growth through the estrogen receptor β-KLF5 pathway. *Mol. Cell. Biol.*, **36**, 144–156.

60. Leung, Y. K., Gao, Y., Lau, K. M., Zhang, X., Ho, S. M. (2006). ICI 182,780-regulated gene expression in DU145 prostate cancer cells is mediated by estrogen receptor-β/NFκB crosstalk. *Neoplasia*, **8**, 242–249.

61. Wu, W. F., Maneix, L., Insunza, J., Nalvarte, I., Antonson, P., Kere, J., Yu, N. Y., Tohonen, V., Katayama, S., Einarsdottir, E., et al. (2017). Estrogen receptor β, a regulator of androgen receptor signaling in the mouse ventral prostate. *Proc. Natl. Acad. Sci. U.S.A.*, **114**, E3816–E3822.

62. Gehrig, J., Kaulfuss, S., Jarry, H., Bremmer, F., Stettner, M., Burfeind, P., Thelen, P. (2017). Prospects of estrogen receptor β activation in the treatment of castration-resistant prostate cancer. *Oncotarget*, **8**, 34971–34979.

63. Bianco, S., Sailland, J., Vanacker, J. M. (2012). ERRs and cancers: Effects on metabolism and on proliferation and migration capacities. *J. Steroid Biochem. Mol. Biol.*, **130**, 180–185.

64. Bianco, S., Lanvin, O., Tribollet, V., Macari, C., North, S., Vanacker, J. M. (2009). Modulating estrogen receptor-related receptor-α activity inhibits cell proliferation. *J. Biol. Chem.*, **284**, 23286–23292.

65. Bernatchez, G., Giroux, V., Lassalle, T., Carpentier, A. C., Rivard, N., Carrier, J. C. (2013). ERRα metabolic nuclear receptor controls growth of colon cancer cells. *Carcinogenesis*, **34**, 2253–2261.

66. Wu, Y. M., Chen, Z. J., Liu, H., Wei, W. D., Lu, L. L., Yang, X. L., Liang, W. T., Liu, T., Liu, H. L., Du, J., et al. (2015). Inhibition of ERRα suppresses epithelial mesenchymal transition of triple negative breast cancer cells by directly targeting fibronectin. *Oncotarget*, **6**, 25588–25601.

67. Zou, C., Yu, S., Xu, Z., Wu, D., Ng, C. F., Yao, X., Yew, D. T., Vanacker, J. M., Chan, F. L. (2014). ERRα augments HIF-1 signalling by directly interacting with HIF-1α in normoxic and hypoxic prostate cancer cells. *J. Pathol.*, **233**, 61–73.

68. Fradet, A., Sorel, H., Bouazza, L., Goehrig, D., Depalle, B., Bellahcene, A., Castronovo, V., Follet, H., Descotes, F., Aubin, J. E., et al. (2011). Dual function of ERRα in breast cancer and bone metastasis formation: Implication of VEGF and osteoprotegerin. *Cancer Res.*, **71**, 5728–5738.

69. Dwyer, M. A., Joseph, J. D., Wade, H. E., Eaton, M. L., Kunder, R. S., Kazmin, D., Chang, C. Y., McDonnell, D. P. (2010). WNT11 expression is induced by estrogen-related receptor α and β-catenin and acts in an autocrine manner to increase cancer cell migration. *Cancer Res.*, **70**, 9298–9308.

70. Fradet, A., Bouchet, M., Delliaux, C., Gervais, M., Kan, C., Benetollo, C., Pantano, F., Vargas, G., Bouazza, L., Croset, M., et al. (2016). Estrogen related receptor α in castration-resistant prostate cancer cells promotes tumor progression in bone. *Oncotarget*, **7**, 77071–77086.

71. Kroiss, A., Vincent, S., Decaussin-Petrucci, M., Meugnier, E., Viallet, J., Ruffion, A., Chalmel, F., Samarut, J., Allioli, N. (2015). Androgen-regulated microRNA-135a decreases prostate cancer cell migration and invasion through downregulating ROCK1 and ROCK2. *Oncogene*, **34**, 2846–2855.

72. Tribollet, V., Barenton, B., Kroiss, A., Vincent, S., Zhang, L., Forcet, C., Cerutti, C., Perian, S., Allioli, N., Samarut, J., et al. (2016). miR-135a inhibits the invasion of cancer cells via suppression of ERRα. *PLoS One*, **11**, e0156445.

73. Lonard, D. M., Smith, C. L. (2002). Molecular perspectives on selective estrogen receptor modulators (SERMs): Progress in understanding their tissue-specific agonist and antagonist actions. *Steroids*, **67**, 15–24.

74. Taneja, S. S., Smith, M. R., Dalton, J. T., Raghow, S., Barnette, G., Steiner, M., Veverka, K. A. (2006). Toremifene—A promising therapy for the prevention of prostate cancer and complications of androgen deprivation therapy. *Expert Opin. Investig. Drugs*, **15**, 293–305.

75. Dutertre, M., Smith, C. L. (2000). Molecular mechanisms of selective estrogen receptor modulator (SERM) action. *J. Pharmacol. Exp. Ther.*, **295**, 431–437.

76. Raghow, S., Hooshdaran, M. Z., Katiyar, S., Steiner, M. S. (2002). Toremifene prevents prostate cancer in the transgenic adenocarcinoma of mouse prostate model. *Cancer Res.*, **62**, 1370–1376.

77. Hariri, W., Sudha, T., Bharali, D. J., Cui, H., Mousa, S. A. (2015). Nano-targeted delivery of toremifene, an estrogen receptor-α blocker in prostate cancer. *Pharm. Res.*, **32**, 2764–2774.

78. Stein, S., Zoltick, B., Peacock, T., Holroyde, C., Haller, D., Armstead, B., Malkowicz, S. B., Vaughn, D. J. (2001). Phase II trial of toremifene in androgen-independent prostate cancer: A Penn cancer clinical trials group trial. *Am. J. Clin. Oncol.*, **24**, 283–285.

79. Price, D., Stein, B., Sieber, P., Tutrone, R., Bailen, J., Goluboff, E., Burzon, D., Bostwick, D., Steiner, M. (2006). Toremifene for the prevention of prostate cancer in men with high grade prostatic intraepithelial neoplasia: Results of a double-blind, placebo controlled, phase IIB clinical trial. *J. Urol.*, **176**, 965–971.

80. Smith, M. R., Malkowicz, S. B., Chu, F., Forrest, J., Price, D., Sieber, P., Barnette, K. G., Rodriguez, D., Steiner, M. S. (2008). Toremifene increases bone mineral density in men receiving androgen deprivation therapy for prostate cancer: Interim analysis of a multicenter phase 3 clinical study. *J. Urol.*, **179**, 152–155.

81. Taneja, S. S., Morton, R., Barnette, G., Sieber, P., Hancock, M. L., Steiner, M. (2013). Prostate cancer diagnosis among men with isolated high-grade intraepithelial neoplasia enrolled onto a 3-year prospective phase III clinical trial of oral toremifene. *J. Clin. Oncol.*, **31**, 523–529.

82. Smith, M. R., Malkowicz, S. B., Chu, F., Forrest, J., Sieber, P., Barnette, K. G., Rodriquez, D., Steiner, M. S. (2008). Toremifene improves lipid profiles in men receiving androgen-deprivation therapy for prostate cancer: Interim analysis of a multicenter phase III study. *J. Clin. Oncol.*, **26**, 1824–1829.

83. Draper, M. W., Flowers, D. E., Huster, W. J., Neild, J. A., Harper, K. D., Arnaud, C. (1996). A controlled trial of raloxifene (LY139481) HCl: Impact on bone turnover and serum lipid profile in healthy postmenopausal women. *J. Bone Miner. Res.*, **11**, 835–842.

84. Shazer, R. L., Jain, A., Galkin, A. V., Cinman, N., Nguyen, K. N., Natale, R. B., Gross, M., Green, L., Bender, L. I., Holden, S., et al. (2006). Raloxifene, an oestrogen-receptor-β-targeted therapy, inhibits androgen-independent prostate cancer growth: Results from preclinical studies and a pilot phase II clinical trial. *BJU Int.*, **97**, 691–697.

85. Ho, T. H., Nunez-Nateras, R., Hou, Y. X., Bryce, A. H., Northfelt, D. W., Dueck, A. C., Wong, B., Stanton, M. L., Joseph, R. W., Castle, E. P. (2017). A study of combination bicalutamide and raloxifene for patients with castration-resistant prostate cancer. *Clin. Genitourin. Cancer*, **15**, 196–202.

86. Chadha, M. K., Ashraf, U., Lawrence, D., Tian, L., Levine, E., Silliman, C., Escott, P., Payne, V., Trump, D. L. (2008). Phase II study of fulvestrant (Faslodex) in castration resistant prostate cancer. *Prostate*, **68**, 1461–1466.

87. Yu, E. Y., Getzenberg, R. H., Coss, C. C., Gittelman, M. M., Keane, T., Tutrone, R., Belkoff, L., Given, R., Bass, J., Chu, F., et al. (2015). Selective estrogen receptor α agonist GTx-758 decreases testosterone with reduced side effects of androgen deprivation therapy in men with advanced prostate cancer. *Eur. Urol.*, **67**, 334–341.

88. Fujimura, T., Takahashi, S., Kume, H., Urano, T., Takayama, K., Yamada, Y., Suzuki, M., Fukuhara, H., Nakagawa, T., Inoue, S., et al. (2015). Toremifene, a selective estrogen receptor modulator, significantly improved biochemical recurrence in bone metastatic prostate cancer: A randomized controlled phase II a trial. *BMC Cancer*, **15**, 836.

89. Howell, A., Osborne, C. K., Morris, C., Wakeling, A. E. (2000). ICI 182,780 (Faslodex): Development of a novel, "pure" antiestrogen. *Cancer*, **89**, 817–825.

90. Carlson, R. W., Allred, D. C., Anderson, B. O., Burstein, H. J., Edge, S. B., Farrar, W. B., Forero, A., Giordano, S. H., Goldstein, L. J., Gradishar, W. J., et al. (2012). Metastatic breast cancer, version 1.2012: Featured updates to the NCCN guidelines. *J. Natl. Compr. Cancer Netw.*, **10**, 821–829.

91. Rothschild, S. I. (2015). Targeted therapies in non-small cell lung cancer-beyond EGFR and ALK. *Cancers*, **7**, 930–949.

92. Choudhury, A. D., Eeles, R., Freedland, S. J., Isaacs, W. B., Pomerantz, M. M., Schalken, J. A., Tammela, T. L., Visakorpi, T. (2012). The role of genetic markers in the management of prostate cancer. *Eur. Urol.*, **62**, 577–587.

93. Markert, E. K., Mizuno, H., Vazquez, A., Levine, A. J. (2011). Molecular classification of prostate cancer using curated expression signatures. *Proc. Natl. Acad. Sci. U.S.A.*, **108**, 21276–21281.

94. Martens-Uzunova, E. S., Jalava, S. E., Dits, N. F., van Leenders, G. J., Moller, S., Trapman, J., Bangma, C. H., Litman, T., Visakorpi, T., Jenster, G. (2012). Diagnostic and prognostic signatures from the small non-coding RNA transcriptome in prostate cancer. *Oncogene*, **31**, 978–991.

95. Cuzick, J., Swanson, G. P., Fisher, G., Brothman, A. R., Berney, D. M., Reid, J. E., Mesher, D., Speights, V. O., Stankiewicz, E., Foster, C. S., et al. (2011). Prognostic value of an RNA expression signature derived from cell cycle proliferation genes in patients with prostate cancer: A retrospective study. *Lancet Oncol.*, **12**, 245–255.

96. Sterbis, J. R., Gao, C., Furusato, B., Chen, Y., Shaheduzzaman, S., Ravindranath, L., Osborn, D. J., Rosner, I. L., Dobi, A., McLeod, D. G., et al. (2008). Higher expression of the androgen-regulated gene PSA/HK3 mRNA in prostate cancer tissues predicts biochemical recurrence-free survival. *Clin. Cancer Res.*, **14**, 758–763.

97. Nonn, L., Vaishnav, A., Gallagher, L., Gann, P. H. (2010). mRNA and micro-RNA expression analysis in laser-capture microdissected prostate biopsies: Valuable tool for risk assessment and prevention trials. *Exp. Mol. Pathol.*, **88**, 45–51.

98. Rogerson, L., Darby, S., Jabbar, T., Mathers, M. E., Leung, H. Y., Robson, C. N., Sahadevan, K., O'Toole, K., Gnanapragasam, V. J. (2008). Application of transcript profiling in formalin-fixed paraffin-embedded diagnostic prostate cancer needle biopsies. *BJU Int.*, **102**, 364–370.

99. Bancroft, E. K., Page, E. C., Castro, E., Lilja, H., Vickers, A., Sjoberg, D., Assel, M., Foster, C. S., Mitchell, G., Drew, K., et al. (2014). Targeted prostate cancer screening in BRCA1 and BRCA2 mutation carriers: Results from the initial screening round of the IMPACT study. *Eur. Urol.*, **66**, 489–499.

100. Na, R., Zheng, S. L., Han, M., Yu, H., Jiang, D., Shah, S., Ewing, C. M., Zhang, L., Novakovic, K., Petkewicz, J., et al. (2017). Germline mutations

in ATM and BRCA1/2 distinguish risk for lethal and indolent prostate cancer and are associated with early age at death. *Eur. Urol.*, **71**, 740–747.

101. Rubicz, R., Zhao, S., Wright, J. L., Coleman, I., Grasso, C., Geybels, M. S., Leonardson, A., Kolb, S., April, C., Bibikova, M., et al. (2017). Gene expression panel predicts metastatic-lethal prostate cancer outcomes in men diagnosed with clinically localized prostate cancer. *Mol. Oncol.*, **11**, 140–150.

102. Takayama, K., Inoue, S. (2013). Transcriptional network of androgen receptor in prostate cancer progression. *Int. J. Urol.*, **20**, 756–768.

103. Takayama, K., Kaneshiro, K., Tsutsumi, S., Horie-Inoue, K., Ikeda, K., Urano, T., Ijichi, N., Ouchi, Y., Shirahige, K., Aburatani, H., et al. (2007). Identification of novel androgen response genes in prostate cancer cells by coupling chromatin immunoprecipitation and genomic microarray analysis. *Oncogene*, **26**, 4453–4463.

104. Takayama, K., Tsutsumi, S., Suzuki, T., Horie-Inoue, K., Ikeda, K., Kaneshiro, K., Fujimura, T., Kumagai, J., Urano, T., Sakaki, Y., et al. (2009). Amyloid precursor protein is a primary androgen target gene that promotes prostate cancer growth. *Cancer Res.*, **69**, 137–142.

105. Takayama, K., Horie-Inoue, K., Ikeda, K., Urano, T., Murakami, K., Hayashizaki, Y., Ouchi, Y., Inoue, S. (2008). FOXP1 is an androgen-responsive transcription factor that negatively regulates androgen receptor signaling in prostate cancer cells. *Biochem. Biophys. Res. Commun.*, **374**, 388–393.

106. Obinata, D., Takayama, K., Urano, T., Murata, T., Kumagai, J., Fujimura, T., Ikeda, K., Horie-Inoue, K., Homma, Y., Ouchi, Y., et al. (2012). Oct1 regulates cell growth of LNCaP cells and is a prognostic factor for prostate cancer. *Int. J. Cancer*, **130**, 1021–1028.

107. Fujimura, T., Takahashi, S., Urano, T., Takayama, K., Sugihara, T., Obinata, D., Yamada, Y., Kumagai, J., Kume, H., Ouchi, Y., et al. (2014). Expression of androgen and estrogen signaling components and stem cell markers to predict cancer progression and cancer-specific survival in patients with metastatic prostate cancer. *Clin. Cancer Res.*, **20**, 4625–4635.

Chapter 29

Precision or Personalized Medicine for Cancer Chemotherapy: Is There a Role for Herbal Medicine?

Zhijun Wang, PhD,[a] Xuefeng Liu, MD,[b] Rebecca Lucinda Ka Yan Ho, MSc,[c] Christopher Wai Kei Lam, PhD,[c] and Moses Sing Sum Chow, PharmD[a]

[a]Center for Advancement of Drug Research and Evaluation,
College of Pharmacy, Western University of Health Sciences, Pomona, California, USA
[b]Department of Pathology and Center for Cell Reprogramming,
Georgetown University Medical Center, Washington, DC, USA
[c]Faculty of Medicine and State Key Laboratory of Quality Research in Chinese Medicines,
Macau University of Science and Technology,
Taipa, Macau, China

Keywords: precision medicine, personalized medicine, herbal medicine, resistant cancer, next-generation sequencing, individual molecular profiles, biomarkers, genomics, proteomics, metabolomics, non-small cell lung cancer, programmed death ligand 1, human epidermal growth factor receptor 2, National Cancer Institute, clear cell renal cell carcinoma, collagen gel droplet embedded culture drug test, conditional reprogramming, conditionally reprogrammed cells, circulating tumor cells, patient-derived xenograft, severe combined immunodeficiency, NOD scid gamma, pan-assay interference compounds, pharmacokinetics, pharmacodynamics, vascular endothelial growth factor receptor, phosphoinositide-3 kinase, chemosensitizing effect, collateral sensitivity, traditional Chinese medicine

The Road from Nanomedicine to Precision Medicine
Edited by Shaker A. Mousa, Raj Bawa, and Gerald F. Audette
Copyright © 2020 Jenny Stanford Publishing Pte. Ltd.
ISBN 978-981-4800-59-4 (Hardcover), 978-0-429-29501-0 (eBook)
www.jennystanford.com

29.1 Introduction: Current Limitations of Cancer Chemotherapy

Although more than 100 chemotherapy drugs (National Cancer Institute, http://www.cancer.gov/about-cancer/treatment/drugs) including alkylating agents, antimetabolites, anti-tumor antibiotics, anthracyclines, topoisomerase inhibitors, mitotic inhibitors, corticosteroids, and other molecularly-targeted agents are available for cancer chemotherapy (either alone or in combination), the overall long term clinical benefit from these agents has been generally disappointing, and modern chemotherapy usually ends in failure due to either severe side effects or loss of effectiveness. One major reason for the loss of effectiveness is the development of chemoresistance [1–5]. To overcome such problems, a new approach called precision medicine or personalized medicine has been proposed and initiated by the National Institute of Health (Bethesda, MD, USA). While the precision medicine approach is likely to yield more effective cancer treatment options in the future, the drug development cost is not expected to be reduced and thus the new drug cost for the patients will likely remain high. According to a recent report from the American Society of Clinical Oncology, the average cost to the patient of newly approved anticancer drugs is $10,000 per month and can be as high as $30,000/month [6]. Such cost is not only difficult to afford for a typical US citizen, it is unlikely to be affordable by more than 80% of the world population living in developing countries [7]. Thus, an alternative approach, such as using herbal medicine, whether alone or in combination with conventional anti-cancer agents, could be a feasible and less costly option. But can herbal medicine meet the rigorous testing required and offer improved personalized cancer chemotherapy over conventional anticancer agents? If it is subjected to rigorous testing, can herbal medicine still be cost effective? We believe the answers to both questions can be an optimistic "yes". This chapter will first review the definition and perspectives of precision medicine versus personalized medicine relevant to research and clinical practice. We shall then examine the existing and future technologies that could speed the development of herbal products for personalized cancer chemotherapy leading to enhanced efficacy and reduced cost for treatment of resistant cancer in individual patients.

In order to quicken the development that could rapidly lead to application in individual patients, this chapter will concentrate in reviewing the phenotypic approach rather than genotypic/proteomic/metabolomic approaches to personalized cancer chemotherapy.

29.2 Precision Medicine vs. Personalized Medicine—Potential Application to Improved Cancer Chemotherapy

29.2.1 Definition of Precision Medicine and Personalized Medicine

Precision medicine refers to the tailoring of medical treatment to the individual characteristics of each patient. Similarly, personalized medicine usually refers to a medical approach that proposes the customization of healthcare—with medical decisions, practices, and/or products being tailored to the individual patient. Thus, personalized medicine is often synonymous with precision medicine [8, 9].

Since precision medicine relies on the comprehensive understanding of individual molecular profiles using genomic, proteomic, metabolomic, as well as bioinformatic approaches to obtain a thorough understanding of the correlation between the regulation of gene(s) (functional protein) and disease status, it depends largely on microarray and next-generation sequencing (NGS) in obtaining the genetic information [10–13].

As genetic variations/mutations are often important factors relating to a particular disease as well as drug response, the first essential step for precision medicine would be the identification of such mutations in specific genes at multiple cell regulatory levels (such as genome, transcription, and epigenetics). Thus, massive genetic screening of genetic variations is necessary in understanding the underlying mechanisms. Based on such information, the unique therapeutic strategy can be made for each individual patient having the genetic variation [14].

The regulation of genes can also occur at the stages of translation and post translation modification. In contrast to the genomic information which is relatively inherently static, dynamic processes of cells cannot be monitored using classical genomic approaches alone. Therefore, proteomic approaches have gained extensive development which can complement the molecular profiles. Proteomics will add additional information to precision medicine by monitoring the cellular function at the protein level and help to identify the quantitative biomarkers which can be essential for characterization of disease course and therapeutic response reliably [15, 16].

Metabolomics is the qualitative identification and quantitative measurement of the dynamic multiparametric metabolic response in a biological system (such as cancer tissue and cells) under the condition of pathophysiological stimuli, genetic modification, or therapeutic treatment [17, 18]. Endogenous metabolites can be determined by using GC-MS/LC-MS profiling [19, 20]. Spectral data can be processed and annotated using commercially available database (NIST), and on-line database (HMDB). Based on such information the metabolite annotation and metabolomic biomarker can be selected.

Thus the "-omics" (genomics/proteomics/metabolomics) technology could enable researchers to further uncover the causative mechanisms or biomarkers and potentially optimize drug efficacy and safety.

29.2.2 Examples of Precision Medicine or Personalized Medicine in Cancer Chemotherapy

Precision medicine can involve a single drug or combination of drugs. A good example of single drug is pembrolizumab for metastatic non-small cell lung cancer (NSCLC) whose tumor cells express programmed death ligand 1 (PD-L1) and progress after platinum-based chemotherapy. When PD-L1 is activated, immune response is inhibited. Pembrolizumab targets PD-L1 and allows the host immune system to recognize and attack tumor cells (www.fda.gov).

A good example of combination therapy is pertuzumab in combination with trastuzumab and docetaxel for the treatment of human epidermal growth factor receptor 2 (HER2) positive metastatic breast cancer. Trastuzumab is an antibody that blocks the function of HER2, a protein produced by a specific gene with cancer-causing potential in HER2-positive metastatic breast cancer. Pertuzumab, another monoclonal antibody which binds to a different epitope on HER2 other than trastuzumab, can inhibit HER2 dimerization. The combination with pertuzumab has been shown to produce a median of 6.1 months longer survival time compared to without it [21].

29.2.3 Precision Medicine vs. Personalized Medicine—Perspectives in Drug Development and Practice

While precision medicine and personalized medicine are often synonymous in cancer chemotherapy, the initiating steps could be different in drug development versus practice. Precision medicine is usually a research initiative followed by practice, whereas personalized medicine often starts with an empiric practice followed by basic research (e.g., identification of biomarkers using genomic/proteomic/metabolomic technology) with subsequent implementation in clinical practice. Thus, personalized medicine approach can begin with the initial step of empiric culture of cancer cells from patient's tumor tissue followed by drug sensitivity testing using relevant techniques/models. Once an effective (high cytotoxicity to cancer cell) but safe (low cytotoxicity to normal cell) agent is found, individualization of its dose to achieve the desired target concentration can be implemented based on individual pharmacokinetics and pharmacodynamics. Additional research steps to identify genomic/proteomic/metabolomic biomarkers in response to drug therapy can be included. At present only, a limited number of drugs have been developed to offer personalized cancer chemotherapy. The potential however is huge for future therapeutic agents including herbal medicines.

29.3 Technologies Relevant to Development of Herbal Medicine for Personalized Cancer Chemotherapy

To implement personalized cancer chemotherapy, a relevant pre-therapeutic diagnostic procedure such as biomarker identification or cancer cell culture and drug sensitivity determination is needed prior to proper therapy.

29.3.1 Identification of Biomarkers

The National Cancer Institute (NCI), in particular, defines biomarker as "a biological molecule found in blood, other body fluids, or tissues that is a sign of a normal or abnormal process, or of a condition or disease". A biomarker can be used to assess the therapeutic response to chemotherapy in cancer patients.

The technologies of "-omics" have been widely used to identify the biomarkers of various cancers. Usually the identification of the biomarkers is initiated by collecting the biologically relevant samples which can be the biopsy tissues, patients' blood and/or urine samples. The biomarker candidates can then be screened using proteomic techniques to identify the potential makers using appropriate protein array or mass spectrometry technologies. This can be incorporated with appropriate filtering criteria followed by validation of the candidate biomarkers using a large independent set of samples with targeted approaches such as SRM and ELISA. Pre-clinical assay development is followed by clinical validation. Final approval of the assay is obtained provided that the assay exceeds the current gold standard, is cost-effective, can easily be integrated into current clinical workflows, and improves patient management [22].

A good example is the quantitative proteomics approach for identifying mitochondrial proteins such as ACAT1 and MnSOD for clear cell renal cell carcinoma (ccRCC) [23]. Another example is the identification of eight new potential markers from formalin fixed paraffin embedded biopsy tissues by Metamark Genetics (Cambridge, MA, USA). Their proteomic profile/signature has been found to predict "favorable" versus "non-favorable" pathology of prostate cancer patients independently and may be used as alternatives for clinical decision [22, 24, 25]. Also, the combination of prostate specific antigen (PSA) concentration with β-microseminoprotein (β-MSMB) level which was identified by using MALDI-MS profiling has increased the diagnostic sensitivity [26].

There are many other reports on identification and application of biomarkers using "-omics" technology. The detailed review of the biomarker is beyond the scope of this chapter. To apply to herbal medicines, the relationship in these biomarkers associated with herbal medicine needs to be established, then the biomarkers can be utilized as a useful monitoring tool for herbal medicine therapy.

29.3.2 Culture of Patient Cancer Cells and Determination of Drug Sensitivity

Another pretherapeutic diagnostic test is the cancer cell culture for determination of drug sensitivity. Over the past 50 years, the worldwide cancer research community has generated between 1000 and 1500 publically available cancer cell lines [27, 28] and the use of such cell lines to determine drug sensitivity has been an accepted technique since the early 1980s [29, 30]. While these conventional cancer cell lines have played an important role in understanding of tumor biology and high-throughput screening for drug development, those generated with traditional methods usually fail to reflect the complex genotypes and phenotypes of the corresponding primary tumors due to *in vitro* selection and the accumulation of genetic and epigenetic alterations during passaging. Further, there is a lack of tumor derived extracellular matrix associated with these cell lines [31].

Thus, more relevant and better techniques of culturing patient cancer cells for drug sensitivity testing can facilitate personalized medicine not only for conventional drugs but also for herbal products. While the present technology of culture and sensitivity is not matured enough for recommendation of its use as a diagnostic assay in clinical practice [32, 33], a number of emerging techniques have provided significant progresses in the understanding of different aspects of cancer initiation, progression, metastasis and tumor microenvironment and can be useful for research and drug development, as described below.

29.3.2.1 Collagen gel droplet embedded culture drug test (CD-DST)

This is a two-dimensional (2D) culture technique of patient cancer cells from tissue samples and it can be utilized for determination of drug sensitivity (e.g., using similar drug concentration exposure as that occurred in patients receiving the same drug). When comparing retrospectively the drug sensitivity from such *in vitro* determination has been found to correspond well to the therapeutic response in patients (sensitivity and specificity of 88.2% and 80.6% respectively with a predictive accuracy of about 84.1%) for a wide variety of cancer cells tested including lung, breast, gastric, esophagus [34]. Furthermore, the CD-DST drug sensitive results have been found to provide a better prediction to clinical response of gastric cancer, non-small cell lung cancer, or ovarian/uterine cancer when compared to those with drug resistant test results [35–37]. One important limitation of this method is the need to obtain $>10^4$ cells from biopsy samples for cell culture.

29.3.2.2 Conditional reprogramming (CR)

Liu et al. [38] developed a method which allows a rapid expansion of primary normal and tumor cells using combination of feeder cells and a Rho kinase inhibitor, Y-27632, termed conditional reprogramming (CR). CR seems to convert adult epithelial cells into a basal or stem-like state [39]. The induction of these conditionally reprogrammed cells (CRC) is reversible, and the removal of feeders and ROCK inhibitor, coupled with their placement in environments that mimic their native environment (Matrigel, air-liquid interface, and the renal capsule of immunodeficient mice) allows cells to differentiate normally. Importantly, the CR technology can generate 2×10^6 cells in a week from small biopsies and can generate cultures from cryopreserved tissue and from fewer than four viable cells. In one case report, a patient with lung tumor induced by a mutant HPV11 was treated with vorinostat based on the *ex vivo* response of their CR-derived tumor sample to this drug, and the patient showed a durable response to treatment [40]. A recent independent study utilized the CR method to initiate cultures from CT-guided lung biopsies from patients with non-small-cell lung cancer (NSCLC) which showed clinical resistance to targeted therapies. These cultures contain the same original driver mutations and maintain resistance to the single agent for which resistance was originally shown in patients. In addition, the use of CR technology has been shown to identify a novel combination of targeted therapies against MEK and ALK to combat resistance to single-agent ALK inhibition in ALK-mutant NSCLC. Such approach may well prove to be a suitable diagnostic test to identify therapeutic strategies for individual patients [41].

29.3.2.3   Organoids and organotypic *ex vivo* technique

Another new technique is the organoids or 3D culture [42]. Organoids allow dissociated patient-derived cells to be expanded in a semi-solid extracellular matrix with growth-factor-enriched medium. Such an *in vitro* model has shown histologic, genomic, and phenotypic characteristics resembling that of the original tumor [43]. Organoids from patients with pancreatic, prostate and colon cancers have been developed and reported [42]. This technique however can require weeks to generate sufficient cells for drug testing, and the success rates of establishment are specific to the tissue of origin.

A recent improvement on the above is the organotypic *ex vivo* technique which incorporates both the tumor matrix proteins and autologous serum in the culturing of the patient's cancer cells. Using this technique as well as an algorithm of responses from drug testing, an impressive predicted drug response rate (87%) has been obtained in 55 patients with head and neck squamous cell carcinoma with the testing being completed in about a week [44].

29.3.2.4   Circulating tumor cells (CTC)

CTC or liquid biopsy is an attractive option for cancer cell culture and subsequent sensitivity testing. These cells are thought to be involved in metastasis and they usually die in the circulation due to circulatory shear stress or the loss of matrix related survival signals. The isolation and cultures of viable CTC are technically challenging, although there are a few reports with low success rates on certain types of metastatic tumor [45]. Its clinical utility is still being investigated at the present time [46].

29.3.2.5   Patient-Derived Xenograft (PDX) Model

The *in vivo* mouse cancer model, especially patient derived cancer xenograft (PDX), has been well accepted for investigation of mechanistic and new therapeutic strategies [47]. The PDX model involves direct implantation of fresh cancer tissue specimens into immunodeficient mice such as nude, SCID (severe combined immunodeficiency), and NSG (NOD scid gamma) mice. It is a more realistic preclinical model which can maintain the heterogeneous architecture and molecular characteristics of the original individual tumor as well as the microenvironment. The fresh tumor specimens can be grafted subcutaneously, orthotopically, and in subrenal capsule. It has been proven that the successful rate is highly dependent on the injection site. Subcutaneous injection is easy to implant. However, this site lacks vascularization and may result in loss of cancer subpopulations. The orthotopic model can best mimic the microenvironment similar to that of the original cancer. However, a complex surgical procedure is required for successful implantation. In addition, the orthotopic site has a limited capacity. Recently, subrenal capsule was found to be a more feasible site for prostate cancer model [48].

The PDX model has been used for prostate cancer and ovarian cancer research [49, 50]. While it has a lot of advantages over the traditional xenograft model, its duration is very long (ranges from few months to years) and is not suitable for routine laboratory evaluation associated with clinical practice. It is more suitable for understanding mechanisms and therapeutic strategies and thus can be very helpful in the initial proof of concept for developing a new product or treatment strategy.

29.3.2.6 Appraisal of cell culture and sensitivity techniques for herbal product evaluation

Among the currently available techniques as described above, the organotypic *ex vivo* test reported by Majumder et al. [44] appears to be the most useful. However, its technique is complicated compared to other *in vitro* tests. The CRC technique appears to be the simplest and fastest test available for determining cytotoxicity of various drugs including herbal compounds. The cells can be cultured in regular cell culture flask or plate. Afterwards, cytotoxicity of the compounds can be tested. The cell viability can also be tested by fluorescent and colorimetric cell viability assays (including tetrazolium-based assay (MTT), MTS cell proliferation assay, and sulforhodamine B assay) [51]. More recently, a technique for real time monitoring cell viability has been developed by detection of luminescent signal generate from the specific luciferase.

If positive results are observed with a given compound, the bio-markers can be further evaluated using real time PCR, western blot or other immunological methods. Such biomarkers can further refine the clinical application when implementing the personalized cancer chemotherapy in the clinical practice [40].

A potential issue about activity of herbal medicines is the possibility of the presence of pan-assay interference compounds (PAINS) which can "function as reactive chemicals rather than discriminating drugs" for a specific target. [52, 53]. However, in resistant cancer, there are likely multiple targets. Will PAINS be actually beneficial for multiple targets? Further research is needed to investigate the relevance of PAINS and herbal therapy. Regardless, the empiric cytotoxic effect of the herbal product (with or without combination with the conventional anticancer agent) in resistant cancer and the safety of the herb are two important issues.

29.3.3 Pharmacokinetic Approach for Optimal Dosing to Achieve Efficacy and Safety

Following the cancer cell culture and determination of drug sensitivity, dose optimization for achieving an effective and safe drug concentration in the patient is also an important step for personalized cancer chemotherapy. Pharmacokinetics (PK) describes the relationship of dosage to drug concentration in body fluids (e.g., plasma), while pharmacodynamics (PD) describes the relationship of drug concentration to the effect. Since the desirable dosage of an anticancer agent should ideally possess high cytotoxicity to the cancer cell but low cytotoxicity to the normal cell, a relevant PK/PD-model can be extremely useful to achieve this goal.

The simplest PK/PD model to use is the linear or log-linear model. An example of its application is for mitotane, an antineoplastic drug used in the treatment of adrenocortical carcinoma. Its minimum effective plasma concentration has been found to be about 14 μg/mL. Because it has a long elimination half-life (18–159 days), the accumulation of plasma concentration is relative slow and thus the onset of therapeutic effect is quite delayed (weeks or even months). A pharmacokinetic model has been established which enabled clinicians to adjust dosing based on a target drug exposure and facilitate personalized therapy with a coefficient of variation of 14%. With the aid of such model, the dose regimen can be adapted based on individual plasma level measurements in prospective setting, which makes it possible to improve the clinical management of mitotane treatment [54].

For predicting or simulating the drug response in cancer chemotherapy, we should also consider the heterogeneous nature of tumor tissues. To address cell heterogeneity, a multi-scale modeling approach has been proposed. A physiologic PK modeling to integrate the data obtained from *in vitro* U87 glioma cells, *in vivo* mice study and cancer patients for the brain disposition of a compound called temozolomide has been described with the DNA adducts serving as the marker for the PD model. This multiscale protocol may be further used for temozolomide PK-PD modeling in various cell populations, thus providing a critical tool to personalize temozolomide based chemotherapy on a cell-type-specific approach [55].

Since cancer chemotherapy often involves drug combination, a unified approach to optimize multidrug chemotherapy using a pharmacokinetic enhanced pharmacodynamic model has been developed. This model is based on the vascular endothelial growth factor receptor (VEGFR) signaling system characterized by ligand-receptor interactions, enzyme recruitment (Grb2-Sos, phospholipase Cγ (PLCγ), and phosphoinositide-3 kinase (PI3K)), and downstream mitogen-activated protein kinase and Akt cascade activation. Drugs targeting these mechanisms (a VEGF inhibitor, a PI3K inhibitor, a PLCγ inhibitor, and a mitogen-activated protein kinase inhibitor) and sunitinib can provide input to optimization-based control analyses. This method can capture the complexities of drug action, tailor cancer chemotherapy, and empower personalized medicine [56].

With more discoveries of genomic/proteomic/metabolomic markers which are being validated for cancer treatment, these markers can serve as the target in PKPD model. For example, the physiologically based PK modeling can simulate the drug disposition in tumor tissues in addition to the normal organs. By linking the drug exposure at the action site (tumor) to response (biomarkers), it is possible to evaluate the dynamic changes in the tumor cells.

Of the various PKPD models, a simple model such as the linear or log-linear model is more practical to empirically generate a desirable dosage to achieve a targeted therapeutic effect. Since the population pharmacokinetics is often already known for a given compound, a simple model is likely to be suitable for rapid simulation of dose concentration relationship of the active drug or active herbal component.

In individualized dosing, individual pharmacokinetics of the active drug can be first determined, then an individualized effective dose generated to achieve a targeted plasma concentration, similar to individualized dosing of aminoglycosides that has been established in the past, by using appropriate pharmacokinetic models [57–59]. However, the application of pharmacokinetics to individualization of herbal extracts (with multiple components) is difficult if not impossible. The identification of a major active component can be helpful to partially describe the dose-concentration-effect relationship. The conventional approaches to determine the effective and safe dose of the herbal extract including maximum tolerated dose of the herbal product in animals, existing human use experience, and dose ranging studies in human subjects can all offer useful consideration in deciding a suitable dose for the herbal product. In addition, information on the biomarker effect in relation to herbal product dosage will be especially useful.

29.4 Special Features of Herbal Medicine for Personalized Cancer Chemotherapy— Implications for Future Therapeutic Advancement

Precision medicine is often linked to the concept of targeting one mechanism with one specific compound. While this approach can be effective in certain single gene inherited diseases, there is no single target involvement in many other diseases, especially with resistant cancer. Consequently, in the latter situation, targeting more than one mechanism would be more "precise", since the single target approach has not been effective for long term efficacy. This is where the role of herbal medicines can be of benefit. The herbs contain multiple active components which are likely to target multiple mechanisms [60]. Recently at least two unique cytotoxic effects have been observed from certain herbs at low concentrations. These are the chemosensitizing effect (CE) and collateral sensitivity (CS). They may offer special utility in resistant cancer and are described below.

29.4.1 Chemosensitizing Effect (CE)

Based on observations in *in vitro* resistant cell lines and patient derived resistant cancer cells as well as animal studies, certain herbal extracts such as *Tripterygium wilfordii*, *Coptis* rhizome, and *Rhei Rhizoma*, or their active single chemical components are capable of inducing a potent CE [3]. CE is the ability of a low concentration of herbal extract or its active component (e.g., using a half or one-quarter of IC_{50} concentration) capable of reversing anticancer drug resistance when combined with a particular anticancer drug which the cancer cell has already developed resistance. Some resistant cancer cells (such as PC3-TxR, a prostate cancer cells resistant to docetaxel) are known to be capable of over-expressing P-glycoprotein (P-gp), an efflux pump capable of pumping the active drug from intracellular site to extracellular site and thus decrease drug intracellular concentration. A well-known mechanism of CE from the herb *Tripterygium wilfordii* is believed to be the suppression of P-gp transporter in addition to several other mechanisms.

29.4.2 Collateral Sensitivity (CS) of Herbs

CS is the ability of the herbal extract or compound to selectively inhibit the growth or kill the resistant cells more than the non-resistant parent cells. The mechanism is believed to be related to inhibition of anti-oxidant effect [61, 62]. Certain flavonoids, verapamil, tiopronin and related compounds as well as extracts of *Tripterygium wilfordii* have been reported to significantly exert such effect in different resistant prostate cancer cell lines, as well as conditional reprogrammed cells from patients with breast, prostate and renal cell carcinoma [61, 62]. In our preliminary test of collateral sensitivity of herbal products, the extracts of several commercially available *Tripterygium wilfordii* tablets (obtained from a Chinese medicine store in Shenzhen, China) were found more toxic in the prostate cancer cells resistant to TRAIL (PC3-TR) than its parent cell line (PC3) (see Fig. 29.1 for one such product).

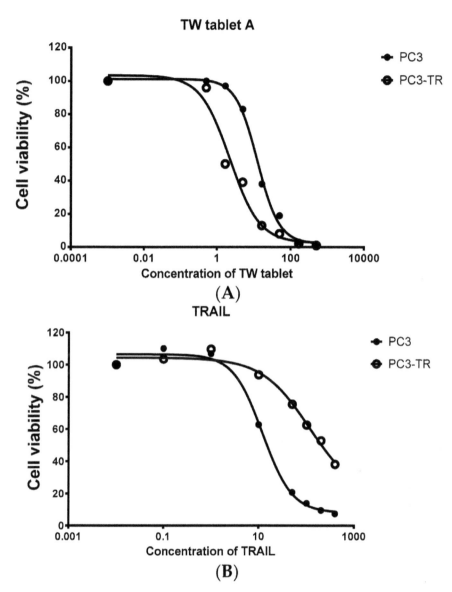

Figure 29.1 Cell viability (measured from triplicate samples) of a prostate cancer cell line (PC3) and its TRAIL resistant cell line (PC3-TR) treated with the extract of *Tripterygium wilfordii* (TW) tablet. (A) The cytotoxicity of TW tablets on PC3-TR was much more potent than PC3 (IC_{50} of 2.3 vs. 12.6 µg/mL for PC3-TR and PC3 respectively); (B) PC3-TR showed resistance to TRAIL (IC_{50} of 512 vs. 18 ng/mL for PC3-TR and PC3, respectively.

In addition, a *Tripterygium wilfordii* extract also showed significant collateral sensitivity in prostate cancer cells resistant to docetaxel (PC3-TxR) than its parent cell line (PC3) (see Fig. 29.2). This could be an exciting area of herbal research. The combined CE and CS effects by herbal medicines could be an exciting potential to retard the development of chemoresistance, as based on clinical experience in antimicrobial chemotherapy and treatment of HIV infections [63, 64]. Such activities may pave the way for further investigations to achieve enhanced benefit of herbal medicine in resistant cancer, which is a major problem at present.

29.5 Perspective and Future Direction of Herbal Medicine—Is There a Role for Personalized Medicine in Cancer Chemotherapy

While precision or personalized medicine involves the use of unique characteristics of a patient's disease status (based on certain biomarkers or genomic/proteomic/metabolomic changes), this basic principle appears to share the same concept as the herbal medicine practice or traditional Chinese medicine (TCM) practice in China. In TCM practice, it has been well recognized that the practice is to provide an individualized therapeutic prescription based on an individual patient's health condition. A typical TCM prescription for a given patient usually consists of a handful of ingredients mixed in a given ratio. Some of these ingredients are referred to as efficacy-enhancing ingredients which are empirically determined.

In this new era of precision medicine, genomic/proteomic/ metabolomic markers may help to understand the mechanisms of enhancing effects of various ingredients of TCM prescriptions, e.g., by using profile (finger printing) of the herbs at "-omics" levels. For example, tanshinone IIA has been shown to target P53 and AKT [65]. Aloe-emodin has been shown to inhibit H460 (non-small lung cancer cell line) by increasing the level of HSP70, 150-kD oxygen-regulated protein and protein disulfide isomerase using proteomics [66]. Ginsenoside Rg3 has been shown to target the apoptosis associated proteins such as Rho GDP dissociation inhibitor, tropomyosin 1 and annexin V and glutathione s-transferase

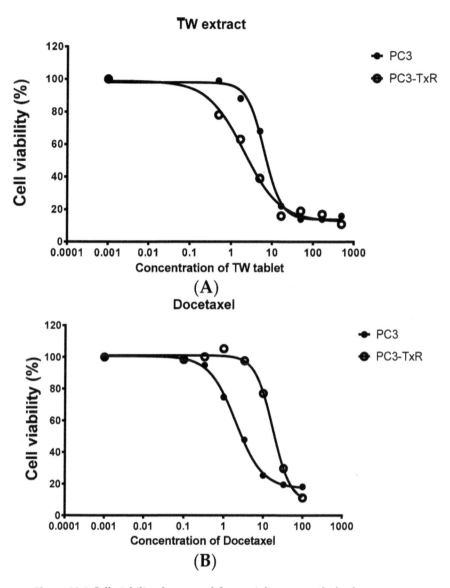

Figure 29.2 Cell viability (measured from triplicate samples) of a prostate cancer cell line (PC3) and its Dtx resistant cell line (PC3-TxR) treated with the extract of *Tripterygium wilfordii* (TW) extract. (A) The cytotoxicity of TW tablets on PC3-TxR was much more potent than PC3 (IC$_{50}$ of 1.9 vs. 6.3 µg/mL for PC3-TxR and PC3 respectively); (B) PC3-TxR showed resistance to TRAIL (IC$_{50}$ of 28.2 vs. 3.4 ng/mL for PC3-TxR and PC3, respectively).

pi-1 [67]. Kampo-derived natural products have been shown to target a panel of genes related to transcriptional processes and nucleic acid interactions [68]). The biomarkers from these "-omics" technologies can help TCM to build a stronger evidence-based practice for personalized medicine and may eventually not only improve cancer chemotherapy but also global health [69].

At present, herbal medicine has already been shown to prevent tumorigenesis, attenuating toxicity and enhancing efficacy as well as reducing tumor recurrence and metastasis [60]. These empiric results are encouraging and may be related to CE and CS effects from the herbs. However further research is needed to demonstrate the benefit from these 2 cytotoxic effects in resistant cancer cells. The fact that 49% of all existing anticancer drugs are either natural products (primarily herbal products) or their derivatives [70], the potential of applying these new cytotoxic actions from the herbal products to overcome resistance is highly attractive and may offer new hope to improve the low success rate (5%) for oncology drug development in the past [71]. While developing herbal product to achieve approval by FDA for resistant cancer with biomarker identification is likely a long process, its application using the personalized medicine approach can be tested as $N = 1$ trials to investigate the benefit of certain specific herbal product in resistant cancer. This can be initiated by empirically culturing resistant cancer cells from individual patient and determining sensitivity of herbal product followed by individualized dosing of the active herb (by utilizing the pharmacokinetics of its major active components). As many of these herbs have already been exposed to human subjects, such an approach should be feasible (after proper patient informed consent) and may offer a more timely therapeutic benefit to many individual patients who are suffering from chemotherapy resistant to conventional drugs.

Abbreviations

β-MSMB: β-microseminoprotein

ccRCC: clear cell renal cell carcinoma

CD-DST: collagen gel droplet embedded culture drug test

CE: chemosensitizing effect

CR: conditional reprogramming
CRC: conditionally reprogrammed cells
CS: collateral sensitivity
CTC: circulating tumor cells
HER2:   human epidermal growth factor receptor 2
NCI: National Cancer Institute
NGS: next-generation sequencing
NSCLC: non-small cell lung cancer
NSG   NOD scid gamma
PAINS: pan-assay interference compounds
PD: pharmacodynamics
PD-L1:   programmed death ligand 1
PDX: patient-derived xenograft
PI3K: phosphoinositide-3 kinase
PK: pharmacokinetics
PLCγ: phospholipase Cγ
PSA: prostate specific antigen
SCID: severe combined immunodeficiency
TCM: traditional Chinese medicine
VEGFR: vascular endothelial growth factor receptor

Disclosures and Conflict of Interest

This chapter was originally published as: Wang, Z., Liu, X., Ho, R.L.K.Y., Lam, C.W.K. and Chow, M.C.C. "Precision or Personalized Medicine for Cancer Chemotherapy: Is there a Role for Herbal Medicine", Molecules **2016**, 21, 889; doi:10.3390/molecules21070889 under the Creative Commons Attribution license (http://creativecommons.org/licenses/by/4.0/). It appears here, with edits and updates, by kind permission of the authors and publisher, MDPI (Basel).

The work described in this chapter was has been partially supported by a research grant from the Macau Science and Technology Development Fund project number FDCT 064/2011/A3. Georgetown University has been awarded a patent by the US

Patent and Trademark Office (US Pat. No. 9,279,106) for conditional cell reprogramming. This technology has been licensed exclusively to a new biotechnology company, Propagenix, for commercialization. Georgetown University and the inventor on this manuscript (X Liu) receives payments and potential royalties from Propagenix. The other authors declare no conflict of interest. No writing assistance was utilized in the production of this chapter and the authors have received no payment for it.

Corresponding Author

Dr. Moses Sing Sum Chow
College of Pharmacy
Western University of Health Sciences
Pomona, CA 91766, USA
Email: mchow@westernu.edu

References

1. Longley, D. B., Johnston, P. G. (2005). Molecular mechanisms of drug resistance. *J. Pathol.*, **205**, 275–292.

2. Thomas, H., Coley, H. M. (2003). Overcoming multidrug resistance in cancer: An update on the clinical strategy of inhibiting p-glycoprotein. *Cancer Control*, **10**, 159–165.

3. Wang, Z., Xie, C., Huang, Y., Lam, C., Chow, M. (2014). Overcoming chemotherapy resistance with herbal medicines: Past, present and future perspectives. *Phytochem. Rev.*, **13**, 323–337.

4. Rottenberg, S., Borst, P. (2012). Drug resistance in the mouse cancer clinic. *Drug Resist. Updates*, **15**, 81–89.

5. Hamilton, G., Rath, B. (2014). A short update on cancer chemoresistance. *Wien. Med. Wochenschr.*, **164**, 456–460.

6. Schnipper, L. E., Davidson, N. E., Wollins, D. S., Tyne, C., Blayney, D. W., Blum, D., Dicker, A. P., Ganz, P. A., Hoverman, J. R., Langdon, R. (2015). American Society of Clinical Oncology statement: A conceptual framework to assess the value of cancer treatment options. *J. Clin. Oncol.*, **33**, 2563–2577.

7. Prasad, S., Tyagi, A. (2015). Traditional medicine: The goldmine for modern drugs. *Adv. Tech. Biol. Med.*, **3**.

8. Roden, D. M., Tyndale, R. F. (2013). Genomic medicine, precision medicine, personalized medicine: What's in a name? *Clin. Pharmacol. Ther.*, **94**, 169–172.

9. Redekop, W. K., Mladsi, D. (2013). The faces of personalized medicine: A framework for understanding its meaning and scope. *Value Health*, **16**, S4–S9.

10. Dong, L., Wang, W., Li, A., Kansal, R., Chen, Y., Chen, H., Li, X. (2015). Clinical next generation sequencing for precision medicine in cancer. *Curr. Genom.*, **16**, 253–263.

11. Zaneveld, J., Wang, F., Wang, X., Chen, R. (2013). Dawn of ocular gene therapy: Implications for molecular diagnosis in retinal disease. *Sci. China Life Sci.*, **56**, 125–133.

12. Guo, Y., Shi, L., Hong, H., Su, Z., Fuscoe, J., Ning, B. (2013). Studies on abacavir-induced hypersensitivity reaction: A successful example of translation of pharmacogenetics to personalized medicine. *Sci. China Life Sci.*, **56**, 119–124.

13. Vizirianakis, I. S. (2002). Pharmaceutical education in the wake of genomic technologies for drug development and personalized medicine. *Eur. J. Pharm. Sci.*, **15**, 243–250.

14. Chen, G., Shi, T. (2013). Next-generation sequencing technologies for personalized medicine: Promising but challenging. *Sci. China. Life Sci.*, **56**, 101.

15. Mesri, M. (2014). Advances in proteomic technologies and its contribution to the field of cancer. *Adv. Med.*, **2014**, 238045.

16. Eckhard, U., Marino, G., Butler, G. S., Overall, C. M. (2016). Positional proteomics in the era of the human proteome project on the doorstep of precision medicine. *Biochimie*, **122**, 110–118.

17. Yu, K.-H., Snyder, M. (2016). Omics profiling in precision oncology. *Mol. Cell. Proteom.*, **15**(8).

18. Klement, G. L., Arkun, K., Valik, D., Roffidal, T., Hashemi, A., Klement, C., Carmassi, P., Rietman, E., Slaby, O., Mazanek, P., et al. (2016). Future paradigms for precision oncology. *Oncotarget*, **19**.

19. Pan, L., Qiu, Y., Chen, T., Lin, J., Chi, Y., Su, M., Zhao, A., Jia, W. (2010). An optimized procedure for metabonomic analysis of rat liver tissue using gas chromatography/time-of-flight mass spectrometry. *J. Pharm. Biomed. Anal.*, **52**, 589–596.

20. Fordahl, S., Cooney, P., Qiu, Y., Xie, G., Jia, W., Erikson, K. M. (2012). Waterborne manganese exposure alters plasma, brain, and liver

metabolites accompanied by changes in stereotypic behaviors. *Neurotoxicol. Teratol.*, **34**, 27–36.

21. Baselga, J., Cortés, J., Kim, S.-B., Im, S.-A., Hegg, R., Im, Y.-H., Roman, L., Pedrini, J. L., Pienkowski, T., Knott, A., et al. (2012). Pertuzumab plus trastuzumab plus docetaxel for metastatic breast cancer. *N. Engl. J. Med.*, **366**, 109–119.

22. Di Meo, A., Pasic, M. D., Yousef, G. M. (2016). Proteomics and peptidomics: Moving toward precision medicine in urological malignancies. *Oncotarget*, **7**(32), 52460–52474.

23. Zhao, Z., Wu, F., Ding, S., Sun, L., Liu, Z., Ding, K., Lu, J. (2015). Label-free quantitative proteomic analysis reveals potential biomarkers and pathways in renal cell carcinoma. *Tumor Biol.*, **36**, 939–951.

24. Blume-Jensen, P., Berman, D. M., Rimm, D. L., Shipitsin, M., Putzi, M., Nifong, T. P., Small, C., Choudhury, S., Capela, T., Coupal, L., et al. (2015). Development and clinical validation of an *in situ* biopsy-based multimarker assay for risk stratification in prostate cancer. *Am. Assoc. Cancer Res.*, **21**, 2591–2600.

25. Shipitsin, M., Small, C., Choudhury, S., Giladi, E., Friedlander, S., Nardone, J., Hussain, S., Hurley, A. D., Ernst, C., Huang, Y. E., et al. (2014). Identification of proteomic biomarkers predicting prostate cancer aggressiveness and lethality despite biopsy-sampling error. *Br. J. Cancer*, **111**, 1201–1212.

26. Flatley, B., Wilmott, K. G., Malone, P., Cramer, R. (2014). MALDI MS profiling of post-DRE urine samples highlights the potential of β-microseminoprotein as a marker for prostatic diseases. *Prostate*, **74**, 103–111.

27. Boehm, J. S., Golub, T. R. (2015). An ecosystem of cancer cell line factories to support a cancer dependency map. *Nat. Rev. Genet.*, **16**, 373–374.

28. Barretina, J., Caponigro, G., Stransky, N., Venkatesan, K., Margolin, A. A., Kim, S., Wilson, C. J., Lehar, J., Kryukov, G. V., Sonkin, D., et al. (2012). The Cancer Cell Line Encyclopedia enables predictive modelling of anticancer drug sensitivity. *Nature*, **483**, 603–607.

29. Frei, E. (1982). The national cancer chemotherapy program. *Science*, **217**, 600–606.

30. Venditti, J. M. (1983). The National Cancer Institute antitumor drug discovery program, current and future perspectives: A commentary. *Cancer Treat. Rep.*, **67**, 767–772.

31. Genovese, L., Zawada, L., Tosoni, A., Ferri, A., Zerbi, P., Allevi, R., Nebuloni, M., Alfano, M. (2014). Cellular localization, invasion, and turnover are differently influenced by healthy and tumor-derived extracellular matrix. *Tissue Eng. Part A*, **20**, 2005–2018.

32. Schrag, D., Garewal, H. S., Burstein, H. J., Samson, D. J., Von Hoff, D. D., Somerfield, M. R. (2004). American Society of Clinical Oncology Technology Assessment: Chemotherapy sensitivity and resistance assays. *J. Clin. Oncol.*, **22**, 3631–3638.

33. Burstein, H. J., Mangu, P. B., Somerfield, M. R., Schrag, D., Samson, D., Holt, L., Zelman, D., Ajani, J. A. (2011). American Society of Clinical Oncology clinical practice guideline update on the use of chemotherapy sensitivity and resistance assays. *J. Clin. Oncol.*, **29**, 3328–3330.

34. Kobayashi, H. (2003). Development of a new *in vitro* chemosensitivity test using collagen gel droplet embedded culture and image analysis for clinical usefulness. *Recent Results Cancer Res. Fortschr. Krebsforsch. Progres Dans Les Recherches Sur Le Cancer*, **161**, 48–61.

35. Naitoh, H., Yamamoto, H., Murata, S., Kobayashi, H., Inoue, K., Tani, T. (2014). Stratified phase II trial to establish the usefulness of the collagen gel droplet embedded culture-drug sensitivity test (CD-DST) for advanced gastric cancer. *Gastric Cancer*, **17**, 630–637.

36. Higashiyama, M., Oda, K., Okami, J., Maeda, J., Kodama, K., Imamura, F., Minamikawa, K., Takano, T., Kobayashi, H. (2010). Prediction of chemotherapeutic effect on postoperative recurrence by *in vitro* anticancer drug sensitivity testing in non-small cell lung cancer patients. *Lung Cancer*, **68**, 472–477.

37. Nagai, N., Minamikawa, K., Mukai, K., Hirata, E., Komatsu, M., Kobayashi, H. (2005). Predicting the chemosensitivity of ovarian and uterine cancers with the collagen gel droplet culture drug-sensitivity test. *Anti-Cancer Drugs*, **16**, 525–531.

38. Liu, X., Ory, V., Chapman, S., Yuan, H., Albanese, C., Kallakury, B., Timofeeva, O. A., Nealon, C., Dakic, A., Simic, V., et al. (2012). ROCK inhibitor and feeder cells induce the conditional reprogramming of epithelial cells. *Am. J. Pathol.*, **180**, 599–607.

39. Suprynowicz, F. A., Upadhyay, G., Krawczyk, E., Kramer, S. C., Hebert, J. D., Liu, X., Yuan, H., Cheluvaraju, C., Clapp, P. W., Boucher, R. C., et al. (2012). Conditionally reprogrammed cells represent a stem-like state of adult epithelial cells. *Proc. Natl. Acad. Sci. USA*, **109**, 20035–20040.

40. Yuan, H., Myers, S., Wang, J., Zhou, D., Woo, J. A., Kallakury, B., Ju, A., Bazylewicz, M., Carter, Y. M., Albanese, C., et al. (2012).

Use of reprogrammed cells to identify therapy for respiratory papillomatosis. *N. Engl. J. Med.*, **367**, 1220–1227.

41. Crystal, A. S., Shaw, A. T., Sequist, L. V., Friboulet, L., Niederst, M. J., Lockerman, E. L., Frias, R. L., Gainor, J. F., Amzallag, A., Greninger, P., et al. (2014). Patient-derived models of acquired resistance can identify effective drug combinations for cancer. *Science*, **346**, 1480–1486.

42. Gao, D., Vela, I., Sboner, A., Iaquinta, P. J., Karthaus, W. R., Gopalan, A., Dowling, C., Wanjala, J. N., Undvall, E. A., Arora, V. K., et al. (2014). Organoid cultures derived from patients with advanced prostate cancer. *Cell*, **159**, 176–187.

43. Gao, D., Chen, Y. (2015). Organoid development in cancer genome discovery. *Curr. Opin. Genet. Dev.*, **30**, 42–48.

44. Majumder, B., Baraneedharan, U., Thiyagarajan, S., Radhakrishnan, P., Narasimhan, H., Dhandapani, M., Brijwani, N., Pinto, D. D., Prasath, A., Shanthappa, B. U., et al. (2015). Predicting clinical response to anticancer drugs using an ex vivo platform that captures tumour heterogeneity. *Nat. Commun.*, **6**.

45. Yu, M., Bardia, A., Aceto, N., Bersani, F., Madden, M. W., Donaldson, M. C., Desai, R., Zhu, H., Comaills, V., Zheng, Z., et al. (2014). Ex vivo culture of circulating breast tumor cells for individualized testing of drug susceptibility. *Science*, **345**, 216–220.

46. Alix-Panabieres, C., Pantel, K. (2014). Challenges in circulating tumour cell research. *Nat. Rev. Cancer*, **14**, 623–631.

47. Friedman, A. A., Letai, A., Fisher, D. E., Flaherty, K. T. (2015). Precision medicine for cancer with next-generation functional diagnostics. *Nat. Rev. Cancer*, **15**, 747–756.

48. Wang, Y., Wang, J. X., Xue, H., Lin, D., Dong, X., Gout, P. W., Gao, X., Pang, J. (2015). Subrenal capsule grafting technology in human cancer modeling and translational cancer research. *Differentiation*, **91**, 15–19.

49. Scott, C. L., Becker, M. A., Haluska, P., Samimi, G. (2013). Patient-derived xenograft models to improve targeted therapy in epithelial ovarian cancer treatment. *Front. Oncol.*, **3**, 1–8.

50. Lin, D., Xue, H., Wang, Y., Wu, R., Watahiki, A., Dong, X., Cheng, H., Wyatt, A. W., Collins, C. C., Gout, P. W. (2014). Next generation patient-derived prostate cancer xenograft models. *Asian J. Androl.*, **16**, 407.

51. Van Tonder, A., Joubert, A. M., Cromarty, A. D. (2015). Limitations of the 3-(4,5-dimethylthiazol-2-yl)-2,5-diphenyl-2*H*-tetrazolium bromide

(MTT) assay when compared to three commonly used cell enumeration assays. *BMC Res. Notes*, **8**, 1–10.

52. Baell, J. B., Holloway, G. A. (2010). New substructure filters for removal of pan assay interference compounds (PAINS) from screening libraries and for their exclusion in bioassays. *J. Med. Chem.*, **53**, 2719–2740.

53. Baell, J., Walters, M. A. (2014). Chemistry: Chemical con artists foil drug discovery. *Nature*, **513**, 481–483.

54. Kerkhofs, T. M., Derijks, L. J., Ettaieb, H., Den Hartigh, J., Neef, K., Gelderblom, H., Guchelaar, H.-J., Haak, H. R. (2015). Development of a pharmacokinetic model of mitotane: Toward personalized dosing in adrenocortical carcinoma. *Ther. Drug Monit.*, **37**, 58–65.

55. Ballesta, A., Zhou, Q., Zhang, X., Lv, H., Gallo, J. (2014). Multiscale design of cell-type–specific pharmacokinetic/pharmacodynamic models for personalized medicine: Application to temozolomide in brain tumors. *CPT Pharmacomet. Syst. Pharmacol.*, **3**, 1–11.

56. Zhang, X. Y., Birtwistle, M., Gallo, J. (2014). A general network pharmacodynamic model–based design pipeline for customized cancer therapy applied to the VEGFR pathway. *CPT Pharmacomet. Syst. Pharmacol.*, **3**, 1–9.

57. Sawchuk, R. J., Zaske, D. E. (1976). Pharmacokinetics of dosing regimens which utilize multiple intravenous infusions: Gentamicin in burn patients. *J. Pharmacokinet. Biopharm.*, **4**, 183–195.

58. Platt, D. R., Matthews, S. J., Sevka, M. J., Comer, J. B., Quintiliani, R., Cunha, B. A., Nightingale, C. H., Chow, M. S. (1982). Comparison of four methods of predicting serum gentamicin concentrations in adult patients with impaired renal function. *Clin. Pharm.*, **1**, 361–365.

59. Burton, M. E., Chow, M. S., Platt, D. R., Day, R. B., Brater, D. C., Vasko, M. R. (1986). Accuracy of Bayesian and Sawchuk-Zaske dosing methods for gentamicin. *Clin. Pharm.*, **5**, 143–149.

60. Ling, C.-Q., Yue, X.-Q., Ling, C. (2014). Three advantages of using traditional Chinese medicine to prevent and treat tumor. *J. Integr. Med.*, **12**, 331–335.

61. Pluchino, K. M., Hall, M. D., Goldsborough, A. S., Callaghan, R., Gottesman, M. M. (2012). Collateral sensitivity as a strategy against cancer multidrug resistance. *Drug Resist. Updates*, **15**, 98–105.

62. Hall, M. D., Marshall, T. S., Kwit, A. D. T., Miller Jenkins, L. M., Dulcey, A. E., Madigan, J. P., Pluchino, K. M., Goldsborough, A. S., Brimacombe, K. R., Griffiths, G. L., et al. (2014). Inhibition of glutathione peroxidase

mediates the collateral sensitivity of multidrug-resistant cells to tiopronin. *J. Biol. Chem.*, **289**, 21473–21489.

63. Imamovic, L., Sommer, M. O. (2013). Use of collateral sensitivity networks to design drug cycling protocols that avoid resistance development. *Sci. Transl. Med.*, **5**, 204ra132.

64. Munck, C., Gumpert, H. K., Wallin, A. I., Wang, H. H., Sommer, M. O. (2014). Prediction of resistance development against drug combinations by collateral responses to component drugs. *Sci. Transl. Med.*, **6**, 262ra156.

65. Lin, L. L., Hsia, C. R., Hsu, C. L., Huang, H. C., Juan, H. F. (2015). Integrating transcriptomics and proteomics to show that tanshinone IIA suppresses cell growth by blocking glucose metabolism in gastric cancer cells. *BMC Genom.*, **16**.

66. Lao, Y., Wang, X., Xu, N., Zhang, H., Xu, H. (2014). Application of proteomics to determine the mechanism of action of traditional Chinese medicine remedies. *J. Ethnopharmacol.*, **155**, 1–8.

67. Lee, S. Y., Kim, G. T., Roh, S. H., Song, J. S., Kim, H. J., Hong, S. S., Kwon, S. W., Park, J. H. (2009). Proteomic analysis of the anti-cancer effect of 20*S*-ginsenoside Rg3 in human colon cancer cell lines. *Biosci. Biotechnol. Biochem.*, **73**, 811–816.

68. Efferth, T., Miyachi, H., Bartsch, H. (2007). Pharmacogenomics of a traditional Japanese herbal medicine (Kampo) for cancer therapy. *Cancer Genom. Proteom.*, **4**, 81–91.

69. Yun, H., Hou, L., Song, M., Wang, Y., Zakus, D., Wu, L., Wang, W. (2016). Genomics and traditional Chinese medicine: A new driver for novel molecular-targeted personalized medicine? *Curr. Pharmacogenomics Pers. Med.*, **10**, 6.

70. Newman, D. J., Cragg, G. M. (2016). Natural products as sources of new drugs from 1981 to 2014. *J. Natl. Prod.*, **79**, 629–661.

71. Kola, I., Landis, J. (2004). Can the pharmaceutical industry reduce attrition rates? *Nat. Rev. Drug Discov.*, **3**, 711–715.

Chapter 30

Personomics: The Missing Link in the Evolution from Precision Medicine to Personalized Medicine

Roy C. Ziegelstein, MD

Johns Hopkins University School of Medicine, Baltimore, Maryland, USA

Keywords: personomics, precision medicine, health care, personalized medicine, evidence-based medicine, randomized controlled trials, systems biology, clinical practical guidelines, estrogen receptor, progesterone receptor, human epidermal growth factor receptor 2, Early Breast Cancer Trialists' Collaborative Group, breast cancer, heterogeneity

There is much discussion today about "precision medicine" and "personalized medicine." Although these terms were not in common use a century ago, physicians in 1919 might well have described their practice using these terms. Physicians generally knew their patients well as people. Physicians and patients often shared the same local community, schools, stores, house of worship, current

The Road from Nanomedicine to Precision Medicine
Edited by Shaker A. Mousa, Raj Bawa, and Gerald F. Audette
Copyright © 2020 Jenny Stanford Publishing Pte. Ltd.
ISBN 978-981-4800-59-4 (Hardcover), 978-0-429-29501-0 (eBook)
www.jennystanford.com

events, and gossip. Although medicine may not have been well-grounded in science, medical care was personally tailored to the individual.

Although physicians a century ago may have practiced a type of personalized medicine, it would be difficult to say that personalized medicine is the norm today, however defined. The last several decades in medicine have been marked by the generation of masses of data from randomized controlled trials (RCTs) that have led to the development of a new field of evidence-based medicine (EBM). EBM has, in turn, led to the generation of clinical practice guidelines for many, if not most, common conditions, not only based on RCTs, but on meta-analyses of those trials. In fact, in many countries, incentives are provided to physicians whose practice is consistent with these guidelines, with penalties to those whose practice is not. When medicine is informed strictly by clinical practice guidelines, the patient is not treated as an individual, but rather a member of a group. Individual variation is certainly not the driver of medical practice; instead, care is based on groups of individuals with similar characteristics who have a similar health condition at a similar stage of progression. In this paradigm, treatment recommendations are often guided by predictions about prognosis and by assessments of risks and benefits that are based on groups of patients felt to be similar to the patient being treated. Indeed, the process of prognostication is—like the creation of an actuarial life table—by definition, treating each individual as if he or she is not unique. It is based on the notion that predictions about an individual can be made based on the prior experience of a group of people felt to have similar characteristics in what is felt to be a similar situation.

As we consider the practice of medicine today and the ideal practice of medicine in the future, we should begin by defining the terms precision medicine and personalized medicine. This commentary proposes an operational definition of these terms and the use of a recently coined term—personomics [1]—to help differentiate the two terms. Herein, precision medicine is defined as the application of recent advances in medical science to characterize individuals based on the unique biological characteristics of the individual or of specimens obtained from that individual. Precision medicine uses information derived from genomics, proteomics, metabolomics, epigenomics, pharmacogenomics, and other "-omics" to derive more precisely tailored diagnostics and therapeutics and thereby improve human health [2]. For the data from precision medicine to be used optimally to improve human health, systems biology will need to collect and integrate data "...from the molecular level, through cells, tissues and organisms, to the population level" [3]. Systems biology will need to be applied to human disease in what has been termed systems medicine, [4] using high throughput technologies to produce and integrate enormous data sets that lead to an improved understanding of human biology. It has been suggested that as digital technology allows these data to be communicated more readily to activated patients and consumers, the possibility of a new healthcare system may be realized that is predictive, preventive, personalized, and participatory (P4) [3].

Personalized medicine adds to that information derived from knowing the patient's unique psychosocial situation, taking into consideration the individual's personal preferences and health beliefs, as well as the individual's values and goals. As much as genomics, proteomics, metabolomics, epigenomics, pharmacogenomics, and other "-omics" are the tools of precision medicine, personomics [1] can be used to describe the tools of personalized medicine. Personomics recognizes that individuals are not only distinguished by their biological variability, but also by their personalities, health beliefs, social support networks, financial resources, and other unique life circumstances that have important effects on how and when a given health condition will manifest in that individual and how it will respond to treatment [1]. The concept of personomics emphasizes that these components of individuality are as critical to patient care as any of the more traditional "-omics" noted previously. Just as these other "-omics" allow for an improved understanding of the pathogenesis and treatment of disease in precision medicine, personomics facilitates the delivery of personalized medicine by enabling physicians to develop a better understanding and appreciation of the patient's environment and life circumstances. Personalized medicine calls for a deeper understanding of each individual's health behaviors, since behavioral habits such as cigarette smoking, unhealthy diet, and lack of exercise are responsible for many important health conditions that result in premature death [5]. Personalized medicine also means that physicians must understand each patient's financial circumstances, since what diagnostic tests and treatments are appropriate for a patient is based not only on the individual's genomics, proteomics, and metabolomics, but also on what tests and treatments that person can afford. Precision medicine may be considered a necessary step in the evolution of medical care to personalized medicine, with personomics as the missing link. The evolution from precision medicine to personalized medicine is the evolution from "health care" to "health caring." As Chochinov noted, "Where health care asks about a problem, health caring asks, 'What do I need to know about you as a person to give you the best care possible?" [6].

To understand this evolution—over the last 25 years—of medical care informed by clinical practice guidelines to a precision medicine approach, and then to think about how personalized medicine might be practiced, it may be useful to consider the treatment of a 72-year-old woman with node-positive, estrogen receptor (ER)-positive breast cancer diagnosed on a routine mammogram. In 1992, this patient's care would have been informed by an important publication from the Early Breast Cancer Trialists' Collaborative Group on the systemic treatment of early breast cancer by hormonal, cytotoxic, or immune therapy [7]. In that report of 133 RCTs involving 75,000 women, systemic chemotherapy was shown to significantly reduce the risk of recurrence among women with early stage breast cancer. However, very little information was presented from women 70 years of age and older, since most RCTs at the time excluded this age group. The effect of systemic chemotherapy in this age group was therefore statistically uncertain. Indeed, the report itself notes that the reviewed trials "...provide virtually no information about women aged over 70" [7]. A physician treating this 72-year-old woman with node-positive, ER-positive breast cancer in 1992 might well have recommended forgoing systemic chemotherapy based on the paucity of data in this age group, even if the patient in question were otherwise healthy.

The situation for this woman would be different today, since precision medicine allows us to base her treatment recommendations on the classification of her tumor into specific molecular subtypes identified from the analysis of genomic alterations in tumor tissue [8]. Predicted survival of women with breast cancer is no longer based solely on the results of RCTs stratified by age and menopausal status, but also based on breast cancer genomic variability and on the status of certain breast cancer receptors, namely the ER, the progesterone receptor (PR), and the human epidermal growth factor receptor 2 (HER2). ER/PR/HER2 status is known to affect patient survival (with so-called triple negative patients having the worst survival) and helps physicians select appropriate treatment regimens [9]. A host of targeted therapeutic options would now be available to be used based on the genomic profiling of this woman's tumor [10]. The physician's recommendation about systemic chemotherapy would be informed by the 21-gene recurrence score assay that may be used to stratify the risk of recurrence of this woman's type of early-stage, ER-positive breast cancer [11]. This is the essence of precision medicine.

Precision medicine would also recognize if this woman's tumor was among the 15–20% of breast cancers that overexpress HER2, a finding that is associated with aggressive tumor behavior and reduced survival [12,13]. If this woman's tumor overexpressed HER2, today's oncologist would be able to employ a treatment such as trastuzumab, a humanized monoclonal antibody that blocks HER2 activity [14], in addition to systemic chemotherapy. It has been shown that women with HER2-positive breast cancer who received trastuzumab have a 44% lower risk of death compared even to women with HER2-negative breast cancer [15]. In another study, the addition of trastuzumab to chemotherapy in the treatment of women with HER2-positive metastatic breast cancer decreased the risk of death at 1 year from 33% to 22% ($p = 0.008$) and increased median survival from 20.3 to 25.1 months ($p = 0.01$) [14]. Today's oncologist would also be able to treat this woman with lapatinib, a small molecule that reversibly inhibits HER1 and HER2, and has been used in women with HER2-positive advanced breast cancer that has progressed after treatment with conventional chemotherapy and trastuzumab [16]. A recent meta-analysis of randomized controlled trials compared the effects of trastuzumab, lapatinib, alone and in combination, in HER2-positive breast cancer. The meta-analysis included 2350 patients, 837 who received lapatinib, 913 who received trastuzumab, and 555 who received the combination. It showed that while trastuzumab is first-line therapy in women with HER2-positive breast cancer, combination therapy increases the pathological complete response rate compared to trastuzumab alone [17].

Despite these advances, there are limitations to precision medicine, even in the field of oncology. Tannock and Hickman recently reviewed trials of cancer therapy guided by genetic sequencing and noted that "the outcomes of these investigations are discouraging" [18]. While it is possible that these limitations will improve as the science of molecular targeted treatments matures, these authors suggested that benefits will be less than expected because of inherent limitations of targeted therapies and because of intratumor heterogeneity. They noted that targeted therapies only partially inhibit critical signaling pathways and that related signaling pathways are also present in normal cells [18]. Given this, currently existing treatments often exhibit significant dose-limiting side effects. Since cancer cells typically develop resistance to agents that target a single signaling pathway, combinations of targeted therapies that inhibit different pathways are often needed. However, side effects often limit a patient's ability to tolerate effective doses of these combination targeted treatments.

How would this woman be treated using a personalized medicine approach? Using the tools of personomics, the physician would get to know this patient as an individual. The physician would certainly consider the patient's age and general health status, and the physician would most definitely inform treatment decisions based on the molecular subtype of this woman's tumor. In addition, though, the physician would ask and learn about this woman's values and goals. The physician would ask and learn about the support this woman would likely receive (or not) from family and friends. This support would be needed regardless of the type of treatment used, and it may be more likely to be needed in the short term if cytotoxic chemotherapy is recommended or if molecular targeted treatments are used that may result in dose-limiting side effects. The physician would ask and learn about this woman's health beliefs, and in particular whether she favors doing whatever possible to prolong life or whether quality of life is more important to her. The physician would ask and learn who is most important in the patient's life and the extent to which she depends on these individuals to make critical health care and life decisions. The physician would ask about this woman's financial circumstances and health insurance coverage and consider her ability to afford the type of medications that might be needed to treat this tumor type, since some of these medications—in particular many of the newer targeted therapies—may be very expensive and might not be covered by some forms of health insurance. The physician who practices personalized medicine would base a recommendation about systemic chemotherapy and other targeted therapies on this woman's age, general health, recurrence score testing, and all the information gleaned from putting personomics into practice. In summary, the evolution from medical care based on clinical practice guidelines to precision medicine to personalized medicine is, in many ways, an evolution from lumpers to splitters. Clinical practice guidelines lump patients together based on certain shared characteristics. As Sacristan and Dilla note, "...guidelines are based on the results of large RCTs and meta-analyses containing information that, in theory, is applicable to 'average patients'. However, doctors do not treat average patients; thus, this 'generalizable knowledge' that may be useful to standardize medical practice is not the most appropriate to treat individual patients" [19]. Precision medicine, by contrast, splits patients into groups defined by biological characteristics and allows for the delivery of more individualized medical care. In the case of cancer treatment, the biological characteristics are based largely on analysis of tumor tissue. As NIH Director Francis Collins noted, "If I had cancer diagnosed today, I would want that cancer to be completely analyzed. I would want to know 'what are the drivers in my tumor that are making it do those bad things,' and then I'd want to compare that to a menu of targeted drugs..." [20]. Personalized medicine splits patients further, recognizing that knowledge of the patient as a person—personomics—is necessary to deliver truly individualized care to every patient based on the unique circumstances of the individual.

Abbreviations

EBM: evidence-based medicine

ER: estrogen receptor

HER2: human epidermal growth factor receptor 2

PR: progesterone receptor

RCTs: randomized controlled trials

Disclosures and Conflict of Interest

This chapter was originally published as: Ziegelstein, R. C. (2017). Personomics: The missing link in the evolution from precision medicine to personalized medicine, *J. Pers. Med.*, **7**, 11, doi: 10.3390/ jpm7040011, under the Creative Commons Attribution license (http://creativecommons.org/licenses/by/4.0/). It appears here, with edits and updates, by kind permission of the authors and publisher, MDPI (Basel).

The author declares that he has no conflict of interest. No writing assistance was utilized in the production of this chapter and the author has received no payment for it.

Corresponding Author

Dr. Roy C. Ziegelstein
Johns Hopkins University School of Medicine
Baltimore, MD 21205, USA
Email: rziegel2@jhmi.edu

References

1. Ziegelstein, R. C. (2015). Personomics. *JAMA Intern. Med.*, **175**, 888–889.

2. Collins, F. S., Varmus, H. (2015). A new initiative on precision medicine. *N. Engl. J. Med.*, **372**, 793–795.

3. Flores, M., Glusman, G., Brogaard, K., Price, N. D., Hood, L. (2013). P4 medicine: How systems medicine will transform the healthcare sector and society. *Per. Med.*, **10**, 565–576.

4. Hood, L., Balling, R., Auffray, C. (2012). Revolutionizing medicine in the 21st century through systems approaches. *Biotechnol. J.*, **7**, 992–1001.

5. Committee on Population; Division of Behavioral and Social Sciences and Education; Board on Health Care Services; National Research Council; Institute of Medicine (2015). *Measuring the Risks and Causes of Premature Death: Summary of Workshops*. National Academies Press, Washington, DC, USA.

6. Chochinov, H. M. (2014). Health care, health caring, and the culture of medicine. *Curr. Oncol.*, **21**, e668–e669.

7. Early Breast Cancer Trialists' Collaborative Group (1992). Systemic treatment of early breast cancer by hormonal, cytotoxic, or immune therapy: 133 randomised trials involving 31,000 recurrences and 24,000 deaths among 75,000 women. *Lancet*, **339**, 71–85.

8. Bettaieb, A., Paul, C., Plenchette, S., Shan, J., Chouchane, L., Ghiringhelli, F. (2017). Precision medicine in breast cancer: Reality or utopia? *J. Transl. Med.*, **15**, 139.

9. Carels, N., Spinasse, L. B., Tilli, T. M., Tuszynski, J. A. (2016). Toward precision medicine of breast cancer. *Theor. Biol. Med. Model.*, **13**, 7.

10. Stover, D. G., Wagle, N. (2015). Precision medicine in breast cancer: Genes, genomes, and the future of genomically driven treatments. *Curr. Oncol. Rep.*, **17**, 15.

11. Friese, C. R., Li, Y., Bondarenko, I., Hofer, T. P., Ward, K. C., Hamilton, A. S., Deapen, D., Kurian, A. W., Katz, S. J. (2017). Chemotherapy decisions and patient experience with the recurrence score assay for early-stage breast cancer. *Cancer*, **123**, 43–51.

12. Valachis, A., Nearchou, A., Lind, P., Mauri, D. (2012). Lapatinib, trastuzumab or the combination added to preoperative chemotherapy for breast cancer: A meta-analysis of randomized evidence. *Breast Cancer Res. Treat.*, **135**, 655–662.

13. Ross, J. S., Slodkowska, E. A., Symmans, W. F., Pusztai, L., Ravdin, P. M., Hortobagyi, G. N. (2009). The HER-2 receptor and breast cancer: Ten years of targeted anti-HER-2 therapy and personalized medicine. *Oncologist*, **14**, 320–368.

14. Slamon, D. J., Leyland-Jones, B., Shak, S., Fuchs, H., Paton, V., Bajamonde, A., Fleming, T., Eiermann, W., Wolter, J., Pegram, M., et al. (2001). Use of chemotherapy plus a monoclonal antibody against HER2 for metastatic breast cancer that overexpresses HER2. *N. Engl. J. Med.*, **344**, 783–792.

15. Dawood. S., Broglio, K., Buzdar, A. U., Hortobagyi, G. N., Giordano, S. H. (2010). Prognosis of women with metastatic breast cancer by HER2 status and trastuzumab treatment: An institutional-based review. *J. Clin. Oncol.*, **28**, 92–98.

16. Geyer, C. E., Forster, J., Lindquist, D., Chan, S., Romieu, C. G., Pienkowski, T., Jagiello-Gruszfeld, A., Crown, J., Chan, A., Kaufman, B., et al. (2006). Lapatinib plus capecitabine for HER2-positive advanced breast cancer. *N Engl. J. Med.*, **355**, 2733–2743.

17. Xin, Y., Guo, W. W., Huang, Q., Zhang, P., Zhang, L. Z., Jiang, G., Tian, Y. (2016). Effects of lapatinib or trastuzumab, alone and in combination, in human epidermal growth factor receptor 2-positive breast cancer: A meta-analysis of randomized controlled trials. *Cancer Med.*, **5**, 3454–3463.

18. Tannock, I. F., Hickman, J. A. (2016). Limits to personalized cancer medicine. *N Engl. J. Med.*, **375**, 1289–1294.

19. Sacristan, J. A., Dilla, T. (2015). No big data without small data: Learning health care systems begin and end with the individual patient. *J. Eval. Clin. Pract.*, **21**, 1014–1017.

20. Bowman, D., Collins, F. (September 15, 2016). privacy a priority for precision medicine initiative. FierceHealthcare. Available at: http://www.fiercehealthcare.com/healthcare/francis-collins-privacy-a-priority-for-precision-medicine-initiative (accessed on January 9, 2019).

Chapter 31

Stem Cell Banking for Regenerative and Personalized Medicine

David T. Harris, PhD

Department of Immunobiology, University of Arizona, Tucson, Arizona, USA

Keywords: stem cells, banking, cord blood, mesenchymal stem cells, cord tissue, adipose tissue, ES cells, human leukocyte antigen, bone marrow, bone marrow transplantation, regenerative medicine, tissue engineering, multiple sclerosis, systemic lupus erythematosus, osteoarthritis, orthopedics, neurology, cardiovascular, age, investigational new device, institutional review board

31.1   Background Introduction

Political and ethical controversy surrounds the use of embryonic stem (ES) cells, and significant biological and regulatory concerns limit their clinical use (the latter concern also applies to induced pluripotent stem cells (iPSC)). Aside from the moral and political

The Road from Nanomedicine to Precision Medicine
Edited by Shaker A. Mousa, Raj Bawa, and Gerald F. Audette
Copyright © 2020 Jenny Stanford Publishing Pte. Ltd.
ISBN 978-981-4800-59-4 (Hardcover), 978-0-429-29501-0 (eBook)
www.jennystanford.com

controversies, ES cells are by their derivation allogeneic in nature and subject to immune rejection if used *in vivo*. Although ES cells may of themselves lack human leukocyte antigen (HLA) expression, the mature tissues that arise from these cells do express HLA antigens and are subject to immune surveillance. Furthermore, ES cells are prone to give rise to teratoma formation when placed *in vivo*, making their clinical use problematic if not impossible.

Although iPS cell generation avoids the allogenicity issue, the teratoma formation problem is still present. Investigators have tried to avoid this issue by deriving mature tissues from iPS cells for clinical use. However, this process is time consuming (4–6 months) and expensive (estimated to be $50,000 or more to derive a single cell line). It should be noted that having tissues that escape immune surveillance is not desirable, as viral infections could go unchecked leading to viremia and serious consequences. Therefore, in most instances regenerative medicine, tissue engineering, gene therapy and most forms of personalized medicine are best served by using autologous sources of stem and progenitor cells. However, the preferred sources must be easily accessible, contain large numbers of stem cells, and economical to utilize. It is our belief that there are only four such sources available: cord blood (CB; generally, 30–120 cc can be collected) and tissue (CT; generally, 4–10 inches are available), bone marrow (BM; generally, 1000–1500 cc is available by surgical harvest) and adipose tissue (AT; generally, 100–3000 cc available using either local or general anesthesia). CB and CT contain hematopoietic stem cells (HSC) (CB only) and mesenchymal stem cells (MSC) (CT primarily) that are available only at the time of birth. BM contains both HSC and MSC although not at high concentrations and these cells may be subject to the detrimental effects of age and health status. AT contains the highest concentration of MSC in the body and is easily accessible (without surgery) in small or large volumes. Of all cellular sources available for use, those containing MSC may be the most useful for clinical applications.

MSCs offer a multitude of potential applications in regenerative medicine, being able to proliferate and differentiate *in vitro* into multiple lineages [1–3]. Low immunoreactivity and high immunosuppressive properties make MSCs a suitable stem cell source for therapy [4, 5]. It has been shown in numerous model systems that MSCs can be used to successfully treat cardiovascular [6, 7], neurological [8] and musculoskeletal disorders [9] either by differentiation into competent cardiomyocytes, neuron-like cells and chondrocytes, respectively; or through a paracrine effect via the secretion of growth, anti-apoptotic and anti-inflammatory factors. In addition, various clinical trials are now underway to assess the effects of these stem cells in patients (see: http://www.clinicaltrials.gov). To date, bone marrow is the best characterized source of MSCs and most clinical data has been based on bone marrow studies. However, there are limitations to the use of bone marrow-derived MSCs (BM-MSCs); e.g., a painful acquisition process, use of extensive anesthesia during the harvest, and low cell yield per cc of tissue. Further, BM-MSCs have been shown to exhibit a decline in MSC numbers, proliferation, angiogenic and wound healing properties, and differentiation, along with enhanced apoptotic and senescent traits with advancing donor age [7, 10, 11]. Recently, other MSC sources have gained clinical interest for use in regenerative medicine; and adipose tissue (AT) represents one of these sources. AT-MSCs possess morphological, phenotypic and functional characteristics similar to BM-MSC [12], are stable over long term culture, expand efficiently *in vitro* and possess multi-lineage differentiation potential [3, 13]. Human adipose tissue may represent a more practical autologous source of MSCs for various tissue engineering strategies. However, the effectiveness of these cells when obtained from any of these sources, and utilized in elderly patients, must be considered when contemplating cell-based therapies (see Section 31.5.5.1 below).

31.2 Practical Stem Cell Sources

Stem cells can be found throughout the body, being present in many tissues and organs (e.g., heart, brain and muscle). In addition, stem cells can be isolated from the heretofore waste products of birth (CB and CT) as well as being created in the laboratory (i.e., ES and iPS cells). When considering the use of stem cells for regenerative medicine and tissue engineering, one must consider the practical aspects of the endeavor (Fig. 31.1). That is, initially and for some time to come, such therapy will not be reimbursed by insurance and must be funded by either the investigator, industry and/or the patients themselves. In addition, in order to be considered for reimbursement any new stem cell based therapy must be as efficacious as standard therapy and cannot be any more expensive. Thus, when considering a source of stem cells for use in these therapies one must identify a source of autologous tissue (to avoid immune rejection issues) that can be readily and inexpensively accessed, which contains large numbers of stem cells (not requiring expansion before clinical use). It is our belief that these constraints limit our choices of stem cell sources to bone marrow, cord blood and tissue, or adipose tissue. Bone marrow, albeit a source of MSC that has been used extensively, is expensive to harvest (with some risk to the donor) and does not contain large numbers of stem cells per cc of tissue. Therefore, this review will focus on banking stem cells collected from cord blood and tissue, or adipose tissue.

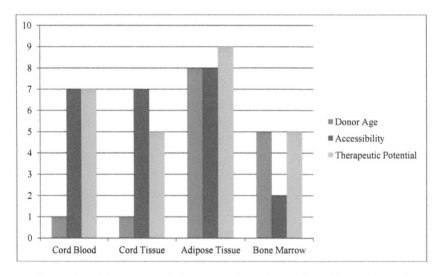

Figure 31.1 Schematic of factors impacting stem cell banking. The most commonly available and practical sources of stem cells are compared *versus* one another for age at acquisition, ease of harvest (accessibility) and therapeutic potential. Donor age refers to total decades of life when one could acquire the stem cell sample; accessibility refers to how easy (10) or difficult (1) the stem cell sample is to collect; while therapeutic potential is used to indicate potential number of uses with (10) representing the greatest potential uses.

The blood in the umbilical cord and placenta after the birth of a child is comparable to bone marrow for use in hematopoietic stem cell transplantation and offers a number of advantages. In the past 20 years, more than 30,000 cord blood transplants have been performed worldwide [14]. Stem cell transplantation for hematological malignancies and genetic disorders however, is an uncommon occurrence. Research performed by several independent laboratories [15–20] has demonstrated that cord blood also contains a mixture of pluripotent stem cells capable of giving rise to cells derived from the endodermal, mesodermal, and ectodermal lineages. In addition, mesenchymal stem cells (MSC), although rare in cord blood, can be easily isolated from the cord tissue (CT) and preserved for later use [3], prompting the development of methods for the collection and cryopreservation of cord tissue [3]. Thus, CB and CT are a readily available stem cell source for use in tissue engineering and regenerative medicine applications, which are hypothesized to be more frequent events than the need for hematopoietic stem cell transplant. It is estimated that almost 1 in 3 individuals in the United States, or 128 million people, could benefit over their lifetime from regenerative medicine, including therapies for cardiovascular, neurological, and orthopedic diseases [21]. However, the absolute numbers of these hematopoietic stem cells (HSC) are limited and to date it has not been possible to successfully replicate HSC. However, MSC can be easily expanded *ex vivo* and CT represents a viable source of such cells. Absolute numbers of CT-MSC again are limited (estimated at 250,000 to 10 million total cells) requiring extended *in vitro* expansion prior to use, which invites additional regulatory oversight, costs and possible culture-induced senescence issues. Fortunately, recent work from our laboratory has demonstrated that CT-MSC and AT-MSC are equivalent in regenerative medicine potential [3]. Although CT-MSC proliferate to a greater extent *in vitro*, the ability to easily acquire 100–1000× more MSC numbers with AT (at almost any time in a patient's life) makes adipose tissue the preferred source for clinical banking of MSC. Additionally, detailed fluorescence activated cell sorter (FACS) analyses showed no differences in phenotypic expression of CD3, CD14, CD19, CD34, CD44, CD45, CD73, CD90 and CD105 molecules between CT-MSCs and

AT-MSCs. Cells from both sources efficiently differentiated into adipose, bone, cartilage and neuronal structures as determined with histochemistry, immunofluorescence and real-time reverse transcriptase polymerase chain reaction. To date, the vast majority of studies have concluded that both tissues are suitable sources of stem cells for potential use in regenerative medicine.

Generally, mesenchymal stem cells (MSCs) are thought of as being non-hematopoietic cells with multi-lineage potential that hold great promise for regenerative medicine. In the past decade, much research has been devoted to bone marrow-derived MSCs. Many clinical and pre-clinical studies have shown that these BM cells can be successfully used for the treatment of various diseases and disorders [22, 23]. Studies have shown differentiation into mesenchymal and non-mesenchymal cell types, including adipocytes, osteoblasts, chondrocytes, muscles and neurons [24–26]. However, bone marrow-derived MSCs are not ideal because of the painful isolation process, the low cell yield and the often older adult age of the cell donor. Lower cell numbers and a reduced proliferative and differentiation capability owing to *in vitro* and *in vivo* aging [10, 27] make bone marrow a less ideal source of MSCs. Use of alternative MSC sources is an important aspect to consider for regenerative medicine applications.

Human adipose tissue may provide the best alternative source of MSCs for tissue engineering and regeneration. Adipose tissue is a convenient, abundant and readily available source of stem cells; the harvest procedure is less invasive than bone marrow aspiration and is associated with little discomfort for the patient. Adipose tissue has 500-fold more stem cells than bone marrow per gram of tissue [28, 29].

Thus, hundreds of millions of MSCs can be easily obtained from a single individual because large adipose samples can be obtained from multiple harvest sites. In addition, the cells proliferate rapidly *in vitro* and demonstrate low levels of senescence after months of *in vitro* culture [29, 30]. Cell-based therapies need viable and ample numbers of cells for regenerative medicine. For autologous therapies, the cells should be available soon after injury, as would be the case for adipose tissue and previously banked cord tissue.

31.3 Stem Cell Banking Approaches

31.3.1 Cord Blood

Most CB banks have adopted the use of small (i.e., pediatric) blood bags (approximately 250 cc in size); although collections can also be made with 60 cc syringes. Bag collections are preferred as collection of the blood (and subsequent processing) occurs in a closed system (preferable for most regulatory stipulations). However, bag collections must not be left unattended in order to prevent unintended contamination or loss of blood flow from occurring. Routinely, collections are completed within 5 min (prior to placental expulsion after clamping and sectioning of the cord) by accessing the umbilical vein. Alternatively, one can wait for delivery of the placenta and collect the blood directly from the expelled placenta.

The vast majority of the cellular constituent in the cord blood collection is red blood cells (RBC), followed by neutrophils (making up 70%–80% of the leukocyte population). In reality only the mononuclear (MNC) fraction (20% of the leukocyte population), which contains the stem cell population, is needed for banking. The stem cells make up approximately 1% of the MNC fraction. CB has a very high hematocrit and RBC can make up more than half of the collection by volume. Thus, to facilitate the banking procedure, the vast majority of CB collections are RBC-depleted or reduced prior to cryopreservation. Several methods are in use to accomplish this goal including Hespan sedimentation to obtain a modified buffy coat [31], density gradient centrifugation (Ficoll method) to obtain enriched MNCs [32], and two automated processes (Sepax® from Biosafe SA, Eysins, Switzerland and the AutoXpress Platform® (AXP) from Thermogenesis, Rancho Cordova, CA; [33, 34]) that result in a buffy coat product. The Hespan, Sepax, and AXP processing methods result in cord blood products containing all nucleated cell populations found in the original collection (MNC, neutrophils, some normal as well as nucleated RBC), while the Ficoll method enriches for the stem cell-containing MNC subpopulation (generally greater than 85% of the final cell composition are MNC with a few neutrophils and nucleated RBC). Total cell counts obtained in the final Ficoll product are generally 50% or less of the cell counts found in the other processes for this reason, although absolute stem cell recovery is similar. The AXP and other automated methods allow for greater sample throughput with fixed personnel numbers (increasing the economy of operations) than manual methods. The AXP device is also an FDA-cleared and functionally closed system (which is recommended under the current regulatory guidelines [35]). In addition, there are a few CB banks that perform plasma reduction as a means of volume reduction prior to banking. It is thought that there may be important components in the non-leukocyte fraction that would be important for clinical use. In addition, the RBC may always be removed later after thawing [36].

An average cord blood collection is generally 70–80 mL of blood from a typical full-term (40 weeks), live birth, containing an average of slightly more (or less) than (850–1100) × 10^6 total nucleated cells (personal experience). The majority of CB banks currently store CB units in multiple aliquots. Generally, this task is accomplished by the use of a freezing bag divided into multiple compartments. Multiple aliquots allow for future use of the stem cells in cell expansion, gene therapy, or for regenerative medicine uses, which may only require a fraction of the frozen unit. Thus, it would not be necessary to thaw the entire unit unless absolutely needed, avoiding the damaging effects of repeated episodes of freezing/thawing. Multiple aliquots also allow for potency testing and confirmation of identify of the unit. Commercially available freezing bags now routinely provide for at least two unit aliquots comprising a 20% and an 80% fraction of the processed unit in separate compartments. Alternatively, one may store small numbers of cells (and plasma if desired) in small 2.5 cc cryovials.

31.3.2 Cord Tissue

An additional source of stem cells that can be simultaneously obtained at birth is the cord tissue (CT) itself, which is a ready source of MSCs. CT can be collected and banked as a future source of stem cells for regenerative medicine and tissue engineering. In addition to MSCs, CT also contains endothelial and epithelial precursor cells that may be useful for these applications [37, 38].

Theoretically, MSCs could be isolated either from cord blood or from cord tissue [37, 38]. However, multiple investigators have reported significant difficulty in isolating sufficient numbers of MSC directly from cord blood (as well as from peripheral venous blood) for clinical use. An example of the difficulty in reproducibly isolating large numbers of MSCs from cord blood was reported recently by Zhang et al. [39]. Their work revealed the specific optimal parameters needed to be met in order to obtain significant numbers of blood-derived MSCs; including CB collections larger than 90 cc and processing of such samples within 2 h of birth. Even when these restrictions were met the absolute numbers of MSCs isolated from CB were too low for immediate use (in the hundreds to thousands of cells), although the MSCs could be rapidly expanded *in vitro* to 1×10^9 MSC in several weeks to several months. Thus, in reality, CT is really the only clinically feasible source of MSCs at the time of birth [40–43].

CT is derived from the human umbilical cord that develops during gestation to support the development of the fetus. The average length of the cord itself is between 30 and 50 cm depending on a variety of factors including maternal age, heath and ethnicity [44]. Stem cells have been identified in various anatomical locations throughout the cord tissue including the amniotic compartment, the Wharton's jelly and the perivascular space surrounding the blood vessels [45]. Numerous papers have described various methods to isolate the stem cells contained within the CT, few of which are relatively facile; including "scraping out" the Wharton's jelly from cut up cord pieces, isolation and digestion of the Wharton's jelly directly, culture of cord tissue explants, culture of amnion explants, digestion of isolated blood vessels from CT, and isolation/culture of venous epithelial stem cell tissues [46–62]. Regardless of the isolation methodology used, one generally finds that the predominant stem cell population obtained is the MSC [45]. However, there are also present endothelial, epithelial and hematopoietic stem cells depending on which isolation protocol is utilized. The total number of MSC that can be obtained varies widely in the literature and may be dependent again on the method used; ranging from as low as 25,000 MSC/cm of CT to as high as 5×10^6 MSC/cm of CT [45, 47, 49, 62–64]. If it is assumed that the average length of CT that could be readily obtained at birth was 30 cm [64] then the number of freshly isolated MSC would range from 750,000 to 150×10^6 cells assuming every cell isolated was actually an MSC [65]. Thus, the limited number of MSC that can be obtained immediately upon isolation without need for *in vitro* culture expansion is a major disadvantage for clinical use of CT-MSC.

Regardless of whether it is actually possible to reproducibly obtain the numbers of MSC claimed at the high end of the CT range noted above, generally insufficient numbers of MSC would be obtained from a typical segment of CT to allow for immediate use in clinical applications (e.g., tissue engineering or regenerative medicine) without prior expansion *in vitro*. Clinical grade MSC expansion would require a minimum of Good Tissue Practice (GTP), if not Good Manufacturing Practice (GMP)-qualified facilities, an investigational approval from the FDA (in terms of an Investigational New Drug (IND) protocol or Biologic License Agreement (BLA)) and additional resources (time and monies) in order to be accomplished successfully. If one could reproducibly obtain 150×10^6 MSC from a 30 cm CT one would have sufficient MSC for some but not many applications (as many applications require $(10-30) \times 10^6$ cells/kg body weight based on our 20 year experience as well as common cell thresholds used for most cord blood transplants). Although the desired regenerative medicine end-use will ultimately determine what cell dose is needed, many applications generally will require some period of *in vitro* MSC expansion. Thus, currently there is no ready-to-use, out of the box, clinical trial-ready CT-MSC methodology available (in contrast to other adult sources of MSC).

Different methods may be used for successful isolation of MSCs from cord tissue including enzymatic [56, 66] and non-enzymatic digestion [58, 59, 67]. However, there have been some conflicting reports regarding the type of MSC and the "multi-lineage differentiation potential" of CT-MSCs [59, 66, 68–70]. Regardless, CT must and can be collected at the time of birth and banked frozen for extended periods of time prior to expected use. To date no study has examined the effects of cryopreservation on the utility of CT-MSCs isolated from thawed cord tissue for use in regenerative medicine and tissue engineering. Cord tissue (CT) is generally obtained from full term deliveries. All samples should be obtained with written consent from the donors. Most collections are 5–10 inch tissue pieces cut from the umbilical cord with sterilized scissors and processed within 24 h. Specifically, the CT is cleaned and sterilized with alcohol and betadine. It is then cut with sterile scissors and a 4–8 inch segment is placed into a sterile container with transport buffer. The transport buffer contains isotonic saline with 10 U/mL heparin, 1% human serum albumin (HAS), Penicillin-Streptomycin, Gentamycin, and Amphotericin. The sample may be held and transported at room temperatures for up to 48 h. Upon receipt the CT sample is placed in isotonic saline followed by a 70% ethanol wash, and a final sterile saline wash. The CT is then cut into small 5 mm ringlets or minced into small pieces using a sterile scalpel. The CT pieces are then placed in isotonic saline containing 0.1 mol/L sucrose, 20% autologous plasma and 1.5 mol/L dimethyl sulfoxide (DMSO) for 30 min at 4°C on a rocking platform. Samples are frozen in 4.5 cc cryovials (~1.0–1.5 grams total/cryovial) using a controlled rate freezer to −180°C. Samples are stored in liquid nitrogen dewars.

When required, CT can be thawed at room temperature for 30 s, followed by a complete thaw in a 37°C water bath (approx. 2 min). Either the tissue strips may then be washed extensively at 4°C to remove DMSO, or a step-down procedure may be used to remove the cryoprotectant. This latter procedure involves the washing of the tissue by agitation for 5 min in each of the following buffers at 4°C:

- 1.5 mol/L DMSO in PBS with 20% autologous plasma and 0.1 mol/L sucrose,
- mol/L DMSO in PBS with 20% autologous plasma and 0.1 mol/L sucrose,
- 0.5 mol/L DMSO in PBS with 20% autologous plasma and 0.1 mol/L sucrose, and
- mol/L DMSO in PBS with 20% autologous plasma and 0.1 mol/L sucrose.

Finally, a last wash in PBS containing 20% autologous plasma is used, followed by re-suspension in PBS/20% plasma for analysis, culture or clinical use. MSCs from cord tissue can then be isolated using a non-enzymatic digestion procedure as described [68–70]. Briefly, pieces of cord tissue are extensively washed with PBS containing penicillin and streptomycin in a

100 mm Petri-plate. The tissue is minced into fine pieces that are placed into a 25 cm^2 culture flask. After 4–6 days the pieces are removed and cultured in a new flask. In 10–14 days cell colonies can be observed. The cells from both flasks are then harvested using trypsin-EDTA and pooled. A total of 25,000 cells are then cultured in each subsequent 25 cm^2 culture flask in alpha-MEM expansion medium.

31.3.3   Adipose Tissue

All adipose samples should be obtained with written consent from the donors and according to any other requirements of the local Institution Review Board (IRB). Adipose tissue samples are generally obtained from scheduled liposuction procedures, or by syringe harvest performed under local anesthesia. The lipoaspirates should be processed and cryopreserved within 24–36 h of collection. For cryopreservation, the tissue is washed extensively with isotonic saline, and the washed tissue slurry is directly placed in a cryo-container (generally a cryobag) and an equal volume of pre-cooled DMSO solution (70% Lactated Ringer's buffer, 20% serum or HSA, 20% DMSO) is added slowly over several minutes at 4°C. The cryo-container is generally mixed at 4°C for 20–30 min to allow for cryoprotectant equilibration. Cryopreservation is performed using a controlled rate freezer to −180°C before final submersion in liquid nitrogen for long term storage [71]. In our experience we have not seen significant differences in AT-MSC harvest by tumescent liposuction, VASER liposuction, power-assisted liposuction or various forms of laser liposuction (although we have not analyzed all possible permutations that are commercially available). In addition, we have not observed any deleterious effects of anesthetic choice on the clinical utility of the harvested AT-MSC (i.e., Lidocaine and Marcaine were equivalent).

When needed, frozen tissues should be thawed rapidly in a 37°C water bath. Immediately after thawing the cryopreservation solution must be diluted with expansion medium (alpha-minimal essential media [α-MEM]) supplemented with FBS, HSA or other source of protein to avoid deleterious effects of the cryoprotectant at elevated temperatures [72]. Adipose tissue may then be utilized as is or further digested as described above. Adipose derived

MSCs (AT-MSC) may be isolated by enzymatic digestion as described [12, 13] if desired. The tissue slurry can be digested with 0.2% type IV collagenase by incubation at 37°C for 15 min. MSC may be expanded by culture in medium consisting of α-MEM media supplemented with 10% FBS (or HSA) and 1% each of non-essential amino acids, sodium pyruvate, L-glutamine and streptomycin/penicillin solution. Expansion medium is also added to the digested tissue to neutralize collagenase. The infranatant is centrifuged at 150 g for 10 min to obtain cells of the stromal vascular fraction (SVF). AT-MSC are then plated in 25 cm^2 culture flasks and maintained in a humidified atmosphere at 37°C with 5% CO^2. After two days of culture, non-adherent cells are removed by changing the medium to leave a homogenous MSC population. Medium is changed twice weekly thereafter.

31.4 Common Stem Cell Uses

Cord blood stem cells have been used in the clinic to treat malignant and genetic blood disorders for more than 20 years now. Over the past decade CB has made its way into multiple clinical trials for use in regenerative medicine applications (http://www.clinicaltrials.gov and see below). It is only recently that CT stem cells have made their way into the clinic; and it is only now that clinical comparisons to bone marrow MSCs can be made. A recent report from Xue et al. examined the use of CT-MSCs in patients with non-healing bone fractures [71]. This study reported significant clinical benefit from intravenously infused MSC. CT-MSCs are also showing positive results in treating GVHD following hematopoietic stem cell transplantation. Two pediatric patients with severe steroid-resistant GVHD were infused with CT-MSCs. The GVHD improved dramatically in both patients following infusion of CT-MSCs, although one patient received multiple infusions of MSCs until the course of treatment was complete [73]. CT-MSCs have also been evaluated for potential therapeutic benefits in autoimmune diseases. Liang et al. reported that CT-MSCs stabilized the disease course of a patient with progressive multiple sclerosis that was not responsive to conventional treatment [74]. A subsequent study from the same group reported dramatic improvements in a patient with systemic lupus erythematosus following intravenous infusion of CT-MSCs [75]. Importantly none of the case reports indicated adverse effects associated with infusion of CT MSCs. Based on promising *in vitro* and *in vivo* results for a wide range of conditions, MSCs, primarily from bone marrow, are currently being investigated in more than 350 planned, ongoing, or recently-completed clinical trials for conditions including ischemic injury (heart attack, stroke, critical limb ischemia), autoimmune diseases (type I diabetes, multiple sclerosis [MS], systemic lupus erythematosus [SLE]), inflammatory conditions (congestive-obstructive pulmonary disease [COPD], Crohn's disease), orthopedic applications (bone fractures, cartilage injuries, osteoarthritis, osteogenesis imperfecta), and transplantation (both stem cells and organs). Specifically, AT-MSC are currently being used in more than 100 ongoing clinical trials for many of the same disease settings.

As might be expected, when it is possible to harvest and bank MSCs from various tissues at different times throughout the lifespan of the donor, the question will always be raised as to why should someone use younger as opposed to older MSC? This topic will be discussed in more detail below. However, numerous studies have indicated that MSC isolated from older donors, as well as from patients with longstanding (chronic) disease conditions are neither as prevalent [76–78] nor as potent [10, 79–82] as those isolated from younger and healthier donors. MSCs collected from older donors and/or donors with chronic diseases (e.g., coronary disease, COPD, etc.) seem to be less able to differentiate into the different cell types needed for tissue engineering [10, 76, 79–82], less able to proliferate and expand to achieve cell concentrations that would allow for multiple treatments [10, 79–82], and are more prone to die off during culture and use [80]. Thus, younger stem cells are likely to be more useful for regenerative medicine applications than older MSCs. Finally, there is anecdotal evidence that increased MSC donor age and disease status negatively impacts clinical utility and successful clinical outcomes. Data on the success rates of treatment of patients with myocardial infarction and chronic heart disease is mixed but seems to be negatively correlated with patient age and chronic disease [83].

31.4.1 Hematologic Settings: CB and CT

Cord blood is unique in that it contains hematologic stem cells and thus may be used to reconstitute the blood and immune system after chemotherapy, radiation and stem cell transplant. Although bone marrow is similar in this regard, neither cord tissue nor adipose tissue has this capability. Cord blood may be used to treat more than 80 malignant and non-malignant hematologic conditions requiring transplant. In addition, CB may also be used for regenerative medicine and tissue engineering applications, as can cord tissue and adipose tissue MSC [14, 21]. In the transplant setting CT-MSCs have shown positive results in the treatment of GVHD following hematopoietic stem cell transplantation. Wu et al. [73] found that CT-MSC had superior proliferative potential and increased immunosuppressive effects as compared to bone marrow MSC. Two pediatric patients with severe steroid-resistant GVHD were infused with *ex vivo* expanded CT MSCs. The GVHD improved dramatically in both patients following infusion of CT-MSCs, although one patient needed to receive multiple infusions of MSCs over the course of treatment [73]. CT-MSCs have also been evaluated for potential therapeutic benefits in autoimmune diseases. Liang et al. [74] reported that CT-MSCs stabilized the disease course of a patient with progressive multiple sclerosis that was not responsive to conventional treatment [74]. A subsequent study from the same group reported dramatic improvements in a patient with systemic lupus erythematosus following intravenous infusion of CT-MSCs [75]. Importantly none of the case reports indicated adverse effects associated with infusion of CT-MSCs.

When contemplating which applications will most likely be first to the clinic as well as ones applicable for most people, we believe that there are three primary stem cell indications: orthopedic (e.g., cartilage repair in the articular joints), cardiovascular (e.g., heart attack), and neurological (e.g., stroke). Most individuals are likely to experience one or more of these categories of problems during their lifetimes and thus be candidates for regenerative medicine therapy. In the following discussion we will focus first on applications in which CB and CT have been used, with a separate section for AT afterwards.

31.4.2 Neurological Settings

31.4.2.1 CB

Cerebrovascular diseases are the third leading cause of death in the United States, not including the multitudes of individuals who survive only to suffer debilitating lifelong issues. Cerebral ischemia (CI) is by far the most prevalent cause of stroke (87%, http://www.americanheart.org [83]). Approximately 700,000 people in the United States are affected by stroke annually; and 1 in 16 Americans who suffer a stroke will die from it [84]. The brain is extremely sensitive to hypoxia and some degree of tissue death is likely from stroke. At a relatively young age the brain loses most of its plasticity, so any significant tissue death can be profoundly devastating. Interestingly, in young children the brain is very plastic and very large portions of the brain can be removed (such as removal of tumors or hemispherectomy for severe seizures) with relatively little long term neurological damage. These facts suggest that younger neural cells might have a greater capacity to regenerate the injured brain.

Nowhere has the potential significance of CB stem cell therapy for the treatment of neurological disease been greater than in this area of stroke therapy. As early as 2001, it was demonstrated that the infusion of CB stem cells into rats in the commonly used MCAO (middle carotid artery occlusion) model of stroke could reverse many of the physical and behavioral deficits associated with this disease [85]. Studies demonstrated that direct injection of the stem cells into the brain was not required [86], and in fact, beneficial effects could be observed even if the stem cells did not actually home into the target organ (probably via the release of growth and repair factors triggered by the anoxia) [87, 88]. The beneficial effects seemed to be dose-dependent and could reduce the size of the infarcted tissue [89]. It appeared that multiple progenitor populations in CB were capable of mediating these effects [90]. Significantly, unlike current pharmacological interventions that require treatment within the first few hours after stroke, CB stem cell therapies were effective up to 48 h after the thrombotic event [91]. In fact, administration of CB stem cells immediately after the ischemic event may be detrimental in that the inflammatory milieu may be toxic to the administered stem cells.

The majority of reported studies [90–97] have shown that CB administration in stroke models resulted in some degree of therapeutic benefit with no adverse effects. Neuroprotective effects [90–92, 95, 96, 98] as well as functional/behavioral improvements [91, 96, 97] from CB therapies have been widely reported. Neurological improvement was accompanied by decreased inflammatory cytokines [89], by neuron rescue/reduced ischemic volume [90–92], as well as by lowered parenchymal levels of granulocytic/monocytic infiltration and astrocytic/microglial activation [91]. Thus, the mechanisms behind the observed beneficial effects afforded by CB therapies included reduced inflammation [92], protection of nervous tissue from apoptosis [90] and nerve fiber reorganization [90]. These observations are particularly encouraging as it implies that CB therapy can mediate both direct restorative effects to the brain as well as tropic neuroprotection. Many of the published studies lend support to this trophic role, in that several investigators reported [89, 90, 97] neural protection with little to no detection of CB cells engrafted in the brain. The level of engraftment in the brain appeared to be a function of the route of CB administration. When CB was administered intravenously [89, 97–99], little or no CB migration to the brain was found. However, when CB was given intraperitoneally [99] there was evidence of neural restorative effects. Early studies have also shown benefit in animal models of hemorrhagic (as opposed to embolic) stroke [96]. For additional information one is referred to the recent review on cell therapies for stroke found in reference [100].

In addition to stroke, CB stem cells have been used in other nervous system injury models, two of which have now instigated clinical trials. Lu et al. [101] demonstrated that intravenous administration of CB mononuclear cells could be used to treat traumatic brain injury in a rat model. In this model the CB cells were observed to enter the brain, selectively migrate to the damaged region of the brain, express neural markers, and reduce neurological damage. Similarly, CB stem cell transplant could also alleviate symptoms of newborn cerebral palsy in a rat model, with improved neurological effects [93]. These observations have now been turned into clinical therapies (see below). Early, albeit anecdotal, reports have indicated beneficial effects from the CB mononuclear cell infusions [102]. Several investigators have begun planning clinical trials to treat children with hypoxic/ischemic and traumatic brain injury utilizing autologous cord blood stem cell infusions

The observation that CB stem cells can become different types of nervous cells extends its utility to other areas of neurological damage, including spinal cord injury. Spinal cord injured rats infused with CB stem cells showed significant improvements five days post-treatment compared to untreated animals. The CB stem cells were observed at the site of injury but not at uninjured regions of the spinal cord [85]. This finding is supported by another study demonstrating that CB stem cells transplanted into spinal cord injured animals differentiated into various neural cells, improved axonal regeneration and motor function [103]. Significantly, in a recently reported clinical use of CB stem cells to treat a patient with a spinal cord injury [104] it was stated that transplantation of CB cells improved her sensory perception and mobility in the hip and thigh regions. Both CT and MRI studies revealed regeneration of the spinal cord at the injury site. Since the CB stem cells were allogeneic in origin it will be significant to determine if immune rejection or other immune-mediated problems occur that might jeopardize the early improvement. Neither additional patients nor additional studies in this area have been reported. However, the use of CB stem cells for spinal cord injury seems to be the next logical clinical trial. Large numbers of children are unfortunate enough to suffer a spinal cord injury at an early age (e.g., diving into a pool, car accidents, falls, etc.) and it would be expected that a significant number would have autologous cord blood banked and available for treatment.

31.4.2.2 CT and AT

CT-MSC can be differentiated into neuron-like cells *in vitro* [3] which may indicate that there could be applications for neurological conditions like stroke, Parkinson's disease and Alzheimer's disease. Animal work in ischemic stroke has also shown promising results [105–108]. In addition, CT-MSCs have shown promising results in intracerebral hemorrhage models [109] and in the treatment of spinal cord injuries [110–113]. Finally, the anti-apoptotic effects of MSC have provided beneficial effects in animal models of Parkinson's disease [47, 114].

AT-MSC can also be differentiated into neural tissue [3] which has led to its introduction into several human clinical trials. Due to a longer history of clinical work with BM-MSC and its similarity, AT-MSC are currently being investigated as a therapeutic approach to spinal cord injury [111], stroke [109, 115, 116], and Parkinson's disease [47].

31.4.3   Cardiovascular Settings

31.4.3.1   CB

Cardiovascular disease is the leading cause of morbidity and mortality for both men and women in the United States. Approximately one million people die of cardiovascular disease annually despite medical intervention. Coronary artery disease comprises approximately half of these deaths. As heart cells have a limited capacity to regenerate after myocardial infarction (MI), application of exogenous stem cells seems a logical alternative for therapy. Recently, numerous pre-clinical and clinical studies examined the use of adult hematopoietic stem cell sources (see ref. [84] for details and additional references). To date, only non-embryonic stem cells have been examined in clinical trials due to political, ethical and biological constraints. There have been no clinical trials using CB stem cells for cardiovascular disease. The lack of clinical trials has been due to the relative youth of the CB banking industry. However, although no clinical trials utilizing CB stem cells for heart failure have been conducted to date, a number of pre-clinical animal studies have been performed [116–122]. Several common observations were noted in these studies regardless of the protocols utilized including selective migration of the CB stem cells to the injured cardiac tissue, increased capillary density at the site of injury, decreased infarct size, improved heart function and a general lack of myogenesis. These observations are thought to be due to the production of angiogenic factors and induction of angiogenesis/vasculogenesis [6, 116, 123]. In fact, work done by Gaballa et al. [84, 116] in myocardial infracted rats showed that CD34+ CB stem cells induced blood vessel formation, reduced infarct size and restored heart function. The effects were thought to be due to the release of angiogenic and growth factors (e.g., vascular endothelial growth factor [VEGF], epithelial growth factor [EGF], and Angiopoietin-1, 2) induced by hypoxia as shown by gene array analyses. This work demonstrated that cord blood stem cells could be induced to become/differentiate into endothelial-like cells. Interestingly, as a prelude to human clinical trials for MI, it has been shown that it is possible to isolate therapeutic cells from CB using a clinical grade apparatus making the transition from bench to bedside a bit more facile. Finally, work from numerous groups seems to indicate that more than one population of pluripotent cells contained in CB is capable of mediating this effect as shown by the ability of CD34+, CD133+, and CD45– cells to induce cardiac repair after MI [116, 117, 121, 122, 124]. Even more important, the numbers and potency of these cells found in CB seem sufficient for adult human applications as shown by work performed in a porcine model [121].

Aside from its application to MI, CB stem cells via the exertion of its angiogenic capability also appear to be useful for the treatment of various ischemic diseases. Many investigators have demonstrated that not only does CB contain cells displaying the phenotypic characteristics of endothelial precursors that are responsible for blood vessel formation, but that these cells are capable of differentiating into endothelial cells and becoming blood vessels [125–132]. These bioengineered blood vessels appeared similar to native blood vessels in terms of their (three layered) tissue organization as well as expression of matrix components [125, 127, 128, 130]. Furthermore, when placed in animal models CB stem cells were able to significantly reverse the effects of ischemia in several model systems [126, 129, 132]. In models of hind limb ischemia, transplantation of CB stem cells or endothelial cells derived from CB stem cells appeared able to reverse surgery-induced ischemia resulting in limb salvage [133–136].

31.4.3.2 AT

The use of MSC in cardiovascular disease therapy has a long history via the use of BM-MSC [84], with mixed results. Due to the phenotypic and functional similarity of BM-MSC and AT-MSC [3], a number of clinical trials using AT-MSC in this arena have been initiated (http://www.clinicaltrials.gov), including Myocardial infarction, Congestive heart failure, Stroke, Critical limb ischemia/peripheral artery disease, and Coronary ischemia. It is still too early to determine how effective this approach will be, and for which indication it will be most efficacious.

31.4.4 Orthopedic Applications

31.4.4.1 CB

The potential of CB stem cells to generate bone and cartilage has been recently examined. It is estimated that more than one million individuals in the USA annually suffer from articular joint injuries involving cartilage, ligaments and/or tendons, as well as difficult to heal bone fractures (see http://www.arthritis.org). CB contains both ES-like and mesenchymal stem cells (MSC) capable of differentiating into both bone and cartilage [137]. In fact, when CB stem cells were placed into animals with fractured femurs there was significant bone healing. Work from the laboratories of Szivek et al. [138] and Harris (unpublished observations) have also examined the ability of cord blood stem cells to become cartilage in comparison to tissues derived from bone marrow MSC and adipose stem cells, with early encouraging results.

31.4.4.2 CT

CT-MSC have made their way into the clinic to treat non-hematological conditions. The potential of CT stem cells to generate bone and cartilage has been recently examined. It is estimated that more than one million individuals in the USA annually suffer from articular joint injuries involving cartilage, ligaments and/or tendons, as well as difficult to heal bone fractures [139]. CT contains mesenchymal stem cells (MSC) capable of differentiating into both bone and cartilage [140]. In fact, when CB stem cells were placed into animals with fractured femurs there was significant bone healing. A recent report from Xue et al. examined the use of CT MSCs in patients with non-healing bone fractures [71]. This study reported significant clinical benefit from intravenously infused MSC.

31.4.4.3 AT

AT-MSC have advanced the most in terms of clinical trials over the past several years. Currently there are human clinical trials using AT-MSC for cartilage replacement, osteonecrosis, osteogenesis imperfect, periodontitis, non-healing bone fractures, to treat benign bone neoplasms, and repair of degenerative spinal discs. Although very promising results have been obtained it is still too early to form definitive conclusions.

31.4.5 Factors Impacting Stem Cell Banking and Use

31.4.5.1 Stem Cell Origin and Age

Multiple factors may impact stem cell use in the clinic including origin and age of the donor stem cells, donor disease status, time from injury to treatment (the "treatment window"), and regulatory guidelines and restrictions. The first factor to consider is the stem cell source. That is, will stem cells be of autologous or allogeneic origin? Stem cells of autologous origin are much easier to implement from a regulatory standpoint (see below) and lessen the concern of disease transmission. In addition, autologous stem cells remove the concern of immune rejection which may be particularly important in instances where multiple stem cell injections are required [141–144]. However, a number of clinical trials have been attempted using allogeneic stem cells. Even in those trials that have been successful one can only treat patients a limited number of times before they become immunized and rejection occurs. The cost of this "off-the-shelf" approach to regenerative medicine is expensive, and generally limited to academic centers where frozen (i.e., banked) allogeneic stem cells can be stored until ready for use. Finally, in instances where the injected stem cells differentiate into long-lasting tissues (e.g., neurons) the use of autologous stem cells is a must. Otherwise, rejection of the newly acquired tissue will eventually occur.

The average human life expectancy has significantly increased due to advances in medical research and improvements in general life style. Unfortunately, however, human aging is associated with many clinical disorders and an inability of the body to maintain tissue turnover and homeostasis. As a result, the number of elderly medical patients have also significantly increased, making them a major target population that could potentially benefit from cell based therapies. As autologous cell sources are preferred for economical and logistical reasons, the effect of donor age on regenerative potential should be determined before clinical use. In recent years, many studies have demonstrated the clinical potential of mesenchymal stem cells (MSCs), both *in vivo* and *in vitro* [7, 145]. However, using MSC collected from the elderly who are most likely to benefit from this technology raises some concerns.

However, before widespread clinical use it is important to determine the potential effect of donor age on the expansion and differentiation capabilities of these cells. There is a logical assumption that organismal aging is linked to diminished organ repair due to reduced functional capacity of tissue resident stem cells. It is believed that such cells residing in the elderly are subjected to age-related changes and thus contribute less to tissue rejuvenation. Similarly, age-related diseases such as diabetes and heart failure also negatively impact the function of endogenous progenitor cells [146]. As stem cells are the basis of tissue regeneration therapies, a diminished functionality of these cells in the elderly may result in reduced efficacy of autologous cell therapies. With an increase in the aging population, cellular therapies are becoming more relevant for aged patients who are the main target population for such therapies.

Analyses have indicated that the overall yield of total nucleated and stem cells were significantly and negatively affected by donor age. Similar observations have been reported in literature by assessing the yield of bone marrow-derived MSCs and circulating endothelial progenitor cells [77, 147]. These results indicated that age-related changes in MSC number should be taken into account whenever these cells are considered for clinical applications in the elderly. Although AT-MSCs from all age groups had the ability to form colonies (an indication of cell function), AT-MSC from younger donors produced more colonies containing larger numbers of cells. Other investigators have reported that the number of cells forming colonies decreased significantly with increasing donor age and is in accordance with the results of our current study [148].

However, recent studies have raised questions about the usefulness of AT-MSCs collected from aged donors [148, 149]. Khan et al. [150] found age-related differences in osteogenic potential of AT-MSCs. These inconsistent results may be due to the different age ranges and the health status of the donors that were studied. Overall, the majority of reports found results similar to our current study; describing an overall decline in osteogenic potential with donor age (regardless of species). Murphy et al. [81] has also reported an age-related decline in chondrogenic potential of MSC similar to the results of our study. In combination, these findings and our osteogenic results indicate that donor age may negatively impact the use of AT-MSC for orthopedic applications which are not uncommon as one grows older.

Numerous studies have indicated that MSC isolated from older donors, as well as from patients with long-standing (chronic) disease conditions are neither as prevalent [76–79] nor as potent [10, 79–82] as those isolated from younger and healthier donors. MSCs collected from older donors and/or donors with chronic diseases (e.g., coronary disease, COPD, etc.) seem to be less able to differentiate into the different cell types needed for tissue engineering [10, 76, 79, 82], less able to proliferate and expand to achieve cell concentrations that would allow for multiple treatments [10, 79–82], and are more prone to die during culture and use [80]. Thus, younger MSC are likely to be more useful for regenerative medicine applications than older MSCs. This hypothesis remains to be proven, however. Finally, there is anecdotal evidence that increased MSC donor age and disease status negatively impact clinical utility and successful clinical outcomes. Data on the success rates of treatment of patients with myocardial infarction and chronic heart disease is mixed but seems to be negatively correlated with patient age and chronic disease [85]. Ultimately, disease status of the MSC donor may prove to be more important than absolute age of the MSC donor.

Finally, there is also some preliminary evidence that age of the stem cell recipient may also impact clinical efficacy [151]. That is, older recipients may require earlier intervention or additional therapies to achieve the same level of clinical success observed with younger recipients, regardless of stem cell donor age. Importantly, limited clinical benefit has been observed when using older stem cells in older recipients, implying that access to banked, younger stem cells may be critical to serving those most in need of regenerative therapies, the elderly.

31.5   Optimal Treatment Windows

Another significant variable impacting clinical outcome is time to therapy. It seems unrealistic to expect to treat almost any condition at almost any time just because one is using stem cells. In many instances, injury will be followed by inflammation and later by fibrosis, scarring and cell death. Inflammation at the beginning of the injury is toxic to cells and could kill the stem cells if administered too early after injury. Once fibrosis and scarring have been established the injured site is essentially "walled off" from therapy as infused stem cells will no longer have access to the damaged tissues. Thus, when is the optimal time to implement stem cell therapy? Most likely it will depend upon the type of injury and the age of the patient. Young patients will be better recipients especially for neurological injuries (as the pediatric brain is still growing until about age 7 years) as their systems are more resilient and less likely to have been damaged by long term inflammatory processes. In fact, the older the patient the more restricted the options will likely be and the more likely the patient will need to be treated sooner than later. Based on published reports as well as our own experience we would estimate that the window of opportunity for treatment of most conditions will be days to months, and most likely will require a waiting period of 48–72 h before infusion to avoid the effects of inflammation. Treatment during the optimal window of time takes advantage of endogenous repair mechanisms that employ resident stem and progenitor cells, rather than needing to construct new tissues *ex vivo* or depend on *in vivo* stem cell differentiation. Thus, prior to tissue death or necrosis but after inflammation has subsided, would seem to be the optimal treatment window. This relatively long treatment window allows an individual that has previously banked young and healthy stem cells to easily retrieve them for use in that time frame from almost any place in the world (and to be sent to almost any place in the world).

31.6 Regulatory Oversight

Despite constant complaints regarding federal regulatory oversight of stem cell banking and clinical use, regulation is necessary to insure standardization and protection from charlatans. Cord blood banking began at the beginning of the stem cell regulatory movement and has been overseen by scientific, state and federal authorities. CB is unique in that the end-user of the stem cells (transplant physicians) is already regulated by insurers, CMS and state medical boards. Thus, regulation of the stem cell providers (i.e., the banks) was easy and straightforward. Regenerative medicine and tissue engineering however, are problematic in that many physician specialties may be involved in therapy. In some instances, these doctors will be at university and other academic institutions where studies will be conducted under the auspices of the local IRB (possibly along with filing a federal IND). However, in many if not most instances the treatments will be conducted at clinics and doctor offices where the basic rules of good tissue practice (GTP) are neither followed nor understood (e.g., disease transmission prevention, sample sterility, stem cell potency, and donor/recipient identity). Many of these therapies will escape federal detection until such time that patient advertising makes them aware, until someone complains, or until a patient is injured or dies. Thus, the FDA is faced with a dilemma. Patients are clamoring for therapies, untrained doctors are often ready to offer such treatments, and many stem cell providers are out for a quick profit. The FDA should (in conjunction with the AATB) regulate the collection, processing and banking of adipose tissue (AT)-derived and cord tissue (CT)-derived stem cells much as it does CB. In addition, a plan must be put in place to oversee the rise of stem cell clinics to protect both patients and the stem cell industry, as well as to capture any useful data that may be derived from such "one-off" therapies. The regulatory plan should not be too onerous or expensive and may consist of nothing more than registration and reporting requirements, with occasional unannounced inspections.

One complicating factor in this area is the general misconception held by many doctors concerning IRB approvals *versus* the need for an IND application. Having served on my institutional IRB and having been involved in IND trials, I can emphatically state that although the IRB and IND complement one another they are definitely not the same thing and one cannot substitute one for the other. IRB approvals (whether institutional or private ones) require descriptions of the proposed clinical study so determinations can be made that patients are well informed, not taken advantage of, and not unnecessarily exposed to risk (i.e., an ethical study will be performed), and can be considered the local approval for a protocol. An IND is a federal approval (based on the HCT/P federal law that gives the FDA the right to regulate stem and progenitor cells as well as tissues) for use of a drug or medical device in a particular application that stipulates how a study will be performed, what disease states can and cannot be treated, what data must be collected and reported, how much one can charge for the procedure, if placebos and blinding of therapy are required, and ultimately will determine if standard of care can be issued so that CMS and insurer reimbursement can occur. If the FDA states that an IND is needed for a particular therapy or use of a particular stem cell source, then IRB approval is no substitute. Unfortunately, many doctors and their medical clinics confuse one for the other and are jeopardizing the entire stem cell field through their earnest but misguided efforts.

When the FDA was given oversight of the stem cell industry they established regulations that determine whether or not an IND application is required, based on patient risk. First, if a stem cell sample is more than minimally manipulated then it needs an IND approval (i.e., in order to qualify for "351 *vs.* 361" regulation standards). Minimal manipulation was defined as processes that did not alter the composition or structure of the tissue containing the stem cells. Thus, with regard to CB, if one isolates the CD34 cells it is considered a manipulated tissue. In terms of adipose tissue-derived stem cells, if one enzymatically isolates the stromal vascular fraction (SVF) it is considered a manipulated tissue. Unfortunately, no definition is yet forthcoming concerning CT although many banks are enzymatically digesting the tissue before banking, which would seem to fit the definition of a manipulated tissue. Without a working definition for CT many stem cell banks are at risk of potentially running afoul of FDA regulations. Second, according to the FDA regulations the proposed use of the tissue needs to be in an autologous setting or for 1st/2nd degree relatives in order to not require an IND. That is, the sample needs to be used in a familial setting where disease transmission is minimized. And finally, and perhaps most confusing for many, the proposed use of the stem cells needs to be homologous in nature. That is, the stem cells need to function in the therapy as they would normally be expected to function in the body. That is, CB stem cells could be used to perform stem cell transplants for cancer without an IND (as these stem cells normally make blood and immune cells), but not to treat brain trauma or stroke. In terms of CT-MSC this definition would seem to preclude ever using these stem cells without an IND as the entire purpose of the stem cells is lost upon birth of the child. For AT-MSC use of the cells in cosmetic and some reconstructive applications would be permitted but use in treating a disease such as multiple sclerosis (MS) would not. AT-MSC regulations here are a bit confusing in that AT-MSC are widespread throughout the body, opening up the definition of homologous function to various interpretations.

This last issue is the most problematic as it could possibly completely shut down the growing field of regenerative medicine. If an IND is required for almost every use, regardless of whether it is an autologous use with an un-manipulated sample, then the field will either die quickly or progress so slowly as not to be terribly useful due to the higher costs involved and the inability to treat large numbers of patients quickly. It would seem more reasonable that as long as patient health was protected, and risk was minimized through meeting the other two requirements, then adequate informed consent should protect the patient from being misled. Perhaps submission of all informed consents to the FDA would satisfy this aspect and allow the field to progress. Obviously something needs to be done and done quickly. I am personally aware of dozens of trials treating hundreds (if not thousands) of patients that have escaped any scrutiny at all, where I am convinced the physicians know little or nothing concerning sterile technique, good manufacturing processes, etc.

It is very surprising that no one has been seriously injured or died as of yet, but without some standardization and oversight, I expect this to occur at any time.

31.7 Conclusions

The ability to bank autologous stem cells for later use has the potential to be a significant linchpin in the development and implementation of regenerative and personalized medicine strategies. CB, CT-MSC and AT-MSC offer the most economical sources of stem cells for almost everyone. Cord blood banking has been available for more than 20 years, is well established and regulated, and has been involved in more than 30,000 stem cell transplants and thousands of regenerative therapies. Cord tissue banking has become available over the past 5–7 years as an adjunct to cord blood banking. Although it appears to be regulated through its association with cord blood entities, in reality it is not. Regardless, its applicability to the clinic seems to be more limited and may eventually be replaced by other MSC sources. Recently, adipose tissue banking has become available and offers one the opportunity to store almost unlimited numbers of stem cells for future use readily and inexpensively. AT-MSCs have been involved in clinical trials for more than 10 years in more than 100 FDA-approved clinical applications. This particular stem cell source may soon replace BM-MSC as the preferred stem cell source for most regenerative and personalized medicine applications.

Clinical trials using cord blood stem cells to treat malignant and non-malignant blood disorders, cerebral palsy and peripheral vascular disease among others have been ongoing for many years now [85, 152]. It is only recently that efforts have focused on the isolation, characterization and utilization of MSC found in CT. In fact, CT stem cells are just now making their way into clinical trials [73–75, 85, 141], albeit to a more limited extent than AT-MSC and after extensive *in vitro* cell expansion. The requirement for cell expansion highlights one of the major disadvantages of CT-MSC; low cell yield upon isolation which requires extensive (and expensive) *ex vivo* expansion before clinical use is possible. AT-MSC however, is well-established and has been involved in more than 100 clinical trials over the past 10 years. AT-MSCs are available from almost any patient and stem cell numbers in the hundreds of millions of cells are easily harvested for immediate clinical use. MSC (from whatever source) are probably the most useful stem cells for regenerative medicine applications, but CB can also be used for stem cell transplants to treat blood (malignant and genetic) and immune disorders, which MSC in general cannot. Regenerative medicine applications will most probably be performed for orthopedic, cardiovascular and neurological applications; meaning that MSC banking will be more important over one's lifetime than other types of stem cell banking. However, stem cells in general need to be banked while young and healthy as older stem cells and those harvested from individuals with longstanding chronic and inflammatory diseases appear to function poorly in these situations. Finally, a lack of logical and clear regulatory oversight for the entire stem cell banking (and use) field is putting patients and the entire field of regenerative medicine at risk. Something needs to be done quickly in order to allow this clinical endeavor to reach its fullest potential and serve those with the greatest and most immediate need.

Abbreviations

AT:	adipose tissue
AT-MSC:	adipose derived MSCs
AXP:	AutoXpress Platform®
BM:	bone marrow
CA:	Rancho Cordova
BLA:	biologic license agreement
BM-MSCs:	bone marrow-derived MSCs
CB:	cord blood
CI:	cerebral ischemia
COPD:	congestive-obstructive pulmonary disease
CT:	cord tissue
DMSO:	dimethyl sulfoxide
EGF:	epithelial growth factor
ES cells:	embryonic stem cells

FACS:	fluorescence activated cell sorter
GMP:	good manufacturing practice
GTP:	good tissue practice
HAS:	human serum albumin
HLA:	human leukocyte antigen
HSC:	hematopoietic stem cells
IND:	investigational new drug
IRB:	Institution Review Board
iPSC:	induced pluripotent stem cells
α-MEM:	alpha-minimal essential media
MCAO:	middle carotid artery occlusion
MI:	myocardial infarction
MNC fraction:	mononuclear fraction
MS:	multiple sclerosis
MSC:	mesenchymal stem cells
RBC:	red blood cells
SLE:	systemic lupus erythematosus
SVF:	stromal vascular fraction
VEGF:	vascular endothelial growth factor

Disclosures and Conflict of Interest

This chapter was originally published as: Harris, D. (2014). Stem cell banking for regenerative and personalized medicine, *Biomedicines* **2**, 50–79, doi:10.3390/biomedicines2010050, under the Creative Commons Attribution license (http://creativecommons.org/licenses/by/4.0/). It appears here, with edits and updates, by kind permission of the author and the publisher, MDPI (Basel). The author declares that he has no conflict of interest and has no affiliations or financial involvement with any organization or entity discussed in this chapter. No writing assistance was utilized in the production of this chapter and the authors have received no payment for it. The author would like to acknowledge all the doctors, nurses, and patients involved over the past two decades in their stem cell banking and therapy efforts.

Corresponding Author

Dr. David T. Harris
AHSC Biorepository
1501 N Campbell Avenue
AHSC 6122, PO Box 245221
University of Arizona
Tucson, AZ 85724, USA
Email: davidh@email.arizona.edu

References

1. Pittenger, M. F., Mackay, A. M., Beck, S.C., Jaiswal, R. K., Douglas, R., Mosca, J. D., Moorman, M. A., Simonetti, D. W., Craig, S., Marshak, D. R. (1999). Multilineage potential of adult human mesenchymal stem cells. *Science*, **284**, 143–147.

2. Lee, C. C., Ye, F., Tarantal, A. F. (2006). Comparison of growth and differentiation of fetal and adult rhesus monkey mesenchymal stem cells. *Stem Cells Dev.*, **15**, 209–220.

3. Choudhery, M. S., Badowski, M., Muise, A., Harris, D. T. (2013). Comparison of human adipose and cord tissue derived mesenchymal stem cells. *Cytotherapy*, **15**, 330–343.

4. Ryan, J. M., Barry, F., Murphy, J. M., Mahon, B. P. (2007). Interferon-γ does not break, but promotes the immunosuppressive capacity of adult human mesenchymal stem cells. *Clin. Exp. Immunol.*, **149**, 353–363.

5. Abumaree, M., Al Jumah, M., Pace, R. A., Kalionis, B. (2012). Immunosuppressive properties of mesenchymal stem cells. *Stem Cell Rev.*, **8**, 375–392.

6. Amado, L. C., Saliaris, A. P., Schuleri, K. H., St. John, M., Xie, J. S., Cattaneo, S., Durand, D. J., Fitton, T., Kuang, J. Q., Stewart, G., et al. (2005). Cardiac repair with intramyocardial injection of allogeneic mesenchymal stem cells after myocardial infarction. *Proc. Natl. Acad. Sci. USA*, **102**, 11474–11479.

7. Choudhery, M. S., Khan, M., Mahmood, R., Mohsin, S., Akhtar, S., Ali, F., Khan, S. N., Riazuddin, S. (2012). Mesenchymal stem cells conditioned with glucose depletion augments their ability to repair-infarcted myocardium. *J. Cell. Mol. Med.*, **16**, 2518–2529.

8. Xin, H., Li, Y., Shen, L. H., Liu, X., Wang, X., Zhang, J., Pourabdollah-Nejad, D. S., Zhang, C., Zhang, L., Jiang, H., et al. (2010). Increasing

tPA activity in astrocytes induced by multipotent mesenchymal stromal cells facilitate neurite outgrowth after stroke in the mouse. *PLoS One*, **5**, e9027.

9. Taylor, S. E., Smith, R. K., Clegg, P. D. (2007). Mesenchymal stem cell therapy in equine musculoskeletal disease: Scientific fact or clinical fiction? *Equine Vet. J.*, **39**, 172–180.

10. Kretlow, J. D., Jin, Y. Q., Liu, W., Zhang, W. J., Hong, T. H., Zhou, G., Baggett, L. S., Mikos, A. G., Cao, Y. (2008). Donor age and cell passage affects differentiation potential of murine bone marrow-derived stem cells. *BMC Cell Biol.*, **9**, 60.

11. Choudhery, M. S., Khan, M., Mahmood, R., Mehmood, A., Khan, S. N., Riazuddin, S. (2012). Bone marrow derived mesenchymal stem cells from aged mice have reduced wound healing, angiogenesis, proliferation and anti-apoptosis capabilities. *Cell Biol. Int.*, **36**, 747–753.

12. Zuk, P. A., Zhu, M., Ashjian, P., de Ugarte, D. A., Huang, J. I., Mizuno, H., Alfonso, Z. C., Fraser, J. K., Benhaim, P., Hedrick, M. H. (2002). Human adipose tissue is a source of multipotent stem cells. *Mol. Biol. Cell*, **13**, 4279–4295.

13. Zuk, P. A., Zhu, M., Mizuno, H., Huang, J., Futrell, J. W., Katz, A. J., Benhaim, P., Lorenz, H. P., Hedrick, M. H. (2001). Multi-lineage cells from human adipose tissue: Implications for cell-based therapies. *Tissue Eng.*, **7**, 211–228.

14. Butler, M. G., Menitove, J. E. (2011). Umbilical cord blood banking: An update. *J. Assist. Reprod. Genet.*, **28**, 669–676.

15. McGuckin, C. P., Forraz, N., Baradez, M. O., Navran, S., Zhao, J., Urban, R., Tilton, R., Denner, L. (2005). Production of stem cells with embryonic characteristics from human umbilical cord blood. *Cell Prolif.*, **38**, 245–255.

16. McGuckin, C. P., Forraz, N., Allouard, Q., Pettengell, R. (2004). Umbilical cord blood stem cells can expand hematopoietic and neuroglial progenitors *in vitro*. *Exp. Cell Res.*, **295**, 350–359.

17. Rogers, I., Yamanaka, N., Bielecki, R., Wong, C. J., Chua, S., Yuen, S., Casper, R. F. (2007). Identification and analysis of *in vitro* cultured CD45-positive cells capable of multi-lineage differentiation. *Exp. Cell Res.*, **313**, 1839–1852.

18. Kucia, M., Halasa, M., Wysoczynski, M., Baskiewicz-Masiuk, M., Moldenhawer, S., Zuba-Surma, E., Czajka, R., Wojakowski, W., Machalinski, B., Ratajczak, M. Z. (2007). Morphological and molecular

characterization of novel population of CXCR4+ SSEA-4+ Oct-4+ very small embryonic-like cells purified from human umbilical cord blood-preliminary report. *Leukemia*, **21**, 297–303.

19. Harris, D. T., He, X., Badowski, M., Nichols, J. C. (2008). Regenerative medicine of the eye: A short review. In: Levicar, N., Habib, N. A., Dimarakis, I., Gordon, M. Y., eds. *Stem Cell Repair & Regeneration*, Imperial College Press, London, UK, Vol. 3, pp. 211–225.

20. Sunkomat, J. N. E., Goldman, S., Harris, D. T. (2007). Cord blood-derived MNCs delivered intracoronary contribute differently to vascularization compared to CD34+ cells in the rat model of acute ischemia. *J. Mol. Cell. Cardiol.*, **42**, S97.

21. Perry, D. (2000). Patient's voices: The powerful sound in the stem cell debate. *Science*, **287**, 1423.

22. Norambuena, G. A., Khoury, M., Jorgensen, C. (2012). Mesenchymal stem cells in osteoarticular pediatric diseases: An update. *Pediatr. Res.*, **71**, 452–458.

23. Pati, S., Gerber, M. H., Menge, T. D., Wataha, K. A., Zhao, Y., Baumgartner, J. A., Zhao, J., Letourneau, P. A., Huby, M. P., Baer, L. A., et al. (2011). Bone marrow derived mesenchymal stem cells inhibit inflammation and preserve vascular endothelial integrity in the lungs after hemorrhagic shock. *PLoS One*, **6**, e25171.

24. See, E. Y., Toh, S. L., Goh, J. C. (2010). Multilineage potential of bone marrow-derived mesenchymal stem cell sheets: Implications for tissue engineering. *Tissue Eng. Part A*, **6**, 1421–1431.

25. Strioga, M., Viswanathan, S., Darinskas, A., Slaby, O., Michalek, J. (2012). Same or not the same? Comparison of adipose tissue-derived *versus* bone marrow-derived mesenchymal stem and stromal cells. *Stem Cells Dev.*, **21**, 2724–2752.

26. Lindroos, B., Suuronen, R., Miettinen, S. (2011). The potential of adipose stem cells in regenerative medicine. *Stem Cell Rev.*, **2**, 269–291.

27. Chen, Y., Xu, X., Tan, Z., Ye, C., Zhao, Q., Chen, Y. (2012). Age-related BMAL1 change affects mouse bone marrow stromal cell proliferation and osteo-differentiation potential. *Arch. Med. Sci.*, **8**, 30–38.

28. Fraser, J. K., Wulur, I., Alfonso, Z., Hedrick, M. H. (2006). Fat tissue: An underappreciated source of stem cells for biotechnology. *Trends Biotechnol.*, **24**, 150–154.

29. Kitagawa, Y. K. M., Toriyama, K., Kamei, Y., Torii, S. (2006). History of discovery of human adipose-derived stem cells and their clinical applications. *J. Plast. Reconstr. Surg.*, **49**, 1097–1104.

30. Vieira, N. M., Brandalise, V., Zucconi, E., Secco, M., Strauss, B. E., Zatz, M. (2010). Isolation, characterization, and differentiation potential of canine adipose-derived stem cells. *Cell Transplant.*, **19**, 279–289.

31. Rubinstein, P., Rosenfield, R. E., Adamson, J. W., Stevens, C. E. (1993). Stored placental blood for unrelated bone marrow reconstitution. *Blood*, **81**, 1679–1690.

32. Badowski, M. S., Harris, D. T. (2011). Collection, processing, and banking of umbilical cord blood stem cells for transplantation and regenerative medicine. *Somat. Stem Cells*, **879**, 279–290.

33. Papassavas, A. C., Gioka, V., Chatzistamatiou, T., Kokkinos, T., Anagnostakis, I., Gecka, G., Redoukas, I., Paterakis, G., Stavropoulos-Giokas, C. (2008). A strategy of splitting individual high volume cord blood units into two half subunits prior to processing increases the recovery of cells and facilitates *ex vivo* expansion of the infused hematopoietic progenitor cells in adults. *Int. J. Lab. Hematol.*, **30**, 124–132.

34. Harris, D. T., McGaffey, A. P., Schwarz, R. H. (2007). Comparing the mononuclear cell (MNC) recovery of AXP and Hespan. *Obstet. Gynecol.*, **109**, 93S.

35. American Association of Blood Banks (AABB) (2012). *Standards for Cellular Therapy Product Services*, 5th ed., AABB Press, Bethesda, MD, USA.

36. Chow, R., Nademanee, A., Rosenthal, J., Karanes, C., Jaing, T.-H., Graham, M. L., Tsukahara, E., Wang, B., Gjertson, D., Tan, P., et al. (2007). Analysis of hematopoietic cell transplants using plasma-depleted cord blood products that are not red blood cell reduced. *Biol. Blood Marrow Transplant.*, **13**, 1346–1357.

37. Lindenmair, A., Hatlapatka, T., Kollwig, G., Hennerbichler, S., Gabriel, C., Wolbank, S., Redl, H., Kasper, C. (2012). Mesenchymal stem or stromal cells from amnion and umbilical cord tissue and their potential for clinical applications. *Cells*, **1**, 1061–1088.

38. Huang, Y.-C., Parolini, O., La Rocca, G., Deng, L. (2012). Umbilical cord *versus* bone marrow-derived mesenchymal stromal cells. *Stem Cells Dev.*, **21**, 2900–2903.

39. Zhang, X., Hirai, M., Cantero, S., Ciubotariu, R., Dobrila, L., Hirsh, A., Igura, K., Satoh, H., Yokomi, I., Nishimura, T., et al. (2011). Isolation and

characterization of mesenchymal stem cells from human umbilical cord blood: Reevaluation of critical factors for successful isolation and high ability to proliferate and differentiate to chondrocytes as compared to mesenchymal stem cells from bone marrow and adipose tissue. *J. Cell. Biochem.*, **112**, 1206–1218.

40. Wexler, S. A., Donaldson, C., Denning-Kendall, P., Rice, C., Bradley, B., Hows, J. M. (2003). Adult bone marrow is a rich source of human mesenchymal stem cells but umbilical cord and mobilized blood adult blood are not. *Br. J. Haematol.*, **121**, 368–374.

41. Lee, O. K., Kuo, T. K., Chen, W. M., Lee, K. D., Hsieh, S. L., Chen, T. H. (2004). Isolation of multipotent mesenchymal stem cells from umbilical cord blood. *Blood*, **103**, 1669–1675.

42. Musina, R. A., Bekchanova, E. S., Belyavskii, A. V., Grinenko, T. S., Sukhikh, G. T. (2007). Umbilical cord blood mesenchymal stem cells. *Bull. Exp. Biol. Med.*, **143**, 127–131.

43. Secco, M., Zucconi, E., Vieira, N. M., Fogaça, L. L., Cerqueira, A., Carvalho, M. D., Jazedje, T., Okamoto, O. K., Muotri, A. R., Zatz, M. (2008). Multipotent stem cells from umbilical cord: Cord is richer than blood. *Stem Cells*, **26**, 146–150.

44. Henry, G., William, P. L., Bannister, L. H. (1995). *Grays Anatomy*, 38th ed., ELBS Churchill Livingstone, London, UK.

45. Bongso, A., Fong, C.-Y. (2013). The therapeutic potential, challenges and future clinical directions of stem cells from Wharton's jelly of the human umbilical cord. *Stem Cell Rev. Rep.*, **9**, 226–240.

46. Jeschke, M. G., Gauglitz, G. G., Phan, T. T., Herndon, D. N., Kita, K. (2011). Umbilical cord lining membrane and Wharton's jelly-derived mesenchymal stem cells: The similarities and differences. *Open Tissue Eng. Regen. Med. J.*, **4**, 21–27.

47. Weiss, M. L., Medicetty, S., Bledsoe, A. R., Rachakatla, R. S., Choi, M., Merchav, S. (2006). Human umbilical cord matrix stem cells: Preliminary characterization and effect of transplantation in a rodent model of Parkinson's disease. *Stem Cells*, **24**, 781–792.

48. Seshareddy, K., Troyer, D., Weiss, M. L. (2008). Methods to isolate mesenchymal-like cells from Wharton's jelly of umbilical cord. *Methods Cell Biol.*, **86**, 101–119.

49. Wang, H. S., Hung, S. C., Peng, S. T., Huang, C. C., Wei, H. M., Guo, Y. J., Fu, Y. S., Lai, M. C., Chen, C. C. (2004). Mesenchymal stem cells in the Wharton's Jelly of the human umbilical cord. *Stem Cells*, **22**, 1330–1337.

50. Fong, C. Y., Richards, M., Manasi, N., Biswas, A., Bongso, A. (2007). Comparative growth behavior and characterization of stem cells from human Wharton's jelly. *Reprod. Biomed. Online*, **15**, 708–718.

51. Fong, C. Y., Gauthaman, K., Bongso, A. (2009). Reproductive stem cells of embryonic origin: Comparative properties and potential benefits of human embryonic stem cells and Wharton's jelly stem cells. In: Simon, C., Pellicer, A., eds. *Stem Cells in Human Reproduction*, 2nd ed., Informa Healthcare, New York, NY, USA, pp. 136–149.

52. Fong, C. Y., Subramanian, A., Biswas, A., Gauthaman, K., Srikanth, P., Hande, M., Bongso, A. (2010). Derivation efficiency, cell proliferation, frozen-thaw survival, 'stemness' properties, and differentiation of human Wharton's jelly stem cells: Their potential for concurrent banking with cord blood for regenerative medicine purposes. *Reprod. Biomed. Online*, **21**, 391–401.

53. Angelucci, S., Marchisio, M., Giuseppe, F. D., Pierdomenico, L., Sulpizio, M., Eleuterio, E. (2010). Proteome analysis of human Wharton's jelly cells during *in vitro* expansion. *Proteome Sci.*, **8**, 18–25.

54. Ding, D. C., Shyu, W. C., Lin, S. Z., Liu, H. W., Chiou, S. H., Chu, T. Y. (2012). Human umbilical cord mesenchymal stem cells support non-tumorigenic expansion of human embryonic stem cells. *Cell Transplant.*, **21**, 1515–1527.

55. Kikuchi-Taura, A., Taguchi, A., Kanda, T., Inoue, T., Kasahara, Y., Hirose, H., Sato, I., Matsuyama, T., Nakagomi, T., Yamahara, K., et al. (2012). Human umbilical cord provides a significant source of unexpanded mesenchymal stromal cells. *Cytotherapy*, **14**, 441–450.

56. Lu, L. L., Liu, Y. J., Yang, S. G., Zhao, Q. J., Wang, X., Gong, W., Han, Z. B., Xu, Z. S., Lu, Y. X., Liu, D., et al. (2006). Isolation and characterization of human umbilical cord mesenchymal stem cells with hematopoiesis-supportive function and other potentials. *Haematologica*, **91**, 1017–1026.

57. Schugar, R. C., Chirieleison, S. M., Wescoe, K. E., Schmidt, B. T., Askew, Y., Nance, J. J., Evron, J. M., Peault, B., Deasy, B. M. (2009). High harvest yield, high expansion, and phenotype stability of CD146 mesenchymal stromal cells from whole primitive human umbilical cord tissue. *J. Biomed. Biotechnol.*, **2009**, 789526.

58. Capelli, C., Gotti, E., Morigi, M., Rota, C., Weng, L., Dazzi, F., Spinelli, O., Cazzaniga, G., Trezzi, R., Gianatti, A., et al. (2011). Minimally manipulated whole human umbilical cord is a rich source of clinical grade human mesenchymal stromal cells expanded in human platelet lysate. *Cytotherapy*, **13**, 786–801.

59. Bosch, J., Houben, A. P., Radke, T. F., Stapelkamp, D., Bünemann, E., Balan, P., Buchheiser, A., Liedtke, S., Kögler, G. (2011). Distinct differentiation potential of 'MSC' derived from cord blood and umbilical cord: Are cord-derived cells true mesenchymal stromal cells? *Stem Cells Dev.*, **21**, 1977–1988.

60. Kita, K., Gauglitz, G. G., Phan, T. T., Herndon, D. N., Jeschke, M. G. (2010). Isolation and characterization of mesenchymal stem cells from the sub amniotic human umbilical cord lining membrane. *Stem Cells Dev.*, **19**, 491–502.

61. Sarugaser, R., Lickorish, D., Baksh, D., Hosseini, M. M., Davies, J. E. (2005). Human umbilical cord perivascular (HUCPV) cells: A source of mesenchymal progenitors. *Stem Cells*, **23**, 220–229.

62. Romanov, Y. A., Svintsitskaya, V. A., Smirnov, V. N. (2003). Searching for alternative sources of postnatal human mesenchymal stem cells: Candidate MSC-like cells from umbilical cord. *Stem Cells*, **21**, 105–110.

63. Fu, Y. S., Shih, Y. T., Cheng, Y. C., Min, M. Y. (2004). Transformation of human umbilical mesenchymal cells into neurons *in vitro*. *J. Biomed. Sci.*, **11**, 652–660.

64. Tsagias, N., Koliakos, I., Karagiannis, V., Eleftheriadou, M., Koliakis, G. G. (2011). Isolation of mesenchymal stem cells using the total length of umbilical cord for transplantation purposes. *Transfus. Med.*, **21**, 253–261.

65. Dominici, M., le Blanc, K., Mueller, I., Slaper-Cortenbach, I., Marini, F., Krause, D., Deans, R., Keating, A., Prockop, D. J., Horwitz, E. (2006). Minimal criteria for defining multipotent mesenchymal stromal cells. The International Society for Cellular Therapy position statement. *Cytotherapy*, **8**, 315–317.

66. Zhang, H., Zhang, B., Tao, Y., Cheng, M., Hu, J., Xu, M., Chen, H. (2012). Isolation and characterization of mesenchymal stem cells from whole human umbilical cord applying a single enzyme approach. *Cell Biochem. Funct.*, **30**, 643–649.

67. Marmotti, A., Mattia, S., Bruzzone, M., Buttiglieri, S., Risso, A., Bonasia, D. E., Blonna, D., Castoldi, F., Rossi, R., Zanini, C., et al. (2012). Minced umbilical cord fragments as a source of cells for orthopaedic tissue engineering: An *in vitro* study. *Stem Cells Int.*, **2012**, 326813.

68. Nekanti, U., Mohanty, L., Venugopal, P., Balasubramanian, S., Totey, S., Ta, M. (2001). Optimization and scale-up of Wharton's jelly-derived mesenchymal stem cells for clinical applications. *Stem Cell Res.*, **5**, 244–254.

69. Pilz, G. A., Ulrich, C., Ruh, M., Abele, H., Schäfer, R., Kluba, T., Bühring, H. J., Rolauffs, B., Aicher, W. K. (2011). Human term placenta-derived mesenchymal stromal cells are less prone to osteogenic differentiation than bone marrow-derived mesenchymal stromal cells. *Stem Cells Dev.*, **20**, 635–646.

70. Girdlestone, J., Limbani, V. A., Cutler, A. J., Navarrete, C. V. (2009). Efficient expansion of mesenchymal stromal cells from umbilical cord under low serum conditions. *Cytotherapy*, **11**, 738–748.

71. Xue, G., He, M., Zhao, J., Chen, Y., Tian, Y., Zhao, B. (2011). Intravenous umbilical cord mesenchymal stem cell infusion for the treatment of combined malnutrition nonunion of the humerus and radial nerve injury. *Regen. Med.*, **6**, 733–741.

72. Choudhery, M. S., Badowski, M., Muise, A., Pierce, J., Harris, D. T. (2014). Cryopreservation of whole adipose tissue for future use in regenerative medicine. *J. Surg. Res.*, **187**(1), 24–35.

73. Wu, K. H., Chan, C. K., Tsai, C., Chang, Y. H., Sieber, M., Chiu, T. H., Ho, M., Peng, C. T., Wu, H. P., Huang, J. L. (2011). Effective treatment of severe steroid-resistant acute graft-versus-host disease with umbilical cord-derived mesenchymal stem cells. *Transplantation*, **91**, 1412–1416.

74. Liang, J., Zhang, H., Hua, B., Wang, H., Wang, J., Han, Z., Sun, L. (2009). Allogeneic mesenchymal stem cells transplantation in treatment of multiple sclerosis. *Mult. Scler.*, **15**, 644–646.

75. Liang, J., Gu, F., Wang, H., Hua, B., Hou, Y., Shi, S., Lu, L., Sun, L. (2010). Mesenchymal stem cell transplantation for diffuse alveolar hemorrhage in SLE. *Nat. Rev. Rheumatol.*, **6**, 486–489.

76. D'Ippolito, G., Schiller, P. C., Ricordi, C., Roos, B. A., Howard, G. A. (1999). Age-related osteogenic potential of mesenchymal stromal stem cells from human vertebral bone marrow. *J. Bone Miner. Res.*, **14**, 1115–1122.

77. Tokalov, S. V., Gruner, S., Schindler, S., Wolf, G., Baumann, M., Abolmaali, N. (2007). Age-related changes in the frequency of mesenchymal stem cells in the bone marrow of rats. *Stem Cells Dev.*, **16**, 439–446.

78. Stolzing, A., Jones, E., McGonagle, D., Scutt, A. (2008). Age-related changes in human bone marrow-derived mesenchymal stem cells: Consequences for cell therapies. *Mech. Aging Dev.*, **129**, 163–173.

79. Bonyadi, M., Waldman, S. D., Liu, D., Aubin, J. E., Grynpas, M. D., Stanford, S. L. (2003). Mesenchymal progenitor self-renewal deficiency leads to age-dependent osteoporosis in Sca-1/Ly-6A null mice. *Proc. Natl. Acad. Sci. USA*, **100**, 5840–5845.

80. Stolzing, A., Scutt, A. (2006). Age-related impairment of mesenchymal progenitor cell function. *Aging Cell*, **5**, 213–224.

81. Murphy, J. M., Dixon, K., Beck, S., Fabian, D., Feldman, A., Barry, F. (2002). Reduced chondrogenic and adipogenic activity of mesenchymal stem cells from patients with advanced osteoarthritis. *Arthritis Rheum.*, **46**, 704–713.

82. Zhang, H., Fazel, S., Tian, H., Mickle, D. A., Weisel, R. D., Fujii, T., Li, R. K. (2005). Increasing donor age adversely impacts beneficial effects of bone marrow but no smooth muscle myocardial cell therapy. *Am. J. Physiol. Heart Circ. Physiol.*, **289**, H2089–H2096.

83. ClinicalTrials.gov. Available at: https://clinicaltrials.gov/ (accessed on February 1, 2019).

84. Furfaro, M. E. K., Gaballa, M. A. (2007). Do adult stem cells ameliorate the damaged myocardium? Is human cord blood a potential source of stem cells? *Curr. Vasc. Pharmacol.*, **5**, 27–44.

85. Chen, J., Sanberg, P. R., Li, Y., Wang, L., Lu, M., Willing, A. E., Sanchez-Ramos, J., Chopp, M. (2001). Intravenous administration of human umbilical cord blood reduces behavioral deficits after stroke in rats. *Stroke*, **32**, 2682–2688.

86. Willing, A. E., Lixian, J., Milliken, M., Poulos, S., Zigova, T., Song, S., Hart, C., Sanchez-Ramos, J., Sanberg, P. R. (2003). Intravenous *versus* intrastriatal cord blood administration in a rodent model of stroke. *J. Neurosci. Res.*, **73**, 296–307.

87. Borlongan, C. V., Hadman, M., Sanberg, C. D., Sanberg, P. R. (2004). Central nervous system entry of peripherally injected umbilical cord blood cells is not required for neuroprotection in stroke. *Stroke*, **35**, 2385–2389.

88. Newman, M. B., Willing, A. E., Manressa, J. J., Sanberg, C. D., Sanberg, P. R. (2006). Cytokines produced by cultured human umbilical cord blood (HUCB) cells: Implications for brain repair. *Exp. Neurol.*, **199**, 201–208.

89. Vendrame, M., Cassady, J., Newcomb, J. J., Butler, T., Pennypacker, K. R., Zigova, T., Sanberg, C. D., Sanberg, P. R., Willing, A. E. (2004). Infusion of human umbilical cord blood cells in a rat model of stroke dose-dependently rescues behavioral deficits and reduces infarct volume. *Stroke*, **35**, 2390–2395.

90. Xiao, J., Nan, Z., Motooka, Y., Low, W. C. (2005). Transplantation of a novel cell line population of umbilical cord blood stem cells ameliorates neurological deficits associated with ischemic brain injury. *Stem Cells Dev.*, **14**, 722–733.

91. Newcomb, J. D., Ajrno, C. T., Sanberg, C. D., Sanberg, P. R., Pennypacker, K. R., Willing, A. E. (2006). Timing of cord blood treatment after experimental stroke determines therapeutic efficacy. *Cell Transplant.*, **15**, 213–223.

92. Vendrame, M., Gemma, C., Pennypacker, K. R., Bickford, P. C., Davis Sanberg, C., Sanberg, P. R., Willing, A. E. (2006). Cord blood rescues stroke-induced changes in splenocyte phenotype and function. *Exp. Neurol.*, **199**, 191–200.

93. Meier, C., Middelanis, J., Wasielewski, B., Neuhoff, S., Roth-Haerer, A., Gantert, M., Dinse, H. R., Dermietzel, R., Jensen, A. (2006). Spastic paresis after perinatal brain damage in rats is reduced by human cord blood mononuclear cells. *Pediatr. Res.*, **59**, 244–249.

94. Chen, S. H., Chang, F. M., Tsai, Y. C., Huang, K. F., Lin, C. L., Lin, M. T. (2006). Infusion of human umbilical cord blood cells protect against cerebral ischemia and damage during heatstroke in the rat. *Exp. Neurol.*, **199**, 67–76.

95. Vendrame, M., Gemma, C., de Mesquita, D., Collier, L., Bickford, P. C., Sanberg, C. D., Sanberg, P. R., Pennypacker, K. R., Willing, A. E. (2005). Anti-inflammatory effects of human cord blood cells in a rat model of stroke. *Stem Cells Dev.*, **14**, 595–604.

96. Nan, Z., Grande, A., Sanberg, C. D., Sanberg, P. R., Low, W. C. (2005). Infusion of human umbilical cord blood ameliorates neurologic deficits in rats with hemorrhagic brain injury. *Ann. N. Y. Acad. Sci.*, **1049**, 84–96.

97. Nystedt, J., Mäkinen, S., Laine, J., Jolkkonen, J. (2006). Human cord blood CD34+ cells and behavioral recovery following focal cerebral ischemia in rats. *Acta Neurobiol. Exp. (Wars)*, **66**, 293–300.

98. Mäkinen, S., Kekarainen, T., Nystedt, J., Liimatainen, T., Huhtala, T., Närvänen, A., Laine, J., Jolkkonen, J. (2006). Human umbilical cord blood cells do not improve sensorimotor or cognitive outcome following transient middle cerebral artery occlusion in rats. *Brain Res.*, **1123**, 207–215.

99. Chang, C. K., Chang, C. P., Chiu, W. T., Lin, M. T. (2006). Prevention and repair of circulatory shock and cerebral ischemia/injury by various agents in experimental heatstroke. *Curr. Med. Chem.*, **13**, 3145–3154.

100. Bliss, T., Guzman, R., Daadi, M., Steinberg, G. K. (2007). Cell transplantation therapy for stroke. *Stroke*, **38**, 817–826.

101. Lu, D., Sanberg, P. R., Mahmood, A., Li, Y., Wang, L., Sanchez-Ramos, J., Chopp, M. (2002). Intravenous administration of human umbilical

cord blood reduces neurological deficit in the rat after traumatic brain injury. *Cell Transplant.*, **11**, 275–281.

102. Kurtzberg, J. (2009). Update on umbilical cord blood transplantation. *Curr. Opin. Pediatr.*, **21**, 22–29.

103. Kuh, S. U., Cho, Y. E., Yoon, D. H., Kim, K. N., Ha, Y. (2005). Functional recovery after human umbilical cord blood cells transplantation with brain derived-neurotropic factor into the spinal cord injured rats. *Acta Neurochir. (Wein.)*, **14**, 985–992.

104. Kang, K.-S., Kim, S. W., Oh, Y. H., Yu, J. W., Kim, K.-Y., Park, H. K., Song, C.-H., Han, H. (2005). A thirty-seven-year old spinal cord-injured female patient, transplanted of multipotent stem cells from human UC blood with improved sensory perception and mobility, both functionally and morphologically: A case study. *Cytotherapy*, **7**, 368–373.

105. Zhang, L., Li, Y., Zhang, C., Chopp, M., Gosiewska, A., Hong, K. (2011). Delayed administration of human umbilical tissue-derived cells improved neurological functional recovery in a rodent model of focal ischemia. *Stroke*, **42**, 1437–1444.

106. Hess, D. C., Borlongan, C. V. (2008). Cell-based therapy in ischemic stroke. *Expert Rev. Neurother.*, **8**, 1193–1201.

107. Ding, D. C., Shyu, W. C., Chiang, M. F., Lin, S. Z., Chang, Y. C., Wang, H. J., Su, C. Y., Li, H. (2007). Enhancement of neuroplasticity through upregulation of beta1-integrin in human umbilical cord-derived stromal cell implanted stroke model. *Neurobiol. Dis.*, **27**, 339–353.

108. Koh, S. H., Kim, K. S., Choi, M. R., Jung, K. H., Park, K. S., Chai, Y. G., Roh, W., Hwang, S. J., Ko, H. J., Huh, Y. M., et al. (2008). Implantation of human umbilical cord-derived mesenchymal stem cells as a neuroprotective therapy for ischemic stroke in rats. *Brain Res.*, **1229**, 233–248.

109. Liao, W., Zhong, J., Yu, J., Xie, J., Liu, Y., Du, L., Yang, S., Liu, P., Xu, J., Wang, J., et al. (2009). Therapeutic benefit of human umbilical cord derived mesenchymal stromal cells in intracerebral hemorrhage rat: Implications of anti-inflammation and angiogenesis. *Cell. Physiol. Biochem.*, **24**, 307–316.

110. Lim, J. H., Byeon, Y. E., Ryu, H. H., Jeong, Y. H., Lee, Y. W., Kim, W. H., Kang, K. S., Kweon, O. K. (2007). Transplantation of canine umbilical cord blood-derived mesenchymal stem cells in experimentally induced spinal cord injured dogs. *J. Vet. Sci.*, **8**, 275–282.

111. Yang, C. C., Shih, Y. H., Ko, M. H., Hsu, S. Y., Cheng, H., Fu, Y. S. (2008). Transplantation of human umbilical mesenchymal stem cells from

Wharton's jelly after complete transection of the rat spinal cord. *PLoS One*, **3**, e3336.

112. Zhang, L., Zhang, H. T., Hong, S. Q., Ma, X., Jiang, X. D., Xu, R. X. (2009). Cografted Wharton's jelly cells-derived neurospheres and BDNF promote functional recovery after rat spinal cord transection. *Neurochem. Res.*, **34**, 2030–2039.

113. Hu, S.-L., Luo, H.-S., Li, J.-T., Xia, Y.-Z., Li, L., Zhang, L.-J., Meng, H., Cui, G.-Y., Chen, Z., Wu, N., et al. (2010). Functional recovery in acute traumatic spinal cord injury after transplantation of human umbilical cord mesenchymal stem cells. *Crit. Care Med.*, **38**, 2181–2189.

114. Fu, Y. S., Cheng, Y. C., Lin, M. Y., Cheng, H., Chu, P. M., Chou, S. C., Shih, Y. H., Ko, M. H., Sung, M. S. (2006). Conversion of human umbilical cord mesenchymal stem cells in Wharton's jelly to dopaminergic neurons *in vitro*: Potential therapeutic application for Parkinsonism. *Stem Cells*, **24**, 115–124.

115. Liao, W., Xie, J., Zhong, J., Liu, Y., Du, L., Zhou, B., Xu, J., Liu, P., Yang, S., Wang, J., et al. (2009). Therapeutic effect of human umbilical cord multipotent mesenchymal stromal cells in a rat model of stroke. *Transplantation*, **87**, 350–359.

116. Copeland, N., Harris, D., Gaballa, M. A. (2009). Human umbilical cord blood stem cells, myocardial infarction and stroke. *Clin. Med.*, **9**, 342–345.

117. Botta, R., Gao, E., Stassi, G., Bonci, D., Pelosi, E., Zwas, D., Patti, M., Colonna, L., Baiocchi, M., Coppola, S., et al. (2004). Heart infarct in NOD-SCID mice: Therapeutic vasculogenesis by transplantation of human CD34+ cells ad low dose CD34+KDR+ cells. *FASEB J.*, **18**, 1392–1394.

118. Henning, R. J., Abu-Ali, H., Balis, J. U., Morgan, M. B., Willing, A. E., Sanberg, P. R. (2004). Human umbilical cord blood mononuclear cells for treatment of acute myocardial infarction. *Cell Transplant.*, **13**, 729–739.

119. Chen, H. K., Hung, H. F., Shyu, K. G., Wang, B. W., Sheu, J. R., Liang, Y. J., Chang, C. C., Kuan, P. (2005). Combined cord blood cells and gene therapy enhances angiogenesis and improves cardiac performance in mouse after acute myocardial infarction. *Eur. J. Clin. Invest.*, **35**, 677–686.

120. Hirata, Y., Sata, M., Motomura, N., Takanashi, M., Suematsu, Y., Ono, M., Takamoto, S. (2005). Human umbilical cord blood cells improve cardiac function after myocardial infarction. *Biochem. Biophys. Res. Commun.*, **327**, 609–614.

121. Kim, B. O., Tian, H., Prasongsukarn, K., Wu, J., Angoulvant, D., Wnendt, S., Muhs, A., Spitkovsky, D., Li, R. K. (2006). Cell transplantation improves ventricular function after a myocardial infarction: A preclinical study of human unrestricted somatic stem cells in a porcine model. *Circulation*, **112**, 196–204.

122. Leor, J., Guetta, E., Feinberg, M. S., Galski, H., Bar, I., Holbova, R., Miller, L., Zarin, P., Castel, D., Barbash, I. M., et al. (2006). Human umbilical cord blood-derived CD133+ cells enhance function and repair of the infracted myocardium. *Stem Cells*, **24**, 772–780.

123. Ma, N., Stamm, C., Kaminski, A., Li, W., Kleine, H. D., Müller-Hilke, B., Zhang, L., Ladilov, Y., Egger, D., Steinhoff, G. (2005). Human cord blood cells induce angiogenesis following myocardial infarction in NOD/scid mice. *Cardiovasc. Res.*, **66**, 45–54.

124. Bonnano, G., Mariotti, A., Procoli, A., Rutella, M., Pessina, S., Scambia, G., Mancuso, G., Pierelli, S., Luca, P. (2007). Human cord blood CD133+ cells imunoselected by a clinical-grade apparatus differentiate *in vitro* into endothelial- and cardiomyocyte-like cells. *Transfusion*, **47**, 280–289.

125. Schmidt, D., Breymann, C., Weber, A., Guenter, C. I., Neuenschwander, S., Zund, G., Turina, M., Hoerstrup, S. P. (2004). Umbilical cord blood derived endothelial progenitor cells for tissue engineering of vascular grafts. *Soc. Thorac. Surg.*, **78**, 2094–2098.

126. Murga, M., Yao, L., Tosato, G. (2004). Derivation of endothelial cells from CD34–umbilical cord blood. *Stem Cells*, **22**, 385–395.

127. Hoerstrup, S. P., Kadner, A., Breymann, C., Maurus, C. F., Guenter, C. I., Sodian, R., Visjager, J. F., Zund, G., Turina, M. I. (2002). Living, autologous pulmonary artery conduits tissue engineered from human umbilical cord cells. *Ann. Thorac. Surg.*, **74**, 46–52.

128. Schmidt, D., Mol, A., Neuenschwander, S., Breymann, C., Gössi, M., Zund, G., Turina, M., Hoerstrup, S. P. (2005). Living patches engineered from human umbilical cord derived fibroblasts and endothelial progenitor cells. *Eur. J. Cardiothorac. Surg.*, **27**, 795–800.

129. Murohara, T., Ikeda, H., Duan, J., Shintani, S., Sasaki, Ki., Eguchi, H., Onitsuka, I., Matsui, K., Imaizumi, T. (2000). Transplanted cord blood-derived endothelial precursor cells augment postnatal neovascularization. *J. Clin. Invest.*, **105**, 1527–1536.

130. Goldberg, J. L., Laughlin, M. J. (2007). UC blood hematopoietic stem cells and therapeutic angiogenesis. *Cytotherapy*, **9**, 4–13.

131. Nieda, M., Nicol, A., Denning-Kendall, P., Sweetenham, J., Bradley, B., Hows, J. (1997). Endothelial cell precursors are normal components of human umbilical cord blood. *Br. J. Hematol.*, **98**, 775–777.

132. Murohara, T. (2001). Therapeutic vasculogenesis using human cord blood-derived endothelial progenitors. *Trends Cardiovasc. Med.*, **11**, 303–307.

133. Ikeda, Y., Fukuda, N., Wada, M., Matsumoto, T., Satomi, A., Yokoyama, S., Saito, S., Matsumoto, K., Kanmatsuse, K., Mugishima, H. (2004). Development of angiogenic cell and gene therapy by transplantation of umbilical cord blood with vascular endothelial growth factor gene. *Hypertens. Res.*, **27**, 119–128.

134. Cho, S. W., Gwak, S. J., Kang, S. W., Bhang, S. H., Won Song, K. W., Yang, Y. S., Choi, C. Y., Kim, B. S. (2006). Enhancement of angiogenic efficacy of human cord blood cell transplantation. *Tissue Eng.*, **12**, 1651–1661.

135. Finney, M. R., Greco, N. J., Haynesworth, S. E., Martin, J. M., Hedrick, D. P., Swan, J. Z., Winter, D. G., Kadereit, S., Joseph, M. E., Fu, P., et al. (2006). Direct comparison of umbilical cord blood *versus* bone marrow-derived endothelial precursor cells in mediating neovascularization in response to vascular ischemia. *Biol. Blood Marrow Transplant.*, **12**, 585–593.

136. Pesce, M., Orlandi, A., Iachininoto, M. G., Straino, S., Torella, A. R., Rizzuti, V., Pompilio, G., Bonanno, G., Scambia, G., Capogrossi, M. C. (2003). Myoendothelial differentiation of human umbilical cord blood-derived stem cells in ischemic limb tissue. *Circ. Res.*, **93**, 51–62.

137. Wang, F. S., Yang, K. D., Wang, C. J. (2004). Shockwave stimulates oxygen radical-mediated osteogenesis of the mesenchymal cells from human umbilical cord blood. *J. Bone Miner. Res.*, **19**, 973–982.

138. Szivek, J. A., Wiley, D., Cox, L., Harris, D. T., Margolis, D. S., Grana, W. A. (2006). Stem cells grown in dynamic culture on micropatterned surfaces can be used to engineer cartilage like tissue. In *Proceedings of the Orthopaedic Research Society Meeting*, Chicago, IL, USA, 19–22 March 2006.

139. Arthritis Foundation. Available at: http://www.arthritis.org (accessed on January 17, 2019).

140. Wagner, W., Wein, F., Seckinger, A., Frankhauser, M., Wirkner, U., Krause, U., Blake, J., Schwager, C., Eckstein, V., Ansorge, W., et al. (2005). Comparative characteristics of mesenchymal stem cells from human bone marrow, adipose tissue, and umbilical cord blood. *Exp. Hematol.*, **33**, 1402–1416.

141. Eliopoulos, N., Stagg, J., Lejeune, L., Pommey, S., Galipeau, J. (2005). Allogeneic marrow stromal cells are immune rejected by MHC class I- and class II-mismatched recipient mice. *Blood*, **106**, 4057–4065.

142. Nauta, A. J., Westerhuis, G., Kruisselbrink, A. B., Lurvink, E. G. A., Willemze, R., Fibbe, W. E. (2006). Donor-derived mesenchymal stem cells are immunogenic in an allogeneic host and stimulate donor graft rejection in a nonmyeloablative setting. *Blood*, **108**, 2114–2120.

143. Zangi, L., Margalit, R., Reich-Zeliger, S., Bachar-Lustig, E., Beilhack, A., Negrin, R., Reisner, Y. (2009). Direct imaging of immune rejection and memory induction by allogeneic mesenchymal stromal cells. *Stem Cells*, **27**, 2865–2874.

144. Buja, L. M., Vela, D. (2010). Immunologic and inflammatory reactions to exogenous stem cells implications for experimental studies and clinical trials for myocardial repair. *J. Am. Coll. Cardiol.*, **56**, 1693–1700.

145. Pérez-Simon, J. A., López-Villar, O., Andreu, E. J., Rifón, J., Muntion, S., Campelo, M. D., Sánchez-Guijo, F. M., Martinez, C., Valcarcel, D., Cañizo, C. D. (2011). Mesenchymal stem cells expanded *in vitro* with human serum for the treatment of acute and chronic graft *versus* host disease: Results of a phase I/II clinical trial. *Haematologica*, **96**, 1072–1076.

146. Dimmeler, S., Leri, A. (2008). Aging and disease as modifiers of efficacy of cell therapy. *Circ. Res.*, **102**, 1319–1330.

147. Scheubel, R. J., Zorn, H., Silber, R. E., Kuss, O., Morawietz, H., Holtz, J., Simm, A. (2003). Age-dependent depression in circulating endothelial progenitor cells in patients undergoing coronary artery bypass grafting. *J. Am. Coll. Cardiol.*, **42**, 2073–2080.

148. Alt, E. U., Senst, C., Murthy, S. N., Slakey, D. P., Dupin, C. L., Chaffin, A. E., Kadowitz, P. J., Izadpanah, R. (2012). Aging alters tissue resident mesenchymal stem cell properties. *Stem Cell Res.*, **8**, 215–225.

149. Zhu, M., Kohan, E., Bradley, J., Hedrick, M., Benhaim, P., Zuk, P. (2009). The effect of age on osteogenic, adipogenic and proliferative potential of female adipose-derived stem cells. *J. Tissue Eng. Regen. Med.*, **3**, 290–301.

150. Khan, W. S., Adesida, A. B., Tew, S. R., Andrew, J. G., Hardingham, T. E. (2009). The epitope characterization and the osteogenic differentiation potential of human fat pad-derived stem cells is maintained with ageing in later life. *Injury*, **40**, 150–157.

151. Harris, D. T., Hilgaertner, J., Simonson, C., Ablin, R. J., Badowski, M. (2012). Cell based therapy for epithelial wounds. *Cytotherapy*, **14**, 802–810.

152. Harris, D. T., Badowski, M., Ahmad, N., Gaballa, M. (2007). The potential of cord blood stem cells for use in regenerative medicine. *Expert Opin. Biol. Ther.*, **7**, 1311–1322.

Chapter 32

Overview of Ethical Issues in Nanomedicine

Raj Bawa, MS, PhD,[a,b,c] and Summer Johnson, PhD[d]

[a]Patent Law Department, Bawa Biotech LLC, Ashburn, Virginia, USA
[b]The Pharmaceutical Research Institute,
Albany College of Pharmacy and Health Sciences, Albany, New York, USA
[c]Guanine Inc., Rensselaer, New York, USA
[d]University of New Haven, School of Health Sciences,
West Haven, Connecticut, USA

Keywords: ethical, ethics, nanotechnology, nanomedicine, nanoscale drug delivery systems, theranostics, therapeutic nanomedicine, diagnostic nanomedicine, nanoimaging, precision medicine, engineered nanodevices, nanodrugs, controlled manipulation, public backlash, US Food and Drug Administration (FDA), patent, research and development (R&D), surgical nanobots, drug-delivery nanoparticles, artificial intelligence, implantable nanoscale medical devices, transhuman, posthuman, blood–brain barrier, privacy, confidentiality, social justice, enhancement, commercialization, primary mode of action (PMOA)

The Road from Nanomedicine to Precision Medicine
Edited by Shaker A. Mousa, Raj Bawa, and Gerald F. Audette
Copyright © 2020 Jenny Stanford Publishing Pte. Ltd.
ISBN 978-981-4800-59-4 (Hardcover), 978-0-429-29501-0 (eBook)
www.jennystanford.com

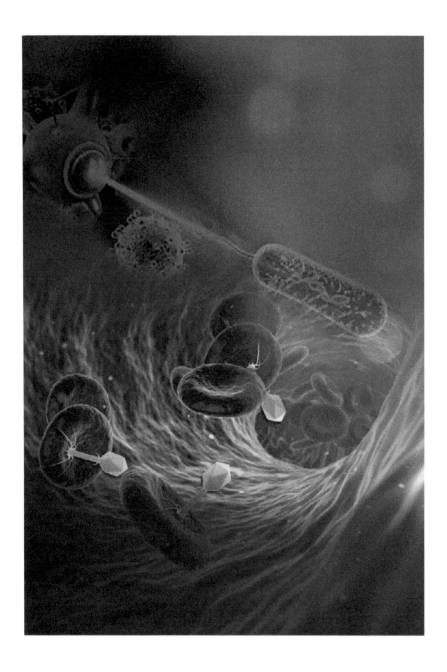

32.1   Introductory Overview

The air is thick with news of nanotechnology[1] breakthroughs, and there is no shortage of excitement and hype when it comes to anything "nano." Optimists tout nano as an enabling technology, a sort of next industrial revolution that could enhance the wealth and health of nations. Pessimists, on the other hand, take a cautionary position, preaching a go-slow approach and pointing to gaps in scientific information on health risks, general failure on the part of regulatory agencies to formulate clear guidelines, and issuance of numerous patents of dubious scope.[2] They highlight that nano is burdened with inflated expectations and hype. Whatever your stance, nano has already permeated virtually every sector of the global economy, with potential applications consistently inching their way into the marketplace.

Medical practice is entering a new era also focused on the nanoscale, more specifically on the practice of "nanomedicine."[3] In fact, in the next decade, many areas within nanomedicine (nanoscale drug delivery systems, nanoimaging, theranostics, etc.) are believed to be a healthcare game-changer by offering patients access to precision medicine. The creation of nanodevices— such as nanobots capable of performing real-time therapeutic plus diagnostic functions *in vivo*—is the major long-term goal

---

[1]Nanotechnology is "the design, characterization, production, and application of structures, devices, and systems by controlled manipulation of size and shape at the nanometer scale (atomic, molecular, and macromolecular scale) that produces structures, devices, and systems with at least one novel/superior characteristic or property." See: Bawa, R. (2007). Patents and nanomedicine. *Nanomedicine (London)*, **2**(3), 351–374.

[2]Nanopatent filings and patent grants have continued unabated since the early 1980s. Universities and industry have jumped into the fray with a clear indication of patenting as much nano as they can grab. Often in this rush to patent anything and everything nano by "patent prospectors," nanopatents of dubious scope and validity are issued by patent offices around the world, thereby generating potential "patent thickets." Since the early 1990s, in light of inadequate search tools/commercial databases available to patent examiners at the US Patent & Trademark Office (PTO) along with an explosion of "prior art" in nanotech, patents of questionable validity and/or scope have dribbled out.

[3]Nanomedicine may be defined as the monitoring, repair, construction, and control of human biological systems at the molecular level, using engineered nanodevices and nanostructures. Nanomedicine is, in a broad sense, the application of nanoscale technologies to the practice of medicine, namely, for diagnosis, prevention, and treatment of disease.

of nanodrug delivery and the Holy Grail of medicine. Advances in nanotherapeutics, miniaturization of analytical tools, improved computational and memory capabilities, advances in genome manipulation, advent of artificial intelligence, higher resolution microscopic and imaging technologies, and developments in remote communications will eventually cross new frontiers in the understanding and practice of medicine.

Nanomedicine is gradually blossoming into a robust industry. Clearly, rapid advances and product development are already in full swing as it continues to influence the pharmaceutical, device, and biotechnology industries [1–8]. The potential impact of nanomedicine on society could be huge [1–8]. Nanomedicine could drastically improve a patient's quality of life, reduce societal and economic costs associated with health care, offer early detection of pathologic conditions, reduce the severity of therapy, and result in improved clinical outcomes for the patient. Numerous companies are actively involved in nanomedicine research and development (R&D), with many nanomedicine-related products (mostly nanodrugs or nanomedicines)[4] already on the market or under development. The global nanomedicine market was reported to be worth $72.8 billion in 2011, $138 billion in 2016 and is predicted to be worth $350 billion by 2025 [9]. Yet, despite all of this R&D in nanomedicine, federal funding related to the research and educational programs on ethical issues has clearly lagged behind. It is critical that ethical, social, and regulatory aspects of nanomedicine be proactively addressed to minimize public backlash similar to that seen with other promising technologies, most notably, genetically modified foods and stem cell research. The public should be properly educated regarding the benefits and risks of nanomedicine. Transparency is essential for greater acceptance and support, and is critical for commercialization.

---

[4]A nanodrug is defined as *"a formulation, often colloidal, containing (1) therapeutic particles (nanoparticles) ranging in size from 1–1,000 nm; and (2) carrier(s) that is/are themselves the therapeutic (i.e., a conventional therapeutic agent is absent), or the therapeutic is directly coupled (functionalized, solubilized, entrapped, coated, etc.) to the carrier(s)."* See: Bawa, R. (2018). Current immune aspects of biologics and nanodrugs: An overview. In: Bawa, R., Szebeni, J., Webster, T. J., Audette, G. F., eds. *Immune Aspects of Biopharmaceuticals and Nanomedicines*, Pan Stanford Publishing, Singapore, chapter 1, pp. 1–82.

Given this backdrop, nanomedicine could be poised to add a profound and complex set of ethical questions for health care professionals. Once nano-based interventions are tested in clinical trials and given US Food and Drug Administration (FDA) approval, it becomes the domain of health care practitioners to use it for the improvement of human health and populations. However, for most physicians and patients, nanomedicine is still an entirely new arena for preventive and diagnostic interventions and curative therapies that will require continuing education, and a heightened awareness of the risks and benefits. We will focus primarily on issues that are likely to emerge once nanomedicine moves out of the preclinical and clinical stages of research and development. In other words, our discussions here will be limited to nanomedicine products as they enter the market and find medical applications in diagnosis, prevention, and treatment.

Nanomedicine raises fundamental questions, such as what is it to be human, how human disease is defined, and how treating disease is approached. Just as with the era of genetics and molecular biology, physicians will have to reconceptualize how they think about the diseases they treat, the means they have to treat them, and the meaning of the phrase "do no harm."

Yet, nanomedicine is not a medical specialty or a single class of medical interventions that can easily be analyzed from an ethical perspective. As discussed earlier, it includes a wide range of technologies that can be applied to medical devices, materials, procedures, and treatment modalities. The simplest way to distinguish categories of nanomedical interventions is to differentiate "diagnostic nanomedicine" from "therapeutic nanomedicine." Diagnostic nanomedicine can include a wide range of interventions, from monitoring changes in blood chemistry, alterations in DNA, or tissue aberrations. It has been postulated that in the near future, clinicians and health care workers at the bedside or in the clinic will be capable of scanning a patient's entire genome in a few minutes and draw remarkably accurate conclusions pertaining to disease potential and corresponding therapies. Therapeutic nanomedicine includes a wide range of interventions—from nanopharmacology to nanobased medical devices, such as nanobots[5]

---

[5]Certain therapeutics submitted to the FDA for regulatory approval are combination products, which consist of two or more regulated components (drug, biologic or device) that are physically, chemically or otherwise combined/mixed to produce a single entity. In such cases, the FDA determines the "primary mode of action (PMOA)"

or nanodrugs to nanomaterials used for bone grafts or other body implants.

Just as different ethical issues exist for preventive medicine versus curative or therapeutic medicine, there exist very different kinds of ethical issues that arise out of diagnostic nanomedicine versus therapeutic nanomedicine. Interventions based on nanotechnologies likely will resurrect old questions about human enhancement, human dignity, and justice that have been asked many times before in the context of pharmaceutic research, stem cell research, artificial life, and gene therapy.

Much of what was discussed or "hyped" in the past two decades as the future of nanomedicine, however, has yet to occur.[6] Therefore, it is difficult for ethicists to predict in advance of the arrival of actual technologies what kinds of issues might arise out of nanomedicine. Yet, on the basis of other kinds of biomedical technologies that have affected health care, it is possible to conjecture what some of the perennial ethical issues and novel ethical problems will be. Therefore, this chapter outlines a range of potential ethical issues for preventive and therapeutic nanomedicine that may occur as these nanotechnologies move from the laboratory to the clinic. Specific focus is on the ethical question of enhancement versus therapy, the risk for and benefits

---

of the product, which is defined as "the single mode of action of a combination product that provides the most important therapeutic action." This process is frequently imprecise because it is not always possible to elucidate a combination product's PMOA. In future, novel "multifunctional/multicomponent" nanobots will be engineered that incorporate both a drug and diagnostic (so called "theranostic"). As these combination products seek regulatory approval, they are sure to present additional challenges for the FDA because the agency's current PMOA regulatory paradigm may prove ineffective.

[6]Nanomedicine's potential benefits are often overstated or inferred to be very close to application when clear bottlenecks to commercial translation exist. Academia, startups and companies are all guilty as they continue to offer inflated promises or exaggerate potential downstream applications based on early-stage preclinical discoveries. Such "spin" or "fake medical news" does great disservice to all stakeholders; it not only pollutes the medical literature but quashes public support for nanomedicine translational activities. This issue is quite serious and often emanates from eminent academic labs perched at distinguished universities or from established industry players. Another common phenomenon observed by us is that many have desperately tagged or thrown around the "nano" prefix to suit their own motives, whether it is for research funding, patent approval, raising venture capital, or running for office.

of nanotechnologies in health care, changing understanding of human disease, and privacy and confidentiality.

32.2   Understanding Human Disease

Diagnostic nanotechnologies eventually will be able to detect and characterize individual cells, subtle molecular changes in DNA, and even minor changes in blood chemistry—scenarios that will likely cause pause and reconsideration of what it means to be a "healthy person" versus a "person who has a disease." In a "nanoworld," we might have to reconsider how to diagnose someone who has, say, cancer. Is the presence of a genetic mutation known to have a predisposition for causing cancer in a single cell a diagnosis? Or is it simply a risk factor? How many cells from the body must be of a cancerous nature for it to be defined as cancer? 1? 50? 1000? The answers to these questions are difficult because no one currently knows exactly how to define, diagnose, or detect disease with this level of sensitivity. Eventually, disease may be able to be detected in this way, but it is important to remember that the development of such diagnostic technologies will require reconceptualizing understanding of disease. Obviously, this will have a significant impact on health care professionals and patients.

The key is that if the slightest abnormality can be discovered, one must ask whether such information will have clinical relevance from a diagnostic, therapeutic, or prognostic point of view. If such knowledge does have clinical relevance, then it seems reasonable to develop technologies, assays, or mechanisms that could detect diseases at their earliest stages with the hope that this early detection would result in fewer side effects, less aggressive treatments, superior patient compliance, and better survival rates.

There may be some cases, however, where more information is simply too much information [10]. Such heightened awareness simply could result in anxious patients and worried family members. One must, therefore, think carefully about which diseases and conditions it would be appropriate to apply such nanotechnologies to so that those interventions are helpful, rather than creating a burden, unnecessary concern, or risk for patients and others. Therefore, the balance of information processed and disseminated versus benefit to society and individual health is a significant consideration for the ethics of nanotech-based diagnostic technologies [10].

32.3   Enhancement versus Therapy

A related distinction for judging the morality of a medical procedure or treatment is whether or not it is regarded as therapeutic or enhancing—a subjective determination that is coupled with the determination of whether or not such action results in a normal or abnormal individual. A little analysis, however, reveals these distinctions to be unavailing because both enhancement and therapy are based on the relative concept of "normal" [11]. Most novel medical technologies that are employed for diagnosis, prevention, or treatment of diseases can also be used to enhance the function of the human body or mind. The traditional distinction between therapy and enhancement lies in the fact that therapy is concerned with maintaining, repairing, or restoring bodily parts or functions that a patient previously had or used. Enhancement, however, is concerned with the creation or improvement of bodily parts or functions that were absent, undamaged, or previously malfunctioning. Using this subjective distinction, the implantation of a nanoscale device that emulates the function of a congenitally absent organ paradoxically would be enhancing rather than therapeutic.

As to this question, a frank prohibition pragmatically is unworkable. There are simply too many potential benefits that nanoscale medical devices offer and policing their use will only be effective when society has reliable methods to detect violations. Rather, the practice of nanomedicine must be governed by a nanomedical ethic that maps the classical principles onto a transhuman and posthuman reality. Of these, the principle of "justice" in access to nanomedical procedures and entitlement to nanomedical treatment likely will be the most contentious. In this context, issues relating to unfair competition, socioeconomic inequality, discrimination, and bias will arise and need to be addressed. At the level of civilization, a morality must be crafted that honors an unprecedented expansion in the meaning of human being and militates against any eugenics agenda.

32.4 Risk versus Benefit

Another important concern for nanomedicine is the need to balance the potentially significant benefits of nanomedical interventions with their potential risks. In the area of therapeutic nanomedicine, for example, it is clear that nanotechnologies will continue to allow chemical compounds, therapeutics, or drugs to be more bioavailable, less toxic, and targeted to specific tissues and even sub-cellular structures. Therefore, these compounds will be needed at lower doses and have fewer side effects in the patient. One likely risk of nanomedicine, however, is that these drugs will receive FDA approval and be on the market long before the long-term risks are conclusive. Nanomedicines have the potential to cross the blood–brain barrier or enter cells easily; therefore, it is a concern that the retention of these molecules in the body may cause long-term or unintentional harm to healthy tissues. Because long-term follow-up data exist for only a handful of nanomedicines, it is important that patients be informed that these drugs may present long-term consequences. Although this is not altogether different from the long-term risks associated with exposure to chemotherapeutic or radiologic agents, it is an important risk factor that must be disclosed to patients taking nanomedicines or any kind of intervention involving nanoparticles or nanomaterials. A similar argument could be extended to nano-nutraceuticals or nano-cosmetics, categories whose definitions may overlap with nanomedicines. In fact, both nano-nutraceuticals or nano-cosmetics may present greater risk in some cases given that neither are subjected to any FDA premarket regulatory approval prior to commercialization.

32.5 Privacy and Confidentiality

Another important ethical issue relates to the protection and maintenance of health information in the era of nanomedicine. Nanotechnologies will make possible the collection of an enormous amount of individual cellular/subcellular level surveillance data of the human body. Nanomedical technology is expected to miniaturize implantable devices so that they function at the subcellular or synaptic level with the ability to monitor or collect data regarding cellular activities and biochemical events within organs, tissues, or individual cells. One application of this technology would be to include a means by which that information could be transmitted remotely.

If and when such technologies are made possible via nanotechnology, a key ethical question arises: Can the health information infrastructure handle, collect, process, and analyze real-time ongoing electronic health data in a secure manner? With healthcare institutions slow to adopt electronic medical record systems and accommodate increasingly large medical files across institutions and time periods, it is of concern that massive amounts of health information is being generated without efficient systems in place to effectively utilize it or that adequate security measures are in place. Clearly, ensuring privacy and confidentiality in such systems would be of utmost importance. Systems without adequate safeguards present serious ethical problems.

32.6 Future Perspectives

Given that nanomedicine is an emerging and evolving arena, it is difficult to precisely predict how ethical issues related to nanomedicine will evolve in the next decade. Nevertheless, ethical considerations will continue to play a significant role in the development and use of nanotechnologic interventions in medicine and healthcare. Initially, some of the important ethical concerns have focused on risk assessment and environmental management. However, in the future, novel ethical issues and unforeseen dilemmas are likely to arise as the field advances further and intercepts other areas of biomedical research, including artificial intelligence, genomics, precision medicine, bioinformatics, and brain science. As with other biomedical and life science advances before it, nanomedicine will face significant challenges as it moves from the lab to the clinic. Along the way, ethical questions regarding social justice, privacy, confidentiality, long-term risks and benefits, and human enhancement are certain to arise. Health care providers must be ready to answer such ethical questions for themselves and be able to address those questions for their patients. Ultimately, it seems likely that nanomedicine will usher in a new area in health care where pharmaceuticals will be more effective, specific, targeted, and less toxic, where disease monitoring will be done on a highly sensitive and specific level, where injections, surgical procedures and a host of other interventions will be made less painful, less toxic, and with fewer side effects than their current versions. It is important to ensure, however, that these advances in medical care do not come at the expense of fairness, safety, transparency, or basic understanding of what it means to be a healthy human being. Ultimately, public and political interest for regulations needs to be carefully balanced with the interests of scientists and businesses for uninhibited science and technological progress. Hype or excitement about nanomedicine should not obscure its important ethical and societal implications. Nanomedicine's future appears brightest if it can be ensured that it also will be a future where such ethical issues are timely, accurately, and transparently addressed via involvement and cooperation of all stakeholders. This will also ensure that the public's desire for novel nanomedical products, investment from venture communities, and big pharma's interest in nanomedicine are not quenched.

Disclosure and Conflicts of Interest

The views in this chapter are those of the authors and do not necessarily those of the organizations they are affiliated with. There is no conflict of interest reported by the authors. This work was supported by Bawa Biotech LLC, Ashburn, VA, USA. Dr. Bawa is a scientific advisor to Teva Pharmaceutical Industries, Ltd., Israel. No writing assistance was utilized in the production of this chapter and no payment was received for its preparation. This chapter is based on the authors' publication titled "The ethical dimensions of nanomedicine" that was published in 2007 in *The Medical Clinics of North America* **91**(5), 881–887. It appears here with kind permission of Elsevier, Netherlands.

Corresponding Author

Dr. Raj Bawa
Bawa Biotech LLC
21005 Starflower Way
Ashburn, VA 20147, USA
Email: bawa@bawabiotech.com

References

1. Bawa, R., Audette, G. F. Rubinstein, I., eds. (2016). *Handbook of Clinical Nanomedicine: Nanoparticles, Imaging, Therapy, and Clinical Applications.* Pan Stanford Publishing, Singapore.

2. Bawa, R., ed.; Audette, G. F., Reese, B. E., asst. eds. (2016). *Handbook of Clinical Nanomedicine: Law, Business, Regulation, Safety and Risk.* Pan Stanford Publishing, Singapore.

3. Bawa, R., Szebeni, J., Webster, T., Audette, G. F. eds. (2018). *Immune Aspects of Biopharmaceuticals and Nanomedicines.* Pan Stanford Publishing, Singapore.

4. Mousa, S. A., Bawa, R., Audette, G. F., eds. (2020). *The Road from Nanomedicine to Precision Medicine.* Pan Stanford Publishing, Singapore.

5. Bawa, R. (2017). A practical guide to translating nanomedical products. In: Cornier, J., et al., eds. *Pharmaceutical Nanotechnology: Innovation and Production*, 1st ed., Wiley-VCH Verlag, chapter 28, pp. 663–695.

6. Bawa, R. (2016). Small is beautiful. In: Bawa, R., Audette, G., Rubinstein, I., eds. *Handbook of Clinical Nanomedicine: Nanoparticles, Imaging, Therapy, and Clinical Applications,* Pan Stanford Publishing, Singapore, pp. xxxvii–xliv.

7. Bawa, R., Bawa, S. R., Mehra, R. (2016). The translational challenge in medicine at the nanoscale. In: Bawa, R., ed.; Audette, G. F., Reese, B. E., asst. eds. *Handbook of Clinical Nanomedicine: Law, Business, Regulation, Safety and Risk,* Pan Stanford Publishing, Singapore, chapter 58, pp. 1291–1346.

8. Tinkle, S., McNeil, S. E., Mühlebach, S., Bawa, R., Borchard, G., Barenholz, Y., Tamarkin, L., Desai, N. (2014). Nanomedicines: Addressing the scientific and regulatory gap. *Ann. N. Y. Acad. Sci.,* **1313**, 35–56.

9. Nanomedicine market size worth $350.8 billion by 2025. Grand View Research, Inc. Available at: https://www.grandviewresearch.com/press-release/global-nanomedicine-market (accessed on August 20, 2019).

10. Bawa, R., Johnson, S. (2008). Emerging issues in nanomedicine and ethics. In: Allhoff, F., Lin, P. eds., *Nanotechnology & Society: Current and Emerging Ethical Issues,* Springer, Dordrecht, pp. 207–223.

11. Best, R., Khushf, G. (2006). The social conditions for nanomedicine: Disruption, systems, and lock-in. *J. Law Med. Ethics,* **34**(4), 733–740.

Chapter 33

Amphiphilic Nanocarrier Systems for Curcumin Delivery in Neurodegenerative Disorders

Miora Rakotoarisoa, MS, and Angelina Angelova, PhD

Institut Galien Paris-Sud CNRS UMR 8612,
LabEx LERMIT, Université Paris-Sud, Université Paris-Saclay, France

Keywords: curcumin, amphiphilic nanoparticles, self-assembled nanoparticles, solid lipid nanoparticles, liposomes, liquid crystalline carriers, nanomedicine, drug delivery, neurodegenerative diseases, Alzheimer's disease, Parkinson's disease, Huntington disease, amyotrophic lateral sclerosis, oxidative stress, neuroprotection, blood–brain barrier, SHSY-5Y cells

33.1 Introduction

Neurodegenerative diseases (Alzheimer's disease (AD), Parkinson's disease (PD), Huntington disease (HD), and amyotrophic lateral sclerosis (ALS)) are disabling chronic disorders characterized by

The Road from Nanomedicine to Precision Medicine
Edited by Shaker A. Mousa, Raj Bawa, and Gerald F. Audette
Copyright © 2020 Jenny Stanford Publishing Pte. Ltd.
ISBN 978-981-4800-59-4 (Hardcover), 978-0-429-29501-0 (eBook)
www.jennystanford.com

the progressive loss of neurons in different areas of the central nervous system. Neuronal cell death leads to cognitive, behavioral, sensory, and motor dysfunctions [1–13]. Currently, age-related neuronal diseases have higher incidences because of increasing life expectancies. Neurodegenerative disorders are caused by multiple factors, such as the accumulation of misfolded proteins, the depletion of endogenous antioxidant enzyme activity, mitochondrial dysfunction, and the deficiency of neurotrophin brain-derived neurotrophic factor (BDNF), neuro-inflammation, as well as various genetic mutations [14–30].

In recent years, several studies have shown that curcumin is a safe natural compound which may prevent the deleterious effects of risk factors causing brain damage as well as slowing down the progressive neuronal loss via different pathways [26–50]. However, clinical studies performed with AD patients with various degrees of progression have reported poor results on the AD symptoms following curcumin treatment [51–54]. This did not allow firm conclusions about the therapeutic or neuroprotective potential of curcumin to be drawn. The obstacles for curcumin utilization as a drug originate from its limited water solubility, poor physicochemical stability, high-grade metabolism, and low plasma concentrations [36, 53–55]. The development of nanoparticulate delivery systems for curcumin has attracted scientific interest in order to improve its bioavailability and stability as a drug compound [56–65]. Curcumin administration to neurodegenerative disease models by nanoparticles has been realized using liposomes, solid lipid nanoparticles, and polymeric particles. Delivery by other carriers such as amphiphilic proteins, e.g., casein, is also possible, but has not been examined as a means of transporting curcumin across the BBB towards neuro-regeneration.

In this review, we briefly summarize the *in vitro* and the *in vivo* evaluations of curcumin, which are linked to multiple risk factors and the multi-target mechanisms of neurodegenerative diseases, and discuss the reported clinical investigations of varying efficacy in humans. Then, we highlight the variety of amphiphilic curcumin-loaded nanocarriers including liposomes, liquid crystalline nanoparticles (cubosomes, hexosomes, and spongosomes), solid lipid nanoparticles, micelles, and polymeric nanoparticles as potential nanomedicine formulations in regeneration therapies of the major neurological disorders.

33.2 Risk Factors for Neurodegenerative Disorders

Alzheimer's disease (AD) is the most common cause of dementia. It currently affects about 10% of the world's population over 60–65 years of age, and about 50% over 85 years of age. More than 30 million people may expect to be affected by AD during the next 20 years due to the increasing lifespan of the world population [1, 2]. Major pathological features of AD include the accumulation of extracellular amyloid plaques and fibrils, intracellular neurofibrillary tangles, and disruption of the cholinergic transmission, including reduced acetylcholine levels in the basal forebrain (Table 33.1). The most common symptom is the short-term memory loss, i.e., difficulty in remembering recent events [2–5]. Other symptoms include disorientation, mood, language, and behavioral issues, and loss of motivation, depending on the progression of the disease. The treatments of AD have employed acetylcholinesterase inhibitors (tacrine, rivastigmine, galantamine, and donepezil) to overcome the decrease of the ACh levels as a result of the death of cholinergic neurons. The NMDA receptor antagonist (memantine) acts by inhibiting the overstimulation by glutamate, which can cause cell death. Atypical antipsychotics have modest efficacy in reducing the aggression and psychosis of AD patients. These medications provide little benefit, and provoke various adverse effects [6, 7].

The second most common disorder, Parkinson's disease (PD), affects more than 1% of the population over 60 years of age and 5% over 85. PD is characterized by progressive impairments in locomotive ability such as tremor, rigidity, and bradykinesia. These symptoms are attributed to the loss of dopaminergic neurons in the substantia nigra and the formation of Lewy bodies in the brain [8, 9]. Treatments are symptomatic and aim at boosting the depleted levels of dopamine (Table 33.1). The most used drug is levodopa. Dopamine agonists are used when the treatment by levodopa becomes less efficient. The inhibitors of MAO-B and COMT (safinamide, selegiline, rasagiline, and tolcapone) are used to inhibit the activity of the enzymes which degrade dopamine. These medications become less effective as the neurons are continuously lost during disease progression. At the same time, they produce complications marked by the involuntary movements of the patients [8, 9].

Huntington disease (HD) is a rare disease which affects about 1/10,000 people (usually between 30 to 50 years of age) in the United States and 1/18,000 people in Europe. It is a poly-glutamine (PolyQ) autosomal genetic disorder characterized by impairments of cognitive, psychiatric, and motor functions [10]. The hallmark of the HD pathology is the abnormal accumulation of misfolded mutated huntingtin protein (mHTT) as intracellular aggregates. They cause selective neuronal loss, primarily in the cortex and in the medium spiny neurons of striatum. Symptoms develop from a general lack of coordination to apparent uncoordinated, jerky body movements [11]. The physical abilities of the patients gradually worsen until coordinated movement becomes difficult. There is no effective cure available to HD (Table 33.1). The only approved medication, tetrabenazine, and other tested drugs (neuroleptics and antipsychotics) help to reduce chorea and psychiatric symptoms.

Amyotrophic lateral sclerosis (ALS) is a severe debilitating disease caused by motors neurons degeneration in the brain and the spinal cord. It is generally characterized by progressive paralysis starting at the limbs and ultimately leading to death caused by respiratory failure within 3 to 5 years after the onset of the symptoms. There is no cure for ALS (Table 33.1). The approved medication, riluzole, may extend life by just a few months [12, 13].

The pathological characteristics, genetic factors, clinical symptoms, and actual medications of these diseases are summarized in Table 33.1. It should be emphasized that the existing therapeutic approaches do not exert disease-modifying effects on the neurodegeneration. The associated economic and societal challenges lead to a growing public health burden.

Table 33.1 Pathological characteristics, genetic factors and clinical symptoms of Alzheimer's disease (AD), Parkinson's disease (PD), Huntington's disease (HD), and amyotrophic lateral sclerosis (ALS) [1–30]

Diseases	Characteristics	Genetics factors	Symptoms	Actual treatments
AD	Senile plaques from extracellular amyloid-Aβ accumulation, Intracellular neurofibrillary tangles, Tau protein aggregation, Irreversible neuronal loss, Brain atrophy	Inherited form (70% of patients): mutations of APP, PSEN1 or PSEN2; Sporadic form (30%): presence of ApoE4 allele in the ApoE gene	Progressive memory loss, Decision judgment loss, Autonomy loss	Anticholinergics (tacrine, rivastigmine, galantamine and donepezil), Memantine, Antipsychotics, NSAIDs
PD	α-Synucleinopathy, Presence of Lewy bodies, Degeneration of dopaminergic neurons in the substance nigra of the brain, Dopamine deficiency	Gene mutations: α-synuclein SNCA, Parkin PRKN, PARK7, PINK1, LRRK2, GBA, DJ-1, VPS35, EIF4G1, DNAJC13 and CHCHD2	Hypokinesia, Bradykinesia, Rigidity, Postural instability, Neuropsychiatric disturbances	Levodopa, Dopamine agonists, MAO-B inhibitors, COMT inhibitors, Anticholinergics
HD	Accumulation of mutant Huntingtin protein in the brain	Expansion of CAG trinucleotide in Huntingtin gene (HTT)	Chorea, Cognitive and neuropsychiatric disorders	Tetrabenazine, Neuroleptics, Antipsychotics
ALS	Progressive degeneration of motor neurons	Sporadic form: 90% of patients; Inherited form: 10% Mutations of SOD1, TARDBP, FUS, UBQLN2, OPTN, and C9ORF72 genes	Spasms, Muscle atrophy, Squelettal muscle paralysis, Cognitive or behavioral dysfunction	Riluzole

Although the etiology and the pathological mechanism of the neurodegenerative diseases are not completely understood, it has been established that the progressive loss of neurons results from the combination of multiple factors (Fig. 33.1). First, genetic factors are involved in the appearance of misfolded amyloid-Aβ protein and other misfolded mutant forms like hyperphosphorylated Tau (p-Tau) and Huntingtin proteins [3–5, 10, 14]. All these mutated proteins aggregate and form deposits. The resulting aggregates can be toxic and can affect the intracellular organelles such as mitochondria [14, 21, 25, 27]. The disruption of the mitochondrial membrane causes neuronal cell death [25, 27].

Second, neurotrophic factors deficiency has been reported in the severe neurodegenerative disorders 33]. Neurotrophins regulate the neuronal survival, differentiation, growth, and regeneration, as well as the synaptic plasticity. Studies have shown that the levels of brain derived neurotrophic factor (BDNF) and its tropomyosin kinase B (TrkB) receptor are decreased in the hippocampus and the cerebral cortex at the beginning of the Alzheimer's disease [11]. In addition, the administration of BDNF mimetics into transgenic mouse models of AD has enhanced learning and memory capacities [31].

Third, oxidative stress is the most common feature of neurodegenerative diseases [15–20]. Reactive Oxygen Species (ROS) such as superoxide anions, hydroxyl radicals, and hydrogen peroxide (H_2O_2) are produced by the mitochondrial transport chain, the endoplasmic reticulum, the Krebs cycle, and the plasma membrane involving the superoxide-generating NADPH oxidase (NOX) macromolecular complex [17]. Oxidative stress occurs under environmental factors and when the generation of ROS exceeds the natural antioxidant defenses of the cell (promoted by superoxide dismutase, catalase, glutathione peroxidase, carotenoids, and vitamins E or C) [15–20]. ROS accumulation attacks proteins, nucleic acids, and membrane lipids, and thus, causes impairments of the neuronal cell functions and integrity [18–20]. Mitochondrial lesions are also mediated by ROS. This leads to the alteration of the neuronal cell bioenergetics, the disruption of the calcium (Ca^{2+}) homeostasis, or the activation of the mitochondrial permeability transition pore (mPTP). Thus, a vicious cycle is formed (Fig. 33.1), which amplifies the cellular dysfunction and triggers neurodegeneration [17–26].

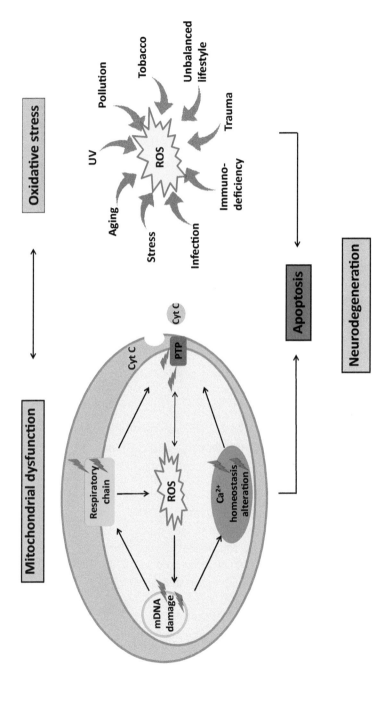

Figure 33.1 Neurodegeneration is triggered and boosted by a vicious circle involving neurotoxic protein accumulation, oxidative stress, mitochondrial damage, DNA damage, and impairment of the calcium (Ca^{2+}) homeostasis, neurotrophin deficiency, neuroinflammation, genetic, and environmental factors.

Fourth, neuro-inflammation is a crucial factor that favors neurodegenerative disease development. Several inflammatory markers (such as chemokines, cytokines, or proteins in the acute phase) are upregulated and cause inflammation. In fact, elevated levels of the inflammatory markers have been found during the progression of the neurodegenerative diseases [27, 28].

33.3 Curcumin Potential for Neuroprotection against Neurodegenerative Diseases

Curcumin is a hydrophobic polyphenolic substance (Fig. 33.2) produced in the root of the plant *Curcuma Longa* L. This antioxidant compound is extensively marketed worldwide as a nutraceutical in various preparations, because it has a very safe nutraceutical profile with low side effects. Curcumin has been reported to be well tolerated at doses up to 8 g per day over short periods in humans [32]. Research on the pharmacological activities of curcumin has attracted strong attention in relation to its multiple actions of therapeutic interest, e.g., the anti-inflammatory, antioxidant, antiviral, antibacterial, antifungal, and antitumor activities [36]. These activities appear to be dose-dependent [33].

33.3.1 *In vitro* and *in vivo* Studies of Curcumin Properties in Neurodegenerative Disease Models

The neuroprotective potential of curcumin (Fig. 33.2) and its antioxidant, anti-inflammatory, and amyloid Aβ binding properties have been highlighted in *in vitro* and *in vivo* investigations of different neurodegenerative disease models [33–42]. Curcumin has been found to increase the levels of glutathione (GSH) and malondialdehyde (MDA), as well as the antioxidant enzyme [superoxide dismutase (SOD), glutathione peroxidase (GPx), glutathione reductase (GR), and catalase (CAT)] activities in the rat brain, thus preventing the stress-induced oxidative damage of brain [37, 39]. The anti-inflammatory properties of curcumin have been characterized by the inhibition of the inflammatory

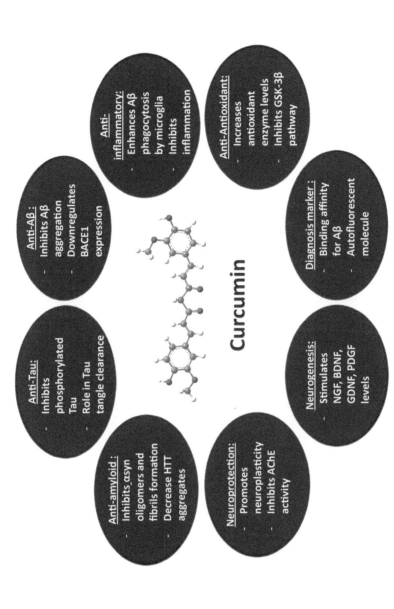

Figure 33.2 Summary of the curcumin activities and suggested mechanisms of action, which can be exploited for treatment of neurodegenerative diseases.

chemokines, by increased levels of the anti-inflammatory cytokines, and by enhanced expression levels of induced nitric oxide synthase (iNOS) and the transcription factor NF-κB [39]. Curcumin has been shown to prevent the fibrillation of α-synuclein at the earliest stage of the aggregation process, as well as the fibrillation of the globular protein hen egg-white lysozyme (HEWL) [40]. Both proteins are known to form amyloid-like fibrils. These results have suggested that curcumin might be a potential therapeutic agent for preventing protein aggregation in Alzheimer's, and Parkinson's diseases [40]. Recent *in vitro* and *in vivo* investigations of curcumin's activities in neurodegenerative disease models [41–50] are summarized in Table 33.2.

33.3.2 Clinical Trials and Curcumin Limits

A serious obstacle to the pharmaceutical application of curcumin has arisen from its limited water-solubility and low bioavailability. In addition, this compound is chemically instable, which may cause a loss of biological activities. The failure of free curcumin in clinical trials is likely due to its limited bioavailability. For instance, curcumin has been delivered in doses between 1 and 4 g/per day as capsules or as powder mixed with food in trials for treatment of Alzheimer's disease patients. The performed 6-month treatment study found no differences in the Aβ-amyloid levels between the treatment groups or in the Mini Mental State Examination (MMSE) scores [51]. Similarly, oral curcumin in a 24-week, randomized, double blind, and placebo-controlled study for AD treatment has shown no detectable differences in the measured biomarkers from the different treatment groups [52]. A clinical study with three single cases of patients receiving curcumin (100 mg/day) reported that only one patient increased his MMSE score from 12/30 to 17/30 after 12 weeks of treatment (improved calculation, concentration, transcription of the figure, and spontaneous writing). Two of the patients were on donepezil treatment before starting the curcumin trial [53]. Based on all performed trials with AD patients, it was, however, difficult to conclude if curcumin has positive effects on the AD symptoms [54].

Table 33.2 Recently reported curcumin (CU) activities in *in vitro* and *in vivo* models of neurodegenerative diseases [41–50]

Disease	Model/administration route	Mechanism	Outcomes
AD	*In vitro*: human neuroblastoma SH-SY5Y and IMR-32 cells	Enhancement of the expression of DNA repair enzymes (APE1, pol β, and PARP1[1]) to halt the oxidative DNA base damage via base excision repair (BER) pathway; Activation of the antioxidant response element (ARE) via Nrf2 upregulation	Revitalization of the neuronal cells from Aβ[2]-induced oxidative stress [41]
AD	*In vitro*: mouse hippocampal clone neuronal cell line HT-22 cells treated with Aβ 1-42, *In vivo*: mice with APP/PS1 transgenes	Decrease of the autophagosomes number; Increase of the lysosomal Ca^{2+} regulation of PI(3,5)P2 and Transient Receptor Potential Mucolipin-1 Expression (TRPME)	Neuronal cell growth, Protective role of CU on memory and cognition impairments [42]
AD	*In vivo*: rat, oral supplementation	Increase of GPx,[3] CAT,[4] GSH[5] activities, and Ach[6] levels	Improving memory and cognitive abilities [43]
PD	*In vivo*: Drosophila model of PD with dUCH[7] knockdown	Effects on dUCH[7] knockdown, a homolog of human UCH-L1	Decrease of ROS levels, Improved locomotive abilities, Reduction of dopaminergic neurons degeneration [44]
PD	*In vivo*: male Sprague-Dawley rats injured by 6-OHDA[8] in the left striatum	Activation of the Wnt/β-catenin signaling pathway, Higher Wnt3a and β-catenin mRNA and protein expressions, c-myc and cyclin D1 mRNA expression, enhanced SOD[9] and GPx[3] contents, decreased MDA[10] content and elevated mitochondrial membrane potential	Protective effect of CU against oxidative stress-induced injury, Enhanced viability, survival, and adhesion, attenuated apoptosis of deutocerebrum primary cells [45]

(Continued)

Table 33.2 *(Continued)*

Disease	Model/administration route	Mechanism	Outcomes
PD	*In vivo*: MPTP[11] mice, intranasal mode of administration of CU (mucoadhesive system)	Increase of dopamine concentration in brain, which improves muscular coordination and gross behavioral activities of the test animal, Significant reduction of the MPTP[11]-mediated dopamine depletion	Improvement in motor performance, Symptomatic neuroprotection against MPTP-induced neurodegeneration in the striatum [46]
HD	*In vivo*: CAG140 mice, a knock-in (KI) mouse model of HD	Decreased huntingtin aggregates, increased striatal DARPP-32 and D1 receptor mRNAs	Partial improvement of transcriptional deficits, partial behavioral improvement [47]
Diazepam-induced cognitive impairment	*In vivo*: diazepam-treated rats, oral supplementation	Downregulation of the extracellular signal-regulated kinase (ERK 1/2)/nuclear transcription factor-(NF-) κB/pNF-κB pathway in the hippocampus and the iNOS[12] expression in the hippocampus and frontal cortex	Improvement of the cognitive performance, Decrease of blood and brain oxidative stress levels [48]
Alcohol-induced neurodegeneration	*In vivo*: rat, oral supplementation	Decrease of the reduced form of GSH,[5] SOD,[9] GPx,[3] GR,[13] change in the Bcl-2 levels, Activation of the CREB-BDNF signaling pathway	Neuroprotection against alcohol-induced oxidative stress, apoptosis and inflammation [49]
Nicotine-induced neurodegeneration	*In vivo*: rat, oral supplementation	Activation of the CREB-BDNF signaling pathway	Neuroprotection against nicotine-induced inflammation, apoptosis and oxidative stress, Reduction of the motor activity disturbances [50]

[1]Poly [ADP-ribose] polymerase 1; [2]Aβ-amyloid; [3]Glutathione Peroxidase; [4]Catalase; [5]Glutathione; [6]Acetylcholine; [7]Ubiquitin carboxy-terminal hydrolase; [8]6-Hydroxydopamine; [9]Superoxide dismutase; [10]Malondialdehyde; [11]1-Methyl-4-phenyl-1,2,3,6-tetrahydropyridine; [12]Induced nitric oxide synthase; [13]Glutathione reductase.

In fact, the major fraction (35–89%) of orally administered curcumin can be lost due to its low bioavailability. The intestinal mucosa and mucus form a physical barrier to curcumin adsorption. The drug cannot reach the circulation in a bioactive form as it undergoes reduction and metabolism/conjugation in the liver. Reductases enzymatically reduce curcumin to dihydrocurcumin, tetrahydrocurcumin, and hexahydrocurcumin. Furthermore, curcumin may be conjugated to sulfates and glucuronides [55–57]. Thus, most of the circulating curcumin is in a conjugated form.

The necessity of the development of a delivery system for the protection of curcumin from rapid metabolism and for the improvement of its bioavailability has become evident [58]. A randomized, double-blind, placebo-controlled clinical trial examined the acute administration (effects 1 h and 3 h after a single dose application), chronic (4 weeks) administration, and acute-on-chronic (1 h or 3 h after a single dose followed by a chronic treatment) effects of solid-lipid-nanoparticle (SLNP) loaded by curcumin. The results of the SLNP formulation of curcumin (400 mg in capsules Longvida®) on cognitive function, mood, and blood biomarkers were obtained with 60 healthy adults (aged 60–85). SLNP-loaded curcumin significantly improved the performance in sustained attention and working memory tasks one hour after its administration (as compared to placebo). Working memory and mood (general fatigue and change in the calmness state, contentedness, and fatigue induced by psychological stress) were essentially improved following chronic treatment. A significant acute-on-chronic treatment effect on alertness and contentedness was also observed [59].

33.4 Nanocarrier-Mediated Curcumin Delivery

Nanotechnology for nanomedicine development employs functional materials with appropriate nanoscale organization that can interact with biological systems and induce desired physiological responses while minimizing undesirable side effects [60]. Nanotechnology-based delivery systems can influence drug capacity to cross the biological barriers (e.g., the BBB) and reach the targeted brain regions [58–61]. Therefore, nanocarriers are promising for the development of personalized medicines for the treatment of neurological disorders [62–67].

Lipid-based nanoparticles, including solid lipid nanoparticles (SLNPs), nanostructured lipid carriers (NLC), liposomes and liquid crystalline nanocarriers (LCN), as well as polymer-based nanoparticles (Fig. 33.3), have been developed to overcome the poor solubility, stability, and bioavailability of curcumin, and to promote its utilization as a drug in disease treatments [68–124].

Lipid-based nanoparticles have the advantage of being the least toxic carriers for *in vivo* applications. The lipids used to prepare biocompatible and biodegradable nanoparticles are usually naturally occurring molecules with low acute and chronic toxicity. In the case of polymeric nanoparticles, the *in vivo* degradation of the polymer matrices might cause toxic effects [94]. The biocompatibility and the physicochemical diversity of lipids and their capacity to enhance the oral bioavailability of drugs have made this kind of nanocarriers very attractive systems for drug delivery. As a matter of fact, lipid-based formulations may positively influence drug absorption in several ways, e.g., by influencing the solubilization properties, preventing the drug precipitation upon intestinal dilution, increasing the membrane permeability, inhibiting the efflux transporters, reducing the CYP enzymes, or enhancing the lymphatic transport [94, 122].

Among the lipid-based nanoparticles, SLNPs have been intensively developed because they combine the advantages of different carrier systems like liposomes and polymeric particles. Similar to liposomes, SLNPs are composed of physiologically biocompatible excipients (lipids and fatty acids). In the same way to polymeric NPs, their solid matrix core can efficiently protect the loaded active pharmaceutical ingredient against chemical degradation under the harsh conditions of biological milieux. Therefore, SLNPs provide controlled release profiles of the encapsulated drugs [95].

In addition to the above advantages, liposomes can encapsulate and transport both lipophilic and hydrophilic drugs. They have a high degree of similarity to cell membranes in terms of lipid composition and organization, which facilitates the bioavailability of the pharmaceutical compounds [102]. Liquid crystalline

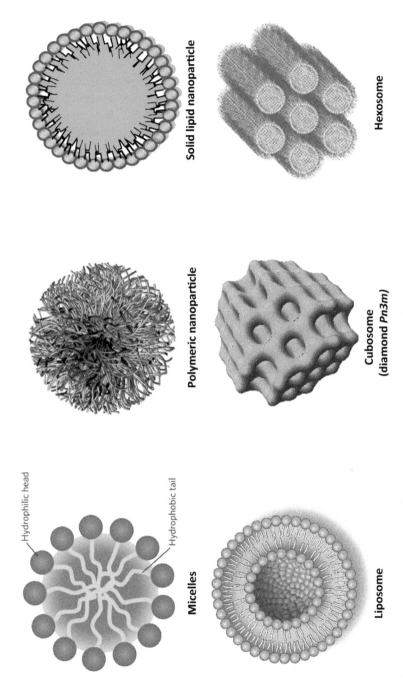

Figure 33.3 Schematic presentation of amphiphilic nanocarriers enabling the encapsulation and protection of hydrophobic and hydrophilic molecules of therapeutic significance.

nanocarriers (LCN) such as cubosomes and hexosomes (Fig. 33.3) involve multiple compartments for encapsulation of either lipophilic or hydrophilic drugs. They display structural advantages which enable high encapsulation efficacy for molecules of various sizes and hydrophilicity [113, 116]. LCNs are formed by self-assembly of lyotropic lipids such as unsaturated monoglycerides, phospholipids, glycolipids, and other amphiphilic molecules. For example, monoolein, which is a nontoxic, biodegradable, and biocompatible lipid, is classified as a GRAS (Generally Recognized As Safe), and ω-3 polyunsaturated fatty acids (n-3 PUFA) have been shown to be highly beneficial in various disease models of neurodegeneration [123, 124].

In the following section, we summarize recently reported works on curcumin delivery to *in vitro* and *in vivo* models of neurodegenerative diseases.

33.4.1 Curcumin Delivery by Polymeric Nanoparticles

Polymeric nanoparticles of a biocompatible and biodegradable nature are of essential interest as drug delivery nanocarriers. The release of the encapsulated drug can be modulated by altering the polymer composition and amphiphilicity. Poly(lactic-co-glycolic acid) (PLGA) is one of the most commonly used biodegradable synthetic polymers. It is a FDA-(US) and EMA-approved platform for the delivery of drugs to humans. PLGA-derived nanoparticles have been successfully used for the encapsulation of different hydrophobic compounds (such as curcumin) by nanoprecipitation or by single emulsion techniques [71]. Hydrophilic molecules can be encapsulated by means of double emulsions or by two-step nanoprecipitation methods. On the other site, polymeric micelles have been studied towards site-specific drug delivery [68–70].

Micelles are formed by amphiphilic macromolecules, which self-assemble into nano-sized (10–100 nm in diameter) core/shell structures in excess aqueous media (Fig. 33.3, top). The core–shell organization facilitates the incorporation of curcumin inside the hydrophobic core, while the water solubility of the nanocarriers is ensured by their hydrophilic corona [68]. Amphiphilic co-polymers self-assemble into micelles in aqueous solutions due to the hydrophobic interactions among their water-insoluble segments. Curcumin-loaded polymeric micelles have received attention due to various features (Table 33.3), like (i) the enhanced solubility of the drug; (ii) the sustained CU release profile; and (iii) the small size of the PEG-decorated carriers (<200 nm), which stabilizes them in biological fluids [64–66]. Micelles formed by the PLGA-PEG-PLGA synthetic copolymer have shown potential in modifying the pharmacokinetics and tissue distribution of curcumin. Evaluation of pharmacokinetics and biodistribution has demonstrated a prolonged half-life of the CU-micelles and a more efficient drug delivery to brain areas as compared to the carrier-free CU administration [71].

33.4.2 Curcumin Delivery by Lipid Nanoparticles

33.4.2.1 Solid lipid nanoparticles and nanostructured lipid carriers

Solid lipid nanoparticles (SLNPs) are submicron colloidal lipid carriers (from 50 nm to 1000 nm in diameter) which maintain a solid, spherical shape at room temperature. They possess a solid lipid core matrix stabilized by emulsifiers that can solubilize lipophilic molecules. The CU-SLNPs are usually small, ranging from 100 to 300 nm in diameter. The total drug content can reach up to 92% when the SLNPs are manufactured using the micro-emulsification technique [87]. In an experimental rat model of cerebral ischemic reperfusion injury, animals fed with CU-loaded SLNPs have had a 90% improvement in their cognitive function along with a 52% inhibition of the acetyl cholinesterase activity [88]. The investigated formula has been shown to increase the levels of superoxide dismutase (SOD), catalase (CAT), glutathione (GSH), and the activities of mitochondrial enzymes, while decreasing the lipid peroxidation and the peroxynitrite levels. Furthermore, this formulation showed a 16.4 to 30-fold improvement in the bioavailability of CU in the brain upon oral and intravenous (IV) administrations, respectively [88]. The product Longvida® (Verdure Sciences Inc.) is a SLNP-formulation of curcumin which can yield from 0.1 to 0.2 µM plasma levels of CU with associated 1–2 µM brain levels of free CU in animals [89–91]. This formula was later optimized as "lipidated Cur," which can yield more than 5 µM CU in mouse brain [93]. Other formulations of CU-loaded SLNPs, tested in Alzheimer's disease models, are outlined in Table 33.4.

Table 33.3 Curcumin-loaded polymeric nanoparticles studied in *in vitro* and *in vivo* models of neurodegenerative diseases

Disease	Nanoformulation type	Model/administration route	Outcomes
AD	PLGA[1] nanoparticles	*In vitro*: SK-N-SH human neuroblastoma cells	Protection against H_2O_2-induced oxidative damage [70]
AD	PLGA nanoparticles	*In vitro*: Neural stem cells, *In vivo*: Aβ[2]-amyloid induced rat model of AD-like phenotypes	Expression of genes involved in neuronal proliferation and differentiation, Reverse learning and memory impairments [73]
AD	PLGA nanoparticles conjugated with Tet-1 peptide	*In vitro*	Anti-amyloid activity unchanged, decrease of aggregates size [74], Diminution of anti-oxidant activity
AD	PLGA nanoparticles functionalized with glutathione	*In vitro*: in SK-N-SH cells	Neuronal uptake, Enhanced curcumin action [75, 76]
AD	PLGA nanoparticles	*In vivo*: Rat, IV, oral	Increased CU bioavailability and plasma concentration [77]
AD	PLGA nanoparticles	*In vivo*: Rat	Prolonged CU retention time in cerebral cortex and hippocampus [78]
AD	Apolipoprotein E3-mediated poly[butyl]cyano acrylate nanoparticles	*In vitro*: SH-SY5Y cells	Protection against Aβ-induced cytotoxicity [79]
AD	Pegylated poly(alkyl cyanoacrylate) nanoparticles with anti-Aβ 1–42 antibody at the surface	*In vitro*	Inhibition of Aβ aggregation [80]

Disease	Nanoformulation type	Model/administration route	Outcomes
AD	Spherical (SPNs) or Discoidal (DPNs) polymeric nanocontructs PLGA, DSPE-PEG[3]	*In vitro*: Raw 264.7 cells, *In vitro* production of Aβ fibers	Decrease of the pro-inflammatory cytokines in macrophages stimulated via Aβ fibers [81]
AD	Polymeric nanoparticles (NanoCurc[TM])	*In vitro*: SK-N-SH differentiated cells, *In vivo*: Mice, parenteral injection	Protection against H_2O_2-induced oxidative stress, Downregulation of caspase 3 and 7 activities, mediators of the apoptotic pathway, Increased glutathione levels [82]
AD	Nanocurcumin CU within polyethylene glycol-polylactide diblock polymer micelles	*In vivo*: AD model Tg2576 mice	Higher curcumin concentration in plasma, 6 times higher area under the curve and mean residence time in brain than ordinary CU, Improved memory function [83]
AD	Nanoemulsion	*In vitro*: SK-N-SH cell line, Sheep nasal mucosa	Safe for intranasal delivery for brain targeting, Higher flux and permeation across sheep nasal mucosa [84]
Mitochondrial dysfunction in brain aging	Micelles	*In vitro*: PC12 cells, *In vivo*: NMRI mice, *Ex vivo*: isolated mouse brain mitochondria	Improved bioavailability of native curcumin around 10- to 40-fold in plasma and brain of mice, Prevention of mitochondrial swelling in isolated mouse brain mitochondria, Protection of PC12 cells from nitrosative stress as compared to free CU [85]
PD	Alginate nanocomposites	*In vivo*: Drosophila, oral	Reduction of oxidative stress and apoptosis in the brain [86]

[1]Poly[lactic-co-glycolic acid]; [2]Aβ-amyloid; [3]Distearoyl phosphatidylethanolamine-polyethylene glycol.

Table 33.4 Curcumin-loaded lipid nanoparticles studied in *in vitro* and *in vivo* models of neurodegenerative diseases

Disease	Nanoformulation type	Model/administration route	Outcomes
AD	Solid lipid nanoparticles	*In vitro*: Mouse neuroblastoma cells after Aβ1 exposure	Decreased ROS production, Prevented apoptotic death, Inhibition of Tau formation [89, 90]
AD	Solid lipid curcumin particle (SLCP), Longvida®	*In vitro*: lipopolysaccharide (LPS)-stimulated RAW 264.7 cultured murine macrophages	Improved solubility over unformulated curcumin, Decreased LPS induced pro-inflammatory mediators NO, PGE2, and IL-6 by inhibiting the activation of NF-kB [92]
AD	Solid lipid particles with CU (SLCP)	*In vivo*: one-year-old 5xFAD-and age-matched wild-type mice, intraperitoneal injections of CU/SLCP	Decrease in Aβ plaque loads in dentate gyrus of hippocampus, More anti-amyloid, anti-inflammatory, and neuroprotective [91]
AD	Solid lipid nanoparticles	*In vivo*: Rat, oral	Effective delivery across the BBB[2] [88]
HD	Solid lipid nanoparticles (CU-SLNs)	*In vivo*: (3-NP)-induced HD in rats	Restored glutathione levels and superoxide dismutase activity, Activation of nuclear factor-erythroid 2 antioxidant pathway, Reduction in mitochondrial swelling, lipid peroxidation, protein carbonyls and reactive oxygen species [89]
CNS disorders	Solid lipid nanoparticles (CU-SLNs) and nanostructured lipid carriers (CU-NLCs)	*In vivo*: male Sprague–Dawley rats 6-8 weeks old, oral	Enhanced curcumin brain uptake, Cur-NLCs enhance the absorption of brain curcumin more than 4-folds in comparison with Cur-SLNs [95]
AD	Lipoprotein (LDL)-mimic nanostructured lipid carrier (NLC) modified with lactoferrin (Lf) and loaded with CU	*In vivo*: Rat, oral	Cellular uptake mediated by the Lf receptor, Permeability through the BBB and preferentially accumulation in the brain, Efficacy in controlling the damage associated with AD [96]

AD	Liposomes functionalized with TAT-peptide	In vitro	Permeability across the BBB enhanced [98]
AD	Liposomes containing cardiolipin	In vitro: SK-N-MC cells	Inhibition of the phosphorylation of p38, JNK, and tau protein, Protection against serious degeneration of Aβ insulted neurons [101]
AD	WGA[3]-conjugated and cardiolipin-incorporated liposomes carrying NGF[4] and CU	In vitro: Human astrocytes and to protect SK-N-MC cells Apoptosis induced by β-amyloid 1–42 (Aβ 1–42) fibrils	Increased entrapment efficiency of NGF and CU, of NGF release and cell viability, Decreased release of CU, Permeability of NGF and CU across the blood–brain barrier [102]
AD	Liposomes	In vivo: Mice, stereotaxic injection in the right hippocampus and neocortex	Decrease in Aβ secretion and toxicity [97]
AD	Liposomes decorated with anti-transferrin receptor mAb	In vivo injection, hippocampus and neocortex	Decrease in Aβ 1–42 aggregation, Internalization in the BBB model [99]
AD	Liposomes functionalized with a curcumin-alkyne derivative TREG	Human biological fluids from sporadic AD patients and down syndrome subjects	Sequestration of Aβ 1–42 [100, 101]
Neuronal loss	Liquid-crystalline lipid nanoparticles carrying curcumin and DHA	In vitro: SH-SY5Y cells	Neuronal viability and neurite outgrowth by activation of the TrkB receptor signaling, and promotion of phosphorylated CREB protein expression [118]
AD	Lipopeptide: a short microtubule-stabilizing peptide conjugated to a hydrophobic palmitic acid chain	In vitro: Neuro-2a cells, PC-12 differentiated cells	Neurite outgrowth in absence of external growth factors, Neural cells morphology and health amelioration [120, 121]

[1]Aβ-amyloid; [2]blood–brain barrier; [3]wheat germ agglutinins; [4]nerve growth factor.

Nanostructured lipid carriers (NLC) are referred to as the "second generation" of SLNPs. NLCs are composed of mixtures of sterically different amphiphilic molecules. Often, mixtures of liquid-phase lipids and solid-phase lipids yield matrices with imperfections, which may incorporate increased quantities of drug molecules as compared to the SLNPs. Despite of the presence of the liquid-phase lipid, the NLC matrix appears to be in a solid state at room and body temperatures. The solid state is controlled by the fraction of the included liquid-phase lipid [94]. Sadegh-Malvajerd et al. have reported an enhanced entrapment efficiency of curcumin in NLCs (94% ± 0.74) as compared to SLNPs (82% ± 0.49).

The pharmacokinetic studies, performed after intravenous (IV) administration of 4 mg/kg dose of curcumin in rats, have indicated that the amount of curcumin available in the brain was significantly higher for curcumin-loaded NLCs (AUC_{0-t} = 505.76 ng/g h) as compared to free curcumin (AUC_{0-t} = 0.00 ng/g h) and curcumin-loaded SLNs (AUC_{0-t} = 116.31 ng/g h) ($P < 0.005$) [95]. The outcomes of other recent investigations of CU-loaded NLCs in models of neurodegenerative diseases are summarized in Table 33.4.

33.4.2.2 Liposomes

Liposomes are lipid bilayer-based, self-assembled, closed colloidal structures, typically 25 nm to 5 μm in diameter [97]. They usually have a spherical shape comprising an aqueous core surrounded by a hydrophobic lipid membrane (Fig. 33.3). The lipid bilayer can be loaded with hydrophobic or amphiphilic molecules, whereas the hydrophilic molecules can be encapsulated in the aqueous reservoir of the liposomes. Often, liposomes are composed of phospholipids (e.g., phosphatidylcholines) or mixtures of phospholipids with co-lipids. Various liposome architectures can form depending on the preparation methods; for instance, multilamellar vesicles (MLV, involving a stack of several lipid bilayers), small unilamellar vesicles (SUV, constituted by a single lipid bilayer), large unilamellar vesicles (LUV), tubular vesicles, and cochleate vesicles.

Curcumin encapsulated in liposomes has been proven to be a safe formulation, which enhances the CU solubility and its cellular uptake [97–101]. Liposomes deliver CU into the cells via membrane fusion or endocytosis process. Liposomal formulations with a PEG surface coating provide a longer circulation time for the encapsulated drug. Biomolecular ligands can be anchored to the liposome surface in order to enhance the receptor targeting capacity, and hence, the permeability across the blood–brain barrier (BBB) [102, 103]. The outcomes of the investigated CU-loaded liposomes in models of neurodegenerative diseases are summarized in Table 33.4.

33.4.2.3 Liquid crystalline nanoparticles with internal structure

Liquid crystalline nanoparticles (LCNPs) are self-assembled architectures of lyotropic lipids, co-lipids (surfactants or oils), and water. They are typically formed upon dispersion and fragmentation of bulk lyotropic liquid crystalline phases (e.g., bicontinuous cubic, sponge, or inverted hexagonal phases) [104–106]. The amphiphilic molecules spontaneously organize into compartments with hydrophobic and hydrophilic domains (Fig. 33.3), which can encapsulate lipophilic or hydrosoluble guest compounds. The structures formed by this self-assembly process are thermodynamically stable. The initial liquid crystalline phases are usually viscous and have a short-range order in comparison to solids, but long-range order in comparison to liquids. A typical example of a lyotropic liquid crystalline phase is the inverted bicontinuous cubic phase formed upon mixing of unsaturated monoglyceride lipids with water [105]. Cubosomes are produced upon dispersion of the bicontinuous cubic liquid crystalline phases in excess aqueous medium. Their periodic structures comprise folded bicontinuous lipid bilayer membranes and periodic networks of aqueous channels (Fig. 33.3). The latter enable high encapsulation capacity for hydrophilic guest macromolecules [106–111]. Lipid nanocarriers of liquid crystalline internal structures have received considerable attention as delivery vehicles through the BBB [111, 112].

Curcumin has been successfully entrapped into monolein-based LCNPs with almost 100% encapsulation efficiency [113]. LCNPs dispersion was very stable in terms of nanocarrier sizes and surface charge upon storage. LCNPs were efficiently taken up by cultures cells following the sustained release of curcumin. In addition, they provided inhibition of the cell proliferation and apoptosis induction in an anticancer study [113]. A recent investigation of an inverse hexagonal (H_{II}) liquid crystalline phase encapsulating curcumin has demonstrated that the release of curcumin was a concentration-diffusion controlled process in the early stages, whereas multiple diffusion mechanisms coexisted in the later stages of drug release. Radical scavenging experiments have shown that curcumin-loaded LCNPs exert the desired antioxidant activity [114]. Thus, curcumin-loaded LCNPs may be promising for neurodegenerative disease treatments using sustained-release nanoformulations for combination therapies [115, 116]. Further results obtained with lipid-based LCNPs in models of neurodegenerative disease are presented in Table 33.4.

33.5 Conclusions

The naturally occurring compound curcumin is increasingly studied in neurodegenerative disease models due to its neurogenesis-stimulating properties and its anti-amyloid and anti-tau potential. Nanotechnology-based delivery systems of curcumin have been developed with the purpose of improving its solubility, stability, and bioavailability in potential treatment strategies of neurodegenerative disorders. We summarized recent advances in research on safe liquid crystalline lipid-based nanocarriers (cubosome, spongosome, hexosome, and liposome particles) and solid lipid nanoparticles, as well as on selected biodegradable polymer-based nanocarriers. The emphasis is given on the observed biological outcomes of the curcumin nanoformulations in *in vitro* and *in vivo* models of the multifactor neurodegenerative diseases (AD, PD, HD and ALS). Despite the difficulty of overcoming biological barriers, promising results on the enhancement of the permeability of the BBB and receptor-mediated delivery across the BBB have been reported with liposome and cubosome nanocarriers. Further investigations will be required in order to understand the involved mechanisms of action of curcumin nanoformulations in the proposed neurodegenerative disease models, and to optimize the delivery systems and strategies towards translation into clinics.

Abbreviations

Aβ:	amyloid beta
ACh:	acetylcholine
AD:	Alzheimer's disease
ALS:	amyotrophic lateral sclerosis
ApoE:	apolipoprotein E
APP:	amyloid beta precursor protein
ARE:	antioxidant response element
BBB:	blood–brain barrier
BDNF:	brain derived neurotrophic factor
Ca^{2+}:	calcium ion
CAT:	catalase
CHCHD2:	coiled-coil-helix-coiled-coil-helix domain 2
C9ORF72:	chromosome 9 open reading frame 72
COMT:	catechol-*O*-methyltransferase
CREB:	cAMP (Cyclic adenosine monophosphate response) element-binding protein
CU:	curcumin
CYP:	cytochrome P450
DARPP:	dopamine and adenosine $3^t,5^t$-monophosphate-regulated phosphoprotein
DHA:	docosahexaenoic acid
DNA:	deoxyribonucleic acid
DNAJC13:	DNA J heat shock protein family (Hsp40) member C13
DSPE:	distearoyl phosphatidylethanolamine
EIF4G1:	eukaryotic translation initiation factor 4 gamma 1
EMA:	European Medicines Agency
ERK:	Extracellular signal regulated kinase
FDA:	Food and Drug Administration
FUS:	RNA binding protein Fused in Sarcoma
GBA:	glucocerebrosidase
GRAS:	generally recognized as safe
GPx:	glutathione peroxidase
GR:	glutathione reductase

GSH:	glutathione
HEWL:	hen egg white lysozyme
HD:	Huntington disease
H_2O_2:	hydrogen peroxide
HTT:	Huntingtin gene
IL-6:	interleukin 6
iNOS:	induced nitric oxide synthase
IV:	intravenous
JNK:	Jun N-terminal kinase
LCNs:	liquid crystalline nanocarriers
LDL:	low density lipoprotein
Lf:	lactoferrin
LPS:	lipopolysaccharide
LRRK1:	leucine-rich repeat kinase 1
LUV:	large unilamellar vesicles
MAO-B:	monoamine oxidase type B
MDA:	malondialdehyde
MLV:	multilamellar vesicles
MPTP:	1-methyl-4-phenyl-1,2,3,6-tetrahydropyridine
MMSE:	mini mental state examination
mRNA:	messenger ribonucleic acid
NF-κB:	nuclear factor kappa beta
NGF:	nerve growth factor
NLC:	nanostructured lipid carriers
NO:	nitric oxide
NPs:	nanoparticles
Nrf2:	nuclear factor erythroid 2–related factor 2
NSAIDs:	non-steroidal anti-inflammatory drugs
6-OHDA:	6 hydroxydopamine
OPTN:	optineurin
PD:	Parkinson disease
PEG:	polyethylene glycol
PINK1:	PTEN-induced putative kinase 1
PLGA:	poly (lactic-*co*-glycolic acid)
PRKN:	parkin

PSEN: presenilin
PUFA: polyunsaturated fatty acids
ROS: reactive oxygen species
SLCP: solid lipid curcumin nanoparticles
SLN: solid lipid nanoparticles
SOD1:   superoxide dismutase 1
SNCA: synuclein alpha
SUV: small unilamellar vesicles
TARDBP: TAR DNA binding protein (TDP-43)
TREG: T regulatory cell
TRPME: transient receptor potential mucolipin-1 expression
TrkB: tropomyosin receptor kinase B
UBQLN2:   ubiquitin 2
UCH: ubiquitin carboxy-terminal hydrolase
VPS35:   vascular protein sorting
WGA: wheat-germ agglutinin

Disclosures and Conflict of Interest

This chapter was originally published as: Rakotoarisoa, M., Angelova, A. (2018). Amphiphilic nanocarrier systems for curcumin delivery in neurodegenerative disorders, *Medicines* **5**, 126, doi:10.3390/medicines5040126, under the Creative Commons Attribution license (http://creativecommons.org/licenses/by/4.0/). It appears here, with edits and updates, by kind permission of the author and the publisher, MDPI (Basel).

This work is supported by the "IDI 2017" project funded by the IDEX Paris-Saclay, ANR-11-IDEX-0003-02. The authors declare no conflict of interest.

Corresponding Author

Dr. Angelina Angelova
Institut Galien Paris-Sud, CNRS UMR 8612
Univ Paris-Sud, Univ Paris-Saclay
5 rue Jean-Baptiste Clément
F-92296 Châtenay-Malabry, France
Email: angelina.angelova@u-psud.fr

References

1. Brookmeyer, R., Abdalla, N., Kawas, C. H., Corrada, M. M. (2018). Forecasting the prevalence of preclinical and clinical Alzheimer's disease in the United States. *Alzheimers Dement.*, **14**, 121–129.

2. Jack, C. R., Albert, M. S., Knopman, D. S., McKhann, G. M., Sperling, R. A., Carrillo, M. C., Thies, B., Phelps, C. H. (2011). Introduction to the recommendations from the National Institute on Aging-Alzheimer's Association workgroups on diagnostic guidelines for Alzheimer's disease. *Alzheimers Dement.*, **7**, 257–262.

3. Haass, C., Selkoe, D. J. (2007). Soluble protein oligomers in neurodegeneration: Lessons from the Alzheimer's amyloid beta-peptide. *Nat. Rev. Mol. Cell Biol.*, **8**, 101–112.

4. Wakasaya, Y., Kawarabayashi, T., Watanabe, M., Yamamoto-Watanabe, Y., Takamura, A., Kurata, T., Murakami, T., Abe, K., Yamada, K., Wakabayashi, K. (2011). Factors responsible for neurofibrillary tangles and neuronal cell losses in tauopathy. *J. Neurosci. Res.*, **89**, 576–584.

5. Stoothoff, W. H., Johnson, G. V. (2005). Tau phosphorylation: Physiological and pathological consequences. *Biochim. Biophys. Acta*, **1739**, 280–297.

6. Wang, J., Wang, Z. M., Li, X. M., Li, F., Wu, J. J., Kong, L. Y., Wang, X. B. (2016). Synthesis and evaluation of multi-target-directed ligands for the treatment of Alzheimer's disease based on the fusion of donepezil and melatonin. *Bioorgan. Med. Chem.*, **24**, 4324–4338.

7. Cole, G. M., Morihara, T., Lim, G. P., Yang, F., Begum, A., Frautschy, S. A. (2004). NSAID and antioxidant prevention of Alzheimer's disease: Lessons from *in vitro* and animal models. *Ann. N. Y. Acad. Sci.*, **1035**, 68–84.

8. Alexander, G. E. (2004). Biology of Parkinson's disease: Pathogenesis and pathophysiology of a multisystem neurodegenerative disorder. *Dialogues Clin. Neurosci.*, **6**, 259–280.

9. Dauer, W., Przedborski, S. (2003). Parkinson's disease: Mechanisms and models. *Neuron*, **39**, 889–909.

10. Landles, C., Bates, G. P. (2004). Huntingtin and the molecular pathogenesis of Huntington's disease: Fourth in molecular medicine review series. *EMBO Rep.*, **5**, 958–963.

11. Zuccato, C., Marullo, M., Vitali, B., Tarditi, A., Mariotti, C., Valenza, M., Lahiri, N., Wild, E. J., Sassone, J., Ciammola, A. (2011). Brain-derived

neurotrophic factor in patients with Huntington's disease. *PLoS One*, **6**, e22966.

12. Zhou, S., Zhou, Y., Qian, S., Chang, W., Wang, L., Fan, D. (2018). Amyotrophic lateral sclerosis in Beijing: Epidemiologic features and prognosis from 2010 to 2015. *Brain Behav.*, **19**, e01131.

13. Glajch, E., Ferraiuolo, L., Mueller, K. A., Stopford, M. J., Prabhkar, V., Gravanis, A., Shaw, P. J., Sadri-Vakili, G. S. (2016). Microneurotrophins improve survival in motor neuron-astrocyte co-cultures but co cot improve disease phenotypes in a mutant SOD1 mouse model of amyotrophic lateral sclerosis. *PLoS One*, **10**, e0164103.

14. Varinderpal, S. D., Michael, F. (2014). Mutations that affect mito-chondrial functions and their association with neurodegenerative diseases. *Mutat Res. Rev. Mutat. Res.*, **759**, 1–13.

15. Barnham, K. J., Masters, C. L., Bush, A. I. (2004). Neurodegenerative diseases and oxidative stress. *Nat. Rev. Drug Discov.*, **3**, 205–214.

16. Jiang, T., Suna, Q., Chena, S. B. (2016). Oxidative stress: A major pathogenesis and potential therapeutic target of antioxidative agents in Parkinson's disease and Alzheimer's disease. *Progr. Neurobiol.*, **147**, 1–19.

17. Gilgun-Sherki, Y., Melamed, E., Offen, D. (2001). Oxidative stress induced-neurodegenerative diseases: The need for antioxidants that penetrate the blood brain barrier. *Neuropharmacology*, **40**, 959–975.

18. Uttara, B., Singh, A. V., Zamboni, P., Mahajan, R. T. (2009). Oxidative stress and neurodegenerative diseases: A review of upstream and downstream antioxidant therapeutic options. *Curr. Neuropharmacol.*, **7**, 65–74.

19. Dumont, M., Beal, M. F. (2011). Neuroprotective strategies involving ROS in Alzheimer disease. *Free Radic. Biol. Med.*, **51**, 1014–1026.

20. Zhong, L., Zhou, J., Chen, X., Lou, Y., Liu, D., Zou, X., Yang, B., Yin, Y., Pan, Y. (2016). Quantitative proteomics study of the neuroprotective effects of B12 on hydrogen peroxide-induced apoptosis in SH-SY5Y cells. *Sci. Rep.*, **6**, 22635.

21. Onyango, I. G., Khan, S. M., Bennett, J. P. (2017). Mitochondria in the pathophysiology of Alzheimer's and Parkinson's diseases. *Front. Biosci.*, **22**, 854–872.

22. Ankarcronaa, M., Mangialascheb, F., Winblada, B. (2010). Thinking Alzheimer's disease therapy: Are mitochondria the key? *J. Alzheimers Dis.*, **20**(Suppl 2), S579–S590.

23. Federico, A., Cardaioli, E., Da Pozzo, P., Formichi, P., Gallus, G. N. (2012). Mitochondria, oxidative stress and neurodegeneration. *J. Neurol. Sci.*, **322**, 254–262.

24. Stahon, K. E., Bastian, C., Griffith, S., Kidd, G. J., Brunet, S., Baltan, S. (2016). Age-related changes in axonal and mitochondrial ultrastructure and function in white matter. *J. Neurosci.*, **36**, 9990–10001.

25. Cho, D. H., Nakamura, T., Lipton, S. A. (2010). Mitochondrial dynamics in cell death and neurodegeneration. *Cell Mol. Life Sci.*, **67**, 3435–3447.

26. Lu, M., Su, C., Qiao, C., Bian, Y., Ding, J., Hu, G. (2016). Metformin prevents dopaminergic neuron death in MPTP/P-induced mouse model of Parkinson's disease via autophagy and mitochondrial ROS clearance. *Int. J. Neuropsychopharmacol.*, **19**, 1–11.

27. Zádori, D., Klivényi, P., Szalárdy, L., Fülöp, F., Toldi, J., Vécsei, L. (2012). Mitochondrial disturbances, excitotoxicity, neuroinflammation and kynurenines: Novel therapeutic strategies for neurodegenerative disorders. *J. Neurol. Sci.*, **322**, 187–191.

28. Rahimifard, M., Maqbool, F., Moeini-Nodeh, S., Niaz, K., Abdollahi, M., Braidy, N., Nabavi, S. M., Nabavi, S. F. (2017). Targeting the TLR4 signaling pathway by polyphenols: A novel therapeutic strategy for neuroinflammation. *Ageing Res. Rev.*, **36**, 11–19.

29. Géral, C., Angelova, A., Lesieur, S. (2013). From molecular to nanotechnology strategies for delivery of neurotrophins: Emphasis on brain-derived neurotrophic factor (BDNF). *Pharmaceutics*, **5**, 127–167.

30. Angelova, A., Angelov, B., Drechsler, M., Lesieur, S. (2013). Neurotrophin delivery using nanotechnology. *Drug Discov. Today*, **18**, 1263–1271.

31. Devi, L., Ohno, M. (2012). 7,8-Dihydroxyflavone, a small-molecule TrkB agonist, reverses memory deficits and BACE1 elevation in a mouse model of Alzheimer's disease. *Neuropsychopharmacology*, **37**, 434–444.

32. Soleimani, V., Sahebkar, A., Hosseinzadeh, H. (2018). Turmeric (*Curcuma longa*) and its major constituent (curcumin) as nontoxic and safe substances. *Rev. Phytother. Res.*, **32**, 985–995.

33. Mahdavi, H., Hadadi, Z., Ahmadi, M. A. (2017). Review of the anti-oxidation, anti-inflammatory and anti-tumor properties of curcumin. *Tradit. Integr. Med.*, **2**, 188–195.

34. Gupta, S. C., Patchva, S., Aggarwal, B. B. (2013). Therapeutic roles of curcumin: Lessons learned from clinical trials. *AAPS J.*, **15**, 195–218.

35. Brondino, N., Re, S., Boldrini, A., Cuccomarino, A., Lanati, N., Barale, F., Politi, P. (2014). Curcumin as a therapeutic agent in dementia: A mini systematic review of human studies. *Sci. World J.*, **2014**, 1–6.

36. Serafini, M. M., Catanzaro, M., Rosini, M., Racchi, M., Lanni, C. (2017). Curcumin in Alzheimer's disease: Can we think to new strategies and perspectives for this molecule? *Pharmacol. Res.*, **124**, 146–155.

37. Maiti, P., Dunbar, G. (2018). Use of curcumin, a natural polyphenol for targeting molecular pathways in treating age-related neurodegenerative diseases. *Int. J. Mol. Sci.*, **31**, 19.

38. Haan, J. D., Morrema, T. H. J., Rozemuller, A. J., Bouwman, F. H., Hoozemans, J. J. M. (2018). Different curcumin forms selectively bind fibrillar amyloid beta in post mortem Alzheimer's disease brains: Implications for in-vivo diagnostics. *Acta Neuropathol. Commun.*, **6**, 75.

39. Samarghandian, S., Azimi-Nezhad, M., Farkhondeh, T., Samini, F. (2017). Anti-oxidative effects of curcumin on immobilization-induced oxidative stress in rat brain, liver and kidney. *Biomed. Pharmacother.*, **87**, 223–229.

40. Ahmad, B., Borana, M. S., Chaudhary, A. P. (2017). Understanding curcumin-induced modulation of protein aggregation. *Int. J. Biol. Macromol.*, **100**, 89–96.

41. Sarkar, B., Dhiman, M., Mittal, S., Mantha, A. K. (2017). Curcumin revitalizes Amyloid beta (25-35)-induced and organophosphate pesticides pestered neurotoxicity in SH-SY5Y and IMR-32 cells via activation of APE1 and Nrf2. *Metab. Brain Dis.*, **32**, 2045–2061.

42. Zhang, L., Fang, Y., Cheng, X., Lian, Y. J., Xu, H. L., Zeng, Z. S., Zhu, H. C. (2017). Curcumin exerts effects on the pathophysiology of Alzheimer's disease by regulating PI(3,5)P2 and transient receptor potential mucolipin-1 expression. *Front. Neurol.*, **8**, 531.

43. Liaquat, L., Batool, Z., Sadir, S., Rafiq, S., Shahzad, S., Perveen, T., Haider, S. (2018). Naringenin-induced enhanced antioxidant defense system meliorates cholinergic neurotransmission and consolidates memory in male rats. *Life Sci.*, **194**, 213–223.

44. Nguyen, T. T., Vuu, M. D., Huynh, M. A., Yamaguchi, M., Tran, L. T., Dang, T. P. T. (2018). Curcumin effectively rescued Parkinson's disease-like phenotypes in a novel drosophila melanogaster model with dUCH knockdown. *Oxid. Med. Cell. Longev.*, **2018**, 1–12.

45. Wang, Y. L., Ju, B., Zhang, Y. Z., Yin, H. L., Liu, Y. J., Wang, S. S., Zeng, Z. L., Yang, X. P., Wang, H. T., Li, J. F. (2017). Protective effect of

curcumin against oxidative stress-induced injury in rats with Parkinson's disease through the Wnt/β-catenin signaling pathway. *Cell. Physiol. Biochem.*, **43**, 2226–2241.

46. Snigdha, D. M., Surjyanarayan, M., Jayvadan, P. (2017). Intranasal mucoadhesive microemulsion for neuroprotective effect of curcumin in mPTP induced Parkinson model. *Braz. J. Pharm. Sci.*, **53**.

47. Hickey, M. A., Zhu, C., Medvedeva, V., Lerner, R. P., Patassini, S., Franich, N. R., Maiti, P., Frautschy, S. A., Zeitlin, S., Levine, M. S., et al. (2012). Improvement of neuropathology and transcriptional deficits in CAG 140 knock-in mice supports a beneficial effect of dietary curcumin in Huntington's disease. *Mol. Neurodegener.*, **7**, 12.

48. Sevastre-Berghian, A. C., Făgărăsan, V., Toma, V. A., Bâldea, I., Olteanu, D., Moldovan, R., Decea, N., Filip, G. A., Clichici, S. V. (2017). Curcumin reverses the diazepam-induced cognitive impairment by modulation of oxidative stress and ERK 1/2/NF-κB pathway in brain. *Oxid. Med. Cell. Longev.*, **2017**, 1–16.

49. Motaghinejad, M., Motevalian, M., Fatima, S., Hashemi, H., Gholami, M. (2017). Curcumin confers neuroprotection against alcohol-induced hippocampal neurodegeneration via CREB-BDNF pathway in rats. *Biomed. Pharmacother.*, **87**, 721–740.

50. Motaghinejad, M., Motevalian, M., Fatima, S., Faraji, F., Mozaffari, S. (2017). The neuroprotective effect of curcumin against nicotine-induced neurotoxicity is mediated by CREB-BDNF signaling pathway. *Neurochem. Res.*, **42**, 2921–2932.

51. Baum, L., Lam, C. W., Cheung, S. K., Kwok, T., Lui, V., Tsoh, J., Lam, L., Leung, V., Hui, E., Ng, C. (2008). Six-month randomized, placebo-controlled, double-blind, pilot clinical trial of curcumin in patients with Alzheimer disease. *J. Clin. Psychopharmacol.*, **28**, 110–113.

52. Ringman, J. M., Frautschy, S. A., Teng, E., Begum, A. N., Bardens, J., Beigi, M., Gylys, K. H., Badmaev, V., Heath, D. D., Apostolova, L. G. (2012). Oral curcumin for Alzheimer's disease: Tolerability and efficacy in a 24-week randomized, double blind, placebo-controlled study. *Alzheimers Res. Ther.*, **4**, 43.

53. Hishikawa, N., Takahashi, Y., Amakusa, Y., Tanno, Y., Tuji, Y., Niwa, H., Murakami, N., Krishna, U. K. (2012). Effects of turmeric on Alzheimer's disease with behavioral and psychological symptoms of dementia. *AYU*, **33**, 499–504.

54. Mazzanti, G., Di Giacomo, S. (2016). Curcumin and resveratrol in the management of cognitive disorders: What is the clinical evidence? *Molecules*, **21**, 1243.

55. Ireson, C. R., Jones, D. J., Orr, S., Coughtrie, M. W., Boocock, D. J., Williams, M. L., Farmer, P. B., Steward, W. P., Gescher, A. J. (2002). Metabolism of the cancer chemopreventive agent curcumin in human and rat intestine. *Cancer Epidemiol. Biomark. Prev.*, **11**, 105–111.

56. Marczylo, T. H., Steward, W. P., Gescher, A. J. (2009). Rapid analysis of curcumin and curcumin metabolites in rat biomatrices using a novel ultraperformance liquid chromatography (UPLC) method. *J. Agric. Food. Chem.*, **57**, 797–803.

57. Marczylo, T. H., Verschoyle, R. D., Cooke, D. N., Morazzoni, P., Steward, W. P., Gescher, A. J. (2007). Comparison of systemic availability of curcumin with that of curcumin formulated with phosphatidylcholine. *Cancer Chemother. Pharmacol.*, **60**, 171–177.

58. Ghalandarlaki, N., Alizadeh, A. M., Ashkani-Esfahani, S. (2014). Nanotechnology-applied curcumin for different diseases therapy. *Biomed. Res. Int.*, **2014**, 1–23.

59. Cox, K. H., Pipingas, A., Scholey, A. B. (2015). Investigation of the effects of solid lipid curcumin on cognition and mood in a healthy older population. *J. Psychopharmacol.*, **29**, 642–651.

60. Modi, G., Pillay, V., Choonara, Y. E. (2010). Advances in the treatment of neurodegenerative disorders employing nanotechnology. *Ann. N. Y. Acad. Sci.*, **1184**, 154–172.

61. Sahni, J. K., Doggui, S., Ali, J., Baboota, S., Dao, L., Ramassamy, C. (2011). Neurotherapeutic applications of nanoparticles in Alzheimer's disease. *J. Control. Release*, **152**, 208–231.

62. Craparo, E. F., Bondì, M. L., Pitarresi, G., Cavallaro, G. (2011). Nanoparticulate systems for drug delivery and targeting to the central nervous system. *CNS Neurosci. Ther.*, **17**, 670–677.

63. Fonseca-Santos, B., Gremiao, D., Palmira, M., Chorilli, M. (2015). Nanotechnology-based drug delivery systems for the treatment of Alzheimer's disease. *Int. J. Nanomed.*, **10**, 4981–5003.

64. Kanwar, J. R., Sun, X., Punj, V., Sriramoju, B., Mohan, R. R., Zhou, S. F., Chauhan, A., Kanwar, R. K. (2012). Nanoparticles in the treatment and diagnosis of neurological disorders: Untamed dragon with fire power to heal. *Nanomedicine*, **8**, 399–414.

65. Gao, H. (2016). Progress and perspectives on targeting nanoparticles for brain drug delivery. *Acta Pharm. Sin. B*, **6**, 268–286.

66. Roney, C., Kulkarni, P., Arora, V., Antich, P., Bonte, F., Wu, A., Mallikarjuana, N. N., Manohar, S., Liang, H. F., Kulkarni, A. R., et al. (2005). Targeted nanoparticles for drug delivery through the blood–brain barrier for Alzheimer's disease. *J. Control. Release*, **108**, 193–214.

67. Modi, G., Pillay, V., Choonara, Y. E., Ndesendo, V. M. K., Du Toit, L. C., Naidoo, D. (2009). Nanotechnological applications for the treatment of neurodegenerative disorders. *Progr. Neurobiol.*, **4**, 272–285.

68. Ma, Z., Haddadi, A., Molavi, O., Lavasanifar, A., Lai, R., Samuel, J. (2008). Micelles of poly(ethyleneoxide)-b-poly(epsilon-caprolactone) as vehicles for the solubilization, stabilization, and controlled delivery of curcumin. *J. Biomed. Mater. Res. A*, **86**, 300–310.

69. Yu, H., Li, J., Shi, K., Huang, Q. (2011). Structure of modified epsilon-polylysine micelles and their application in improving cellular antioxidant activity of curcuminoids. *Food Funct.*, **2**, 373–380.

70. Podaralla, S., Averineni, R., Alqahtani, M., Perumal, O. (2012). Synthesis of novel biodegradable methoxypoly (ethylene glycol)-zein micelles for effective delivery of curcumin. *Mol. Pharm.*, **9**, 2778–2786.

71. Song, Z., Feng, R., Sun, M., Guo, C., Gao, Y., Li, L., Zhai, G. (2011). Curcumin-loaded PLGA-PEG-PLGA triblock copolymeric micelles: Preparation, pharmacokinetics and distribution *in vivo*. *J. Colloid Interface Sci.*, **354**, 116–123.

72. Doggui, S., Sahni, J. K., Arseneault, M., Dao, L., Ramassamy, C. (2012). Neuronal uptake and neuroprotective effect of curcumin-loaded PLGA nanoparticles on the human SK-N-SH cell line. *J. Alzheimers Dis.*, **30**, 377–392.

73. Tiwari, S. K., Agarwal, S., Seth, B., Yadav, A., Nair, S., Bhatnagar, P., Karmakar, M., Kumari, M., Chauhan, L. K., Patel, D. K., et al. (2014). Curcumin-loaded nanoparticles potently induce adult neurogenesis and reverse cognitive deficits in Alzheimer's disease model via canonical Wnt/β-catenin pathway. *ACS Nano*, **8**, 76–103.

74. Mathew, A., Fukuda, T., Nagaoka, Y., Hasumura, T., Morimoto, H., Yoshida, Y., Maekawa, T., Venugopal, K., Kumar, D. S. (2012). Curcumin loaded-PLGA nanoparticles conjugated with Tet-1 peptide for potential use in Alzheimer's disease. *PLoS One*, **7**, e32616.

75. Paka, G. D., Doggui, S., Zaghmi, A., Safar, R., Dao, L., Reisch, A., Klymchenko, A., Roullin, V. G., Joubert, O., Ramassamy, C. (2016). Neuronal uptake and neuroprotective properties of curcumin-loaded nanoparticles on SK-N-SH cell line: Role of poly(lactide-coglycolide) polymeric matrix composition. *Mol. Pharm.*, **13**, 391–403.

76. Paka, G. D., Ramassamy, C. (2017). Optimization of curcumin-loaded PEG-PLGA nanoparticles by GSH functionalization: Investigation of the internalization pathway in neuronal cells. *Mol. Pharm.*, **14**, 93–106.

77. Tsai, Y. M., Jan, W. C., Chien, C. F., Lee, W. C., Lin, L. C., Tsai, T. H. (2011). Optimised nano-formulation on the bioavailability of hydrophobic polyphenol, curcumin, in freely-moving rats. *Food Chem.*, **127**, 918–925.

78. Tsai, Y. M., Chien, C. F., Lin, L. C., Tsai, T. H. (2011). Curcumin and its nano-formulation: The kinetics of tissue distribution and blood–brain barrier penetration. *Int. J. Pharm.*, **416**, 331–338.

79. Mulik, R. S., Mönkkönen, J., Juvonen, R. O., Mahadik, K. R., Paradkar, A. R. (2010). ApoE3mediated poly (butyl) cyanoacrylate nanoparticles containing curcumin: Study of enhanced activity of curcumin against beta amyloid induced cytotoxicity using *in vitro* cell culture model. *Mol. Pharm.*, **7**, 815–825.

80. Le Droumaguet, B., Nicolas, J., Brambilla, D., Mura, S., Maksimenko, A., DeKimpe, L., Salvati, E., Zona, C., Airoldi, C., Canovi, M., et al. (2012). Versatile and efficient targeting using a single nanoparticulate platform: Application to cancer and Alzheimer's disease. *ACS Nano*, **6**, 5866–5879.

81. Ameruoso, A., Palomba, R., Palange, A., Cervadoro, A., Lee, A., Di Mascolo, D., Decuzzi, P. (2017). Ameliorating amyloid-β fibrils triggered inflammation via curcumin-loaded polymeric nano-constructs. *Front. Immunol.*, **8**, 1411.

82. Ray, B., Bisht, S., Maitra, A., Maitra, A., Lahiri, D. K. (2011). Neuroprotective and neurorescue effects of a novel polymeric nanoparticle formulation of curcumin (NanoCurcTM) in the neuronal cell culture and animal model: Implications for Alzheimer's disease. *J. Alzheimers Dis.*, **23**, 61–77.

83. Cheng, K. K., Yeung, C. F., Ho, S. W., Chow, S. F., Chow, A. H., Baum, L. (2013). Highly stabilized curcumin nanoparticles tested in an *in vitro* blood–brain barrier model and in Alzheimer's disease Tg2576 mice. *AAPS J.*, **15**, 324–336.

84. Sood, S., Jain, K., Gowthamarajan, K. (2014). Optimization of curcumin nanoemulsion for intranasal delivery using design of experiment and its toxicity assessment. *Colloids Surf. B Biointerfaces*, **113**, 330–337.

85. Hagl, S., Kocher, A., Schiborr, C., Kolesova, N., Frank, J., Eckert, G. P. (2015). Curcumin micelles improve mitochondrial function in neuronal PC12 cells and brains of NMRI mice—Impact on bioavailability. *Neurochem. Int.*, **89**, 234–242.

86. Siddique, Y. H., Wasi Khan, S. B. R., Naqvi, A. H. (2013). Synthesis of alginate-curcumin nanocomposite and its protective role in transgenic drosophila model of Parkinson's disease. *ISRN Pharmacol.*, **2013**, 1–8.

87. Kakkar, V., Muppu, S. K., Chopra, K., Kaur, I. P. (2013). Curcumin loaded solid lipid nanoparticles: An efficient formulation approach for cerebral ischemic reperfusion injury in rats. *Eur. J. Pharm. Biopharm.*, **85**, 339–345.

88. Kakkara, V., Mishrab, A. K., Chuttanib, K., Kaura, I. P. (2013). Proof of concept studies to confirm the delivery of curcumin loaded solid lipid nanoparticles (C-SLNs) to brain. *Int. J. Pharm.*, **448**, 354–359.

89. Maiti, P., Dunbar, G. L. (2017). Comparative neuroprotective effects of dietary curcumin and solid lipid curcumin particles in cultured mouse neuroblastoma cells after exposure to Abeta42. *Int. J. Alzheimers Dis.*, **2017**, 1–13.

90. Maiti, P., Hall, T. C., Paladugu, L., Kolli, N., Learman, C., Rossignol, J., Dunbar, G. L. (2016). A comparative study of dietary curcumin, nanocurcumin, and other classical amyloid-binding dyes for labeling and imaging of amyloid plaques in brain tissue of 5x-familial Alzheimer's disease mice. *Histochem. Cell Biol.*, **146**, 609–625.

91. Maiti, P., Paladugu, L., Dunbar, G. L. (2018). Solid lipid curcumin particles provide greater anti-amyloid, anti-inflammatory and neuroprotective effects than curcumin in the 5xFAD mouse model of Alzheimer's disease. *BMC Neurosci.*, **19**, 7.

92. Pragati, P., Nahar, A., Slitt, L., Navindra, P. (2015). Anti-inflammatory effects of novel standardized solid lipid curcumin formulations. *J. Med. Food*, **18**, 786–792.

93. Dadhaniya, P., Patel, C., Muchhara, J., Bhadja, N., Mathuria, N., Vachhani, K., Soni, M. G. (2011). Safety assessment of a solid lipid curcumin particle preparation: Acute and subchronic toxicity studies. *Food Chem. Toxicol.*, **49**, 1834–1842.

94. Ganesan, P., Narayanasamy, D. (2017). Lipid nanoparticles: Different preparation techniques, characterization, hurdles, and strategies for the production of solid lipid nanoparticles and nanostructured lipid carriers for oral drug delivery. *Sustain. Chem. Pharm.*, **6**, 37–56.

95. Sadegh-Malvajerd, S., Azadi, A., Izadi, Z., Kurd, M., Dara, T., Dibaei, M., Sharif-Zadeh, M., Akbari-Javar, H., Hamidi, M. (2019). Brain delivery of curcumin using solid lipid nanoparticles and nanostructured

lipid carriers: Preparation, optimization, and pharmacokinetic evaluation. *ACS Chem. Neurosci.*, **10**(1), 728–739.

96. Meng, F., Asghar, S., Gao, S., Su, Z., Song, J., Huo, M., Meng, W., Ping, Q., Xiao, Y. (2015). A novel LDL-mimic nanocarrier for the targeted delivery of curcumin into the brain to treat Alzheimer's disease. *Colloids Surf. B Biointerfaces*, **134**, 88–97.

97. Lazar, A. N., Mourtas, S., Youssef, I., Parizot, C., Dauphin, A., Delatour, B., Antimisiaris, S. G., Duyckaerts, C. (2013). Curcumin-conjugated nanoliposomes with high affinity for Aβ deposits: Possible applications to Alzheimer disease. *Nanomed. Nanotechnol. Biol. Med.*, **9**, 712–721.

98. Sancini, G., Gregori, M., Salvati, E., Cambianica, L., Re, F., Ornaghi, F., Canovi, M., Fracasso, C., Cagnotto, A., Colombo, M., et al. (2013). Functionalization with TAT-peptide enhances blood–brain barrier crossing *in vitro* of nanoliposomes carrying a curcumin-derivative to bind amyloid-β peptide. *J. Nanomed. Nanotechnol.*, **4**, 1–8.

99. Mourtas, S., Canovi, M., Zona, C., Aurilia, D., Niarakis, A., La Ferla, B., Salmona, M., Nicotra, F., Gobbi, M., Antimisiaris, S. G. (2011). Curcumin-decorated nanoliposomes with very high affinity for amyloid beta1–42 peptide. *Biomaterials*, **32**, 1635–1645.

100. Airoldi, C., Zona, C., Sironi, E., Colombo, L., Messa, M., Aurilia, D., Gregori, M., Masserini, M., Salmona, M., Nicotra, F., et al. (2011). Curcumin derivatives as new ligands of apeptides. *J. Biotechnol.*, **156**, 317–324.

101. Conti, E., Gregori, M., Radice, I., Da Re, F., Grana, D., Re, F., Salvati, E., Masserini, M., Ferrarese, C., Zoia, C. P., et al. (2017). Multifunctional liposomes interact with Abeta in human biological fluids: Therapeutic implications for Alzheimer's disease. *Neurochem. Int.*, **108**, 60–65.

102. Kuo, Y. C., Lin, C. Y., Li, J. S., Lou, Y. I. (2017). Wheat germ agglutinin-conjugated liposomes incorporated with cardiolipin to improve neuronal survival in Alzheimer's disease treatment. *Int. J. Nanomed.*, **12**, 1757–1774.

103. Kuo, Y. C., Lin, C. C. (2015). Rescuing apoptotic neurons in Alzheimer's disease using wheat germ agglutinin-conjugated and cardiolipin-conjugated liposomes with encapsulated nerve growth factor and curcumin. *Int. J. Nanomed.*, **10**, 2653–2672.

104. Valldeperas, M., Wis´niewska, M., Ram-On, M., Kesselman, E., Danino, D., Nylander, T., Barauskas, J. (2016). Sponge phases and nanoparticle dispersions in aqueous mixtures of mono- and diglycerides. *Langmuir*, **32**, 8650–8659.

105. Milak, S., Zimmer, A. (2015). Glycerol monooleate liquid crystalline phases used in drug delivery systems. *Int. J. Pharm.*, **478**, 569–587.

106. Angelova, A., Angelov, B., Mutafchieva, R., Lesieur, S., Couvreur, P. (2011). Self-assembled multicompartment liquid crystalline lipid carriers for protein, peptide, and nucleic acid drug delivery. *Acc. Chem. Res.*, **44**, 147–156.

107. Angelov, B., Angelova, A., Filippov, S. K., Drechsler, M., Štěpánek, P., Lesieur, S. (2014). Multicompartment lipid cubic nanoparticles with high protein upload: Millisecond dynamics of formation. *ACS Nano*, **8**, 5216–5226.

108. Zerkoune, L., Lesieur, S., Putaux, J. L., Choisnard, L., Gèze, A., Wouessidjewe, D., Angelov, B., Vebert-Nardin, C., Doutch, J., Angelova, A. (2016). Mesoporous self-assembled nanoparticles of bio-transesterified cyclodextrins and nonlamellar lipids as carriers of water-insoluble substances. *Soft Matter*, **12**, 7539–7550.

109. Han, S., Shen, J. Q., Gan, Y., Geng, H. M., Zhang, X. X., Zhu, C. L., Gan, L. (2010). Novel vehicle based on cubosomes for ophthalmic delivery of flurbiprofen with low irritancy and high bioavailability. *Acta Pharmacol. Sin.*, **31**, 990–998.

110. Zou, A., Li, Y., Chen, Y., Angelova, A., Garamus, V. M., Li, N., Drechsler, M., Angelov, B., Gong, Y. (2017). Self-assembled stable sponge type nanocarriers for *Brucea javanica* oil delivery. *Colloids Surf. B*, **153**, 310–319.

111. Azhari, H., Strauss, M., Hook, S., Boyd, B. J., Rizwan, S. B. (2016). Stabilising cubosomes with tween 80 as a step towards targeting lipid nanocarriers to the blood–brain barrier. *Eur. J. Pharm. Biopharm.*, **104**, 148–155.

112. Wu, H., Li, J., Zhang, Q., Yan, X., Guo, L., Gao, X., Qiu, M., Jiang, X., Lai, R., Chen, H. (2012). A novel small odorranalectin-bearing cubosomes: Preparation, brain delivery and pharmacodynamic study on amyloid-b25–35-treated rats following intranasal administration. *Eur. J. Pharm. Biopharm.*, **80**, 368–378.

113. Baskaran, R., Madheswaran, T., Sundaramoorthy, P., Kim, H. M., Yoo, B. K. (2014). Entrapment of curcumin into monoolein-based liquid crystalline nanoparticle dispersion for enhancement of stability and anticancer activity. *Int. J. Nanomed.*, **9**, 3119–3130.

114. Wei, L., Li, X., Guo, F., Liu, X., Wang, Z. (2018). Structural properties, *in vitro* release and radical scavenging activity of lecithin based curcumin-encapsulated inverse hexagonal (HII) liquid crystals. *Colloids Surf. A*, **539**, 124–131.

115. Angelova, A., Angelov, B. (2017). Dual and multi-drug delivery nanoparticles towards neuronal survival and synaptic repair. *Neural. Regen. Res.*, **12**, 886–889.

116. Angelova, A., Drechsler, M., Garamus, V. M., Angelov, B. (2018). Liquid crystalline nanostructures as pegylated reservoirs of omega-3 polyunsaturated fatty acids: Structural insights toward delivery formulations against neurodegenerative disorders. *ACS Omega*, **3**, 3235–3247.

117. Sandhir, R., Yadav, A., Mehrotra, A., Sunkaria, A., Singh, A., Sharma, S. (2014). Curcumin nanoparticles attenuate neurochemical and neurobehavioral deficits in experimental model of Huntington's disease. *Neuromol. Med.*, **16**, 106–118.

118. Guerzoni, L. P., Nicolas, V., Angelova, A. (2017). *In vitro* modulation of TrkB receptor signaling upon sequential delivery of curcumin-DHA loaded carriers towards promoting neuronal survival. *Pharm. Res.*, **34**, 492–505.

119. Chen, Y., Angelova, A., Angelov, B., Drechsler, M., Garamus, V. M., Willumeit-Römere, R., Zou, A. (2015). Sterically stabilized spongosomes for multi-drug delivery of anticancer nanomedicines. *J. Mater. Chem. B*, **3**, 7734–7744.

120. Biswas, A., Kurkute, P., Jana, B., Laskar, A., Ghosh, S. (2014). An amyloid inhibitor octapeptide forms amyloid type fibrous aggregates and affects microtubule motility. *Chem. Commun.*, **50**, 2604–2607.

121. Adak, A., Das, G., Barman, S., Mohapatra, S., Bhunia, D., Jana, B., Ghosh, S. (2017). Biodegradable neuro-compatible peptide hydrogel promotes neurite outgrowth, shows significant neuroprotection, and delivers anti-Alzheimer drug. *ACS Appl. Mater. Interfaces*, **9**, 5067–5076.

122. Shrestha, H., Bala, R., Arora, S. (2014). Lipid-based drug delivery systems. *J. Pharm.*, **2014**, 1–10.

123. Ganem-Quintanar, A., Quintanar-Guerrero, D., Buri, P. (2000). Monoolein: A review of the pharmaceutical applications. *Drug Dev. Ind. Pharm.*, **26**, 809–820.

124. Bousquet, M., Saint-Pierre, M., Julien, C., Salem, N., Cicchetti, F., Calon, F. (2008). Beneficial effects of dietary omega-3 polyunsaturated fatty acid on toxin-induced neuronal degeneration in an animal model of Parkinson's disease. *FASEB J.*, **22**, 1213–1225.

Chapter 34

Peptide-Based Drug Delivery Systems: Future Challenges, Perspectives, and Opportunities in Nanomedicine

Diego Tesauro, PhD,[a] Antonella Accardo, PhD,[a] Carlo Diaferia, PhD,[a] Vittoria Milano, PhD,[a,b] Jean Guillon, PhD,[b] Luisa Ronga, PhD,[c] and Filomena Rossi, PhD[a]

[a]Department of Pharmacy and CIRPeB, Università Federico II, Naples, Italy
[b]ARNA, INSERM U1212/UMR CNRS 5320, UFR des Sciences Pharmaceutiques, Université de Bordeaux, Bordeaux, France
[c]Institute of Analytical Sciences, IPREM, UMR 5254, CNRS-University of Pau, Pau, France

Keywords: peptide, peptide backbone structures, drug delivery, peptide self-assembling carriers, active targeting receptors, diphenylalanine, binding peptides, hydrogels, contrast agents, nanomedicine, supramolecular chemistry, bioimaging, micelles, liposomes, integrin receptors, nanoaggregates, octreotide, bombesin receptors, aromatic interactions, amphiphilic peptides, β-sheet peptides

34.1   Introduction

Peptides of natural and synthetic origin are compounds involved in a wide variety of biological roles. They act as hormones, enzyme

The Road from Nanomedicine to Precision Medicine
Edited by Shaker A. Mousa, Raj Bawa, and Gerald F. Audette
Copyright © 2020 Jenny Stanford Publishing Pte. Ltd.
ISBN 978-981-4800-59-4 (Hardcover), 978-0-429-29501-0 (eBook)
www.jennystanford.com

substrates and inhibitors, antibiotics, biological regulators, and so on. Therefore, peptides play an essential role in biotechnological applications as therapeutic and diagnostic agents. Their advantages depend on the strategy applied to produce them and include biocompatibility, low cost, tunable bioactivity, chemical variety, and specific targeting. Moreover, they are easily synthesized, for example, by using solid-phase peptide methodologies where the amino acid sequence can be exactly selected at the molecular level by tuning the basic units [1]. Although the drawbacks related to their use are referred to as metabolic instability via protease degradation, an improved metabolic stability can be pursued through several chemical approaches aimed to modify the original peptide sequences. Some examples are the introduction of specific coded or un-coded amino acids, D-counterparts, cyclization, and DNA recombinant technology. Recently, peptides achieved resounding success in drug delivery and in nanomedicine smart applications, thanks to these innovative approaches. These applications are among the most significant challenges of recent decades in transporting drugs only to pathological tissues whilst other districts in the body are preserved from side effects. This specific feature allows the reduction of unwanted drug effects and increases the drug efficacy [2].

In peptide-containing aggregates, peptide sequence can fulfill a structural or a bioactive role. In detail, peptides play a structural role when the primary amino acid sequence drives or affects the molecular self-assembly by adding remarkable weak non-covalent bonds, electrostatic interactions, hydrogen bonds, hydrophobic and Van der Waals interactions, and π–π stacking between the side chains. Furthermore, peptides play a bioactive role when the full sequence, or a part of it, is deputed to recognize specific receptors, such as those overexpressed by pathological cells. In this review, we will focus on the peptide ability to self-assemble and on potential applications of peptide based nanosystems for nanomedicine. In addition, we report recent examples of peptides employed as delivery systems of anticancer drugs and/or contrast agents for the imaging of tumor pathologies. Finally, we will describe peptide nanosystems able to actively address the active pharmaceutical ingredients (APIs) toward specific biological targets.

34.2   Peptide Self-Assembled Nanostructures

Peptides are able to gather into assorted nanostructures, including nanotubes, nanofibers, nanospheres, and nanovesicles, supported by their device and self-assembly conditions [3]. Different types and structures of peptides, including cyclic and linear peptides, amphiphilic peptides, and α-helical and β-sheet peptides, can self-assemble into nanostructures (see Fig. 34.1).

Figure 34.1 Different classes of peptides can be arranged in supramolecular structures handling the self-assembling phenomena. Various morphologies can be generated according to the rational design of the primary sequence.

34.2.1   α-Helical and β-Sheet Peptides

The primary feature in the design and synthesis of peptide-based biomolecules regards the peptide backbone arrangement in α-helical and β-sheet secondary structures. This is a consequence of the hydrogen bonding pattern interactions through the amide and carbonyls groups in the peptide backbone. After that, the β-strands turn into a β-sheet self-assembled structure that could be rearranged in parallel or antiparallel arrays, according to the direction of the peptide sequences. The peptide is typically designed to contain repeating amino acid residues and distinct hydrophobic and hydrophilic regions. Consequently, the hydrophobic moiety could be hidden within the self-assembled nanostructure while the hydrophilic area could be better exposed to the solvent (water) environment [4]. Unlike β-sheets, α-helices are formed by single peptide chains, where backbone amide components are intramolecularly hydrogen bonded. This arrangement leads to the exposition of side chains of amino acids on the surface of each helix. Thus, their positioning further facilitates the accessibility of the peptide in the solvent.

The regular α-helical peptides with 2,5 helices are shown to aggregate around each other and their structure evolves in nanofibers [5, 6]. These α-helical peptides can also self-assemble into nanofibers if they have at least 30 amino acid residues, through helical coiled-coil structures [7]. The hydrophobic residues could promote the helix oligomerization through hydrophobic collapse.

The β-sheet secondary structures are the naturally occurring motifs most similar to those which carry on into pay peptide self-assembly [8, 9]. The β-sheet determines regular alternating hydrophilic and hydrophobic regions in the peptide sequence. The same structure provides the amphiphilic property to the peptide that drives the self-assembly of β-sheets. For instance, β-sheet peptides (namely the QQR holding sequences) can self-assemble into a pH-responsive hydrogel by means of the side chain's ion affinity for acidic residues of Glu and Arg. These peptides are soluble in neutral pH conditions and they switch into a hydrogel material, in acidic pH surroundings. This behavior can be rationalized in terms of the rearrangement of antiparallel β-sheet tapes. Those β-sheet tapes are obtained at lower pH values and, afterward, stacked together to form nanofibrils in hydrogels [10]. The intramolecular folding β-hairpin peptides are well represented in the self-assembled sequences in various nanostructures, both in water and in space boundaries. Indeed, the self-assembly of β-hairpins in proteins is carried on by the arrangement of two β-sheets in antiparallel plans. The modulation of pH values reproduces the status in which these materials could be gained and tailored. It is well described that the fundamental mechanism in hydrogels engendered by self-assembly of the β-sheet hairpin structure strongly depends on the increase of the pH values [11].

The peptide self-assembly processes are also kept up by non-covalent interactions that sometime show the key role in the overall configuration. The non-covalent interactions should be taken into high consideration for these grounds, especially when designing peptide self-assembled nanostructures for drug delivery. Indeed, non-covalent interactions shall be rationally applied in the strategies. These non-covalent interactions are effortlessly unfair by external environments, for instance, pH values, temperature array, and the solvent polarity. Indeed, pH values are critical in peptide sequences richer in charged amino acids, such as Glu, Asp, Lys, His, and Arg, as stated above. As a consequence of the rank, these peptides can exhibit negative or positive shell charges. Then, those peptides can self-assemble into different nanostructures, according to the pH values.

34.2.2 Linear Peptides

Recently, data in the literature have reported that short (below six residues) and ultra-short (referred to dipeptides and tripeptides) linear peptides have the ability to self-assemble into many different nanostructures. This aspect, particularly interesting, allows to minimize the synthesis and purification steps and to reduce the cost of the production process [12].

One of the most studied prototypes of self-assembling linear peptides is the ultra-short homodipeptide Phe-Phe (FF) (Fig. 34.2) identified by Gazit and co-workers in 2003 as the smallest region of the Aβ-amyloid peptides ($A\beta_{1-40}$ and $A\beta_{1-42}$) prone to the aggregation [13]. The characterization of FF assemblies via single-crystal X-ray diffraction studies showed that this dipeptide is able to generate ring-like networks with a hexagonal symmetry, promoted by head-to-tail interactions established by its charged N- and C-terminus. This association trend is further stabilized by "T-shaped" contacts between the phenyl aromatic side chains [14]. Molecular dynamic simulations (MD) corroborated these observations, suggesting the ability of this system to form open ring-like peptide networks in aqueous solution [15]. Additional studies highlighted the structural versatility of this motif by showing that, despite its molecular simplicity, FF homodipeptide is able to form more complex supramolecular architectures [16]. Interestingly, simple modifications of the charged state at the C- or N-terminus of FF, strongly affected its self-assembling pathway. Indeed, the introduction of a thiol group or of a Fmoc-fluorenylmethyloxycarbonyl group can alter the self-assembling phenomena, for example [16].

Nanotubes, nanowires, nanofibrils, spherical vesicles, and organogels are just a few examples of the new peptide materials, based on FF self-assembly. These materials exhibit mechanical properties [17], electrical properties [18], electrochemical properties [19], or optical properties (photoluminescence [20, 21] and optical waveguide [22]) properties. In this contest, it is significant to highlight the main property of FF self-assembled nanotubes referring their thermal stability, which is the distinctive skill in bioinspired materials [23]. All these physicochemical characteristics make them suitable for several applications in nanomedicine (tissue engineering, drug delivery, and bioimaging) [18, 24, 25] and in nanofabrication fields (biosensors, nanodevices, and conducting nanomaterials) (Fig. 34.2) [26, 27].

In 2013, Alves et al. suggested for the first time FF-based microtubes (FF-MTs) as potential drug delivery vehicles. In their studies, the authors used Rhodamine B (RhB), a common dye, as a model drug. Data suggested the low *in vitro* toxicity of the FF-MTs and the potential of these carriers to deliver drugs at constant rates [28]. Beyond the low *in vitro* toxicity, FF-MTs units showed very high thermal stability (up to 120°C) and stability towards to the protease degradation. Self-assembled nanotubes obtained for aggregation of unnatural fluorinated-peptides containing two aryl units were found able to penetrate the cultured primary human smooth muscle cells and to locate in their cytoplasmic/perinuclear region [29]. Successively, FF-micro and nanotubes, covalently conjugated to folic acid/magnetic nanoparticles (FA/MNPs), were also evaluated as a potential delivery systems of the anti-cancer therapeutic 5-fluorouracil (5-FU), and of the anti-inflammatory cargo flufenamic acid (FFA) [30].

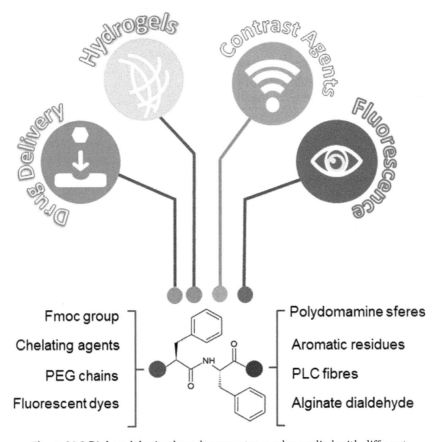

Figure 34.2 Diphenylalanine based aggregates can be applied with different biotechnological scopes, producing drug delivery systems, hydrogels matrices, supramolecular contrast agents, and fluorescent aggregates. Chemical and functional decorations (like sequences modification, incorporation of fluorescent dyes and conjugation with chelating agents and polymers) at N- and C-terminus of the primary sequence produce innovative nanostructurated tools.

Analogously to micro- and nanotubes, FF peptide-based nanofibers have also been recently proposed as vehicles of hydrophobic drugs, like hydroxycamphothecin (HCPT), for cancer therapy. In this study, the replacement of some L-amino acids with D-counterparts permitted to further improve the *in vitro* and *in vivo* biostability of these peptides against the hydrolysis catalyzed by endogenous peptidases. The protease stability allows for prolonged therapeutic effect with reduction of a tumor mass in a rat model [31]. At the same time, nanofibers of D-peptides generated via enzymatic dephosphorylation were also investigated for the controlled release of the anticancer drug taxol, and of a fluorophore (e.g., 4-nitro-2,1,3-benzoxadiazole) used as imaging agents *in vivo* [32]. Moreover, the combination of FF with others organic/inorganic molecules brought to the formation of novel hybrid smart materials responsive to the external stimuli, such as pH, enzyme, and oxidative stress. For example, aldehyde molecules can induce cationic diphenylalanine to assemble into biocompatible and biodegradable enzyme-responsive nanocarriers. These nanocarriers loaded with doxorubicin have been proposed as intelligent antitumor agents [33]. Another example of hybrid stimuli-responsive biomaterials is represented by magnetic hydrogel generated for co-assembly under mild conditions of FF with polydopamine spheres coated with Fe_3O_4 magnetic nanoparticles [34]. Successively, in 2016 Alves et al. reported the formulation of hybrid materials obtained by conjugation of electrospun polycaprolactone (PCL) fibers and micro/nanotubes of L,L-diphenylylalanine (FF-MNTs). This biodegradable matrix allows the achievement of a stable release of lipophilic anesthetic benzocaine over periods of up to ≈13 h, much higher than commercially available scaffolds [35].

Concurrently, Wu et al. reported FF-based hybrid nanospheres responsive to pH- and GSH-stimuli. In these spheres, natural alginate dialdehyde (ADA)was employed as cross-linker to induce self-assembly of FF and *in situ* reducer of Au^{3+} ions into Au nanoparticles (Au NPs). These biocompatible spheres were proposed as drug loading and delivery systems. Indeed, they were found able to encapsulate more than 95% of hydrophobic chemotherapeutic drug (camptothecin, CPT). CPT-loaded spheres exhibited satisfactory stability under normal physiological conditions and excellent pH- and GSH-responsive release at pH 5.0 with 10 mM GSH, which is similar to the tumor microenvironment. Moreover, these nanocarriers can be taken up by cancer cells and have greater cytotoxicity than free drugs [36].

In 2006, Ulijn and co-workers identified Fmoc-FF as one of the first dipeptides able to form a homogeneous, transparent, self-supporting hydrogel with fibrous nanostructure under physiological conditions [37]. The supramolecular nature of Fmoc-FF aggregates suggested its potential use in biological applications, such as controlled drug release, tissue engineering and cell culture. Many examples of multicomponent hydrogels as biocompatible drug delivery systems have been reported until now. FF based hydrogels have been also proposed as vehicle for the delivery of two complementary anticancer drugs, dexamethasone, and either taxol or dehydro-CPT. These peptide hydrogels showed a high *in vitro* biocompatibility for concentrations up to 100 µM over 48 h. Moreover, their principal advantages are the improved stability of the drugs over time at 37°C (i.e., 48 h for the taxol-derivative, and over two weeks for the others) and the slow drug release [38]. Beyond the encapsulation of anticancer drugs, peptide hydrogels can encapsulate non-steroidal anti-inflammatory drugs (NSAIDs) for local use [39] or SPECT tracers [40]. Recently, Fmoc-FF has been utilized in combination with plasmonic gold nanorods (AuNRs) for the development of near-infrared laser-activatable microspheres. These AuNR-embedded dipeptide microspheres, loaded with the anticancer agent, doxorubicin (DOX), were proposed as a smart drug-delivery platform for native, continuous and pulsatile drug release. Results of the study demonstrated the capability to achieve a sustained and on-demand DOX release from the microspheres by manipulating the laser exposure time [41].

Hybrid hydrogel encapsulating docetaxel were also prepared via the calcium-ion-triggered co-assembly of Fmoc-FF peptide and alginate. Due to the synergic effect of these two components, the final material presented much better stability than the single components in both water and a phosphate-buffered solution. Controlled drug release was obtained by varying the concentration ratio between the peptide and the polysaccharide [42]. Fmoc-FF dipeptide has been also utilized to confer mechanical rigidity and stability to natural polymers like hyaluronic acid (HA), a major component of the extracellular matrix. Fmoc-FF/HA composite hydrogels showed a sustained release of curcumin, a hydrophobic polyphenol showing antioxidant, anti-inflammatory, and antitumor activities. Additionally, in this study, it was observed a direct relationship between the rate of curcumin released and the concentration of the Fmoc-FF peptide within the hydrogel matrix [43]. Fmoc-FF/poly-L-lysine (PLL) injectable multicomponent hydrogels, encapsulating the photosensitive drug Chlorin e6 (Ce6), were also proposed as promising delivery platform in the photodynamic antitumor therapy. *In vivo* studies indicated an efficient inhibition of the tumor growth with no detectable toxicity or damages to normal organs during the treatment [44]. Inspired by Fmoc-FF, recently Adler-Abramovic et al. reported the synthesis of the peptide 6-nitroveratryloxycarbonyl-diphenylylalanine (Nvoc-FF) containing an ultraviolet (UV)-sensitive phototrigger [45]. The UV irradiation of the self-supporting hard hydrogel obtained by the self-assembling of this ultra-short peptide prompts the controlled release of the encapsulated insulin-fluorescein isothiocyanate (insulin-FITC), used as a drug model.

Over the years, multidisciplinary studies have revealed that the self-assembly of short linear peptides and still of single amino acids can make a broad range of diversified materials. It was observed that the conjugation of the polyethylene glycol (PEG) to short homopeptides, containing aromatic amino acids permits a substantial increase of their water solubility. The aromatic residues that can be such are, for example: phenylalanine (Phe, F), tyrosine (Tyr, Y) [46], tryptophan (Trp, W) [47], and naphthylalanine (Nal). Recently, Diaferia et al. proposed a polymer-peptide (PEGylated-F4) functionalized with Gd-DTPA and Gd-DOTA as contrast agent for potential diagnostic applications in magnetic resonance imaging (MRI) [48]. Single peptides in the aggregates are in a β-sheet conformation with an antiparallel alignment along the fiber axis. Each Gd-complex in the nanostructure exhibits a relaxivity value around 15 $mM^{-1}s^{-1}$ at 20 MHz, approximately three-fold higher than the classical contrast agents at low molecular weight (4.7 $mM^{-1}s^{-1}$). These relaxometric parameters are in line with other examples of Gd(III) based supramolecular (micelles or liposomes) contrast agents [49]. Due to the significant internalization efficiency and due to the high relaxivity values, these nanostructures are able to enhance the MRI cellular response on J774A.1 mouse macrophages cell line. In detail, within those cells the cytotoxicity of the fibril nanoaggregates was negligible with an incubation time of 3 h in the 0.5–5.0 mg/mL concentration range.

The same authors also evaluated the effect of the Gd-complex position in the aromatic framework and of the replacement of the phenylalanine with the non-coded amino acid 2-Nal [50, 51]. They observed that the different positioning of the chelating agent into the aromatic framework (at the center or at the N-terminus of the F4-motif) causes a drastic loss of the tendency to self-assemble and of the relaxivity value (11 mM^{-1}·s^{-1}). The decrease of the latter is related to the major flexibility of the Gd-complex on the supramolecular aggregate. On the other hand, the replacement of Phe in the homodimer with non-coded 2-Nal amino acid permits to restore in dipeptide π–π interactions and to prompt self-assembly of Gd-conjugates. This happened thanks to an extended aromatic ring in its side chain. Above 20 mg/mL, Gd-2Nal$_2$ peptide derivative gels and it was found able to encapsulate anticancer drugs like DOX, thus suggesting a potential use as theranostic systems.

Recently, Rosenman et al. [21, 52] reported the capability of FF and FFF NSs to emit photoluminescence (PL) in the blue and in the green spectral regions, upon thermally induced reconstructive phase transition. Kaminski et al. [53] also observed the same phenomenon in proteins and in amyloid-like fibrils rich in β-sheet structures. According to the literature, in 2016 Diaferia et al., synthesized series of PEGylated oligo-phenylalanines (PEG$_8$-F6, PEG$_{12}$-F6, PEG$_{18}$-F6, PEG$_{24}$-F6) able to generate supramolecular systems rich in β-organization [54, 55]. Due to the presence of their β-sheet structures, these polymer-peptides have been proposed as promising bioimaging agents. These peptide nanostructures keep their optoelectronic properties both in solution and at the solid state, upon excitation at 370, 410, and 460 nm, respectively. From a comparison along PEG-series arises that the PEG length and its composition could alter both the structural and the optoelectronic properties of the final material. The differences in the optoelectronic properties were attributed to the extension of the electron delocalization via hydrogen bonds along the cross β-structure of the peptide spine. This effect is due to the number of amide bonds along the PEG chain. With the aim to develop novel biocompatible peptide nanostructures as bioimaging tools, the same authors tried to improve the performance in terms of PL intensity and of wavelength range compatible with *in vivo* applications. They demonstrated that the intrinsic PL of these peptides nanostructures can be transferred to an acceptor dye like 4-chloro-7-nitrobenzofurazan (NBD) confined in proximity of the nanofiber. Then, the entrapment of NBD in these NSs caused a red-shift from 460 to 530 nm. This evidence was the proof of concept that PL can be red-shifted towards the infrared region of the visible spectrum [56].

34.2.3 Cyclic Peptides

In 1974, theoretical analysis suggested the possible arrangement of a cyclic peptide in a hollow tubular structure [57]. Twenty years later, Ghadiri and coauthors solved the first crystalline structure of nanotube structure, by ring stacking of cyclic peptides incorporating alternating D and L amino acids: cyclo-(L-Gln-D-Ala-L-Glu-D-Ala)$_2$ [58]. The peptide side chains were devised on the external area. It is observed that they were organized in the typical nanotube structures, as a consequence of the alternating D and L amino acids. The nanotubes are self-assembled and stabilized by hydrogen bonds between amide groups of the cyclic backbone.

In addition to alternating D- and L-type α-amino acids, several cyclic peptide sequences can make the self-assembly by alternating α- and β-amino acids, β-amino acids, and δ-amino acids by molecular stacking and H-bonds between backbones [59–62]. The size of the cavity depends on the length of the cyclic peptide, from 2 to 13 Å, increasing from a tetramer to a dodecamer. This parameter with charges on the side chain is essential for the use in biotechnological applications. By tailoring the chemical structure of the cyclic peptide, supramolecular self-assembled architectures can be accustomed to meet the requirements of applications, including stimuli-responsive nanomaterials antibacterial agents, for ion channeling and ion sensing and gene delivery.

Despite a large number of cyclic peptide nanotubes (cPNT) designed, their use as carriers of anti-cancer drugs is very poor. Zhang and co-workers [63] designed an eight-residue cyclic peptide containing Glu and Cys amino acids able to self-organize in a micro-scaled aggregate. PEGylated aggregates loaded with

DOX showed a high drug encapsulation ratio. Compared to free DOX, the PEG-modified DOX-loaded CPNT bundles demonstrated higher cytotoxicity, increased DOX uptake and altered intracellular distribution of DOX in human breast cancer MCF-7/ADR cells *in vitro*.

34.2.4 Amphiphilic Peptides (PAs)

Nature has elected amphiphilic molecules to generate life, by using them to circumscribe portions of the environment. Instead, membranes are able to confine biomolecules and to promote the transport of molecules and ions. Imitating the Nature, amphiphilic peptides self-assemble into different nanostructures, including vesicles, micelles, nanofibers, and nanotubes, thus playing a pivotal role in the production of nanomaterials for biotechnological applications [64, 65]. These molecules contain distinct hydrophobic and hydrophilic segments. The most simple peptides able to self-aggregate are constituted by short or long homo-chains of hydrophobic amino acids, like Val, Ala, Gly, and Phe, followed by one or more electrostatic charged residues (such as Asp, Glu, or His). The driving forces of the aggregation are electrostatic (for Asp, Glu, and His residues) and hydrophobic interactions (for Val, Ala, Gly, and Phe residues) that address into a wide variety of nanostructures, depending on their physical and chemical properties.

In 2002, Zhang and co-workers investigated peptide sequences containing an hydrophobic moiety (valine [66], glycine [67] and alanine residues [68]) and an hydrophilic head (one or two Asp residues or one of Lys). These molecules self-assemble into various nanotubes or nanovesicles. In particular, the Lys cationic residues on the head of these peptides could favor their own conjugation with negatively charged DNA and RNA opening the possibility of application in gene drug delivery. Later, Hamley's group has investigated a cationic peptide in which the hydrophobic tail consists in six Ala residues with an Arg head group. At low concentration, this peptide self-assembles in ultrathin sheets, whereas at higher concentrations the sheets wrap around themselves to form nanotubes and helical ribbons. This structure shows antimicrobial properties [69]. It is noteworthy that another sequence able to aggregate is obtained by alternating hydrophobic amino acids with residues bearing on the side chain positive and negative charges. In this case the EAK16 model peptide aggregates into nanofibers [70]. The methyl groups of alanines form a sheet structure inside, whilst the charge residues are exposed on the external wall. This structure is able to delivery ellipticine, an anticancer drug. The UV analysis and the fluorescence demonstrated electrostatic interactions and the conjugation method between the drug and the nanofibers [71].

Aliphatic peptides and lipopeptides were also proposed as building blocks for self-assembling. The feature of lipopeptides is the presence of different short or long alkylic chains as hydrophobic moiety, in the monomer structure. In this case, the aggregation is supported by van der Waals forces. The simplest lipopeptides are also able to form nanostructures. L-dodecanoylserine monomer forms: nanotubes, partially wrapped nanotubes and helical ribbon structures [72]. A peptide amphiphile comprising a single Lys residue, an alpha-(L-Lys),omega-(amino)bolaamphiphile, it was shown to form nanotubes in acidic aqueous solution [73]. Many lipophilic PAs can self-assemble into cylindrical nanofibers, as a consequence of H-bonds among peptide moieties and hydrophobic collapse of alkyl tails [74, 75]. The induction of self-assembly in these cylindrical structures could be obtained in aqueous media, in presence of suitable stimuli such as pH [24, 40]. The tetrapeptide sequence, composed by hydrophobic and negatively charged residues (Val-Glu-Val-Glu), it grafted to an alkyl tail at sixteen carbon atoms, self-assembled into monodispersed nanobelts in an aqueous solution at a concentration of 0.1 wt% [76]. In this regard, Hartgerink et al. described two different self-assembling modes [74]: acid-induced self-assembly and Ca^{2+} induction. For the acid-induced self-assembly, PA including $C_{16}H_{31}O$ grafted to $C_4G_3S(PO_4)RGD$, they can aggregate in nanofibers after dissolution in water and exposition to gaseous HCl. On the other hand, the treatment of a solution of these conjugates with Ca^{2+} instantly caused the gel formation in solution. This Ca^{2+}-induced self-assembly may be particularly helpful for medical applications at physiological pH, where formation of a gel is mandatory.

Drug delivery applications of hydrogelation based on PAs were studied by the Stupp's group [75]. In their studies, the chemical structure of the PA molecule ($C_{16}V_2A_2E_2$) is composed of three segments: an hydrophobic tail (palmytic acid), the well-established β-sheet amino acid sequence V_2A_2 and two negatively charged glutamates able to induce cross-linking in presence of Ca^{2+} ions in solution. Properties of this peptide as controlled drug release tool were investigated linking prodan, a fluorescent lipophilic tag used as a dielectric probe for cell membranes. This probe was linked to the peptide through a hydrazine bond inserting a Lys residue at different positions along the backbone of the peptide. This pH-sensitive bond can be broken in acidic conditions of a cell compartment. Hydrogel formation was induced by adding a 100 mM Ca^{2+} solution after dissolving PA in NaOH.

Lipopeptides with similar chemical structure can form self-assembled micelles. Stearyl-H_3CR_5C lipopeptide, that are crosslinked by disulfide bonds (SHRss), have been used to form DOX-loaded micelles with an average diameter of 233 nm by nanoprecipitation and probe-based ultrasonication methods. DOX and microRNA are then loaded into the micelles through hydrophobic interactions [77]. The DOX release profile strongly depends on the pH value. Indeed, a larger amount of drug was released at pH 5.5 (82.6%) (corresponding to the endolysomial pH) than at physiological pH 7.4 (63.3%). Cellular uptake has also been investigated by flow cytometry and confocal laser scanning microscopy (CLSM) on DU145 (human prostate cancer) cells. Results indicated that the uptake of micelles were time-dependent with an intracellular uptake rate higher after 4 h of incubation (89.50±0.99%) than after 1 h of incubation (82.56±1.55%). Cationic micelles can be obtained by self-assembly of PAs in which the hydrophobic moiety is represented by cholesterol (Chol) and the hydrophilic head contains a variable number of positively charged residues, such as histidine and arginine (Chol-CH_5R_5, Chol-CH_3R_3, Chol-CR_5, and Chol-CR_3,) [78]. These aggregates are able to adsorb on their surface different amount of DNA depending on the ratios between the arginine residues and DNA phosphate bases. Formation of micelles with palmitoyl-p53$_{14-29}$ has been studied by Missirlis et al. [79]. This PA self-assembled in 10 mM phosphate buffer; the hydrophobic interactions induced the simultaneous formation of micelles with a hydrodynamic diameter of 319 nm and the formation of elongated micelles with a diameter of 10 nm having a length of a few hundred nanometers.

34.3 Self-Assembling PAs for Targeting in Nanostructures

In PA-based nanostructures, the main goal of peptides is to drive the self-aggregation and to regulate the loading and the release of the encapsulated drug. In the wide category of PAs, the sequence of the amino acids is responsible for the targeting and delivery features. For this purpose, the sequence is selected on the ability to cross the cell membrane or to bind overexpressed receptors on the cell membranes.

34.3.1 Cell Penetrating Peptide (CPPs) and Smart Sequences

Cell penetrating peptides (CPPs) are a large class containing more than 1700 different experimentally validated sequences [80, 81]. Most common CPPs are cationic and they are widely used. A class of CPPs are derived from the α-helical domain of the Tat protein, covering residues from 48 to 60. Those residues are mainly composed of basic amino acids, such as the TAT dodecapeptide: GRKKRRQRRRPQ [82]. The CPP role as a nanovector has been described in many reviews [83, 84].

Despite high cellular uptake efficiency, CPPs lack cancer cell specificity. To overcome this drawback stimuli-responsive CPPs have been developed recently to enhance the cellular uptake of therapeutic cargo only in the tumor tissue. As previously reported, these stimuli can respond to pH variation or to enzyme activity or to oxidative stress. CPPs can be considered as responsive molecules when containing residues able to vary the net charge depending on the pH. One residue able to tune the net charge of the peptide is His. Therefore, Zhang et al. designed an α-helical CPP to obtain a pH-responsive peptide, by replacing all its lysines with histidines (THAGYLLGHINLHHLAHL(Aib)HHIL). This peptide (TH) showed a neutral charge at physiological pH, but the net charge became positive under acidic conditions, so that its cell penetration capacity was activated [85]. PEGylated liposomes functionalized with TH peptide showed a more efficient internalization into C26 colon cancer cells, when compared to non-targeted liposomes. Moreover, PTX-loaded liposomes suppressed C26 colon tumors *in vivo* with high apoptosis levels where the tumor inhibition rate reached 86.3%. Proteases abundant in tumor tissue can constitute internal stimuli for activating CPPs. Liu et al. developed a liposome able to carry DOX labeled with the sequence AAN-Tat [86]. The AAN sequence is a substrate of Legumain endoprotease. The addition of the AAN moiety to the fourth lysine in the TAT generates a branched peptide moiety, which leads to a decrease in the transmembrane transport capacity of TAT up to 72.65%. The action of the enzyme allows restoring the penetrating capacity of TAT. *In vivo* assays carried out on nude mice showed inhibition of the tumor growth significantly higher in mice administered with AAN-TAT liposomal DOX than control. Simultaneously, for the group treated with targeted DOC liposomes, it was also observed a prolonged survival period. Another stimulus could be derived externally, such as UV or NIR irradiation. In these cases, the CPP peptide is modified by the action of a residue with a photolabile protecting group. Upon reaching the tumor site, the peptide-functionalized liposomes are irradiated by UV or NIR light that cleaves the protective groups and in this way allows the CPP to play its role. Following this methodology, a CPP (CKRRMKWKK or CGRRMKWKK) enhances the efficiency of the translocation, derived from the penetration. This CPP is designed to turn into an inactive form by neutralizing the positive changes of the lysines, which are caged with photoresponsive groups [87].

34.3.2 Peptide Able to Interact with Overexpressed Receptors

In certain PAs, the receptor-targeting peptides are able to induce high levels of internalization within tumor cells due to a receptor-mediated endocytosis mechanism. The peptide sequence can be composed in this manner [88]. These strategies could allow the intracellular delivery of the payload. Some known endogenous proteins are able to bind the target receptor with high affinity. A significant topic of research is about how to preserve the affinity for the overexpressed receptors, especially after the conjugation to the hydrophobic moiety. Furthermore, evidence showed how all the residues which are involved in the receptor binding are well-exposed on the nanostructure surface. Those residues maintained a conformation suitable to the interaction with the receptor [89]. Further studies are aimed to preserve the *in vivo* chemical stability, due to the high sensitivity of peptides to the protease degradation. Improved metabolic stability and pharmacokinetics can be achieved by modifying peptide sequences with either specific coded amino acids or un-coded amino acids, or, as well, with amino acids in the D configuration. Alternative strategies consist of: the cyclization between the *N*- and *C*-terminals, the cyclization between the *N*- or *C*-terminal and a side-chain, or the cyclization between two side-chains.

Peptide sequences can act as cell surface receptor antagonists if molecules are modeled, allowing selective targeting towards to receptors. Antagonist peptides show a dual advantage if compared with their agonist counterparts: On the one hand, they do not act in the biological pathways following receptor binding; on the other hand they present higher binding capacities [90, 91]. However, the strategy of rational design of new compounds has some limits, one of the most significant being the limit of the knowledge requirement related to the structure of ligand/ receptor interaction [92]. A further possibility to identify novel peptide sequences is the use of the phage display technique, concerning recognizing tumor-associated proteins [93]. Next to the identification of a peptide sequence, some suitable spacers (charged or neutral) can be inserted between the hydrophobic region and the peptide. One of the most used spacers in this context is uncharged PEG. Indeed, the presence of one or more ethoxilic units permits an increase in the blood circulation time of the supramolecular aggregates (enhanced permeability retention effect). In addition, the lack of charge prevents interactions with the residues on the bioactive portion which could induce unnatural and incorrect conformations. Furthermore, the spacer allows the maintenance of the molecule's flexibility, mobility, and increases, in some cases, the solubility. The formulation of supramolecular aggregates, like micelles and liposomes, externally functionalized with bioactive peptides, may be obtained by using different approaches, such as pre-functionalization and the post-functionalization strategies [94–97]. In the pre-funtionalization, the peptide sequence is placed on the aggregate during the nanostructure preparation: Figure 34.3 (left panel) shows an amphiphilic peptide derivative added during the formulation step. In the post-functionalization strategy (Fig. 34.3, right panel), the peptide is chemically conjugated on the aggregate surface after nanostructure organization. The advantages of the first method are a defined quantity of bioactive molecules in the aggregate and the avoidance of impurities, but it requires as input a well-purified amphiphilic peptide molecule. However, in the case of liposomes, the bioactive peptide is located on both the external surface and in the inner aqueous compartment, after liposomial formulation. In the post-functionalization approach, peptide coupling concerns the introduction of suitable functional groups (already activated) onto the external side of liposomes

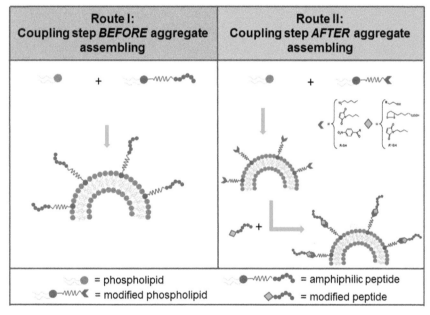

Figure 34.3 Schematic representation of two approaches used for the synthesis of peptide containing liposomes. In route I (pre-liposomal functionalization) a PA is inserted directly during the liposome formulation. In route II (post-liposomal functionalization) a peptide is anchored on the external surfaces after liposome formulation by a selective reaction between two functional groups displayed on the peptide and on the liposome, respectively.

or nanoparticles for covalent or non-covalent peptide binding. In order to make sure about the proper orientation of the targeting ligand, biorthogonal, and site-specific surface, it is necessary to choose the appropriate reactions. In this sense, the most used chemical approaches are: enzymatic ligation, Cu-free chemistry, the amine in case of the N-Hydroxysuccinimide coupling method; thiol for maleimide; Michael addition; azide for Cu(I)-catalyzed Huisgen cycloaddition (CuAAC); biotin for non-covalent interaction with avidin; triphosphines for Staudinger ligation; and hydroxylamine for the oxime bond [98].

In the literature, targeting peptides are tailored toward three broad types of receptors which are overexpressed or exclusively expressed in cancer vasculature and/or cancer cells: integrins; growth factor receptors (GFRs); and G-protein-coupled receptors (GPCRs). Several examples of supramolecular systems loaded with therapeutic or diagnostic agents and externally decorated with homing peptides, able to selective recognize integrin receptors or membrane receptors belonging to the GPCR superfamily, are reported in Table 34.1.

34.3.2.1 Peptide Target for Integrin Receptors

Integrins are heterodimers transmembrane receptors related to the cell-extracellular matrix (ECM) adhesion. Upon ligand binding, integrins activate cellular signals such as regulation of the cell cycle, organization of the intracellular cytoskeleton, and movement of new receptors to the cell membrane. Integrins are one of the most important receptors that can be used in active targeting strategies [122]. Among the different subfamilies of these heterodimeric transmembrane proteins, integrins $\alpha_v\beta_3$ and $\alpha_v\beta_5$ have prominent roles in angiogenesis and metastatic disseminations. The integrin $\alpha_v\beta_3$ plays a very domineering role in angiogenesis and is overexpressed in endothelial cells of the tumor. Recently a large exploration in the field of $\alpha_v\beta_3$ integrin-mediated bioactive targeting for cancer treatment has been reported. All designed peptide sequences contain the RGD motif.

Table 34.1 Supramolecular systems decorated with homing peptides able to selectively recognize integrin receptors or membrane receptors belonging to the GPCR superfamily

Receptor	Peptide Sequence	Peptide Derivative	Drug	Ref.
Integrin receptor Avβ3	c(RGDfK)	c(RGDfK)-NHS-PEG-PLA	CA4	[99]
	c(RGDfK)	c(RGDfK)-SH post liposome modification	CDDP	[100]
	c(RGDfC)	MBPE-c(RGDfC) post-insertion	DOX	[101]
	c(RGDf[N-Met]K)	c(RGDf[N-Met]K (Ac-SCH2CO))	DOX	[102]
	c(RGDyK)	DSPE-PEG-c(RGDyK)	CDDP	[103]
	cAbaRGD cAmpRGD	DSPE-PEG-cAbaRGD DSPE-PEG-cAmpRGD	DOX DOX	[104]
	iRGD	iRGD-HES-SS-C18 NCs	DOX/sorafenib	[105]
	Octreotide	OCA-DOTA/OCA-DTPAGlu	Gd-complex	[106]
	Octreotide	$(C18)_2(AdOO)_5OCT$	Gd-complex	[107]
	Octreotide	$(C18)_2(AdOO)_5OCT$	CDDP/DOX	[108]
	Octreotide	OCT-(PTX)-PEG-b-PCL	PTX	[109]
	Octreotide	Oct-Phe-PEG-SA	DOX	[110]
	Octreotide	H40-PLA-PEG-OCT	TDP-A	[111]
	Octreotide	SAMA-TOC post liposome modification	^{111}In-DTPA	[112]
	Octreotide	HSPE-PEG4000-OCT	DOX	[113]
G-Protein coupled receptor	[Tyr^3]-Octreotate	Maleimido-TATE	^{64}Cu-DOTA	[114]
	KE108	KE108 post micelle modification via NHS	TDP-A AB3	[115] [116]
	[7–14]BN wild-type	$(C18)_2$-L$_5$-[7–14]BN $(C18)_2$-PEG3000-[7–14]BN	^{111}In-DOTA	[117]
	[7–14]BN-AA1 analogue	MonY-BN-AA1	DOX	[118]
			AUL12	[119]
			DOX	[120]
	CCK8	$(C18)_2$-L$_5$CCK8	Gd-DOTA/ Gd-DTPA	[121]

Note: The bioactive peptide, the peptide conjugation, the encapsulated API, and the corresponding references are reported.

In most of the cases, the cyclization is commonly employed to improve the binding properties, conferring rigidity to the structure. In linear peptides, the fourth amino acid alters the binding specificity and the nature of residues, by flanking the RGD sequence. The fourth amino acid could influence receptor affinity, receptor selectivity, and other biological properties [123]. Therefore, nanoaggregates grafted with cRGD sequence have been widely evaluated for the treatment of different cancers, such as ovarian cancer, melanoma, and breast carcinoma [124, 125]. The first examples of aggregates functionalized with RGD containing peptides they were formulated only in half of the last decade. In fifteen years, more than 450 articles were published on RGD-labeled liposomes or micelles delivering hydrophilic drugs like DOX [126, 127], but also with many others anticancer drugs, such as cisplatin (CDDP) [128], paclitaxel (PTX) [129, 130], docetaxel (DTX) [131], combretastatin A4 (CA4) [132], and 5-fluorouracil (5-FU) [133, 134].

Generally, the RGD sequence is inserted in a five residue cycle in which one of the amino acids is in the D-configuration. Therefore, in the case of a single D-amino acid and four L-amino acids, the homodetic cyclic pentapeptide prefers a II′/conformation with the D amino acid in the $i + 1$ position of the II′-turn [135]. Most aggregates were grafted with the c(RGDfK) cyclic peptide (Fig. 34.4a). This sequence, developed by Kessler et al. [136], is able to target the $\alpha_v\beta_3$ and $\alpha_v\beta_5$ integrin receptors [137]. For example, c(RGDfK) was coupled to poly(L-lactide)-block-poly(ethylene glycol)-succinic ester (NHS-PEG-PLA) [138] to obtain polymeric micelles able to deliver hydrophobic drugs like CA4 [99].

The density of cRGD on the micelle surface can affect the amount of drug delivered into the cells. This issue was studied by Kataoka's group varying the quantity the monomer functionalized by cRGD ligand from 5% up to 40%. cRGDfK-labeled micelles were prepared by "post conjugating" Cys-containing cRGD peptides onto maleimide-functionalized DACHPt/micelles obtained from a mixture of poly(ethylene glycol)-b-poly(L-glutamic acid) (MeO-PEG-b-P(Glu)) and maleimide-conjugated poly(ethylene glycol)-b-poly-(L-glutamic acid) (Mal-PEG-b-P(Glu)) [100]. The better results in terms of uptake and cytotoxicity were observed for cis-platinum-loaded micelles functionalized with an amount of peptide ranging between 20% and 40%.

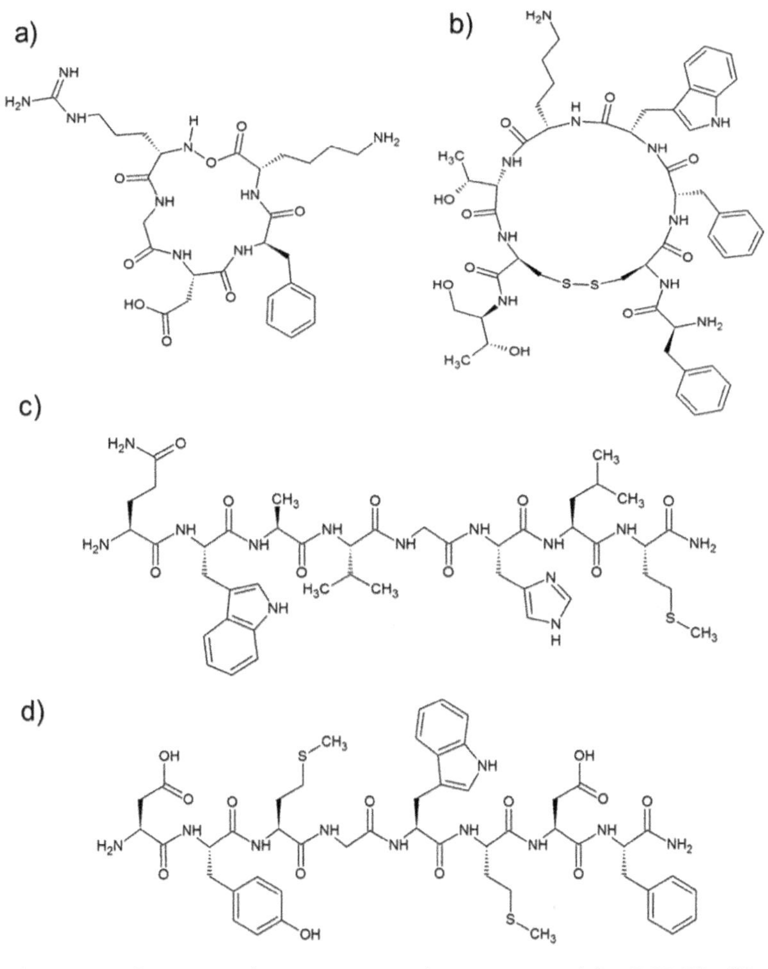

Figure 34.4 Structure of targeting peptide sequences: (**a**) c(RGDfK); (**b**) Otreotide[c(2–7)-FCFWKTCTol]; (**c**) [7–14]-Bombesin (H-QWAVGHLM-NH₂), and (**d**) CCK8(H-DYMGWMDF-NH₂).

Others analogue cyclic RGDs were also analyzed for drug delivery applications. One of those studies was performed by Tao et al., who formulated DOX loaded liposomes labeled with c(RGDfC): the peptide was conjugated to the liposomal surface by a thiol-maleimide coupling reaction with MBPE lipid [MBPE-c(RGDfC)] and the PEG coating of liposomes was obtained by using the post-insertion method [101]. Key factors in tumor therapy are biodistribution and clearance of the aggregates affected by peptide hydrophilicity. In view to assess the effect of the peptide hydrophilicity, PEGylated liposomal DOX has been tested *in vitro* on integrin-expressing HUVEC cells, after the insertion of three cyclic RGD analogues: RGDyC, RGDf[N-Met]K, and RGDfK. Liposomal systems grafted with the RGDf[N-Met]K sequence were than compared to the other two analogue sequences (RGDyC and RGDfK). They showed the lowest undesired (not) specific interactions with other integrin-presenting sites, localization in tumor, and lower DOX side effects [102]. Moreover, a further cyclic sequence c(RGDyk) was then put in place as a labeling cisplatin delivery system for therapeutic applications against bone metastasis derived from prostate cancer in a mouse model [103]. CDDP-loaded targeted liposomes showed a higher cytotoxicity (IC_{50} = 1.83 µM) than free drug (IC_{50} = 15.4 µM) or untargeted liposomal drug (IC_{50} = 10.0 µM). The capability of these targeted liposomes to be selectively accumulated in metastatic tumor bones was tested during several *in vivo* assays. They showed a clear tumor egression.

Moreover, other analogues were obtained introducing in the peptide cycle of the cyclo azabicycloalkane and aminoproline residues. Zanardi et al. arranged targeted liposomal doxorubicin by incorporating a 5% molar ratio of DSPE-PEGcAbaRGD or DSPE-PEG-cAmpRGD amphiphiles into cAbaRGD-LP or cAmpRGD-LP, respectively [104]. They also studied their *in vitro* behavior on three different cell lines (MCF7, HUVECs, and HepG2). Results showed how both targeted liposomes (cAbaRGD-LP or cAmpRGD-LP) possess higher kinetics of nuclei internalization and a higher percentage of cell death when compared to the free drug.

In the last decade, a new cyclic peptide (CRGDKGPDC) iRGD was identified for peptides hosting tumor metastases [139]. This peptide was found to bind αβ integrin overexpressed on the surface of cancer cells and on tumor-vessel cells, but not in normal vessel cells. In delivery applications, the peptide was anchored on the surface by a post-insertion method, in turn, to develop the iRGD properties of several aggregates transporting the isoliquiritigenin (ISL), a natural anti-breast cancer dietary compound [140] or the doxorubicin and sorafenib [105]. However, self-assembling iRGD-based amphiphilic molecules have rarely been reported. The targeting motif was chemically modified with a hydrophilic arginine-rich sequence and hydrophobic alkyl chains sequentially able to self-assemble in a nanostructure.

This adjustment aimed to deliver photosensitizer hypocrellin B for photodynamic application the iRGD [141]. Moreover, very recently, a new PA containing iRGD and a hydrocarbon chain, in addition to hydroxyethyl starch (HES), a semi-synthetic polysaccharide (iRGD-HES-SS-C18 NCs), was formulated [142].

34.3.2.2 GPR Target Peptide

A wide number of nanostructures were functionalized with peptides able to recognize GPCRs, in particular to target receptors for somatostatin (SST), cholecystokinin (CCK), gastrin-releasing peptides (GRP/Bombesin), lutein, and neurotensin.

34.3.2.2.1 Somatostatin receptors

Nanoaggregates directed toward somatostatin receptors have been widely exploited for diagnostic and therapeutic applications. Instead, a side effect is highly frequent in the expression of SSTR in human tumors of neuroendocrine origin, mostly affecting the expression in normal tissues. In general, SSTR2 is the most common SSTR subtype found in human tumors, followed by SSTR1, with SSTR3, that are four and five times less common. Due to the very low *in vivo* half-life of the wild-type SST tetradecapeptide, researchers have preferred to label aggregates with more stable somatostatin analogues. The most renowned selected analogue is the octreotide (OCT) shown in Fig. 34.4b. The OCT is a cyclic peptide containing eight amino acids in L and D configuration, developed in 1992 by Sandoz (now Novartis) [143]. This cyclic peptide is able to cross the cell membranes via endocytosis by binding to SSTR2 with high affinity and inhibitory concentration better than wild-type SST (IC_{50} = 2 nM). The OCT is able to bind also SSTR3 (IC_{50} = 376 nM) and SSTR5 (IC_{50} = 299 nM), but with a lower degree if compared to SSTR2. Moreover, the OCT peptide binding to receptors is not affected by chemical modifications on its *N*-terminus.

Octreotide-labeled aggregates may be obtained following both previously reported approaches and employing opportune strategies aimed to avoid possible undesired compounds. Many studies demonstrated that the β-like turn, formed by Phe-Lys-DTrp-Thr residues, is involved in receptor binding [144]. Therefore, it is essential to verify the retention of the amino acid configuration and the exposition of the tryptophan residue on the external aggregate surface. Hence, after their formulation, peptide properties on the liposomes have to be fully characterized. In 2009, Morisco et al. synthetized OCT amphiphlic molecules able to self-assemble in micelles for the selective delivery of MRI contrast agents [106]. These PAs contain three different regions: a hydrophobic moiety based on two stearyl chains, a chelating agent (DTPAGlu or DOTA) able to coordinate Gd^{3+} ions as a contrast agent, and the bioactive peptide. Fluorescence studies indicate for all micelles a complete exposure of OCT on the surface. CD measurements show the predominant presence of a β-sheet peptide conformation, characterized by a β-like turn.

The majority of aggregates in the literature are not formulated only by PA self-assembling, but they are obtained by mixing PAs with other surfactant molecules. Morelli's group has studied mixed aggregates formulated by co-assembling of the OCT lipopeptide with a second monomer containing in the hydrophilic head: a metal complex acting as diagnostic or therapeutic agent. In diagnostic aggregates (Gd-DTPAGlu, Gd-DTPA, and Gd-DOTA complexes), the metal chelate is covalently bound through a lysine residue to two eighteen-carbon chains [107]. Structural characterization of the aggregates indicates a shape and size of the supramolecular aggregates suitable for *in vivo* use. Therapeutic aggregates were formulated by co-assembling, at a 10/90 molar ratio, of the OCT lipopeptide with a second amphiphilic monomer containing a cytotoxic platinum complex anchored to the lipophilic tails, $(C18)_2PKAG-Pt$ [108]. The $(C18)_2$-PKAG-Pt/$(C18)_2(AdOO)_5$-OCT mixed aggregates generate large liposomes with an average diameter of 168 nm. These liposomes were further loaded in their inner aqueous compartment with the hydrophilic DOX drug. Indeed, platinum complexes are frequently used as chemotherapeutics, in combination with other drugs such as paclitaxel, bleomycin, vinblastine, and in several trials with DOX. This represents the proof of concept of combined therapy based on DOX and platinum complexes.

Targeted OCT aggregates were also largely investigated as carriers for the delivery of hydrophobic anticancer drugs, such as paclitaxel (PTX): a mitotic inhibitor used to treat patients with lung, ovarian, breast, head and neck cancers, and advanced forms of Kaposi's sarcoma. Zhang et al. loaded PTX in polyethylene glycol-polycaprolactone (PEG-PCL) polymeric micelles, they obtained the

OCT-(PTX)-PEG-b-PCL (OCT-M-PTX) and the salinomycin (SAL)-loaded PEG-b-PCL (M-SAL). The OCT was coupled to NHS-PEG-b-PCL through the activated NHS group [109]. These micelles had a diameter of approximately 25–30 nm and the encapsulation efficiency of the drug was 90%. Moreover, by adding free OCT, the interaction was inhibited, then it was confirmed that cellular uptake occurs through a receptor-mediated mechanism.

Zou et al. coupled OCT to hydrophobilized chitosan polymer [110]. This peptide derivative was able to self-assemble in micelles having very low cytotoxicity, an excellent biocompatibility, and biodegradability. In detail, the authors formulated *N*-octyl-*O,N*-carboxymethyl chitosan (OCC) and *N*-deoxycholic acid-*O,N*-hydroxyethylation chitosan (DAHC) micelles. Then, they conjugated the OCT on the *N*-terminal moiety of free carboxylic groups of OCC. The coupling had an extremely low (about 3%) yield, which is largely due to the high molecular weights of OCT and chitosan derivatives, due to the strong hydrogen bonds in the chitosan backbone, and due to poor solubility of chitosan derivatives in organic solvent. This result pushed toward alternative mixed aggregates, adding to DAHC a ligand-PEG-lipid conjugate able to guarantee same long circulation time in blood and ligand targeting. Both micelle types showed good DOX loading capability, with a drug loading content (DLC) in the 22–30% range. As an alternative strategy, the same authors anchored *N*-terminal peptides in solution to a PEG fragment and this moiety was conjugated to an aliphatic chain or to the deoxycholic acid obtaining the OCT(Phe)-PEG-SA (OPS) monomer or the OCT(Phe)PEG-DOCA (OPD), respectively [110].

More recently, OCT-functionalized unimolecular micelles were exploited to delivery thailandepsin-A (TDP-A) toward neuroendocrinal tumor cells. TDP-A is a relatively new naturally produced histone deacetylase (HDAC) inhibitor. Target selective micelles were obtained by the self-assembling of individual hyperbranched polymer molecules, consisting of a hyperbranched polymer core (Boltorn®H40) and approximately 25 amphiphilic polylactide-poly(-ethlyene glycol) (PLA-PEG) block copolymer arms covalently bound through the succinimidyl group (NHS) to octreotide (H40-PLA-PEG-OCT) b [111].

Helboketal. [112] synthesized an amphiphilic OCT derivative by cross-linking the *S*-acetyl-mercaptopropionic acid peptide

(SAMA-TOC) with the Mal-DSPE-PEG2000 phospholipid. Next mixed liposomes were obtained by adding to the OCT derivative adequate amounts of palmitoyloleoyl-phosphatidylcholine (POPC), lyso-stearyl-phosphatidylglycerol (Lyso-PG), distearyl phosphatidylcholine–polyethyleneglycol-2000 (DSPE-PEG2000), and dimyristoylphosphoethanolamine-DTPA (DMPE-DTPA) in a molar ratio of 0.1:11:7.5:0.9:2, respectively. These aggregates are usually employed in nuclear medicine applications radiolabeling with indium-111.

Octreotide-targeted liposomal doxorubicin was constructed with different ligand density by post-inserting HSPE-PEG4000-Octreotide into pre-formed liposomes. The octreotide ligand insertion was confirmed by the activity detection of octreotide in HSPE-PEG4000-Octreotide with synchronous fluorescence. Results indicated that an octreotide density around 1% could achieve the best uptake efficiencyon NCI-H-446 and SMMC-7721 cell lines among the studied liposomes [113].

Similar properties were shown by the somatostatin [Tyr^3]-octreotate (TATE) analogue. Petersenet al. [114] conjugated this peptide to maleimide, covalently attached to the distal end of DSPE-PEG2000 via a thioether bond. Targeted liposomes (DSPC/Chol/DSPE-PEG2000/DSPE-PEG2000-TATE in a molar ratio of 50:40:9:1, respectively), they encapsulated a positron emitter ^{64}Cu, as diagnostic agent for positron emission tomography (PET) imaging. Peptide-labeled liposomes displayed significantly higher tumor-to-muscle (T/M) ratio (12.7±1.0) compared to control-liposomes without TATE (8.9±0.9) and to the ^{64}CuDOTA-TATE peptide (7.2±0.3). These results reveal the feasibility of utilizing somatostatin analogs for specific targeting of the above-described aggregates to tumors overexpressing somatostatin receptors.

Very recently, TDP-A and AB3, new histone deacetylase inhibitors, they were encapsulated in the hydrophobic core of self-assembling micelles labeled with a somatostatin analog KE108 (PAMAM–PVL–PEG–OCH$_3$/Cy5/KE108) [115, 116]. This nonapeptide analogue contains the Phe–D-Trp–Lys–Thr motif, crucial for high-affinity somatostatin receptor binding like octreotide. Being formed by eight residues, the cycle size of this analog is larger than octreotide. It possesses high affinity to all five subtypes of SSTR. KE108 exhibited superior targeting ability in medullary thyroid cancer (MTC) cells, if compared to octreotide.

34.3.2.2.2 Bombesin receptors

The four receptor subtypes which are associated with the Bombesin-like peptides (BLP) family have been identified and found to be overexpressed in prostate, breast, small cell lung, [145] ovarian, and gastrointestinal stromal tumors [146]. A peptide able to bind these receptors is the bombesin (BN), which is constituted by 14 aminoacid residues. Its eight-residue *C*-terminal peptide sequence ([7–14]BN), reported in Fig. 34.4c, and many other BN analogs have been modified to selectively carry diagnostic or therapeutic agents to their receptors. They act both as agonists or antagonists. Many studies demonstrate that the [7–14]BN fragment and its analogues conjugated on the N-terminus with amino acid linkers, aliphatic or hydrophilic moiety, they all keep the affinity for receptors [147–149].

Despite the large overexpression of these receptors only few aggregates were developed. In the first example, Accardo et al. prepared mixed liposomes composed by two amphiphilic derivatives $(C18)_2$-L5-[7–14]BN (or $(C18)_2$-PEG3000-[7–14]BN) and $(C18)_2$-DOTA(^{111}In), both of them containing the same hydrophobic portion (two stearyl tails) and alternatively BN peptide or indium complex. The presence of a metal complex could allow to detect the *in vitro* fate of the liposome and its binding capability. Peptide was anchored to the alkyl chains trough different ethoxylic spacers (L5 or PEG3000). These spacers permit to improve the hydrophilicity of the final monomer and to increase the bioavalability of the peptide sequence on the external surface of the liposome. It is worth noting how to perform a really active targeting, as it is relevant to consider the length of the ethoxylic region. In fact, a long chain hides the bioactive sequence [117]. Successively, the same authors synthetized DSPC-based liposomes derivatized by the pre-functionalization approach with the MonY-BNAA1 monomer containing [7–14]BN analogue, DOTA chelating agent and the alkyl chains in the same molecule. Specific binding capability and cytotoxicity of these targeted liposomes, loaded with DOX, were carried out in PC-3 xenograft-bearing mice. An inhibition of the tumor growth in mice treated with DSPC/MonY-BN/DOX targeted liposomes [118] was observed.

More recently, the same sequence was grafted to cholesterol by a click chemistry, following a post-insertion method. The liposome obtained by mixing this monomer with DPPC was able to load subphthalocyanines (SubPc), an interesting hydrophobic probe for optical imaging, with a geometry that prevent aggregation [150]. An amphiphilic derivative of the [7–14]BN peptide was also used to prepare sterically stabilized mixed micelles (SSMMs) as drug delivery systems for gold(III) complexes (AUL12). The latter is already known for its *in vitro* and *in vivo* high antitumor activity, even in the CDDP-resistant cell lines. These micelles were able to encapsulate the hydrophobic metal complex with high loading efficiency while maintaining the gold (III) complex table in the +3 oxidation state over a period of 72 h. The *in vitro* binding ability and cytotoxicity of this target selective micelles were assessed in PC-3 cells overexpressing the GRP/bombesin receptors [119].

Anyway, circulation time *in vivo* of the [7–14]BN wild-type ($t_{1/2}$ = 15.5 h) remains relatively short. This evidence led the same authors to develop a new peptide analog, BNAA1, in which Sta^{13}-Leu^{14} and the Gly^{11} residue with the *N*-methyl-glycine replaced Leu^{13}-Met^{14} residues. These changes were finalized to increase the resistance towards the aminopeptidase and carnitine enzymes. The labeled DSPC/MonY-BN-AA1/DOX liposomes reduce the tumor volume showing value reductions superior to 20% when compared to DSPC/MonY-BN/DOX liposomes [120].

34.3.2.2.3 CCK receptors

In neuroendocrine origin tumors, such as medullary thyroid cancers, it was found that both CCK1 and CCK2 receptors were overexpressed. The same phenomenon was found in small cell lung cancers and in gastroenteropancreatic (GEP) tumors. The peptide CCK8 is able to recognize both receptors.

In Fig. 34.4d, one can see the eight residue C-terminus sequence of the endogenous hormone cholecystokinin. The CCK8 can be tailored on *N*-terminus without affecting receptor binding. This feature is essentially due to the interaction of receptor *N*-terminal extra domain with the amino acid side chains. The latter lies on the *C*-terminal moiety of the peptide ligand, as demonstrated by solution NMR [151] and theoretical studies [152]. Based on these data, in the last 10 years, Accardo et al. developed a wide class of CCK8-decorated supramolecular aggregates (namely Naposomes), by anchoring the bioactive moiety through the *N*-terminus [107]. The exposition of the CCK8 peptide was assessed by fluorescence measurements [153]. However, the peptide availability on surface aggregates is not an exclusive requirement for the receptor binding: the correct peptide conformation is crucial to assure high affinity and selectivity in ligand/protein binding processes. In this case, the CCK8 peptide needs to adopt a pseudo-α-helix conformation to give high binding affinity towards to the CCK1-R and CCK2-R receptors, according to the membrane-bound pathway theory [151]. The authors demonstrated that only peptide amphiphiles having an initial random coil conformation were able to adopt the pseudo-α-helix conformation in the presence of the receptor. Unlike them, peptides like $(C18)_2$-L_5CCK8, in which the peptide displays a β–sheet conformation, do not show *in vitro* cellular binding. Closing that chemical modification on the CCK8, the *N*-terminus seems to play an important role in stabilizing the peptide active conformation in self-assembling. The CCK8 amphiphilic monomers were combined with a second monomer containing the DOTA or DTPA chelating agent (general formula: $(C18)_2$-L_5CCK8 and $(C18)_2$-CA, respectively). The morphology and the size of the resulting aggregates (micelles, liposomes or open bilayers) are determined by several parameters, such as ionic strength, pH, monomer structure (length of polioxiethylene spacers), composition, and formulation procedure (dissolution in buffered solution or well-assessed procedures based on sonication and extrusion) [107, 121, 154]. Moreover, these aggregates can play a double role as theranostic delivering contrast agents and drugs [155].

34.3.2.3 Supramolecular System Based on Disordered Linear Peptides

The design of supramolecular systems could drive the disordered peptides to fold into a stable structure. This structural modification could be a promising route to develop a new class of bio-molecules for processes in which a specific conformational rearrangement is required [156]. These considerations deserve an in-depth study of the intrinsic disorder of peptide behavior in solution and their performance on surface of nanostructures [157]. Recently, the authors have studied the structural preferences of linear synthetic peptides with CPC-containing sequences (chemokine receptor CXCR4) characterized by the presence of some unordered amino acids [158]. In particular, these studies showed the conformational flexibility of both peptides, tested on the CXCR4 receptor through an indirect binding assay. Additionally, the authors tested the inhibition of CXCL12-induced migration and cAMP reduction. In addition, they proved how disordered peptides possess a stronger inhibitory capability on the adenilate cyclase, if compared to the AMD3100, which is, nowadays, the best characterized CXCR4 inhibitor. Trial evidence highlights that short, flexible peptides with no regular secondary structure can dynamically explore some conformational ensembles by targeting the chemokine receptor CXCR4. The employment of intrinsically disordered peptides could lie in the skill to control the transition between different structural states, especially as biosensors and in molecular recognition [159].

34.4 Conclusions

In this review, we have reported the most recent evidence on peptide-based drug-delivery systems in biotechnological applications. During the examination of the very rich literature data, several very remarkable and significant aspects have come into sight. Without a doubt, the extensive use of peptides to build more complex molecular constructs for biotechnological applications is well known. This is mostly due to their ease of achieving, and automation in, the synthesis of ad hoc designed sequences. In addition, peptides are also suitable for modification and control to gain desired biostructures in different aggregates. Additionally, several of their specific features allow operative research groups to obtain a broad variety of biotechnological materials. Furthermore, the option to arrange them in both linear and cyclic peptide sequences is worth mentioning; the likelihood to draw on side chains of the amino acid residues; the possibility to load on them charges and functional groups; and, finally, the intrinsic opportunity to arrange predictable physical and chemical patterns, suitable for biotechnological modular applications. In structured and/or disordered peptides, we can also consider the option of using conformational preferences: we can always put up micelles, liposomes, and gels based on peptides with preferential secondary dimensions and structures.

All these characteristics can also be engaged by means of bioactive sequences and/or through the recognition of post-translation moieties between biosystems. Therefore, it seems evident that the concrete possibilities that these biomaterials open up many sectors of peptide research, which can be engineered for specific applications in the various biotechnology sectors.

As said above, nature itself has elected amphiphilic molecules to generate life, by using them to circumscribe a portion of the environment. Indeed, surfactants constitute membranes able to contain biomolecules inside cells and they can select and transport molecules and ions. While imitating nature, amphiphilic peptides self-assemble into different nanostructures, such as vesicles, micelles, nanofibers, and nanotubes. In our guess, this is the way can play a key role in the production of new nanomaterials designed for biotechnological applications.

Abbreviations

5-FU:	5-fluorouracil
ADA:	alginate dialdehyde
APIs:	active pharmaceutical ingredients
Au NPs:	Au nanoparticles
AuNRs:	gold nanorods
BLP:	Bombesin-like peptides
BN:	bombesin
CA4:	combretastatin A4
CCK:	cholecystokinin
CDDP:	cisplatin
CLSM:	confocal laser scanning microscopy
cPNT:	cyclic peptide nanotubes
CPPs:	cell penetrating peptide
CPT:	camptothecin
CuAAC:	Cu(I)-catalyzed Huisgen cycloaddition
DAHC:	*N*-deoxycholic acid-*O,N*-hydroxyethylation chitosan
DLC:	drug loading content
DMPE-DTPA:	dimyristoylphosphoethanolamine-diethylenetriaminepentaacetic acid
DTPA:	diethylenetriaminepentaacetic acid

DOTA:	1,4,7,10-tetraazacyclododecane-1,4,7,10-tetraacetic acid
DOX:	doxorubicin
DTX:	docetaxel
ECM:	extracellular matrix
FA/MNPs:	folic acid/magnetic nanoparticles
FF:	Phe-Phe
FF-MTs:	FF-based microtubes
FFA:	flufenamic acid
GEP:	gastroenteropancreatic
GFRs:	growth factor receptors
GPCRs:	G-protein-coupled receptors
GRP:	gastrin-releasing peptides
HA:	hyaluronic acid
HDAC:	histone deacetylase
HES:	hydroxyethyl starch
HCPT:	hydroxycamphothecin
insulin-FITC:	insulin-fluorescein isothiocyanate
Lyso-PG:	lyso-stearyl-phosphatidylglycerol
Mal-PEG-b-P(Glu):	maleimide-conjugated poly(ethylene glycol)-b-poly-(L-glutamic acid)
MD:	molecular dynamic simulations
MRI:	magnetic resonance imaging
MTC:	medullary thyroid cancer
NBD:	4-chloro-7-nitrobenzofurazan
NSAIDs:	non-steroidal anti-inflammatory drugs
OCC:	*N*-octyl-*O*,*N*-carboxymethyl chitosan
OCT:	octreotide
OPD:	OCT(Phe)PEG-DOCA
OPS:	OCT(Phe)-PEG-SA
PAs:	amphiphilic peptides
PCL:	polycaprolactone
PEG:	polyethylene glycol

PEG-PCL:	polyethylene glycol-polycaprolactone
PET:	positron emission tomography
PL:	photoluminescence
PLA-PEG:	polylactide-poly(-ethlyene glycol)
PLL:	poly-L-lysine
POPC:	palmitoyloleoyl-phosphatidylcholine
PTX:	paclitaxel
SAL:	salinomycin
SSMMs:	sterically stabilized mixed micelles
SST:	somatostatin
TDP-A:	thailandepsin-A
UV:	ultraviolet

Disclosures and Conflict of Interest

This chapter was originally published as: Tesauro, D., Accardo, A., Diaferia, C. Milano, V., Guillon, J. Ronga, L. and Rossi, F. (2019). Peptide-based drug-delivery systems: Recent advances and perspectives, *Molecules*, **24**, 351, doi:10.3390/molecules24020351, under the Creative Commons Attribution license (http://creativecommons. org/licenses/by/4.0/). It appears here, with edits and updates, by kind permission of the authors and publisher, MDPI (Basel). The authors declare that they have no conflict of interest. No writing assistance was utilized in the production of this chapter and the authors have received no payment for it. This research was received external funding by Bando Vinci 2016, C4-4. Vittoria Milano thanks the Università Italo-Francese (UIF) for financial post-doc support (project Bando VINCI 2016, C4-4).

Corresponding Author

Dr. Filomena Rossi
Department of Pharmacy and CIRPeB
Università Federico II, 80134 Naples, Italy
Email: filomena.rossi@unina.it

References

1. Ruber Perez, C. M., Stephanopoulos, N., Sur, S., Lee, S. S., Newcomb, C., Stupp, S. I. (2015). The powerful functions of peptide-based bioactive matrices for regenerative medicine. *Ann. Biomed. Eng.*, **43**, 501–514.

2. Sahoo, S. K., Labhasetwar, V. (2003). Nanotech approaches to drug delivery and imaging. *Drug Discov. Today*, **8**, 1112–1120.

3. Panda, J. J., Chauhan, V. S. (2014). Short peptide based self-assembled nanostructures: Implications in drug delivery and tissue engineering. *Polym. Chem.*, **5**, 4418–4436.

4. Brack, A., Orgel, L. E. (1975). ß structures of alternating polypeptides and their possible prebiotic significance. *Nature*, **256**, 383–387.

5. Potekhin, S. A., Melnik, T. N., Popov, V., Lanina, N. F., Vazina, A. A., Rigler, P., Verdini, A. S., Corradin, G., Kajava, A. V. (2001). De novo design of fibrils made of short α-helical coiled coil peptides. *Chem. Biol.*, **8**, 1025–1032.

6. Wagner, D. E., Philips, C. L., Ali, W. M., Nybakken, G. E., Crawford, E. D., Schwab, A. D., Smith, W. F., Fairman, R. (2005). Toward the development of peptide nanofilaments and nanopores as smart materials. *Proc. Natl. Acad. Sci. U.S.A.*, **102**, 12656–12661.

7. Moutevelis, E., Woolfson, D. N. (2009). A periodic table of coiled-coil protein structures. *J. Mol. Biol.*, **385**, 726–732.

8. Aggeli, A., Nyrkova, I. A., Bell, M., Harding, R., Carrick, L., McLeish, T. C. B., Semenov, A. N., Boden, N. (2001). Hierarchical self-assembly of chiral rod-like molecules as a model for peptide β-sheet tapes, ribbons, fibrils, and fibers. *Proc. Natl. Acad. Sci. U.S.A.*, **98**, 11857–11862.

9. Fishwick, C. W. G., Beevers, A. J. L., Carrick, M., Whitehouse, C. D., Aggeli, A., Boden, N. (2003). Structures of helical β-tapes and twisted ribbons: The role of side-chain interactions on twist and bend behavior. *Nano Lett.*, **3**, 1475–1479.

10. Aggeli, A., Bell, M., Carrick, L. M., Fishwick, C. W. G., Harding, R., Mawer, P. J., Radford, S. E., Strong, A. E., Boden, N. (2003). pH as a trigger of peptide β-sheet self-assembly and reversible switching between nematic and isotropic phases. *J. Am. Chem. Soc.*, **125**, 9619–9628.

11. Schneider, J. P., Pochan, D. J., Ozbas, B., Rajagopal, K., Pakstis, L., Kretsinger, J. (2002). Responsive hydrogels from the intramolecular

folding and self-assembly of a designed peptide. *J. Am. Chem. Soc.*, **124**, 15030–15037.

12. Veiga, A. S., Sinthuvanich, C., Gaspar, D., Franquelim, H. G., Castanho, M. A. R. B., Schneider, J. P. (2012). Arginine-rich self-assembling peptides as potent antibacterial gels. *Biomaterials*, **33**, 8907–8916.

13. Reches, M., Gazit, E. (2003). Casting metal nanowires within discrete self-assembled peptide nanotubes. *Science*, **300**, 625–627.

14. Görbitz, C. H. (2006). The structure of nanotubes formed by diphenylalanine, the core recognition motif of Alzheimer's β-amyloid polypeptide. *Chem. Comm.*, **22**, 2332–2334.

15. Tamamis, P., Adler-Abramovich, L., Reches, M., Marshall, K., Sikorski, P., Serpell, L., Gazit, E., Archontis, G. (2009). Self-assembly of phenylalanine oligopeptides: Insights from experiments and simulations. *Biophys. J.*, **96**, 5020–5029.

16. Yan, X., Zhu, P., Li, J. (2010). Self-assembly and application of diphenylalanine-based nanostructures. *Chem. Soc. Rev.*, **39**, 1877–1890.

17. Adler-Abramovich, L., Kol, N., Yanai, I., Barlam, D., Shneck, R. Z., Gazit, E., Rousso, I. (2010). Self-assembled organic nanostructures with metallic-like stiffness. *Angew. Chem. Int. Ed.*, **49**, 9939–9942.

18. Wang, M., Du, L., Wu, X., Xiong, S., Chu, P. K. (2011). Charged diphenylalanine nanotubes and controlled hierarchical self-assembly. *ACS Nano*, **5**, 4448–4454.

19. Vasilev, S., Zelenovskiy, P., Vasileva, D., Nuraeva, A., ShurYa, V., Kholkin, A. L. (2016). Piezoelectric properties of diphenylalaninemicrotubes prepared from the solution. *J. Phys. Chem. Solids*, **93**, 68–72.

20. Nikitin, T., Kopyl, S., Shur, V. Y., Kopelevich, Y. V., Kholkin, A. L. (2016). Low-temperature photoluminescence in self-assembled diphenylalanine microtubes. *Phys. Lett. A*, **380**, 1658–1662.

21. Handelman, A., Kuritz, N., Natan, A., Rosenman, G. (2005). Reconstructive phase transition in ultrashort peptide nanostructures and induced visible photoluminescence. *Langmuir*, **32**, 2847–2862.

22. Handelman, A., Apter, B., Turko, N., Rosenman, G. (2016). Linear and nonlinear optical waveguiding in bio-inspired peptide nanotubes. *Acta Biomater.*, **30**, 72–77.

23. Adler-Abramovich, L., Reches, M., Sedman, V. L., Allen, S., Tendler, S. J. B., Gazit, E. (2006). Thermal and chemical stability of diphenylalanine peptide nanotubes: Implications for nanotechnological applications. *Langmuir*, **22**, 1313–1320.

24. Marchesan, S., Vargiu, A. V., Styan, K. E. (2015). The Phe-Phe motif for peptide self-assembly in nanomedicine. *Molecules*, **20**, 19775–19788.

25. Adler-Abramovich, L., Aronov, D., Beker, P., Yevnin, M., Stempler, S., Buzhansky, L., Rosenman, G., Gazit, E. (2009). Self-assembled arrays of peptide nanotubes by vapour deposition. *Nat. Nanotechnol.*, **4**, 849–854.

26. Scanlon, S., Aggeli, A. (2008). Self-assembling peptide nanotubes. *Nano Today*, **3**, 22–30.

27. Hendler, N., Sidelman, N., Reches, M., Gazit, E., Rosenberg, Y., Richter, S. (2007). Formation of well-organized self-assembled films from peptide nanotubes. *Adv. Mater.*, **19**, 1485–1488.

28. Silva, R. F., Araujo, D. R., Silva, E. R., Ando, R. A., Alves, W. A. (2013). L-diphenylalanine microtubes as a potential drug-delivery system: characterization, release kinetics, and cytotoxicity. *Langmuir*, **29**, 10205–10212.

29. Bonetti, A., Pellegrino, S., Das, P., Yuran, S., Bucci, R., Ferri, N., Meneghetti, F., Castellano, C., Reches, M., Gelmi, M. L. (2015). Dipeptide nanotubes containing unnatural fluorine-substituted-diarylamino acid and L-alanine as candidates for biomedical applications. *Org. Lett.*, **17**, 4468–4471.

30. Emtiazi, G., Zohrabi, T., Lee, L. Y., Habibi, N., Zarrabi, A. (2017). Covalent diphenylalanine peptide nanotube conjugated to folic acid/magnetic nanoparticles for anti-cancer drug delivery. *J. Drug Deliv. Sci. Technol.*, **41**, 90–98.

31. Liu, J., Liu, J., Chu, L., Zhang, Y., Xu, H., Kong, D., Yang, Z., Yang, C., Ding, D. (2014). Self-assembling peptide of D-amino acids boosts selectivity and antitumor efficacy of 10-hydroxycamptothecin. *ACS Appl. Mater. Interfaces*, **6**, 5558–5565.

32. Li, J., Gao, Y., Kuang, Y., Shi, J., Du, X., Zhou, J., Wang, H., Yang, Z., Xu, B. (2013). Dephosphorylation of D-peptide derivatives to form biofunctional, supramolecular nanofibers/hydrogels and their potential applications for intracellular imaging and intratumoral chemotherapy. *J. Am. Chem. Soc.*, **135**, 9907–9914.

33. Zhang, H., Fei, J., Yan, X., Wang, A., Li, J. (2015). Enzyme-responsive release of doxorubicin from monodisperse dipeptide-based nanocarriers for highly efficient cancer treatment *in vitro*. *Adv. Funct. Mater.*, **25**, 1193–1204.

34. Das, P., Yuran, S., Yan, J., Lee, P. S., Reches, M. (2015). Sticky tubes and magnetic hydrogels co-assembled by a short peptide and melanin-like nanoparticles. *Chem. Commun.*, **51**, 5432–5435.

35. Liberato, M. S., Kogikoski, S., da Silva, E. R., de Araujo, D. R., Guha, S., Alves, W. A. (2016). Polycaprolactone fibers with self-assembled peptide micro/nanotubes: A practical route towards enhanced mechanical strength and drug delivery applications. *J. Mater. Chem. B*, **4**, 1405–1413.

36. Li, Q., Chen, M., Chen, D., Wu, L. (2016). One-pot synthesis of diphenylalanine-based hybrid nanospheres for controllable pH- and GSH-responsive delivery of drugs. *Chem. Mater.*, **28**, 6584–6590.

37. Jayawarna, V., Ali, M., Jowitt, T. A., Miller, A. E., Saiani, A., Gough, J. E., Ulijn, R. V. (2006). Nanostructured hydrogels for three-dimensional cell culture through self-assembly of fluorenylmethoxycarbonyl-dipeptides. *Adv. Mater.*, **18**, 611–614.

38. Mao, L. N., Wang, H. M., Tan, M., Ou, L. L., Kong, D. L., Yang, Z. M. (2012). Conjugation of two complementary anti-cancer drugs confers molecular hydrogels as a co-delivery system. *Chem. Commun.*, **48**, 395–397.

39. Li, J., Kuang, Y., Gao, Y., Du, X., Shi, J., Xu, B. (2013). D-amino acids boost the selectivity and confer supramolecular hydrogels of a nonsteroidal anti-inflammatory drug (NSAID). *J. Am. Chem. Soc.*, **135**, 542–545.

40. Liang, G., Yang, Z., Zhang, R., Li, L., Fan, Y., Kuang, Y., Gao, Y., Wang, T., Lu, W. W., Xu, B. (2009). Supramolecular hydrogel of a D-amino acid dipeptide for controlled drug release *in vivo. Langmuir*, **25**, 8419–8422.

41. Erdogan, H., Yilmaz, M., Babur, E., Duman, M., Aydin, H. M., Demirel, G. (2016). Fabrication of plasmonic nanorod-embedded dipeptide microspheres via the freeze-quenching method for near-infrared laser-triggered drug-delivery applications. *Biomacromolecules*, **17**, 1788–1794.

42. Xie, Y., Zhao, J., Huang, R., Qi, W., Wang, Y., Su, R., He, Z. (2016). Calcium-ion-triggered co-assembly of peptide and polysaccharide into a hybrid hydrogel for drug delivery. *Nanoscale Res. Lett.*, **11**, 184.

43. Aviv, M., Halperin-Sternfeld, M., Grigoriants, I., Buzhansky, L., Mironi-Harpaz, I., Seliktar, D., Einav, S., Nevo, Z., Adler-Abramovich, L. (2018). Improving the mechanical rigidity of hyaluronic acid by integration of a supramolecular peptide matrix. *ACS Appl. Mater. Interfaces*, **10**(49), 41883–41891.

44. Abbas, M., Xing, R., Zhang, N., Zou, Q., Yan, X. (2018). Antitumor photodynamic therapy based on dipeptide fibrous hydrogels with

incorporation of photosensitive drugs. *ACS Biomater. Sci. Eng.*, **4**, 2046–2052.

45. Roth-Konforti, M. E., Comune, M., Halperin-Sternfeld, M., Grigoriants, I., Shabat, D., Adler-Abramovich, L. (2018). UV light-responsive peptide-based supramolecular hydrogel for controlled drug delivery. *Macromol. Rapid Commun.*, **39**(24), 1800588.

46. Diaferia, C., Balasco, N., Sibillano, T., Ghosh, M., Adler-Abramovich, L., Giannini, C., Vitagliano, L., Morelli, G., Accardo, A. (2018). Amyloid-like fibrillary morphology originated by tyrosine-containing aromatic hexapeptides. *Chem. Eur. J.*, **24**, 6804–6817.

47. Diaferia, C., Balasco, N., Sibillano, T., Giannini, C., Vitagliano, L., Morelli, G., Accardo, A. (2018). Structural characterization of self-assembled tetra-tryptophan based nanostructures: variations on a common theme. *Chem. Phys. Chem.*, **19**, 1635–1642.

48. Diaferia, C., Gianolio, E., Palladino, P., Arena, F., Boffa, C., Morelli, G., Accardo, A. (2015). Peptide materials obtained by aggregation of polyphenylalanine conjugates as gadolinium-based magnetic resonance imaging contrast agents. *Adv. Funct. Mater.*, **25**, 7003–7016.

49. Accardo, A., Tesauro, D., Aloj, L., Pedone, C., Morelli, G. (2009). Supramolecular aggregates containing lipophilic Gd(III) complexes as contrast agents in MRI. *Coord. Chem. Rev.*, **253**, 2193–2213.

50. Diaferia, C., Gianolio, E., Accardo, A., Morelli, G. (2017). Gadolinium containing telechelic PEG-polymers end-capped by di-phenylalanine motives as potential supramolecular MRI contrast agents. *J. Pept. Sci.*, **23**, 122–130.

51. Diaferia, C., Gianolio, E., Sibillano, T., Mercurio, F. A., Leone, M., Giannini, C., Balasco, N., Vitagliano, L., Morelli, G., Accardo, A. (2017). Cross-beta nanostructures based on dinaphthylalanine Gd-conjugates loaded with doxorubicin. *Sci. Rep.*, **7**, 307.

52. Handelman, A., Natan, A., Rosenman, G. (2014). Structural and optical properties of short peptides: Nanotubes-to-nanofibers phase transformation. *J. Pept. Sci.*, **20**, 487–493.

53. Pinotsi, D., Buell, A. K., Dobson, C. M., Kaminski, G. S., Kaminski, C. F. (2013). A label-free, quantitative assay of amyloid fibril growth based on intrinsic fluorescence. *ChemBioChem*, **14**, 846–850.

54. Diaferia, C., Sibillano, T., Balasco, N., Giannini, C., Roviello, V., Vitagliano, L., Morelli, G., Accardo, A. (2016). Hierarchical analysis of self-assembled PEGylated hexaphenylalanine photoluminescent nanostructures. *Chem. Eur. J.*, **22**, 16586–16597.

55. Diaferia, C., Sibillano, T., Altamura, D., Roviello, V., Vitagliano, L., Giannini, C., Morelli, G., Accardo, A. (2017). Structural characterization of PEGylated hexaphenylalanine nanostructures exhibiting green photoluminescence emission. *Chem. Eur. J.*, **23**, 14039–14048.

56. Diaferia, C., Sibillano, T., Giannini, C., Roviello, V., Vitagliano, L., Morelli, G., Accardo, A. (2017). Photoluminescent peptide-based nanostructures as FRET donor for fluorophore dye. *Chem. Eur. J.*, **23**, 8741–8748.

57. De Santis, P., Morosetti, S., Rizzo, R. (1974). Conformational analysis of regular enantiomeric sequences. *Macromolecules*, **7**, 52–58.

58. Ghadiri, M. R., Granja, J. R., Milligan, R. A., McRee, D. E., Khazanovich, N. (1993). Self-assembling organic nanotubes based on a cyclic peptide architecture. *Nature*, **366**, 324–327.

59. Chapman, R., Danial, M., Koh, M. L., Jolliffe, K. A., Perrier, S. (2012). Design and properties of functional nanotubes from the self-assembly of cyclic peptide templates. *Chem. Soc. Rev.*, **41**, 6023–6041.

60. Fernandez-Lopez, S., Kim, H. S., Choi, E. C., Delgado, M., Granja, J. R., Khasanov, A., Kraehenbuehl, K., Long, G., Weinberger, D. A., Wilcoxen, K. M., et al. (2001). Antibacterial agents based on the cyclic D,L-α-peptide architecture. *Nature*, **412**, 452–455.

61. Ishihara, Y., Kimura, S. (2010). Nanofiber formation of amphiphilic cyclic tri-β-peptide. *J. Pept. Sci.*, **16**, 110–114.

62. Hartgerink, J. D., Granja, J. R., Milligan, R. A., Ghadiri, M. R. (1996). Peptide-amphiphile nanofibers: A versatile scaffold for the preparation of self-assembling materials. *J. Am. Chem Soc.*, **118**, 43–50.

63. Wang, Y., Yi, S., Sun, L., Huang, Y., Lenaghan, S. C., Zhang, M. (2014). Doxorubicin-loaded cyclic peptide nanotube bundles overcome chemoresistance in breast cancer cells. *J. Biomed. Nanotechnol.*, **10**, 445–454.

64. Hamley, I. W. (2011). Self-assembly of amphiphilic peptides. *Soft Matter.*, **7**, 4122–4138.

65. Versluis, F., Marsden, H. R., Kros, A. (2010). Power struggles in peptide-amphiphile nanostructures. *Chem. Soc. Rev.*, **39**, 3434–3444.

66. Vauthey, S., Santoso, S., Gong, H., Watson, N., Zhang, S. (2002). Molecular self-assembly of surfactant-like peptides to form nanotubes and nanovesicles. *Proc. Natl. Acad. Sci. U.S.A.*, **99**, 5355–5360.

67. Santoso, S., Hwang, W., Hartman, H., Zhang, S. (2002). Self-assembly of surfactant-like peptides with variable glycine tails to form nanotubes and nanovesicles. *Nano Lett.*, **2**, 687–691.

68. Von Maltzahn, G., Vauthey, S., Santoso, S., Zhang, S. (2003). Positively charged surfactant like peptides self-assemble into nanostructures. *Langmuir*, **19**, 4332–4337.

69. Dehsorkhi, A., Castelletto, V., Hamley, I. W., Seitsonen, J., Ruokolainen, J. (2013). Interaction between a cationic surfactant-like peptide and lipid vesicles and its relationship to antimicrobial activity. *Langmuir*, **29**, 14246–14253.

70. Zhabìng, S., Holmes, T., Lockshin, C., Rich, A. (1993). Spontaneous assembly of a self complementaryoligopeptide to form stable microscopic membrane. *Proc. Natl. Acad. Sci. U.S.A.*, **90**, 3334–3338.

71. Liu, E., Wang, H., Shang, Y., Liu, M., Chen, P. (2012). Molecular binding self assembling peptide EAK16-II with anticancer agent EPT and its implication in cancer cell inhibition. *J. Control. Release*, **160**, 33–40.

72. Boettcher, C., Schade, B., Fuhrhop, J. H. (2001). Comparative cryo-electron microscopy of noncovalent *N*-dodecanoyl-(D- and L-) serine assemblies in vitreous toluene and water. *Langmuir*, **17**, 873–877.

73. Fuhrhop, J. H., Spiroski, D., Boettcher, C. (1993). Molecular monolayer rodsandtubulesmadeofα-(L-lysine),omega.-(amino)bolaamphiphiles. *J. Am. Chem. Soc.*, **115**, 1600–1601.

74. Hartgerink, J. D., Beniash, E., Stupp, S. I. (2002). Peptide-amphiphilenanofibers: A versatile scaffold for the preparation of self-assembling materials. *Proc. Natl. Acad. Sci. U.S.A.*, **99**, 5133–5138.

75. Matson,J.B.,Newcomb,C.J.,Bitton,R.,Stupp,S.I.(2012).Nanostructure-templated control of drug release from peptide amphiphile nanofiber gels. *Soft Matter*, **8**, 3586–3595.

76. Cui, H., Muraoka, T., Cheetham, A. G., Stupp, S. I. (2009). Self-assembly of giant peptide. *Nano Lett.*, **9**(3) 945–951.

77. Yao, C., Liu, J. Y., Wu, X., Tao, Z. G., Gao, Y., Zhu, Q. G., Li, J. F., Zhang, L. J., Hu, C. L., Gu, F. F., et al. (2016). Reducible self-assembling cationic polypeptide-based micelles mediate co-delivery of doxorubicin and microRNA-34a for androgen-independent prostate cancer therapy. *J. Control. Release*, **232**, 203–214.

78. Tang, Q., Cao, B., Wu, H., Cheng, G. (2013). Cholesterol-peptide hybrids to form liposome-like vesicles for gene delivery. *PLoS One*, 8.

79. Missirlis, D., Krogstad, D. V., Tirrell, M. (2010). Subsequent endosomal disruption results in SJSA-1. *Mol. Pharm.*, **7**, 2173–2184.

80. Pujals, S., Fernandez-Carneado, J., Lopez-Iglesias, C., Kogan, M. J., Giralt, E. (2006). Mechanistic aspects of cell-penetrating peptide-

mediated intracellular drug delivery: Relevance of CPP self-assembly. *Biochim. Biophys. Acta Biomembr.*, **1758**, 264–279.

81. Agrawal, P., Bhalla, S., Usmani, S. S., Singh, S., Chaudhary, K., Raghava, G. P., Gautam, A. (2016). CPPsite 2.0: Arepository of experimentally validated cell-penetrating peptides. *Nucleic Acids Res.*, **44**, D1098–D1103.

82. Borrelli, A., Tornesello, A. L., Tornesello, M. L., Buonaguro, F. M. (2018). Cell penetrating peptides as molecular carriers for anti-cancer agents. *Molecules*, **23**, 295.

83. Sun, H., Dong, Y., Feijen, J., Zhong, Z. (2018). Peptide-decorated polymeric nanomedicines for precision cancer therapy. *J. Control. Release*, **290**, 11–27.

84. Gallo, M., Defaus, S., Andreu, D. (2019). 1988–2018: Thirty years of drug smuggling at the nano scale. Challenges and opportunities of cell-penetrating peptides in biomedical research. *Arch. Biochem. Biophys.*, **661**, 74–86.

85. Zhang, Q., Tang, J., Fu, L., Ran, R., Liu, Y., Yuan, M., He, Q. (2013). A pH-responsive α-helical cell penetrating peptide-mediated liposomal delivery system. *Biomaterials*, **34**, 7980–7993.

86. Liu, Z., Xiong, M., Gong, J., Zhang, Y., Bai, N., Luo, Y., Li, L., Wei, Y., Liu, Y., Tan, X. (2014). Legumain protease-activated TAT-liposome cargo for targeting tumours and their microenvironment. *Nat. Commun.*, **5**, 4280.

87. Yang, Y., Yang, Y., Xie, X., Cai, X., Mei, X. (2014). Preparation and characterization of photo-responsive cell-penetrating peptide-mediated nanostructured lipid carrier. *J. Drug Target.*, **22**, 891–900.

88. Reubi, J. C. (2003). Peptide receptors as molecular targets for cancer diagnosis and therapy. *Endocr. Rev.*, **24**, 389–427.

89. Accardo, A., Ringhieri, P., Palumbo, R., Morelli, G. (2014). Influence of PEG length on conformational and binding properties of CCK peptides exposed by supramolecular aggregates. *Pept. Sci.*, **102**, 304–312.

90. Ginj, M., Zhang, H., Waser, B., Cescato, R., Wild, D., Wang, X., Erchegyi, J., Rivier, J., Macke, H. R., Reubi, J. C. (2006). Radiolabeled somatostatin receptor antagonists are preferable to agonists for *in vivo* peptide receptor targeting of tumors. *Proc. Natl. Acad. Sci. U.S.A.*, **103**, 16436–16441.

91. Chan, K.Y., Vermeersch, S., de Hoon, J., Villalón, C. M., Maassenvandenbrink, A. (2011). Potential mechanisms of prospective antimigraine drugs: A focus on vascular (side) effects. *Pharmacol. Ther.*, **129**, 332–351.

92. Allen, F. H., Pitchford, N. A. (1998). Conformational analysis from crystallographic data. In: Codding, P. W., ed. *Structure Based Drug Design*, Kluwer Academic, Dordrecht, The Netherlands, pp. 15–26.

93. Pande, J., Szewczyk, M. M., Grover, A. K. (2010). Phage display: Concept, innovations, applications and future. *Biotechnol. Adv.*, **28**, 849–858.

94. Ringhieri, P., Mannucci, S., Conti, G., Nicolato, E., Fracasso, G., Marzola, P., Morelli, G., Accardo, A. (2017). Liposomes derivatized with multimeric copies of KCCYSL peptide as targeting agents for HER-2-overexpressing tumor cells. *Int. J. Nanomed.*, **12**, 501–514.

95. Ringhieri, P., Diaferia, C., Galdiero, S., Palumbo, R., Morelli, G., Accardo, A. (2015). Liposomal doxorubicin doubly functionalized with CCK8 and R8 peptide sequences for selective intracellular drug delivery. *J. Pept. Sci.*, **21**, 415–425.

96. Accardo, A., Ringhieri, P., Tesauro, D., Morelli, G. (2013). Liposomes derivatized with tetrabranchedneurotensin peptides via click chemistry reactions. *New J. Chem.*, **37**, 3528–3534.

97. Accardo, A., Morelli, G. (2015). Review peptide-targeted liposomes for selective drug delivery: Advantages and problematic issues. *Pept. Sci.*, **104**, 462–479.

98. Feldborg, L. N., Jølck, R. I., Andresen, T. L. (2012). Quantitative evaluation of bioorthogonal chemistries for surface functionalization of nanoparticles. *Bioconjug. Chem.*, **23**, 2444–2450.

99. Wang, Y., Yang, T., Wang, X., Wang, J., Zhang, X., Zhang, Q. (2010). Targeted polymeric micelle system for delivery of combretastatin A4 to tumor vasculature *in vitro*. *Pharm. Res.*, **27**, 1861–1868.

100. Miura, Y., Takenaka, T., Toh, K., Wu, S., Nishihara, H., Kano, M. R., Ino, Y., Nomoto, T., Matsumoto, Y., Koyama, H., et al. (2013). Cyclic RGD-linked polymeric micelles for targeted delivery of platinum anticancer drugs to glioblastoma through the blood-brain tumor barrier. *ACS Nano*, **7**, 8583–8592.

101. Chen, Z., Deng, J., Zhao, Y., Tao, T. (2012). Cyclic RGD peptide-modified liposomal drug delivery system: Enhanced cellular uptake *in vitro* and improved pharmacokinetics in rats. *Int. J. Nanomed.*, **7**, 3803–3811.

102. Amin, M., Badiee, A., Jaafari, M. R. (2013). Improvement of pharmacokinetic and antitumor activity of PEGylated liposomal doxorubicin by targeting with N-methylated cyclic RGD peptide in mice bearing C-26 colon carcinomas. *Int. J. Pharm.*, **458**, 324–333.

103. Wang, F., Chen, L., Zhang, R., Chen, Z., Zhu, L. (2014). RGD peptide conjugated liposomal drug delivery system for enhance therapeutic efficacy in treating bone metastasis from prostate cancer. *J. Control. Release*, **196**, 222–233.

104. Battistini, L., Burreddu, P., Sartori, A., Arosio, D., Manzoni, L., Paduano, L., D'Errico, G., Sala, R., Reia, L., Bonomini, S., et al. (2014). Enhancement of the uptake and cytotoxicactivity of doxorubicin in cancercells by novel cRGD-semipeptide-anchoring liposomes. *Mol. Pharm.*, **11**, 2280–2293.

105. Zhang, J., Hon, J. H., Chan, F., Skibba, M., Liang, G., Chen, M. (2016). RGD decorated lipid-polymer hybrid nanoparticles for targeted co-delivery of doxorubicin and sorafenib to enhance anti-hepatocellular carcinoma efficacy. *Nanomed. Nanotechnol. Biol. Med.*, **12**, 1303–1311.

106. Morisco, A., Accardo, A., Gianolio, E., Tesauro, D., Benedetti, E., Morelli, G. (2009). Micelles derivatized with octreotide as potential target-selective contrast agents in MRI. *J. Pept. Sci.*, **15**, 242–250.

107. Accardo, A., Morisco, A., Tesauro, D., Pedone, C., Morelli, G. (2011). Naposomes: A new class of peptide-derivatized, target-selective multimodal nanoparticles for imaging and therapeutic applications. *Ther. Deliv.*, **2**, 235–257.

108. Accardo, A., Mangiapia, G., Paduano, L., Morelli, G., Tesauro, D. (2013). Octreotide labeled aggregates containing platinum complexes as nanovectors for drug delivery. *J. Peptsci.*, **19**, 190–197.

109. Zhang, Y., Zhang, H., Wang, X., Wang, J., Zhang, X., Zhang, Q. (2012). The eradication of breast cancer and cancer stem cells using octreotide modified paclitaxel active targeting micelles and salinomycin passive targeting micelles. *Biomaterials*, **33**, 679–691.

110. Zou, A., Chen, Y., Huo, M., Wang, J., Zhang, Y., Zhou, J., Zhang, Q. (2013). *In vivo* studies of octreotidemodified N-octyl-O, N-carboxymethyl chitosan micelles loaded with doxorubicin for tumor-targeted delivery. *J. Pharm. Sci.*, **102**, 126–135.

111. Jaskula-Sztul, R., Xu, W., Chen, G., Harrison, A., Dammalapati, A., Nair, R., Cheng, Y., Gong, S., Chen, H. (2016). Thailandepsin A-loaded and octreotide-functionalized unimolecular micelles for targeted neuroendocrine cancer therapy. *Biomaterials*, **91**, 1–10.

112. Helbok, A., Rangger, C., von Guggenberg, E., Saba-Lepek, M., Radolf, T., Thurner, G., Andreae, F., Prassl, R., Decristoforo, C. (2012). Targeting properties of peptide-modified radiolabeled liposomal nanoparticles. *Nanomedicine*, **8**, 112–118.

113. Li, H., Yuan, D., Minjie, S., Ping, Q. (2016). Effect of ligand density and PEG modification on octreotide-targetedliposome via somatostatin receptor *in vitro* and *in vivo*. *Drug Deliv.*, **23**, 3562–3572.

114. Petersen, A. L., Binderup, T., Jølck, R. I., Rasmussen, P., Henriksen, J. R., Pfeifer, A. K., Kjær, A., Andresen, T. L. (2012). Positron emission tomography evaluation of somatostatin receptor targeted ^{64}Cu-TATE-liposomes in a human neuroendocrine carcinoma mouse model. *J. Control. Release*, **160**, 254–263.

115. Chen, G., Jaskula-Sztul, R., Harrison, A., Dammalapati, A., Chen, H., Gong, S., Xube, W., Cheng, Y. (2016). KE108-conjugated unimolecular micelles loaded with a novel HDAC inhibitor thailandepsin-A for targeted neuroendocrine cancer therapy. *Biomaterials*, **97**, 22–33.

116. Jaskula-Sztul, R., Chen, G., Dammalapati, A., Harrison, A., Tang, W., Gong, S., Chen, H. (2017). AB3-loaded and tumor-targeted unimolecular micelles for medullary thyroid cancer treatment. *J. Mater. Chem. B*, **5**, 151–159.

117. Accardo, A., Mansi, R., Morisco, A., Mangiapia, G., Paduano, L., Tesauro, D., Radulescu, A., Aurilio, M., Aloj, L., Arra, C., et al. (2010). Peptide modified nanocarriers for selective targeting of bombesin receptors. *Mol. Biosyst.*, **6**, 878–887.

118. Accardo, A., Salzano, G., Morisco, A., Aurilio, M., Parisi, A., Maione, F., Cicala, C., Tesauro, D., Aloj, L., De Rosa, G., et al. (2012). Peptide-modified liposomes for selective targeting of bombesin receptors overexpressed by cancer cells: A potential theranostic agent. *Int. J. Nanomed.*, **7**, 2007–2017.

119. Ringhieri, P., Iannitti, R., Nardon, C., Palumbo, R., Fregona, D., Morelli, G., Accardo, A. (2014). Target selective micelles for bombesin receptors incorporating Au(III)-dithiocarbamato complexes. *Int. J. Pharmaceut.*, **473**, 194–202.

120. Accardo, A., Mansi, R., Salzano, G., Morisco, A., Aurilio, M., Parisi, A., Maione, F., Cicala, C., Ziaco, B., Tesauro, D., et al. (2013). Bombesin peptide antagonist for target-selective delivery of liposomal doxorubicin on cancer cells. *J. Drug Target.*, **21**, 240–249.

121. Accardo, A., Tesauro, D., Morelli, G., Gianolio, E., Aime, S., Vaccaro, M., Mangiapia, G., Paduano, L., Schillen, K. (2007). High-relaxivity supramolecular aggregates containing peptide and Gd complexes agents in MRI. *J. Biol. Inorg. Chem.*, **12**, 267–276.

122. Gasparini, G., Brooks, P. C., Biganzoli, E., Vermeulen, P. B., Bonoldi, E., Dirix, L. Y., Ranieri, G., Miceli, R., Cheresh, D. A. (1998). Vascular integrin

$\alpha_v\beta_3$: A new prognostic indicator in breast cancer. *Clin. Cancer Res.*, **4**, 2625–2634.

123. Liu, S. (2006). Radiolabeled multimeric cyclic RGD peptides as integrin alphavbeta3 targeted radiotracers for tumor imaging. *Mol. Pharm.*, **3**, 472–487.

124. Xiong, X. B., Huang, Y., Lu, W. L., Zhang, X., Zhang, H., Nagai, T., Zhang, Q. (2005). Enhanced intracellular delivery and improved antitumor efficacy of doxorubicin by sterically stabilized liposomes modified with a synthetic RGD mimetic. *J. Control. Release*, **107**, 262–275.

125. Danhier, F., Le Breton, A., Préat, V. (2012). RGD-based strategies to target alpha(v) beta(3) integrin in cancer therapy and diagnosis. *Mol. Pharm.*, **9**, 2961–2973.

126. Schiffelers, R. M., Koning, G. A., ten Hagen, T. L. M., Fens, M. H. A. M., Schraa, A. J., Janssen, A. P. C. A., Kok, R. J., Molema, G., Storm, G. (2003). Anti-tumor efficacy of tumor vasculature-targeted liposomal doxorubicin. *J. Control. Release*, **91**, 115–122.

127. Murphy, E. A., Majeti, B. K., Barnes, L. A., Makale, M., Weis, S. M., Lutu-Fuga, K., Wrasidlo, W., Cheresh, D. A. (2008). Nanoparticle-mediated drug delivery to tumor vasculature suppresses metastasis. *Proc Natl. Acad. Sci. U.S.A.*, **105**, 9343–9348.

128. Guan, X., Hu, X., Liu, S., Sun, X., Gai, X. (2008). Cyclic 6RGD targeting cisplatin micelles for near-infrared imaging-guided chemotherapy. *RSC Adv.*, **6**, 1151–1157.

129. Zhao, H., Wang, J.-C., Sun, Q.-S., Luo, C.-L., Zhang, Q. (2009). RGD-based strategies for improving antitumor activity of paclitaxel-loaded liposomes in nude mice xenografted with human ovarian cancer. *J. Drug Target.*, **17**, 10–18.

130. Meng, S., Su, B., Li, W., Ding, Y., Tang, L., Zhou, W., Song, Y., Li, H., Zhou, C. (2010). Enhanced antitumor effect of novel dual-targeted paclitaxel liposomes. *Nanotechnology*, **21**, 415103.

131. Li, Y., Zheng, X., Sun, Y., Ren, Z., Li, X., Cui, G. (2014). RGD-fatty alcohol-modified docetaxel liposomes improve tumor selectivity *in vivo. Int. J. Pharm.*, **468**, 133–141.

132. Pattillo, C. B., Sari-Sarraf, F., Nallamothu, R., Moore, B. M., Wood, G. C., Kiani, M. F. (2005). Targeting of the antivascular drug combretastatin to irradiated tumors results in tumor growth delay. *Pharm. Res.*, **22**, 1117–1120.

133. Dubey, P. K., Mishra, V., Jain, S., Mahor, S., Vyas, S. P. (2004). Liposomes modified with cyclic RGD peptide for tumor targeting. *J. Drug Target.*, **12**, 257–264.

134. Garg, A., Tisdale, A. W., Haidari, E., Kokkoli, E. (2009). Targeting colon cancer cells using PEGylated liposomes modified with a fibronectin-mimetic peptide. *Int. J. Pharm.*, **366**, 201–210.

135. Kessler, H., Kutscher, B., Klein, A. (1986). Peptidkonformationen, 39. NMR-studien zur konformation von cyclopentapeptidanalogen des thymopoietins. *Liebigs Ann. Chem.*, **1986**, 893–913.

136. Kessler, H., Diefenbach, B., Finsinger, D., Geyer, A., Gurrath, M., Goodman, S. L., Hölzemann, G., Haubner, R., Jonczyk, A., Müller, G., et al. (1995). Design of superactive and selective integrin receptor antagonists containing the RGD sequence. *Lett. Pep. Sci.*, **2**, 155–166.

137. Haubner, R., Gratias, R., Diefenbach, B., Goodman, S., Jonczyk, A., Kessler, H. (1996). Structural and functional aspects of RGD-containing cyclic pentapeptides as highly potent and selective integrin v 3 antagonists. *J. Am. Chem. Soc.*, **118**, 7461–7472.

138. Wang, Y., Wang, X., Zhang, Y., Yang, S., Wang, J., Zhang, X., Zhang, Q. (2009). RGD-modified polymeric micelles as potential carriers for targeted delivery to integrin-overexpressing tumor vasculature and tumor cells. *J. Drug Target.*, **17**, 459–467.

139. Kazuki, N., TambetTeesalu, S., Karmali, P. P., Kotamraju, V. R., Agemy, L., Girard, O. M., Hanahan, D., Mattrey, R. F., Ruoslahti, E. (2009). Tissue-penetrating delivery of compounds and nanoparticles into tumors. *Cancer Cell*, **6**, 510–520.

140. Gao, F., Zhang, J., Fu, C., Xie, X., Peng, F., You, J., Tang, H., Wang, Z., Li, P., Chen, J. (2017). iRGD-modified lipid–polymer hybrid nanoparticles loaded with isoliquiritigenin to enhance anti-breast cancer effect and tumor-targeting ability. *Int. J. Nanomed.*, **12**, 4147–4162.

141. Jiang, Y., Pang, X., Liu, R., Xiao, Q., Wang, P., Leung, A. W., Luan, Y., Xu, C. (2018). Design of an amphiphilicRGD peptide and self-assembling nanovesicles for improving tumor accumulation and penetration and the photodynamic efficacy of the photosensitizer. *ACS Appl. Mater. Interfaces*, **10**, 31674–31685.

142. Hu, H., Wan, J., Huang, X., Tang, Y., Xiao, C., Xu, H., Yang, X., Li, Z. (2018). iRGD-decorated reduction-responsive nanoclusters for targeted drug delivery. *Nanoscale*, **10**, 10514–10527.

143. Lamberts, S. W. (1998). *Octreotide: The Next Decade*, BioScientifica: Bristol, UK.

144. Melacini, G., Zhu, Q., Goodman, M. (1997). Multiconformational NMR analysis of sandostatin (octreotide): Equilibrium between beta-sheet and partially helical structures. *Biochemistry*, **36**, 1233–1241.

145. Cuttitta, F., Carney, D. N., Mulshine, J., Moody, T. W., Fedorko, J., Fischler, A., Minna, J. D. (1985). Bombesin-like peptides can function as autocrine growth factors in human small-cell lung cancer. *Nature*, **16**, 823–826.

146. Patel, O., Shulkes, A., Baldwin, G. S. (2006). Gastrin-releasing peptide and cancer. *Biochim. Biophys. Acta*, **1766**, 23–41.

147. Smith, C. J., Volkert, W. A., Hoffman, T. J. (2005). Radiolabeled peptide conjugates for targeting of the bombesin receptor superfamily subtypes. *Nucl. Med. Biol.*, **32**, 733–740.

148. Parry, J. J., Kelly, T. S., Andrews, R., Rogers, B. E. (2007). *In vitro* and *in vivo* evaluation of ^{64}Cu-labeled DOTA-linker-bombesin(7–14) analogues containing different amino acid linker moieties. *Bioconjug. Chem.*, **18**, 1110–1117.

149. Jamous, M., Tamma, M. L., Gourni, E., Waser, B., Reubi, J. C., Maecke, H. R., Mansi, R. (2014). RPEG spacers of different length influence the biological profile of bombesin-based radiolabeled antagonists. *Nucl. Med. Biol.*, **41**, 464–470.

150. Bernhard, Y., Gigot, E., Goncalves, V., Moreau, M., Sok, N., Richard, P., Decréau, R. A. (2016). Direct subphthalocyanine conjugation to bombesin vs. indirect conjugation to its lipidic nanocarrier. *Org. Biomol. Chem.*, **14**, 4511–4518.

151. Pellegrini, M., Mierke, D. F. (1999). Molecular complex of cholecystokinin-8 and N-terminus of the cholecystokinin A receptor by NMR spectroscopy. *Biochemistry*, **38**, 14775–14783.

152. Morelli, G., De Luca, S., Tesauro, D., Saviano, M., Pedone, C., Dolmella, A., Visentin, R., Mazzi, U. (2002). CCK8 peptide derivatized with diphenylphosphine for rhenium labelling: Synthesis and molecular mechanics calculations. *Pept. Sci.*, **8**, 373–381.

153. Accardo, A., Morisco, A., Palladino, P., Palumbo, R., Tesauro, D., Morelli, G. (2011). Amphiphilic CCK peptides assembled in supramolecular aggregates: Structural investigations and *in vitro* studies. *Mol. Biosyst.*, **7**, 862–870.

154. Tesauro, D., Accardo, A., Gianolio, E., Paduano, L., Teixeira, J., Schillen, K., Aime, S., Morelli, G. (2007). Peptide derivatizedlamellar aggregates as target-specific MRI contrast agents. *ChemBioChem*, **8**, 950–995.

155. Accardo, A., Tesauro, D., Aloj, L., Tarallo, L., Arra, C., Mangiapia, G., Vaccaro, M., Pedone, C., Paduano, L., Morelli, G. (2008). Peptide-containing aggregates as selective nanocarriers for therapeutics. *Chem. Med. Chem.*, **3**, 594–602.

156. Vincenzi, M., Accardo, A., Costantini, S., Scala, S., Portella, L., Trotta, A., Ronga, L., Guillon, J., Leone, M., Colonna, G., et al. (2015). Intrinsically disordered amphiphilic peptides as potential targets in drug delivery vehicles. *Mol. Biosyst.*, **11**, 2925–2932.

157. Accardo, A., Leone, M., Tesauro, D., Aufiero, R., Bénarouche, A., Cavalier, J. F., Longhi, S., Carriere, F., Rossi, F. (2013). Solution conformational features and interfacial properties of an intrinsically disordered peptide coupled to alkyl chains: A new class of peptide amphiphiles. *Mol. Biosyst.*, **9**, 1401–1410.

158. Vincenzi, M., Costantini, S., Scala, S., Tesauro, D., Accardo, A., Leone, M., Colonna, G., Guillon, J., Portella, L., Trotta, A., et al. (2015). Conformational ensembles explored dynamically from disordered peptides targeting chemokine receptor CXCR4. *Int. J. Mol. Sci.*, **16**, 12159–12173.

159. Banta, S., Megeed, Z., Casali, M., Rege, K., Yarmush, M. L. (2007). Engineering protein and peptide building blocks for nanotechnology. *J. Nanosci. Nanotechnol.*, **7**, 387–401.

Chapter 35

Hydrogels and Their Applications in Targeted Drug Delivery

Radhika Narayanaswamy, MSc, and Vladimir P. Torchilin, PhD, DSc

Center for Pharmaceutical Biotechnology and Nanomedicine,
Northeastern University, Boston, Massachusetts, USA

Keywords: hydrogels, targeted drug delivery, drug release, hydrophobic drug delivery, clinical translation, administration routes, supramolecular, bio-inspired, multi-functional, stimuli-responsive, polymeric networks, cancer therapy, cardiac delivery, ophthalmic delivery, bio-compatible, smart carriers, porosity, cross-linking, shape memory, gastroretentive dosage forms, biomimetic delivery, cancer immunotherapy, cyclodextrins, mag-bot, drosera-inspired, reactive oxygen species, temperature-sensitive, pH-responsive

35.1 Introduction

Hydrophilic polymeric networks that are capable of imbibing huge volumes of water and undergoing swelling and shrinkage suitably to facilitate controlled drug-release are called hydrogels. Their

The Road from Nanomedicine to Precision Medicine
Edited by Shaker A. Mousa, Raj Bawa, and Gerald F. Audette
Copyright © 2020 Jenny Stanford Publishing Pte. Ltd.
ISBN 978-981-4800-59-4 (Hardcover), 978-0-429-29501-0 (eBook)
www.jennystanford.com

porosity and compatibility with aqueous environments make them highly attractive biocompatible drug delivery vehicles. Their applications are manifold and for several biomedical needs as they are moldable into varied physical forms such as nanoparticles, microparticles, slabs, films and coatings. The United States has been the largest producer of hydrogels and is expected to remain so for a few more years [1]. Hydrogels are promising, trendy, intelligent, and "smart" drug delivery vehicles that cater to the specific requirements for targeting drugs to the specific sites and controlling drug release. Enzymatic, hydrolytic or environmental stimuli often suffice to manipulate the hydrogels for the drug release at the desirable site [2]. Like the two sides of a coin, there are also the disadvantages associated with their use. The primary disadvantage in drug delivery would be the hydrophobicity of most drugs. The water-loving polymeric core is probably not very ideal to hold incompatible hydrophobic drugs, which is a challenge since many of the drugs that are currently used and effective in disease therapy are hydrophobic. The tensile strength of these hydrogels is weak, and this sometimes causes early release of the drug before arrival at the target site. The following review discusses on how hydrogels are being manipulated presently for improved targeted drug delivery. The modern trends in which the hydrogels are exploited for drug delivery are covered.

The attractive physical properties of hydrogels, especially their porosity, offer tremendous advantages in drug delivery applications such as sustained release of the loaded drug. A high local concentration of the active pharmaceutical ingredient is retained over a long period of time via a suitable release mechanism controlled by diffusion, swelling, chemical means or some environmental stimuli.

Diffusion-controlled drug delivery with hydrogels uses reservoir or matrix devices that allow diffusion-based drug release through a hydrogel mesh or pores filled with water. In the reservoir delivery system the hydrogel membrane is coated on a drug-containing core, producing capsules, spheres or slabs that have a high drug concentration in the very center of the system to facilitate a constant drug-release rate. While the reservoir delivery system produces time-independent and constant drug release, the matrix system works via the macromolecular pores or mesh.

This type of release is time-dependent drug release wherein the initial release rate is proportional to the square root of time, rather than being constant (Fig. 35.1).

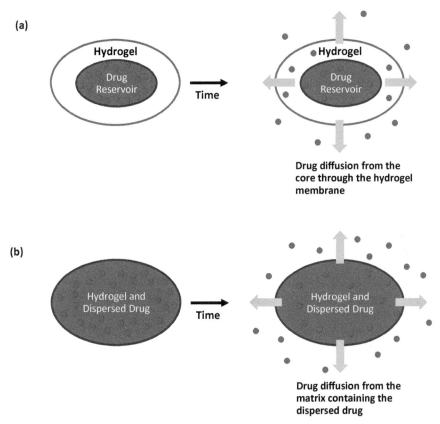

Figure 35.1 (a). Drug-containing core is coated with hydrogel membrane and the drug concentration is higher in the center of the system to allow constant release rate of the same in reservoir delivery system. **(b).** Uniform dissolution or dispersion of the drug throughout the 3D structure of the hydrogel is achieved using matrix delivery. Adapted with permission from [3]; the article is open access and the content reusable.

The swelling-controlled drug release from hydrogels uses drugs dispersed within a glassy polymer which when in contact with a biofluid begins swelling. The expansion during swelling occurs beyond its boundary facilitating the drug diffusion along with the polymer chain relaxation. The process, otherwise referred to as Case II transport, supports time-independent, constant drug release kinetics. Since the gradient between the dispersed drug in the hydrogel and its surrounding environment allows the active ingredient diffusion from a region of higher concentration within the hydrogel to a lower one, the process is also referred to as anomalous transport as it combines both the processes of diffusion and swelling for enabling drug release.

Ocular drug delivery carriers have been developed using hydrogels that are covalently cross-linked. These soft, biodegradable hydrogels with high swelling capacity remain *in situ* in the lacrimal canal offering greater comfort for the patient. Collagen or silicone that may be used decides if the punctal-plug (tear duct plug) system could be used temporarily or permanently, respectively. Poly(ethylene glycol) hydrogels are commonly used for producing ophthalmic drug delivery systems for blocking tear drainage and preventing dry eye.

Drug release in response to environmental changes would be an ideal delivery system as the release becomes very controlled and non-specific side effects at off-target sites are alleviated. Thus, sensitive drug delivery devices responsive to changes in pH, temperature, ionic strength or glucose concentration have been developed that are advantageous in the therapy of diseases such as cancer, and diabetes, characterized by local physiological changes specific to the various disease stages. The polymer composition of the hydrogel responsive to the environmental stimuli is manipulated to make it responsive to the environment [3].

Hydrogels considerably enhance the therapeutic outcome of drug delivery and have found enormous clinical use. The temporal and spatial delivery of macromolecular drugs, small molecules, and cells have greatly improved through hydrogel use for drug delivery [2]. Drug delivery using hydrogels, however, has not been free from challenges, but constant improvements are being made to identify the hydrogel design best suited for specific drug delivery purposes. Therefore, this paper discusses the recent trends in drug delivery applications using hydrogels, including their translation to the clinic and their applications to successfully deliver hydrophobic drugs.

35.2 Current Trend in Hydrogel-Based Targeted Drug Delivery

35.2.1 Supramolecular Hydrogels

The supramolecular hydrogel system is composed of intermolecular interactions that are non-covalent and has two or more molecular entities held together. The non-covalent cross-linking is a very attractive aspect of these hydrogels as it helps circumvent the problems of limited drug loading potential and drug incorporation for use only as implantables, which would be the only possibility with a covalently cross-linked network. Apart from offering the right physical stability for the hydrogels, these achieve drug loading and gelation simultaneously in an aqueous environment without the need for a covalent cross-linking. Recent progress has been made with supramolecular hydrogels using self-assembled inclusion complexes between cyclodextrins (CD) and biodegradable block copolymers that provide sustained and controlled release of macromolecular drugs [4].

Natural cyclic oligosaccharides composed of six, seven or eight D-(+)-glucose units linked by D-(+)-1,4-linkages (termed α-, β-, and γ-CD, respectively), are called cyclodextrins and are well-suited for use in supramolecular systems. They offer hydrophobic internal cavities with a suitable diameter and their ability to generate supramolecular inclusion complexes with various polymers make them ideal drug delivery vehicles.

A recent study involved development of a glycoconjugate prepared by amidation of homopoly-L-guluronic acid block obtained from *D. antarctica* sodium alginate with mono-6-amino-β-CD. The study was aimed at treating Chagas disease caused by *Trypanosoma cruzi*. Lipophilic non-hydroxylated coumarins were loaded into the hydrophobic core of the β-cyclodextrin to render them with trypanocidal activity. Interaction between the carboxylate groups of unconjugated α.0-L-glucuronate residues with calcium ions produced supramolecular hydrogels of glycoconjugate of homopoly-L-guluronic block fraction (GG) with 6-NH$_2$-β-CD (6-amino-β-cyclodextrin). As the *T. cruzi* parasites have only one mitochondrion, it is an ideal target for drugs to manipulate its energy process and apoptosis. Mitochondrial membrane potential studies revealed that the cyclodextrin complex with the drugs produced significant oxidative stress to destroy the parasites. The drug in the complex had increased solubility, showed improved bioavailability, controlled drug release and improved trypanocidal activity in comparison to the corresponding free amidocoumarins [5].

Cyclodextrin-functionalized polyhydrazines were used to prepare hydrogels *in situ* via hydrazine bond formation with aldehyde groups on dextran aldehyde. No toxicity was observed *in vitro* with these hydrogels and they could accommodate nicardipine as hydrophobic drug into the cyclodextrin cavities. Steady release of nicardipine over 6 days was observed with the hydrogel preparation having higher hydrazine linkages. Thus, a gel capable of hydrophobic drug release in an *in situ* formed device over extended periods was generated [6].

Bleeding control and wound healing by bioadhesive hydrogels find enormous biomedical applications. *In situ* forming hydrogels are used to heal injured tissues based on their ability to accumulate and produce a fibrin bridge that permit fibroblast migration and collagen secretion for healing tissue injury. β-Cyclodextrins are non-toxic adjuvants for pharmaceutical and mucoadhesive applications. Partly oxidized β-cyclodextrin was used in a recent study to exploit aldehyde groups on a hydrogel matrix for favorable reaction with amines in the tissue to result in an imine bond (Schiff's base reaction) in order to adhere to the skin and to provide improved cyclodextrin solubility in order to improve loading efficiency. Blending gelatin (the common extracellular component) with the β-cyclodextrin partly oxidized with oxidation in the presence of H_2O_2 (Hydrogen Peroxide)/ horseradish peroxidase, resulted in very rapid formation of gelatin-β-cyclodextrin hydrogels (Fig. 35.2). Hydrophobic drugs such as dexamethasone could be released with 2.7 fold higher efficacy when delivered in presence of the cyclodextrin relative to the gelatin-only hydrogels [7].

Curcumin has been shown to have several therapeutic benefits and found enormous applications in conventional therapy. The challenging aspect of its delivery is the extremely low aqueous solubility. However, a glycyrrhetinic acid (GA) molecule-modified curcumin-based hydrogel has been developed to address the problem of delivery of the insoluble drug for hepatocellular carcinoma. The GA molecule-modified curcumin supplied in the pro-gelator form could produce a supramolecular hydrogel *in vitro* due to disulfide reduction by glutathione (GSH) and increase curcumin bioavailability and solubility as reported in HepG2 cells. Higher cellular uptake and potent anti-cancer activity were observed with the hydrogel *in vitro* relative to an already known curcumin-targeting compound that was tested [8].

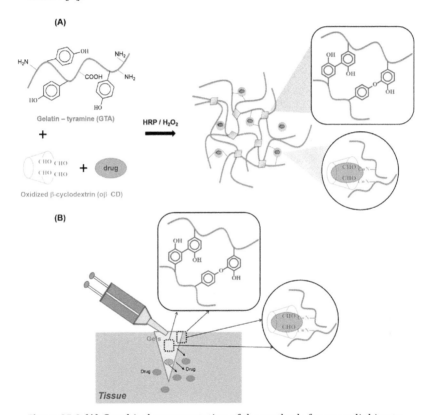

Figure 35.2 (A) Graphical representation of the methods for cross-linking to obtain gelatin-β-cyclo-dextrin (GTA–ob-CD) hydrogels to load hydrophobic drugs. (B). Schematic representation of adhesive GTA–ob-CD hydrogels *in situ* formed by combining HRP catalysis and the Schiff base reaction with therapeutic release. Reprinted from [7] with permission from The Royal Society of Chemistry. The article is licensed by Creative Commons and the link to the license is https://creativecommons.org/licenses/by/3.0/.

35.2.2 DNA-Hydrogels

Hybrid bionanomaterials could be developed using DNA as the building block. Predictable two- or three-dimensional structures are formed from DNA molecules. Highly structured networks are formed by hybridizing complementary DNA molecules and the resultant hydrogel structures expand upon encounter with an aqueous environment that result in swelling. Not only do these materials append to any other type of nucleic acid molecules (such as siRNA, miRNA), but they can also load DNA binding drugs. High solubility, biocompatibility, versatility and responsiveness are key features of such hydrogels. Apart from these features they can also be tagged with suitable fluorescent molecules for tracking biological studies *in vitro* [9].

An interesting application of hydrogels has been made with the development of multi-functional quantum dot (QD) DNA hydrogels. DNA hydrogels are composed of complementary strands of DNA hybridized to form a cross-linked network that swells in an aqueous environment. However, in order for biological studies to be more easily and effectively performed there has to be a tracer or a fluorescent molecule attached to the hydrogel which would then be a superlative option offering both targeted delivery and imaging. With this aspect in mind Zhang et al. recently developed a QD-based DNA hydrogel that had highly tunable size, spectral and delivery properties and bound to the DNA binding drug doxorubicin (Fig. 35.3). The drug targeted cancer cells and the QD-DNA hydrogel increased the potency of the drug *in vitro*. The single-step assembling zinc sulfide QD doxorubicin DNA hydrogels showed increased tumor accumulation *in vitro*, high biocompatibility, was threefold more efficacious than free DOX and served as an excellent tool for *in vivo* bioimaging in monitoring tumor growth over time. Aptamers such as siRNA were used to target specific cell types to deliver drug specifically and modulate protein expression in various cell types [9] (Fig. 35.3).

Drug-loaded cytosine-phosphate-guanine (CpG)-DNA hydrogels have been used for cancer immunotherapy. These CpG sequences containing hydrogels elicit immune responses. As an example, these DNA hydrogels containing CpG nucleotides stimulated innate immunity through Toll-like receptor 9 and promoted immune responses of ovalbumin (OVA) incorporated into the gel by acting as an adjuvant. Adverse reactions were considerably reduced with the use of the CpG-DNA hydrogel in comparison to the OVA injected with alum or complete Freund's adjuvant [10, 11].

a

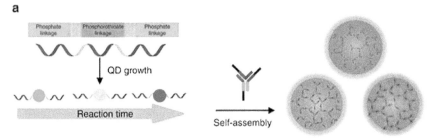

Quantum dot DNA Hydrogels (QDHs)

b

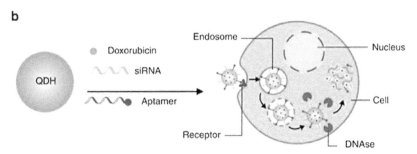

Figure 35.3 (a). Schematic view of the synthesis process of the DNA functionalized QDs followed by the formation of the QD hydrogel through hybridization with DNA. (b). Schematic view of the modification process on the QD hydrogel for specific targeting of the cell using aptamer for drug delivery. Release of Doxorubicin and siRNA happens after uptake into the cell via endocytosis. Reprinted (adapted) with permission from [9]; the article is open access and the content reusable. Creative Commons International License 4.0 for the article is available from http://creativecommons.org/licenses/by/4.0/ [9].

There has been significant advancement in hydrogel modifications for enhancing the shape memory and the reversibility of the hydrogel shape after any prolonged stress on them (e.g., external stimuli such as temperature). This advances the thermal responsivity behavior of the hydrogels and an example is the construction of supramolecular hydrogel with tunable mechanical properties and multi-shape memory effects. These hydrogels make use of agar that is physically cross-linked to form the hydrogel network and a supramolecular network that is cross-linked by suitable chemical bonds that fix shapes temporarily and produces a multi-shape memory effect made possible by reversible interactions (Fig. 35.4). The supramolecular hydrogel possesses enormous biocompatibility and biodegradability characteristics and rely on non-covalent interactions to drive the self-assembly of small molecules in water. The formed structures thus have supramolecular architectures and encapsulate water [12, 13].

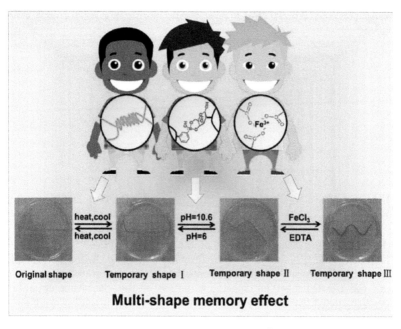

Figure 35.4 Schematic depiction of a novel Fe^{3+}-, pH-, and thermo-responsive hydrogel with tunable mechanical properties easily changed by adjusting cross-linking densities of polymers. Moreover, several co-ordination interactions along with supporting stimuli could be exploited to stabilize temporary shapes in order to realize shape memory behavior. Programmable multi-shape memory effect can be realized by combining the three reversible switches. Reprinted with permission from [12], copyright (2017) American Chemical Society.

A programmed temporary shape can turn into the memorized original shape when placed in an appropriate environment or exposed to a trigger. Such shape-memory hybrid hydrogels could also be synthesized using DNA cross-linkers. These hydrogels not only undergo phase transitions in the trigger of the stimulus, but also possess memory code to recover to the original matrix shape. Guo et al. developed a pH-controlled shape memory DNA hydrogel that was formed by co-polymerization of acrylamide residues with acrydite modified with (1) cytosine-rich sequences (forming i-motif subunits) and (2) nucleic acids exhibiting self-complementarity. Self-assembly of cytosine-rich nucleic acid strands into an i-motif structure occurred at pH 5.0 and the disassembly to a random coil form happened at pH 8.0, leading to a quasi-liquid state for the hydrogel. Re-acidification to pH 5.0 restored the original structure of the gel [14, 15].

35.2.3 Bioinspired Hydrogels

A newer variety of hydrogels used for drug delivery applications are the bioinspired hydrogels. These 3D materials recapitulate the biological micro-environment relevant to the disease condition and support studies on how the targeted drug delivery process could be optimized, how the therapy behaved *in vivo*, how the disease progressed, and so on. These are particularly useful in cancer therapy as the disease is particularly complex and normally associated with intricate cellular and physiological changes that require progressive monitoring. Engineering such microenvironments would thus be a very useful approach to promote research and study the disease condition and therapeutic process better. The stiffness of the 3D model used for studying liver cancer is a critical attribute in regulating molecular diffusivity and malignancy. The elastic moduli of the collagen gels were increased by stiffening interconnected collagen fibers with varied amounts of poly(ethylene glycol) di(succinic acid *N*-hydroxysuccinimidyl ester). The softer gels produced malignant cancer spheroids while the stiffer ones showed suppressed malignancy. The model provided better understanding and regulation of the emergent behaviors of cancer cells [16].

Contact lenses that are bioinspired have been developed recently with improved drug delivery properties suited to perfectly match the eye condition in the diseased state, especially of the anterior-eye segment.

Inner layer of ethyl cellulose and Eudragit S 100 hydrogels encapsulated with diclofenac sodium salt showed sustained drug release in simulated tear film. Sustained drug levels in tear fluid for a prolonged period of time in relation to eye drops were observed. The drug timolol was released in response to lysozyme in the medium and not in PBS. Enzyme cleavable polymers in hydrogels enabled this.

Ergosterol-liposome grafted silicone materials loaded with nystatin (an anti-fungal agent) resembled a fungal infection and triggered nystatin release through a competitive mechanism. In the absence of the ergosterol (important sterol in the cell membranes of fungi) in the medium, the drug release was very negligible [17–20].

With recent progress in cancer therapy using immunology, the research has progressed in the area with the use of cells such as erythrocytes, macrophages, stem cells, dendritic cells and bacteria as building blocks to create targeted delivery systems. Stem cell membrane coated gelatin nanogels that were high tumor targeting and biocompatible drug delivery systems were developed recently. Those gelatin nanogels had excellent stability and tumor-targeting ability *in vitro* and *in vivo*. Targeting doxorubicin (DOX) enhanced anti-tumor therapeutic efficacy significantly higher than the gelatin-DOX of free DOX nanogels. Evident side effects were absent in the tissues of heart, liver, spleen, lung and kidney as was revealed with histopathology staining of treated mice [21].

A similar biomimetic drug delivery carrier is the endosomal membrane-coated nanogel extracted from the source cancer cells for specific delivery of the small molecule drug. Hyaluronic acid nanogel composed of SiO_2/Fe_3O_4 nanoparticles formed the inner core of the hydrogel and the endosome membrane outer shell was used. The hyaluronic acid could target CD44 receptors overexpressed in a variety of tumor types. Core–shell mesoporous silica nanoparticles with Fe_3O_4 nanocrystals were used as the core to encapsulate photo-initiators and cross-linkers, followed by hyaluronic acid coating on the surface via electrostatic interaction. Photo-polymerization upon UV irradiation resulted in the *in situ* formed nanogel in endosomes following incubation of resultant nanoparticles with the source cells. In the presence of the loaded anti-cancer drug doxorubicin, the endosomal membrane nanogels could specifically interact with the target source cells followed by internalization to release doxorubicin. High targeting specificity, cytotoxicity and uptake of the prepared endosomal membrane coated nanogels with DOX over the bare DOX-nanogels was observed *in vitro* [22].

Detoxification of pore-forming toxins resulting from animal bites/stings using a 3D bioinspired hydrogel matrix resembling the 3D structure of liver is a recent achievement. The 3D matrix resembling the liver is created by the hydrogel while the polydiacetylene nanoparticles installed in the hydrogel matrix serve to attract, capture and sense toxins. The 3D printed biomimetic detoxification device with installed nanoparticles would be a breakthrough innovation replacing vaccines, monoclonal antibodies, antisera to eliminate toxins from blood. Commonly used conventional antidote molecules target specific epitope structures on the pore forming toxins warranting specific treatments for the various toxins. IV nanoparticle administration for binding and removal of toxins may cause secondary risk by poisoning from nanoparticle accumulation in liver. While challenges do not limit nanoparticle use for toxin clearance, the 3D printed *in vitro* devices have been clinically approved for toxin removal [23].

Adhesive bioinspired gels comprising of sodium alginate, gum arabic, and calcium ions have been developed in order to mimic properties of natural sundew-derived adhesive hydrogels. Mouse adipose-derived stem cells were used in combination with the sundew-inspired hydrogel to confer interesting properties such as improved wound healing, enhanced wound closure, less noticeable toxicity and inflammation. Sundew plant's natural adhesive hydrogels were quite promising, but the researchers were quite intrigued about collecting the desirable quantity from the natural source. This inspired development of the above described hydrogel type with improved properties [24].

Magnetically actuated gel-bot (Mag-bot) is yet another recent discovery that facilitates remote control of the motion of magnetically actuated hydrogels. These hydrogels adopt a crawling motion resembling that of a maggot and move in a confined space of 3D porous media including, foams, tissues, fibers and so on, opening up broader options for targeted drug delivery. Pattern structures and lubrication effects on hydrogel mobility in the confined 3D space have been experimented in the paper [25].

It is not just the maggot that has inspired development of a new generation of hydrogels with applications in targeted drug delivery but also the drosera. The drosera-inspired model of hydrogel for targeted drug delivery with a bifunctional attribute to it has been developed recently. These hydrogels use the "catch and kill prey" mechanism as a drosera would. They have a bottom layer functionalized with double stranded DNA that can bind the anti-cancer drug doxorubicin and there are aptamers on top that offer targeting advantages to these particles. Though there is requirement for *in vivo* studies, the sustained expression of specific aptamer on surface and the continuous killing of cancer cells by sustained drug release make these hydrogels an ideal targeted drug delivery vehicle for killing cancer cells [26].

35.2.4 Multi-Functional and Stimuli-Responsive Hydrogels

Hydrogels that were multi-functional and carriers of anti-cancer drugs are a typical example of the versatility of these delivery vehicles and their amenability to chemical modifications to enhance their therapeutic effects. Magnetite nanoparticles that supported increased intracellular uptake by HeLa cells also had folate ligand on them to enable targeted delivery. Moreover, the hydrogel polymers were thermally responsive and DOX loaded. Those modified hydrogels offered advantages such as increased cellular uptake and apoptotic activity *in vitro* [27].

Another stimuli-responsive hydrogel developed very recently made use of biocompatible thermally responsive polymers that facilitated rupture of cancer cells. The study was conducted *in vitro* with an external source of heat and showed successful cell rupture. Those hydrogel particles had RGD (arginine, glycine, aspartate) peptides attached to their surface that could bind to cells (Fig. 35.5). The paper discussed experiments performed to confirm firmness of the RGD peptides with the cells (ex: MDAMB21) and their rupture using external heat stimulation.

Since *in vivo* conditions in tumors normally produce a higher temperature than the surrounding areas, and also that it has been proved that the physical force generated from the expansion of the cells when they were heated was sufficient to kill them, that type of thermally responsive hydrogels proved to be highly advantageous options for achieving targeted cell killing [28].

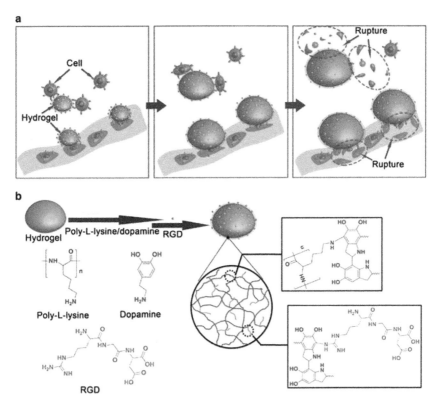

Figure 35.5 (a) Schematic illustration of the cancer cell attachment on stimuli-responsive hydrogel surface. External stimulus such as temperature causes expansion of the hydrogels and rupture of the cancer cells. (b) Schematic depiction of the surface modification of hydrogel. The first layer of coating was polydopamine (PDA) and poly-L-lysine (PLL) followed by a layer of RGD peptides for binding cancer cells. Reprinted (adapted) from [28]; the article is open access and the content reusable. Creative Commons International License 4.0 for the article is available in https://creativecommons.org/licenses/by/4.0/ [28].

Hydrogen peroxide trigger-mediated release of drugs could be achieved using hydrogels. This has been demonstrated previously with the use of ABC-type triblock copolymer poly [(propylenesulfide)-*b*-(*N*,*N*-dimethylacrylamide)-*b*-(*N*-isopropylacrylamide)] [29, 30] (PPS-*b*-PDMA-*b*-PNIPAM). A model hydrophobic drug, the dye red Nile was encapsulated in the hydrogel in order to demonstrate this. The gelation to form the cross-linked hydrogel happened above 37°C and the ambient 25°C temperature was sufficient to encapsulate the hydrophobic drug through the formation of micelles. Thus, the temperature-responsive polymer could cross-link to be a stable hydrogel carrying the hydrophobic drug based on the temperature modification and also showed a release of the red colored dye based on the H_2O_2 release in the environment (Fig. 35.6).

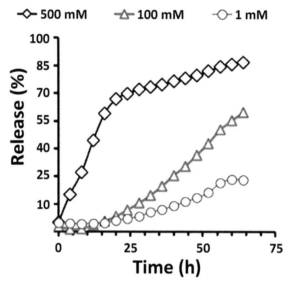

Figure 35.6 Drug release dependent on Reactive Oxygen Species concentration as displayed by PPS_{60}-b-PDMA 150-b-PNIPAAM150 triblock polymer-based thermos-responsive Nile red-loaded hydrogels. Nile red-loaded hydrogels (5 wt% triblock copolymer concentration) in PBS (pH 7.4) have been used to demonstrate Hydrogen Peroxide dependent drug release kinetics *in vitro* at 37°C. 1, 100 and 500 mM concentrations of H_2O_2 over a 64 h time course were incubated with the hydrogel samples to study ROS-dependent drug release. Reprinted with permission from [31], copyright (2014) American Chemical Society.

Apart from the polymer solubility switch mechanism that helped release based on the presence of the reactive oxygen species in the environment, cleavable units of polymer that underwent scission in presence of ROS have also been developed. These are also biodegradable and help avoid any toxic build up or unwanted immune response in the system from the presence of the hydrogels [31, 32].

Diselenide-containing block copolymer and a peptide amphiphile were used as building blocks for the construction of a

UV responsive hydrogel. The gel sol transition of the hydrogel occurred in presence of gamma irradiation. The peptide amphiphile was composed of naproxen, the drug and a hexapeptide sensitive to UV irradiation, the radiation trigger was sufficient to cause release of the drug from hydrogel. The stimuli responsive hydrogel system was thus capable of providing both chemo- and radiotherapy benefits on application [13].

Covalently cross-linked hydrogels are typically stable and elastic while the physically cross-linked hydrogels produced through ionic interactions are less stable and exhibit reduced mechanical properties. The dynamic covalent chemistry is an option to acquire stability, elasticity and also shear-thinning and self-healing characteristics for the hydrogel. Based on this, Volkan et al. reported the synthesis of hydrogel networks from reversible interactions between phenylboronic acid and cis-diols. The gel strength was dependent on pH and the hydrogel was evaluated for protein delivery *in vitro*. The pKa of phenylboronic acid and the pH of the environment were the determining factors for establishing the extent of hydrogel cross-linking. Soft, moldable hydrogels were produced that could be injected using standard syringe needles and demonstrated shear thinning and healing properties in addition to their injectable nature. Protein encapsulation was performed with the hydrogel, and the release effect monitored. Size-dependent protein release could be attributed to the mesh size of the hydrogel network. Insulin and IgG release from the hydrogel network was monitored in presence of glucose. Glucose-responsiveness of the hydrogel was confirmed, and the release kinetics could be controlled by the mesh size of the hydrogel. 3T3 fibroblast cells showed no significant toxicity as quantified using MTT (3-(4,5-dimethylthiazol-2-yl)-2,5-diphenyltetrazolium bromide) assay for 24 h. No chronic inflammation was observed *in vivo* and the materials were quite biocompatible [33].

The temperature sensitive hydrogels respond by undergoing sol-gel transitions with changes in temperature from room ones to the physiological ones. Recently developed hydrogels were that were thermo-responsive and accommodated hydrophobic drugs efficiently were built from amphiphilic triblock co-polymers, poly(N-isopropylacrylamide)-*b*-poly(4-acryloylmorpholine)-*b*-poly(2-((((2-nitrobenzyl)oxy)carbonyl)amino)ethylmethacrylate)

(PNIPAM-*b*-PNAM-*b*-PNBOC). The hydrogel carried the hydrophilic drug gemcitabine and the hydrophobic drug doxorubicin. The triblock co-polymers first assembled into micelles that had the hydrophobic and temperature responsive components formed below the lower critical solution temperature, while higher polymer concentration and temperature above the critical gelation temperature were used to form the hydrogels of physically cross-linked micellar nanoparticles. The study demonstrated the synthesis, characterization and the temperature and UV irradiation triggered synergistic release of both the hydrophobic and hydrophilic drugs *in vitro* [34, 35].

Hydrogels formed by covalent cross-linking of polymers to facilitate targeted drug delivery have been achieved in recent years. The nanoparticle-hydrogel is a hybrid system that is formed in three different ways. The first method is to entrap a hydrogel within a nanoparticle, the second one is to form a 3D hydrogel network with the nanoparticles by cross-linking of the latter using hydrophobic interactions or mixing up nanoparticles of opposite charges and the final one is to covalently couple the nanoparticle with the hydrogel [36]. A liposome cross-linked hybrid hydrogel has been developed recently. Glutathione-triggered release from the stimuli responsive polymer favored drug release. Arylthioether succinimide cross-links were introduced between the peg polymers and the liposome nanoparticles to produce the 3D hydrogel network. In presence of glutathione, the matrix was degraded, and the encapsulated drug molecules were released. Malemimide-functionalized liposomes cross-linked using peg polymers were constructed for co-delivery of doxorubicin and cytochrome-C (apoptotic cascade initiator) and their release testing *in vitro* was monitored in presence of glutathione. The presence of glutathione triggered drug release from the hydrogel in a reducing microenvironment such as a tumor [37].

35.3 Diverse Physical Attributes of Hydrogels for Drug Delivery

Structural-modification amenability of the hydrogels renders them in various shapes and sizes. This feature is particularly interesting for drug delivery applications in order to design the hydrogels as per the target sites to which the drugs have to be delivered. The hydrogel-based dosage forms can have different designs and shapes depending on the route of drug administration (Table 35.1).

Table 35.1 Different types of hydrogel products administered via different routes of administration [38–50]

Route of administration	Shape	Typical dimensions
Peroral	Spherical beads	1 µm to 1 mm
	Discs	Diameter of 0.8 cm and thickness of 1 mm
	Nanoparticles	10–1000 nm
Rectal	Suppositories	Conventional adult suppositories dimensions (length ~32 mm) with a central cavity of 7 mm and wall thickness of 1.5 mm
Vaginal	Vaginal tablets	Height of 2.3 cm, width of 1.3 cm and thickness of 0.9 cm
	Torpedo-shaped pessaries	Length of 30 mm and thickness of 10 mm
Ocular	Contact lens	Conventional dimensions (typical diameter ~12 mm)
	Drops	Hydrogel particles present in the eye drops must be smaller than 10 µm
	Suspensions/Ointments	N/A
	Circular inserts	Diameter of 2 mm and total weight of 1 mg (round shaped)
Transdermal	Dressings	Variable
Implants	Discs	Diameter of 14 mm and thickness of 0.8 mm
	Cylinders	Diameter of 3 mm and length of 3.5 cm

35.4   Specific Therapeutic Areas Using Hydrogels for Drug Delivery at Present

35.4.1   Ophthalmic

Conventional eye drops have problems with sustained drug delivery and there is huge wastage of drug immediately following application through eye drainage. Dextenza is a very recently FDA approved (December 3, 2018) ocular therapeutic hydrogel formulation for human use. This is used for ocular pain following ophthalmic surgery and is the first intracanalicular implant developed for drug delivery by the company Ocular Therapeutix (Bedford, MA, USA) [51].

Thermo-responsive polymer developed by mixing poly-(acrylic acid-graft-N-isopropylacrylamide) (PAAc-graft-PNIPAAm) with PAAc-co-PNIPAAm geL and incorporating epinephrine was used in the *in vitro* evaluation of ophthalmic drug release. The approach augmented the effect of intraocular pressure reduction from 8 h with the traditional drops to 36 h. The cross-linking density of the hydrogel affected the capillary network formation and offered a convenient controlled drug release method for ophthalmic drug delivery [52].

Intraocular pressure (IOP) elevates during glaucoma and alleviating this pressure has been quite challenging. Hydrogels could be used to resolve this problem by using them in order to prepare soft contact lenses composed of polymers to form networks. The highly hydrated polymer networks of hydrogels cause the drug to elute out very rapidly and this is not favorable for glaucoma therapy, which mainly uses hydrophilic drugs. However, with suitable modifications, soft contact lenses have been developed using polymers of N,N-diethylacrylamide and methacrylic acid, which delivered the hydrophilic drug timolol for about 24 h, thereby opening up ways to allow sustained hydrophilic drug delivery using hydrogels. Storing the contact lenses in a hydrated state can leach out drug and to wear them all the time are the limitations though [53].

Inner layer-embedded contact lenses have been investigated for the sustained release of highly water-soluble drug betaxolol hydrochloride on the ocular surface. Cellulose acetate and Eudragit S-100 were selected as the inner layer of the contact lenses which showed a promising sustained drug release for over 240 h in tear fluid of rabbits *in vivo* to create a controlled-release drug-carrier in ophthalmic drug delivery [19].

Controlled drug release behavior from hydrogels was also evaluated using nepafenac as the model drug. 3D cross-linked thermos and pH sensitive hydrogel was designed that was composed of carboxymethylchitosan (CMC) and poloxamer with glutaraldehyde as the cross-linking agent. The hydrogel was found to undergo reversible sol-gel transition at temperature and/or pH alteration at a very low concentration. Sustained release of the drug nepafenac was observed in the *in vitro* model and maximum release was observed at 35°C and pH 7.4. Cytocompatibility of the hydrogel with human corneal epithelial cells was high [54].

35.4.2 Oral, Intestinal

Gastroretentive drug dosage forms (GRDDFs) are particularly attractive for drugs that are absorbed in the proximal part of gastrointestinal tract. Enhancing the retention time of the drugs in the GI tract is very important in order to improve their bioavailability and enhance their therapeutic effects. These dosage forms could be exploited for their mucoadhesion to the gastric mucosa, modified to float or sink in order to prevent leaving the stomach or increase their swelling behavior and make them as large to prevent passage through pylorus for prolonged periods. Based on these ideas, polyionic complex hydrogels of chitosan with ring-opened PVP (polyvinyl pyrrolidone) have been developed for Osteoporosis therapy. The formulation was used to release alendronate in the upper GI tract. Enhanced mucoadhesion, delayed clearance from swelling, minimal localized irritation, improved bioavailability and slower release of the active ingredients are the interesting aspects of the preparation. Also, *in vivo* experimentation showed that these hydrogels could provide optimized PK properties that maintained the drug in the therapeutic levels for a sustained period of time, minimizing fluctuations in therapeutic levels, hence also the possible side effects [55].

Inflammatory diseases such as irritable bowel syndrome have been recently treated using hydrogels. These provided safer alternatives to delivery methods that may cause systemic toxicity. Zhang et al. developed negatively charged hydrogels that preferentially accumulated in the positively charged inflamed colon and acted as carriers of the corticosteroid drug dexamethasone (Dex). The hydrogel was prepared from ascorbyl palmitate which had labile bonds responsive to inflammatory conditions and was Generally Regarded as Safe (GRAS) for administration. Enema administration to the colon of inflammation targeting (IT) hydrogel microfibers not only reached the target site but also stayed there owing to charge interaction. The formulation was therapeutically very efficacious and revealed lesser systemic drug exposure than with free Dex in the IBS mice model *in vivo* [56].

Complexation hydrogel prepared from poly(methacrylic acid-g-ethylene glycol) (P(MAA-g-EG)) has been described. The targeting ligand used was the octarginine cell-penetrating peptide that caused specific delivery of insulin to the intestine. This method facilitated ideal targeting, absorption at target and allowed immediate release of insulin from absorption site. Great hypoglycemic responses were achievable and increased insulin absorption was noted from diabetic rat models used for testing. 18% glucose reduction was observed immediately on administration of the hydrogel containing insulin [57].

35.4.3 Cardiac Illness and Cancer

Myocardial infarction is a leading cause of death and disability in the world. Intramyocardial administration of biomaterials such as hydrogels along the perimeter region of myocardial infarction has proven to be beneficial.

Chen et al. proposed the use of a combination of curcumin (known for its anti-oxidant, anti-inflammatory and anti-oxidation properties) and nitric oxide (known as an anti-angiogenesis agent) in a hydrogel to treat myocardial infarction. The mixed component hydrogel created with the combination drugs improved therapeutic efficacy synergistically. Protective effects such as myocytic apoptotic death alleviation, reduced collagen deposition, increased vessel density (attributable to NO in the combination) and upregulated Silent Information Regulator 1 (SIRT-1), a histone deacetylase that conferred resistance to the heart from ischemic injury were observed in diseased mice models *in vivo*. The hydrogel was prepared using peptide derivatives of curcumin and NO in a ratio of 4:1 and showed sustained curcumin release at a low concentration of 2.5 μg per ml per 24 h. NO was released in presence of the enzyme β-galactosidase that could break glucosidic bonds to release NO [58].

Growth factors and cytokines (paracrine factors) secreted by stem cells have been proven to be effective in repairing damaged myocardial tissue. The whole cocktail of the paracrine factors is referred to as a secretome and is isolated *in vitro*. The biomolecular composition of the secretome can be manipulated suitably by varying stem cell culture conditions. An injectable hydrogel to deliver to peri-infarct myocardium has been recently developed using secretome from human adipose derived stem cell secretome. Nanocomposite hydrogel was formed from a combination of gelatin and laponite carrying the secretome and tested both *in vitro* and *in vivo* for their therapeutic effects via monitoring angiogenesis, scar formation and heart function. Significantly reduced scar area and improved cardiac function were observed *in vivo* in the secretome loaded hydrogel group in relation to the control [59].

The very recent development of a paintable hydrogel to serve as cardiac patch for treating myocardial infarction is worth mentioning in this context. The hydrogel eliminates the damage to tissue through suture or light triggered reactions as it is paintable. It has been constructed by a Fe^{3+} triggered polymerization reaction wherein the covalently linked pyrrole and dopamine undergo simultaneous polymerization with the trigger and the conductive polypyrrole produced uniquely cross-links the network further. The functional patch is both adhesive and conductive and forms a suture-free alternative for reconstruction of cardiac function and revascularization. Bonding within 4 weeks to the beating heart boosts the transmission of electrophysiological signals with conductivity profiles equivalent to that of the normal myocardium [60].

Biomaterial-based immunotherapy platforms for targeted drug delivery to cancers are the latest trend observable in cancer therapy. Based on this idea, novel STINGels have been developed by Leach et al. that are peptide hydrogels to show controlled delivery of cyclic dinucleotides (CDN). Dramatic improvement in survival was observed in murine models of head and neck cancer in comparison to CDN alone or CDN delivered from a collagen hydrogel [61].

Thyroid cancer treatment using local drug delivery system formed of glycol chitosan (GC) hydrogel and doxorubicin hydrochloride (DOX·HCl) called GC10/dox has been recently developed (Fig. 35.7). Visible light regulated the storage and swelling aspects of the hydrogel and a controlled sustained release followed the initial burst release within 18 h. Potent antitumor effects were observed *in vivo* and *in vitro* in comparison to free DOX·HCl and this is a promising research direction for thyroid cancer therapy [62].

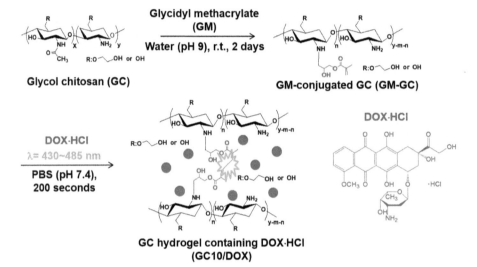

Figure 35.7 To glycol chitosan solution glycidyl methacrylate (GM) was added in water (adjusted to pH 9) and maintained for 2 days at room temperature. The white solid conjugate of GM was dissolved in water and riboflavin added. DOX·HCl was added and the mixture was irradiated using blue visible light (430–485 nm, 2100 mW/cm^2) for 10 min in order to promote hydrogelation. Reprinted with permission from [62]; the article is open access and the content reusable. Creative Commons International License 4.0 for the article is available in https://creativecommons.org/licenses/by/4.0/.

Injectable hydrogels responsive to Reactive Oxygen Species (ROS) that degrade in the presence of ROS and promote immunogenic tumor phenotype via local gemcitabine delivery is a recent discovery. The PVA cross-linked hydrogel with ROS-labile linkers enhance anti-tumor response with a localized release of immune checkpoint blocking antibody (anti-PD-L1 blocking antibody (aPDL-1)) in *in vitro* and immunogenic *in vivo* mice models. Tumor recurrence prevention after primary resection is the therapeutic advantage of this chemo-immunotherapy [63].

35.5 Translation to the Clinic

With enormous potential for therapeutic applications, several hydrogel formulations have crossed the barriers of *in vitro*/pre-clinical studies and found their way into the market. Some of them are still in the clinical study phases. Hydrogels have evolved over time to be one of the best and the most versatile drug delivery platforms. Table 35.2 lists the widespread practical applications of the hydrogel concept that have been translated to the clinical level.

Table 35.2 Examples of hydrogels translated to clinical use [64–68]

Product	Type of hydrogel	Therapeutic application	Drug delivered
Sericin	Dextran	Optically trackable drug delivery system for malignant melanoma	Doxirubicin
Hyalofemme/ Hyalo Gyn	Carbomer propylene glycol, Hyaluronic acid derivative	Vaginal dryness, estrogen alternative	Hyaluronic acid derivative
Destenza	Polyethylene glycol	Intra-canalicular delivery for post-operative ophthalmic care	Dexamethasone
Regranex	Carboxymethyl cellulose	Diabetic foot ulcer	Recombinant human derived growth factor
muGard	Mucoadhesive	Oral lichen planus	—
—	2% Poloxamer	Cervical cancer recurrence	Carboplatin

35.6 Conclusions

Hydrogels offer a versatile platform for the therapy of several diseases including cancer and diabetes. The water-loving nature of hydrogels and the ability to shrink and swell depending on several environmental cues or the mere presence of water is attractive for drug delivery applications. They have a high degree of porosity and the polymers building them could be cross-linked to varying degrees by adjusting their densities. With a physical structure highly amenable to modification in several ways, the hydrogel applications are not just limited to targeted drug delivery. They also find applications in hygiene products, wound dressings, contact lenses and tissue engineering.

Recent developments of hydrogels in the field of targeted drug delivery have been tremendous. They are modified with targeting ligands and diverse polymer types that confer very interesting properties on them for drug delivery. Ophthalmic drug delivery is an area seeing significant impact in therapy from hydrogels. From comfortable contact lenses to biodegradable drug delivery the applications in eye care have been enormous. They are 90% water, provide steady state drug release over days or months, deliver small molecules or large proteins, are fully absorbed in delivery and remain visible during monitoring [51].

Noteworthy is the application of pH responsive hydrogels for cancer therapy and glucose responsive hydrogels for diabetes. The use of modified stem cell membranes for targeted delivery is a very recent and attractive strategy for drug delivery. These membranes coated on hydrogels (nanogels) loaded with drugs are highly specific to the disease site in cancer and are highly biocompatible.

Immunotherapy platforms using hydrogels are very significant in cancer therapy. Hydrogels enabling localized delivery of antibodies and other immune-regulatory molecules at cancer sites are promising drug delivery vehicles for cancer therapy. Gastro-retentive drug dosage forms are versatile drug delivery platforms for intestine and they offer the advantages of adjusting the nanoparticle size to facilitate retention of the active ingredient in the GI tract for as long as required.

Though the hydrogel-based drug delivery was originally influenced by the hydrophobicity of the drugs, several improvements have been made recently including development of cyclodextrins modified to accommodate the hydrophobic drug sufficiently. Adhesive and conductive patches developed using hydrogels are useful in cardiac repair and vascularization. Remotely controlled motility of hydrogel (mimicking motion of a maggot) and the QD DNA hydrogels are novel ideas to facilitate targeted drug delivery.

As discussed in the paper, though there are several hydrogel formulations in clinical use, there is always scope for improvement and modification of hydrogels to enhance their applications. With subtle modifications to the existing ones, the hydrogels could become superlative drug delivery vehicles surpassing the disadvantages and current limitations with the use of several conventional delivery forms and provide promising results for therapy of several illnesses.

Abbreviations

aPDL-1:	anti-PD-L1 blocking antibody
CD:	cyclodextrin
CDN:	cyclic dinucleotides
CMC:	carboxymethylchitosan
CpG:	cytosine-phosphate-guanine
DOX:	doxorubicin
DOX·HCl:	doxorubicin hydrochloride
GA:	glycyrrhetinic acid
GC:	glycol chitosan
GG:	glycoconjugate of homopoly-L-guluronic block fraction
GM:	glycidyl methacrylate
GRAS:	Generally Regarded as Safe
GRDDFs:	gastroretentive drug dosage forms
GSH:	glutathione
GTA–ob-CD:	gelatin-β-cyclo-dextrin
H_2O_2:	hydrogen peroxide
IOP:	intraocular pressure
IT:	inflammation targeting

MTT:	3-(4,5-dimethylthiazol-2-yl)-2,5-diphenyltetrazolium bromide
OVA:	ovalbumin
P(MAA-g-EG):	poly(methacrylic acid-g-ethylene glycol)
PAAc-graft-PNIPAAm:	poly(acrylic acid-graft-*N*-isopropylacrylamide)
PDA:	polydopamine
PLL:	poly-L-lysine
PNIPAM-*b*-PNAM-*b*-PNBOC:	poly(N-isopropylacrylamide)-*b*-poly(4-acryloylmorpholine)-*b*-poly(2-((((2-nitrobenzyl)oxy)carbonyl)amino)ethylmethacrylate)
PPS-*b*-PDMA-*b*-PNIPAM:	poly[(propylenesulfide)-*b*-(*N*,*N*-dimethylacrylamide)-*b*-(*N*-isopropylacrylamide)]
PVP:	polyvinyl pyrrolidone
QD:	quantum dot
ROS:	reactive oxygen species
SIRT-1:	Silent Information Regulator 1

Disclosures and Conflict of Interest

This chapter was originally published as: Narayanaswamy, R., and Torchilin, V. P. (2019). Hydrogels and their applications in targeted drug delivery, *Molecules*, **24**, 603, doi:10.3390/molecules24030603, under the Creative Commons Attribution license (http://creativecommons.org/licenses/by/4.0/). It appears here, with edits and updates, by kind permission of the copyright holders. The authors declare no conflict of interest.

Radhika Narayanaswamy would like to extend her sincere gratitude to all who offered their valuable support for successful completion of this review article. In particular, she wholeheartedly thanks Vladimir P. Torchilin for providing her the opportunity to contribute the review to the journal *Molecules*, and to Jianxun Ding, the guest editor for the special-issue "Smart and functional polymers," where this chapter was originally published. R.N. also gratefully acknowledges the contributions of the reviewers for their useful suggestions during the publication process, and to her friends and colleagues for their kind support during the publication process of this chapter.

Corresponding Author

Prof. Vladimir P. Torchilin
Center for Pharmaceutical Biotechnology & Nanomedicine
Bouvé College of Health Sciences, 206TF
Northeastern University
360 Huntington Avenue, Boston, MA 02115, USA
Email: v.torchilin@neu.edu

References

1. Coinlogitic (2018). Hydrogel consumption market analysis by current industry status and growth opportunities. Available at: https://coinlogitic.com/hydrogel-consumption-market-research-report/51472/ (accessed on October 11, 2018).

2. Hoare, T. R., Kohane, D. S. (2008). Hydrogels in drug delivery: Progress and challenges. *Polymer*, **49**, 1993–2007.

3. Caló, E., Khutoryanskiy, V. V. (2015). Biomedical applications of hydrogels: A review of patents and commercial products. *Eur. Polym. J.*, **65**, 252–267.

4. Li, J. (2010). Self-assembled supramolecular hydrogels based on polymer–cyclodextrin inclusion complexes for drug delivery. *NPG Asia Mater.*, **2**, 112–118.

5. Moncada-Basualto, M. (2019). Supramolecular hydrogels of β-cyclodextrin linked to calcium homopoly-L-guluronate for release of coumarins with trypanocidal activity. *Carbohydr. Polym.*, **204**, 170–181.

6. Jalalvandi, E. (2016). Cyclodextrin-polyhydrazine degradable gels for hydrophobic drug delivery. *Mater. Sci. Eng. C*, **69**, 144–153.

7. Thi, T. T. H. (2017). Oxidized cyclodextrin-functionalized injectable gelatin hydrogels as a new platform for tissue-adhesive hydrophobic drug delivery. *RSC Adv.*, **7**, 34053–34062.

8. Chen, G. (2017). A glycyrrhetinic acid-modified curcumin supramolecular hydrogel for liver tumor targeting therapy. *Sci. Rep.*, **7**, 44210.

9. Zhang, L. (2017). Multifunctional quantum dot DNA hydrogels. *Nat. Commun.*, **8**, 381.

10. Nishikawa, M. (2014). Injectable, self-gelling, biodegradable, and immunomodulatory DNA hydrogel for antigen delivery. *J. Control. Release*, **180**, 25–32.

11. Shahbazi, M. A., Bauleth-Ramos, T., Santos, H. A. (2018). DNA hydrogel assemblies: Bridging synthesis principles to biomedical applications. *Adv. Ther.*, **1**, 1800042.

12. Le, X. (2017). Fe^{3+}-, pH-, Thermoresponsive supramolecular hydrogel with multishape memory effect. *ACS Appl. Mater. Interfaces*, **9**, 9038–9044.

13. Cao, W. (2013). γ-Ray-responsive supramolecular hydrogel based on a diselenide-containing polymer and a peptide. *Angew. Chem.*, **125**, 6353–6357.

14. Guo, W. (2015). pH-Stimulated DNA hydrogels exhibiting shape-memory properties. *Adv. Mater.*, **27**, 73–78.

15. Gehring, K., Leroy, J. L., Gueron, M. (1993). A tetrameric DNA-structure with protonated cytosine.cytosine base-pairs. *Nature*, **363**, 561–565.

16. Liang, Y. (2011). A cell-instructive hydrogel to regulate malignancy of 3D tumor spheroids with matrix rigidity. *Biomaterials*, **32**, 9308–9315.

17. Alvarez-Lorenzo, C. (2019). Bioinspired hydrogels for drug-eluting contact lenses. *Acta Biomater.*, **84**, 49–62.

18. Zhu, Q. (2018). Inner layer-embedded contact lenses for pH-triggered controlled ocular drug delivery. *Eur. J. Pharm. Biopharm.*, **128**, 220–229.

19. Zhu, Q. (2018). Sustained ophthalmic delivery of highly soluble drug using pH-triggered inner layer-embedded contact lens. *Int. J. Pharm.*, **544**, 100–111.

20. Segura, T. (2014). Materials with fungi-bioinspired surface for efficient binding and fungi-sensitive release of antifungal agents. *Biomacromolecules*, **15**, 1860–1870.

21. Gao, C. (2016). Stem cell membrane-coated nanogels for highly efficient *in vivo* tumor targeted drug delivery. *Small*, **12**, 4056–4062.

22. Yu, J. (2016). Endosome-mimicking nanogels for targeted drug delivery. *Nanoscale*, **8**, 9178–9184.

23. Gou, M. (2014). Bio-inspired detoxification using 3D-printed hydrogel nanocomposites. *Nat. Commun.*, **5**, 3774.

24. Sun, L. (2016). Sundew-inspired adhesive hydrogels combined with adipose-derived stem cells for wound healing. *ACS Appl. Mater. Interfaces*, **8**, 2423–2434.

25. Shen, T. (2017). Remotely triggered locomotion of hydrogel magbots in confined spaces. *Sci. Rep.*, **7**, 16178.

26. Li, S. (2015). A Drosera-bioinspired hydrogel for catching and killing cancer cells. *Sci. Rep.*, **5**, 14297.

27. Kim, H. (2017). Synergistically enhanced selective intracellular uptake of anticancer drug carrier comprising folic acid-conjugated hydrogels containing magnetite nanoparticles. *Sci. Rep.*, **7**, 41090.

28. Fang, Y. (2018). Rupturing cancer cells by the expansion of functionalized stimuli-responsive hydrogels. *NPG Asia Mater.*, **10**, e465.

29. Schlegel, P. N. (2001). Effective long-term androgen suppression in men with prostate cancer using a hydrogel implant with the GnRH agonist histrelin. *Urology*, **58**, 578–582.

30. Wang, X. (2016). Vaginal delivery of carboplatin-loaded thermosensitive hydrogel to prevent local cervical cancer recurrence in mice. *Drug Deliv.*, **23**, 3544–3551.

31. Gupta, M. K. (2014). Cell protective, ABC triblock polymer-based thermoresponsive hydrogels with ROS-triggered degradation and drug release. *J. Am. Chem. Soc.*, **136**, 14896–14902.

32. Saravanakumar, G., Kim, J., Kim, W. J. (2017). Reactive-oxygen-species-responsive drug delivery systems: Promises and challenges. *Adv. Sci.*, **4**, 1600124.

33. Yesilyurt, V. (2016). Injectable self-healing glucose-responsive hydrogels with pH-regulated mechanical properties. *Adv. Mater.*, **28**, 86–91.

34. Wang, C. (2017). Photo-and thermo-responsive multicompartment hydrogels for synergistic delivery of gemcitabine and doxorubicin. *J. Control. Release*, **259**, 149–159.

35. Larrañeta, E. (2018). Hydrogels for hydrophobic drug delivery. Classification, synthesis and applications. *J. Funct. Biomater.*, **9**, 13.

36. Gao, W. (2016). Nanoparticle-hydrogel: A hybrid biomaterial system for localized drug delivery. *Ann. Biomed. Eng.*, **44**, 2049–2061.

37. Liang, Y., Kiick, K. L. (2016). Liposome-cross-linked hybrid hydrogels for glutathione-triggered delivery of multiple cargo molecules. *Biomacromolecules*, **17**, 601–614.

38. Lee, P. I., Kim, C.-J. (1991). Probing the mechanisms of drug release from hydrogels. *J. Control. Release*, **16**, 229–236.

39. Ahmed, E. T., Maayah, M. F., Asi, Y. O. M. A. (2017). Anodyne therapy versus exercise therapy in improving the healing rates of venous leg ulcer. *Int. J. Res. Med. Sci.*, **1**(3), 198–203.

40. Hamidi, M., Azadi, A., Rafiei, P. (2008). Hydrogel nanoparticles in drug delivery. *Adv. Drug Deliv. Rev.*, **60**, 1638–1649.

41. Bilia, A. (1996). *In vitro* evaluation of a pH-sensitive hydrogel for control of GI drug delivery from silicone-based matrices. *Int. J. Pharm.*, **130**, 83–92.

42. Rahimi, M. (2016). Preemptive morphine suppository for postoperative pain relief after laparoscopic cholecystectomy. *Adv. Biomed. Res.*, **5**, 57.

43. Karasulu, H. Y. (2004). Efficacy of a new ketoconazole bioadhesive vaginal tablet on Candida albicans. *Il Farm.*, **59**, 163–167.

44. Mandal, T. K. (2000). Swelling-controlled release system for the vaginal delivery of miconazole. *Eur. J. Pharm. Biopharm.*, **50**, 337–343.

45. Hu, X. (2011). Hydrogel contact lens for extended delivery of ophthalmic drugs. *Int. J. Polym. Sci.*, **2011**, 814163.

46. Ludwig, A. (2005). The use of mucoadhesive polymers in ocular drug delivery. *Adv. Drug Deliv. Rev.*, **57**, 1595–1639.

47. Hornof, M. (2003). Mucoadhesive ocular insert based on thiolated poly(acrylic acid): Development and *in vivo* evaluation in humans. *J. Control. Release*, **89**, 419–428.

48. Brazel, C. S., Peppas, N. A. (1996). Pulsatile local delivery of thrombolytic and antithrombotic agents using poly(N-isopropylacrylamide-co-methacrylic acid) hydrogels. *J. Control. Release*, **39**, 57–64.

49. Omidian, H., Park, K. (2012). Hydrogels. In Siepmann, J., Siegel, R. A., Rathbone, M. J., eds. *Fundamentals and Applications of Controlled Release Drug Delivery*, Springer, Boston, MA, USA, pp. 75–105.

50. Naveed, S., Farooq, S., Abbas, S. S., Jawed, S. H., Qamar, F., Hussain, M. Y., Ali, I. (2015). Contemporary trends in novel ophthalmic drug delivery system: An overview, *Sci. Res.*, **2**(5), 1–10.

51. Ocular Therapeutix (2017). Engineered for ocular innovation. Available at: https://www.ocutx.com/about/hydrogel-technology/ (accessed on May 29, 2019).

52. Prasannan, A., Tsai, H.-C., Hsiue, G.-H. (2018). Formulation and evaluation of epinephrine-loaded poly (acrylic acid-co-N-

isopropylacrylamide) gel for sustained ophthalmic drug delivery. *React. Funct. Polym.*, **124**, 40–47.

53. Lavik, E., Kuehn, M., Kwon, Y. (2011). Novel drug delivery systems for glaucoma. *Eye*, **25**, 578–586.

54. Yu, S. (2017). A novel pH-induced thermosensitive hydrogel composed of carboxymethyl chitosan and poloxamer cross-linked by glutaraldehyde for ophthalmic drug delivery. *Carbohydr. Polym.*, **155**, 208–217.

55. Su, C.-Y. (2018). Complex hydrogels composed of chitosan with ring-opened polyvinyl pyrrolidone as a gastroretentive drug dosage form to enhance the bioavailability of bisphosphonates. *Sci. Rep.*, **8**, 8092.

56. Zhang, S. (2015). An inflammation-targeting hydrogel for local drug delivery in inflammatory bowel disease. *Sci. Transl. Med.*, **7**, ra128–ra300.

57. Fukuoka, Y. (2018). Combination strategy with complexation hydrogels and cell-penetrating peptides for oral delivery of insulin. *Biol. Pharm. Bull.*, **41**, 811–814.

58. Chen, G. (2017). A mixed component supramolecular hydrogel to improve mice cardiac function and alleviate ventricular remodeling after acute myocardial infarction. *Adv. Funct. Mater.*, **27**, 1701798.

59. Waters, R. (2018). Stem cell-inspired secretome-rich injectable hydrogel to repair injured cardiac tissue. *Acta Biomater.*, **69**, 95–106.

60. Liang, S. (2018). Paintable and rapidly bondable conductive hydrogels as therapeutic cardiac patches. *Adv. Mater.*, **30**, 1704235.

61. Leach, D. G. (2018). STINGel: Controlled release of a cyclic dinucleotide for enhanced cancer immunotherapy. *Biomaterials*, **163**, 67–75.

62. Yoo, Y. A. (2018). Local drug delivery system based on visible light-cured glycol chitosan and doxorubicinhydrochloride for thyroid cancer treatment *in vitro* and *in vivo*. *Drug Deliv.*, **25**, 1664–1671.

63. Wang, C. (2018). *In situ* formed reactive oxygen species-responsive scaffold with gemcitabine and checkpoint inhibitor for combination therapy. *Sci. Transl. Med.*, **10**, eaan3682.

64. Liu, J. (2016). Sericin/dextran injectable hydrogel as an optically trackable drug delivery system for malignant melanoma treatment. *ACS Appl. Mater. Interfaces*, **8**, 6411–6422.

65. Chen, J. (2013). Evaluation of the efficacy and safety of hyaluronic acid vaginal gel to ease vaginal dryness: A multicenter, randomized,

controlled, open-label, parallel-group, clinical trial. *J. Sex. Med.*, **10**, 1575–1584.

66. Blizzard, C., Desai, A., Driscoll, A. (2016). Pharmacokinetic studies of sustained-release depot of dexamethasone in beagle dogs. *J. Ocul. Pharmacol. Ther. Off. J. Assoc. Ocul. Pharmacol. Ther.*, **32**, 595–600.

67. Allison, R. R. (2014). Multi-institutional, randomized, double-blind, placebo-controlled trial to assess the efficacy of a mucoadhesive hydrogel (MuGard) in mitigating oral mucositis symptoms in patients being treated with chemoradiation therapy for cancers of the head and neck. *Cancer*, **120**, 1433–1440.

68. Li, J., Mooney, D. J. (2016). Designing hydrogels for controlled drug delivery. *Nat. Rev. Mater.*, **1**, 16071.

Chapter 36

Emergence of Real World Evidence in Precision Medicine

Neil A. Belson, MS, JD

Law Office of Neil A. Belson, LLC, Port Tobacco, Maryland, USA

Keywords: real world evidence, real world data, health economics, HEOR, drug development, drug approval, clinical evidence, drug regulation, US Food and Drug Administration, randomized clinical trials, natural history studies, rare diseases, precision medicine, personalized medicine, patient registries, disease registries, product registries, 21st Century America Cures Act, electronic health records, research and development, guidance documents, Expanded Access Program

36.1   Introduction

In June 2017, the US Food and Drug Administration (FDA) approved a new indication for a medical device without requiring the device manufacturer to conduct any new clinical trials. Instead, the Agency relied on the manufacturer's "Real World Evidence" (RWE) demonstrating safety and efficacy. Using RWE dramatically accelerated the time to FDA approval: Relying on traditional

The Road from Nanomedicine to Precision Medicine
Edited by Shaker A. Mousa, Raj Bawa, and Gerald F. Audette
Copyright © 2020 Jenny Stanford Publishing Pte. Ltd.
ISBN 978-981-4800-59-4 (Hardcover), 978-0-429-29501-0 (eBook)
www.jennystanford.com

clinical trials the United States had been only the 42nd nation to approve the original use of the device, a transcatheter aortic valve replacement (TAVR). In contrast, using RWE in its regulatory decision-making process allowed the United States to *become* the first country to approve a new valve-in-valve procedure for the device [1, 2].

This device approval demonstrated the emergence of RWE as a potentially transformative new tool in personalized medical care and drug development and regulation. RWE can provide the information which informs precision medicine decisions. As one commentator has been stated [3]:

> As the pharmaceutical industry shifts to value-based, personalized health care, RWE can help is answer the hard questions in health care, such as what works, for whom, why does it work, and in what context. All of these questions are at the heart of value-based personalized medicine...

> RWE is becoming essential to decisions across every aspect of the pharmaceutical value chain, from the early research and development stage.

RWE could substantially reduce the time and cost of commercializing new drugs and medical devices. RWE is already impacting the life sciences industry: A Deloitte 2018 survey found that nearly 90% of biopharmaceutical companies surveyed that they have either already established or plan to invest in RWE capabilities [4].

RWE is also impacting the FDA. Former FDA Commissioner Dr. Scott Gottlieb stated in December 2018 that RWE and Real World Data (RWD, which is used to generate RWE) are "a top strategic priority for the FDA" [5]. The FDA centers responsible for drugs, biologics and medical devices have all issued guidance documents relating to the use of RWE for regulatory purposes. RWE could supplement or even replace traditional randomized clinical trials (RCTs) in many situations.

The need for new methods of drug and device regulation is clear. Director of the FDA's Center for Drug Evaluation (CDER), Dr. Janet Woodcock, has repeatedly referred to the FDA's traditional system for approving new drugs, using RCTs, as "broken" and not serving the interests of patients [6]. RCTs represent approximately 60% of the approximately $2.6 billion to develop a new drug [7, 8].[1] Yet they are in many ways an inefficient and incomplete way of measuring drug and device safety and efficacy. Only about 12% of drug candidates which enter Phase I clinical studies ultimately receive FDA commercial marketing approval [9].

The deficiencies of traditional RCTs are well known. While RCTs are useful in evaluating the baseline effectiveness of a drug under controlled conditions, they often fail to accurately predict the effectiveness of a drug in real-life conditions. RCTs generally have strict inclusion and exclusion criteria, which may not accurately reflect the likely patient population. For example, RCTs are often limited to subjects between the ages of 18 to 65 years old, even though a substantial portion of the target population may be outside this age range. Similarly, RCTs often exclude patients with comorbidities (the presence of multiple chronic diseases), although many prospective patients suffer from multiple disease conditions. RCTs may not detect uncommon side effects due to inherent limitations in the number of patients who can participate in a study. Additionally, 80% of clinical trials fail to meet their initial subject enrollment projections [10]. RCTs are also generally unsuitable for studies involving rare diseases, due to the unavailability of adequate numbers of study subjects.

36.2 What Is "Real World Evidence"?

The FDA differentiates between "Real World Evidence" and "Real World Data." It considers RWD to consist of information relating to patient health status or patient health care treatment, including electronic health records (EHRs), insurance claims data, disease, or product registries, or home-use patient monitoring devices and mobile devices. RWD may also include data on socio-economic factors or environmental exposures. As is evident

---

[1]The authors of these publications [7, 8] estimated an average out-of-pocket cost per approved new compound was $1395 million in 2013 US dollars. This estimate linked costs of compounds discontinued prior to commercialization to the costs of compounds which received marketing approval. The authors calculated a total estimated cost of $2558 million by capitalizing the out-of-pocket amount at a real discount rate of 10.5%.

from this list, many sources of RWD are generated for non-regulatory purposes, such as documenting patient care or for submission of insurance claims. Not all RWD is therefore suitable for regulatory purposes such as product approvals.

Precision Medicine Committee, American Bar Association, Section of Science & Technology Law

The Precision Medicine Committee examines legal issues affecting the rapidly emerging field of precision medicine (previously known as "personalized medicine"). Precision medicine utilizes information on an individual's genetics and biomarkers, and environmental and lifestyle factors, to inform decisions on disease prevention and treatment. The PMC Committee's focus includes such diverse topics as drug and device development and regulation, payment and reimbursement, health IT, data privacy, ethical issues, intellectual property and business and investment, as they relate to precision medicine.

Co-Chairs: Neil A. Belson, Rouget F. Henschel
Vice Chairs: Raj Bawa, Roger Klein

In contrast, the FDA defines RWE as "*clinical* evidence about the usage and potential benefits or risks of a medical product derived from analysis of RWD" (emphasis in original text) [11]. The Duke-Margolis Center for Health Policy stated in a 2017 white paper prepared with FDA funding, that RWE is "evidence derived from RWD *through the application of research methods*" (italics added) [12]. RWE has also been described as "information that is generated in health care systems outside of a controlled trial" [13]. RWE is not merely passively collected or anecdotal data. Rather, RWE results from careful study designs for assessing treatment effects on patient outcomes [12].

Studies which generate RWE can be prospective, retrospective, or both—they can utilize pre-existing data, newly generated data, or a combination of both. In each case, though, RWE requires a careful clinical design to measure the effect of the study drug, device, or treatment on patient health. RWE-based studies must meet the same "substantial evidence" standards as RCTs for FDA regulatory purposes.

Used appropriately, RWE could advance medical and regulatory science in many respects, in both pre-approval and post-approval contexts. Pre-approval RWE could supplement RCTs, reducing the time and costs of drug and device development. RWE could help generate research hypotheses in clinical trials and help identify more appropriate clinical trial subjects. RWE could make development of treatments for rare diseases more feasible, particularly when it is not practical to recruit enough clinical subjects to support traditional RCTs. Post-approval RWE analyzing patient uses of a drug or device could help in identifying and approving new indications and in identifying factors in safety, effective clinical treatment practices, and personalized care, which may not be apparent in traditional clinical trials [14]. While RWE is unlikely to completely replace RCTs in the near future, RWE can become an important complementary source of information where RCTs are appropriate and a valuable alternative where they are not.

36.3   Emergence of Real World Evidence

21st Century Cures Act. The 21st Century Cures Act, enacted in December 2016 [15], marked a turning point in the emergence of RWE. Section 3022 of that Act directed the Secretary of Health and Human Services, through the FDA, to establish a program to evaluate potential uses of RWE for two purposes:

(i) helping to support approval of new indications for already approved drugs;

(ii) helping to satisfy post-approval study requirements [16].

In August 2017, the FDA's Center for Devices and Radiological Health (CDRH) and Center for Biologics Evaluation and Research (CBER) published a final guidance document describing situations in which they could accept RWE in support of regulatory decisions involving medical devices. CDER and CBER have since followed with a series of draft guidances and other publications relating to RWE. They also published a framework for evaluating RWE in relation to drugs and biologics in December 2018 [11].

36.3.1 Use of RWE for Medical Devices

Although the 21st Century Cures Act focused on drugs, the Agency's first major final guidance on RWE following the Act addressed medical devices. On August 31, 2017, CDRH and CBER issued a guidance document entitled "Use of real-world evidence to support regulatory decision-making for medical devices" [17].

The FDA noted that there often exists no system for systematically characterizing aggregating and analyzing data from all uses of a medical device. The Agency hoped to create incentives for systematically collecting and organizing information from routine use of devices in medical care by expressing a willingness to consider RWE in its regulatory decisions.

The FDA stated [17] that "under the right conditions, data derived from real world sources can be used to support regulatory decisions" in both premarket and postmarket regulatory contexts (Table 36.1). The Agency noted that use of RWD for regulatory purposes requires a careful study design. Such studies should analyze elements similar to those that would be included in a traditional RCT. The FDA recommended using the Agency's pre-submission process for prospective RWD studies, just as when preparing to conduct an RCT [17].

Since RWD is often developed for non-regulatory purposes, the FDA must determine whether such data is useful for regulatory purposes. When relying on RWD, it is important to have a predefined common set of data elements and a common definitional framework, together with prespecified time intervals for data element collection and analysis. The FDA may also consider the ability to supplement RWD with linkages to other data such as EHRs or claims data.

Table 36.1 Examples of purposes for which FDA will consider use of RWD relating to medical devices

---

- Generating hypotheses to be tested in a prospective clinical study

- A mechanism for collecting data related to a clinical study to support device approval or clearance where a registry or other means of systematic data collection exists

- Evidence to identify, demonstrate, or support clinical validity of a biomarker

- Evidence to support approval or granting of a Humanitarian Device Exemption, Premarket Approval Application (PMA), or *de novo* request

- Support for reclassification of a medical device

- Evidence for expanding the labeling of a device to include additional indications for use or to update the labeling to include new information on safety and effectiveness

- Public health surveillance efforts, if signals suggest there may be a safety issue with a medical device

- Conducting post approval studies imposed as a condition of device approval or to potentially preclude the need for postmarket surveillance studies

---

36.3.2   Use of RWE for Drugs and Biologics

In December 2018, the FDA published a "Framework for FDA's Real World Evidence Program." The Framework covers use of RWE in relation to both drugs and biologics, although it does not cover medical devices. The purposes of the Framework are to

- evaluate potential use of RWE to help support FDA approval for new indications for which drugs which are already commercial;

- support or satisfy drug post-approval study requirements [11].

The RWE program will include demonstration projects, stakeholder engagement, and input from senior leadership and promote shared learning and consistent application, with issuance of guidance documents to assist drug developers.

Since RWE already has a significant history of use in evaluating product safety, the FDA will focus on use of RWE to demonstrate product effectiveness. The FDA will evaluate the potential of RWE to support labeling changes, such as adding a new indication, changing a dosage regimen or route of administration, adding a new population, or adding safety information (Table 36.2).

Table 36.2 Factors that FDA will consider in evaluating use of RWE to demonstrate drug product effectiveness [11]

1. Whether the underlying RWD is fit for use
2. Whether the study design used to generate RWE provides adequate scientific evidence to help answer the regulatory question at issue
3. Whether the study conduct meets FDA requirements (e.g., for study monitoring and data collection)

Any RWD selected should be suitable for addressing specific regulatory questions. The strength of any RWE will depend on the clinical study methodology and the reliability and relevance of underlying data.

The FDA has considerable experience assessing electronic health care data (e.g., EHRs, medical claims data, registries) through experience with the Agency's Sentinel System, which is a national electronic system for monitoring the safety of drugs, devices, and other products on the market, as well as other data systems [11]. In fact, the FDA's use of the RWD and RWE, derived from the Sentinel System, has eliminated the need for postmarketing studies when potential safety issues arose involving five products [5]. The FDA plans to use this experience in assessing sources of RWD used to generate RWE for purposes of drug product effectiveness.

The Agency notes that the specific elements to consider in evaluating different sources of RWD may vary depending on the type of RWD used and its intended purpose. For example, the FDA considers the strengths and limitations of medical claims data as RWD to be well understood based on experience within government agencies, health care insurers, and medical researchers. Conversely, EHR data, which may provide more detailed patient data than medical claims, is at present not usually standardized or collected in structured fields which are readily extractable comparable across systems. Additionally, EHRs and medical claims data may not consistently capture certain co-variables (such as obesity, smoking, or alcohol use) and outcomes (such as mortality or symptomatic changes).

Patient registries are another potentially significant source of RWD. A patient registry systemically collects uniform data from a population with a particular disease or condition, or who are receiving a particular drug or other medical treatment, in order to evaluate outcomes. The FDA considers that the fitness of patient registries for use in generating RWE depends on whether there are adequate processes for gathering follow-up information as needed, to minimize missing or incomplete data and ensure data quality.

Filling gaps in information, which may be difficult to capture in the context of EHRs and medical claims data, will be part of the FDA's RWE evaluation program. Another component of the FDA's program will be addressing the lack of interoperability among different health care systems, and the difficulty of linking data sources for a single patient across different providers and health care systems, while maintaining patient privacy [11].

36.4   Case Studies Using RWE

Both the FDA and the private sector had used RWE and RWD even before enactment of the 21st Century Cures Act, in both regulatory and non-regulatory contexts. The discussion below provides case examples, both prior to and after the Act, in which parties used RWE and RWD in a regulatory context:

(i) *Postmarket Monitoring of a Drug*: The FDA has long used RWE for postmarket monitoring and evaluating the safety of drugs after they have been approved, through the Agency's "Sentinel System." The Agency's primary sources of information for such studies include electronic health data such as medical claims and pharmacy dispensing data. The Sentinel System had data on over 100 million individuals as of August 2018 [11].

(ii) *Natural History "Controls" in Studies of Treatments for Rare Diseases*: One of the most frequent uses of RWE in a drug approval context has been in natural history studies as a "control," particularly in cases of evaluating drugs to treat rare diseases. A "natural history" study follows the progression of a disease or condition in the absence of a treatment from just before its onset until its final outcome (i.e., death, disability, or patient recovery).

In evaluating potential treatments for rare diseases, the FDA has often relied on "single-arm" studies of a drug, using a natural history study of the target disease as a non-treatment "control." Many of these cases have involved rare genetic disorders, where no FDA-approved treatment exists, and death or severe disability is imminent in the absence of treatment. In such instances, ethical considerations preclude the use of actual patient control subjects.

For example, the FDA approved Brineura (Cerliponase alfa) to treat a form of Batten disease based on single-arm study compared with a natural history control. No prior FDA-approved treatment existed for that form of disease, known as late-infantile neuronal ceroid lipofuscinosis type 2 (CLN2), a rare genetic disorder which often causes children to lose walking ability and die in their teens. The FDA approved Brineura to treat CLN2 following an efficacy study involving 22 symptomatic pediatric patients, compared with a natural history cohort "control" or comparator consisting of 42 untreated patients. The study showed patients treated with Brineura suffered fewer declines in walking ability than the untreated patients in the natural history cohort [18].

The FDA plans to expand its use of natural history studies. Former FDA Commissioner Gottlieb stated that the Agency has "been working overtime to develop models that can simulate the behavior of placebo arms in the setting of rare diseases [19]. In conjunction with his comment, the FDA announced it would fund natural history studies relating, respectively, to Friedreich's ataxia, pregnancy and lactation-associated osteoporosis, sickle-cell anemia, Angelman syndrome, and myotonic muscular dystrophy type 1 [19].

(iii) *Utilization of the FDA's Expanded Access Program to generate RWE*: Most current discussion on the use of RWE in the context of regulatory approvals seems to focus new indications for existing products. Apart from its use in natural history studies, there appear to be limited options for utilizing RWE to obtain an initial product approval—after all, how can a Sponsor obtain evidence from real-life drug usage regarding a product which is not approved for commercial use in "real-life?" One such option which may exist is the FDA's "Expanded Access Program" (EAP, also known as the "Compassionate Use" program), which grants patients access to investigational drugs when they have exhausted all approved treatments and cannot participate in clinical trials. The FDA has expressed openness to Sponsors' leveraging data from the EAP program to generate RWE [20].

One example where the FDA relied on RWE from an EAP study to support a drug approval involved Lutathera (lutetium Lu 177 dotatate), a radioactive drug (or "radiopharmaceutical") for treatment of somatostatin receptor-positive instances of a type of cancer known as gastroenteropancreatic neuroendocrine tumors (GEP-NETs) that affects the pancreas or gastrointestinal tract. GEP-NETs are a rare group of cancers for which there are limited treatment options if initial therapy is unsuccessful.

Lutathera's approval was supported by two studies. One was an RCT with 229 patients. The second study was based on data from a single-arm, open-label study of 1,214 patients with somatostatin receptor-positive tumors, including GEP-NETS, who received Lutathera at a site in the Netherlands. Complete or partial tumor shrinkage was reported in 16 percent of a subset of 360 patients with GEP-NETs who were evaluated for response by the FDA. Patients initially enrolled in the study received Lutathera as part of an expanded access program [21, 22].

(iv) *Use of Patient Registries to Obtain Approvals for New Indications*: The manufacturer of the TAVR device described at the start of this article is an example of a company using a patient registry to avoid having to conduct RCTs for a new indication. The manufacturer obtained its initial marketing approval from the FDA in 2011. Upon obtaining this approval, the manufacturer then proactively established a product registry, which generated a database containing records of over 100,000 TAVR records. Among these records were 600 records relating to the then off-label use of a new valve-in-valve procedure. Based on this registry data from these 600 records, the FDA approved the use of the new procedure without requiring any further RCTs [1, 2].

(v) *Private sector use of RWE and RWD*: Private sector drug development companies have used RWD as a tool in traditional RCTs to target their studies more precisely on patient groups and sub-groups that are most likely to show a positive response to drug candidates. RWD can be useful in generating study hypotheses for RCTs, identifying potential biomarkers, and identifying prognostic indicators or patient baseline characteristics to support clinical trial enrichment or stratification of clinical trial subjects. The FDA generally encourages such efforts and regards such uses of RWD as well established [11]. Some biopharma companies systemically collect data from their EAP programs, which they use as RWD in support of regulatory filings [23].

36.5   Future Challenges for the Use of RWE

RWE as a regulatory tool is in its early stages. Our understanding of prospective uses of RWE, and the methods for generating it, is still evolving. The list below identifies some challenges affecting the use of RWE for regulatory purposes:

- *Lack of institutionalized methods of obtaining RWD*: Many sources of RWD, such as those generated in EHRs and medical claims, are not collected for regulatory or research purposes. Some data is generated in formats (such as PDFs) which are not efficient for data analyses. The National Evaluation System for Health Technology Coordinating Center (NESTcc), an organization which evaluates use of RWE in relation of medical devices and which was established through a 2016 grant from the FDA to the Medical Device Innovation Consortium, has stated that the "current fragmented health care ecosystem does not support the seamless, near real-time, cost-effective use of health data to generate high-quality evidence for medical devices needed for regulatory decision-making in both the pre- and postmarket spaces" [24].

- *Lack of Generally Accepted Standards*: The FDA and others are still in the process of evaluating which data is useful or adequate to support regulatory decisions.
- *Data Privacy Concerns*: Recurring breaches of data privacy may discourage individuals from allowing their health data to be used for research purposes.

36.6   Conclusions

The use of RWE for obtaining drug and device approvals is in its early stage, and our understanding of its prospective applications is still evolving. Many potentially useful sources of RWD, such as EHRs and medical claims, are not currently generated with a focus facilitating research. Additionally, there is not yet a common understanding as to what kind of RWE is necessary to adequately demonstrate product safety and efficacy. These factors, combined with lack of interoperability across different health care systems, makes the systemic collection, aggregation, and analysis of RWE a challenge.

Nonetheless, broad agreement exists regarding the limitations of traditional RCTs, and the need to develop new approaches for evaluating drug and device product safety. The biopharma industry is rapidly expanding its investment in RWE. Appropriately utilized, RWE could play an important role in reducing the time and cost of developing new drugs and medical devices and reducing the cost and improving the quality of health care.

Disclosures and Conflict of Interest

The author declares that he has no conflict of interest. The author received no payment for preparing this chapter. The author did not receive any assistance in preparing this chapter, apart from editorial assistance from Dr. Raj Bawa, the Series Editor.

The author is a Contract Attorney Consultant for Axovant Gene Therapies (New York, NY) and also the owner of the Law Office of Neil A. Belson, LLC (Port Tobacco, MD). The author is also a Manager and Membership Interest owner of Leafpro, LLC (Wilson, NC), a company that is developing naturally occurring leaf proteins from plants. The author is not aware of any affiliations, memberships, or funding that might be perceived as affecting the objectivity of this chapter.

Corresponding Author

Neil A. Belson
Law Office of Neil A. Belson, LLC
6225 Hampstead Ct.
Port Tobacco, MD 20677, USA
Email: nabelsonlaw@hotmail.com

References

1. Shuren, J. Zuckerman, B. (2017). How creative FDA regulation led to first-in-the-world approval of a cutting-edge heart valve. *FDA Voice*. Available at: https://www.medicaldesignandoutsourcing.com/creative-fda-regulation-led-first-world-approval-cutting-edge-heart-valve/ (accessed on August 20, 2019).

2. Mezher, M. (2017). FDA used real-world evidence in heart valve approval. *Regulatory Focus*. Available at: https://www.raps.org/regulatory-focus%E2%84%A2/news-articles/2017/6/fda-used-real-world-evidence-in-heart-valve-approval (accessed on August 20, 2019).

3. Snyder, M. (2017). Using real world evidence to propel personalized treatments and improve clinical trial outcomes—Interview with Brett Davis, Deloitte. *R&D Mag.*, May 22. Available at: https://www.rdmag.com/article/2017/05/using-real-world-evidence-propel-personalized-treatments-improve-clinical-trial-outcomes (accessed on August 20, 2019).

4. Deloitte insights: Second annual real world evidence (RWE) benchmarking survey (2018). Available at: https://www2.deloitte.com/content/dam/insights/us/articles/4354_Real-World-Evidence/DI_Real-World-Evidence.pdf?elqTrackId=cba898115de54fb5a9dc04206f4de617&elqaid=453&elqat=2 (accessed on August 20, 2019).

5. FDA (2018). Statement from FDA Commissioner Scott Gottlieb, M.D., on FDA's new strategic framework to advance use of real-world evidence to support development of drugs and biologics. Available at: https://www.fda.gov/news-events/press-announcements/statement-fda-commissioner-scott-gottlieb-md-fdas-new-strategic-framework-advance-use-real-world (accessed on August 20, 2019).

6. Dunn, A. (2018). FDA's Woodcock: The clinical trial system is broken. *Biopharma Dive*. Available at: https://www.biopharmadive.com/news/fdas-woodcock-the-clinical-trial-system-is-broken/542698/ (accessed on August 20, 2019).

7. Viceconti, M., Cobelli, C., Haddad, T., Himes, A., Kovatchev, B., Palmer, M. (2017). *In silico* assessment of biomedical products: The conundrum of rare but not so rare events in two case studies. *Proc IMechE Part H: J Eng. Med.,* **231**(5) 455–466. Available at: http://eprints.whiterose.ac.uk/113971/1/pap%20ISCT%20x%20imeche%20OA%20version.pdf (accessed on August 20, 2019).

8. DiMasi, J. A., Grabowski, H. G., Hansen, R. W. (2016). Innovation in the pharmaceutical industry: New estimates of R&D costs. *J. Health Econ.,* **47**, 20–33 (abstract). Available at: https://www.ncbi.nlm.nih.gov/pubmed/26928437/ (accessed on August 20, 2019).

9. DiMasi, J. A., Grabowski, H. G., Hansen, R. W. (2015). The cost of drug development. Correspondence. *NEJM*. Available at: https://www.researchgate.net/profile/Ronald_Hansen2/publication/276361868_The_Cost_of_Drug_Development/links/55b7738708ae9289a08be6ae.pdf (accessed on August 20, 2019).

10. Deloitte (2017). Getting real with real world evidence: 2017 RWE benchmark survey." Available at: https://www2.deloitte.com/us/en/pages/life-sciences-and-health-care/articles/real-world-evidence-benchmarking-survey.html (accessed on August 20, 2019).

11. FDA (2018). Framework for FDA's real-world evidence program. Available at: https://www.fda.gov/media/120060/download (accessed on August 20, 2019).

12. Duke Margolis Center for Health Policy (2017). A framework for regulatory use of real-world evidence. Available at: https://healthpolicy.duke.edu/sites/default/files/atoms/files/rwe_white_paper_2017.09.06.pdf (accessed on August 20, 2019).

13. Snyder, M. (2017). Using real world evidence to propel personalized treatments and improve clinical trial outcomes. *R&D Mag.*, May 22. Available at: https://www.rdmag.com/article/2017/05/using-real-world-evidence-propel-personalized-treatments-improve-clinical-trial-outcomes (accessed on August 20, 2019).

14. NEHI (Network of Excellence in Health Innovation) (2015). A new era for health care innovation. Available at: https://www.nehi.net/writable/publication_files/file/rwe_issue_brief_final.pdf (accessed on August 20, 2019).

15. Public Law 114–255—December 13, 2016, 130 STAT. 1033.

16. 21 U.S.C. § 355g.

17. FDA (2017). Use of real world evidence to support regulatory decision-making for medical devices. Available at: https://www.fda.gov/media/99447/download (accessed on August 20, 2019).

18. FDA (2017). FDA approves first treatment for a form of Batten's disease. Available at: https://www.fda.gov/news-events/press-announcements/fda-approves-first-treatment-form-batten-disease (accessed on August 20, 2019).

19. Luxner, L. (2018). Rare disease groups welcome FDA's embrace of 'real world' data clinical trials. *SMA News Today*. Available at: https://smanewstoday.com/2018/03/12/rare-disease-groups-welcome-fdas-embrace-real-world-data-clinical-trials/ (accessed on August 20, 2019).

20. Opening remarks from the FDA commissioner. In: *Public Meeting Report: Leveraging Real-World Treatment Experience from Expanded Access Protocols*, November 19, 2018. Reagan-Udall Foundation for the FDA. Available at: http://navigator.reaganudall.org/sites/default/files/Leveraging%20Real-World%20Treatement%20Experience%20from%20Expanded%20Access%20Protocols%20Report.pdf (accessed on August 20, 2019).

21. FDA (2018). FDA approves new treatment for certain digestive tract cancers. FDA news release. Available at: https://www.fda.gov/NewsEvents/Newsroom/PressAnnouncements/ucm594043.htm (accessed on August 20, 2019).

22. Advanced accelerator applications receives US FDA approval for LUTATHERA® for treatment of gastroenteropancreatic neuroendocrine tumors. *Globe Newswire*. January 26, 2018. Available at: https://globenewswire.com/news-release/2018/01/26/1313140/0/en/Advanced-Accelerator-Applications-Receives-US-FDA-Approval-for-LUTATHERA-for-Treatment-of-Gastroenteropancreatic-Neuroendocrine-Tumors.html (accessed on August 20, 2019).

23. Aliu, P. (2018). Data collection in EAPs: The Novartis experience. Presented at Reagan-Udall Foundation Public Meeting on "Leveraging Real-World Treatment Experience from Expanded Access Protocols," November 19 (personal communication).

24. NESTcc Web site. Who we are. Available at: https://nestcc.org/about/who-we-are/ (accessed on August 20, 2019).

Index